AF606778

# Data Migration for SAP®

SAP PRESS is a joint initiative of SAP and Rheinwerk Publishing. The know-how offered by SAP specialists combined with the expertise of Rheinwerk Publishing offers the reader expert books in the field. SAP PRESS features first-hand information and expert advice, and provides useful skills for professional decision-making.

SAP PRESS offers a variety of books on technical and business-related topics for the SAP user. For further information, please visit our website: *www.sap-press.com*.

Prathyusha Garimella, Shashidhar Garimella
SAP Landscape Transformation Replication Server: The Practical Guide
2024, 317 pages, hardcover and e-book
*www.sap-press.com/5821*

Dogra, Isacc, Kalwachwala, Radhakrishnan, Srinivasan, Chahal, Cheekoti, Khambhampati, Lodha, Quirk
SAP Master Data Governance: The Comprehensive Guide (3rd Edition)
2023, 761 pages, hardcover and e-book
*www.sap-press.com/5580*

Martin Koch, Siegfried Zeilinger
Cloud Connector for SAP
2023, 352 pages, hardcover and e-book
*www.sap-press.com/5683*

Densborn, Finkbohner, Höft, Rubarth, Klöß, Mathäß
Migrating to SAP S/4HANA: Operating Models, Migration Scenarios, Tools, and Implementation (3rd Edition)
2024, 633 pages, hardcover and e-book
*www.sap-press.com/5816*

Bilay, Singh, Singh, Gutsche, Krimmel
Cloud Integration with SAP Integration Suite: The Comprehensive Guide (4th Edition)
2024, 940 pages, hardcover and e-book
*www.sap-press.com/5760*

Frank Finkbohner, Martina Höft, Michael Roth, Jonas Kinold,
Wolfgang Kuchelmeister, Lukas Widera

# Data Migration for SAP®

SAP S/4HANA® and Cloud Solutions

**Editor** Rachel Gibson
**Acquisitions Editor** Hareem Shafi
**German Edition Editor** Eva Tripp
**Translation** Winema Language Services, Inc.
**Copyeditor** Yvette Chin
**Cover Design** Bastian Illerhaus
**Photo Credit** Shutterstock: 1315131909/©Roberto Rizzi
**Layout Design** Vera Brauner
**Production** Graham Geary
**Typesetting** III-Satz, Germany
**Printed and bound in** Canada, on paper from sustainable sources

**ISBN 978-1-4932-2522-4**

1st edition 2024
1st German edition published 2023 by Rheinwerk Verlag, Bonn, Germany

**Library of Congress Cataloging-in-Publication Data**
Names: Finkbohner, Frank, 1969- author.
Title: Data migration for SAP : SAP S/4HANA and cloud solutions / by Frank Finkbohner, Martina Ho?ft, Michael Roth, Jonas Kinold, Wolfgang Kuchelmeister, and Lukas Widera.
Other titles: Datenmigration in SAP-Systeme. English
Description: 1st edition. | Bonn ; Boston : Rheinwerk Publishing, 2024. | Translation of: Datenmigration in SAP-Systeme : SAP S/4HANA und Cloud-Lösungen.
Identifiers: LCCN 2024009893 | ISBN 9781493225224 (hardcover) | ISBN 9781493225231 (ebook)
Subjects: LCSH: Systems migration. | SAP HANA (Electronic resource) | Cloud computing.
Classification: LCC QA76.9.S9 F5613 2024 | DDC 004.67/82--dc23/eng/20240328
LC record available at https://lccn.loc.gov/2024009893

# Contents at a Glance

## PART I Basic Principles



## PART II Data Migration in SAP Solutions



## PART III Alternative Methods

# Contents

## 8 IDoc and ALE Distribution 365



## 9 SAP Legacy System Migration Workbench 381

## 12 Related Areas 531



## Appendices 539

# Introduction

No two data migrations are alike

Data migration is a topic that keeps coming up over and again. Every successor system replacement for an existing software solution means that data must be transferred. But as a rule, simply transferring your data does not solve the problem permanently. Changing economic or legal conditions can repeatedly lead to organizational restructuring within a company, which also requires a reorganization of the data inventory in an existing IT system. In the recent past, companies invested in many different areas. Now, they are focusing on their core competencies again. For this purpose, a company may sell or spin off parts of itself (a carve-out) and buy companies that fit better into the portfolio. These actions often require migrating new data. You could say:

*Each data migration is like a snowflake—unique.*

New products require new solutions

Since the last edition the predecessor book *Data Migration with SAP*, a lot has happened in the world of SAP products. SAP ERP was replaced by SAP S/4HANA as the new comprehensive business suite. Currently, SAP S/4HANA is offered as on-premise, private cloud, and public cloud solutions. In addition to these products, SAP has expanded its product portfolio to include other cloud products such as SAP Ariba and SAP SuccessFactors, for which the topic of data migration is also of great importance.

The problem of data transfers is now increasingly highlighted in the literature. Our motivation for this book is to integrate not only the classic data migration solutions for SAP ERP and SAP Business ByDesign, but also the new migration solutions for SAP S/4HANA, SAP Ariba, and SAP SuccessFactors into one book that provides a comprehensive overview.

Data migration—the underestimated challenge

Data migrations are an often-underestimated topic, as they can tie up a large part of your budget and project resources, depending on the scope of the project, and thus often cause personnel bottlenecks. As a rule, data migrations are carried out by consultants or programmers, who, however, often cause personnel and financial bottlenecks due to their overlapping fields of application. To avoid such bottlenecks, this book presents data migration techniques that can be used not only by professional programmers, but also by consultants without profound technical knowledge. Your specialist departments should be enabled both to trigger data transfers or restructuring processes and to map them in IT terms. For this purpose, easy-to-learn migration techniques are presented that do not require any programming as much as possible. In specific situations that do require it, programming should only be used as a last resort, that is, when all other provided solutions for data migration do not lead to the desired results. Our focus in this book is therefore on the tools that do not require programming whenever possible.

**Flexible data migration without programming**

If data must be migrated to an SAP system, solutions are available that do not require any programming at all and are not limited to individual applications. These solutions can be used in a wide variety of applications, such as accounting, logistics, or human resources (HR). What type of data needs to be transferred doesn't matter. The tools and solutions are flexible and can optionally be supplemented with customer-specific coding as required for the data migration, which reinforces their universal applicability. Data migration tools and solutions are also available for SAP cloud solutions such as SAP Business ByDesign, SAP Ariba, and SAP SuccessFactors Employee Central. In the individual chapters of the book, we'll reference the availability of migration solutions in the respective SAP products.

**Who is the target group?**

This book is primarily intended for people who come into contact with data migrations, but do not necessarily need to have programming skills. These individuals include, for example, SAP consultants who are responsible for data migrations on the customer side or in their own organization, as well as project managers who want an overview of the different available methods so as to critically analyze them. This book might also be of interest to SAP developers who want to learn efficient migration techniques.

**How much prior knowledge is required?**

We expect you have some basic understanding of the ABAP platform and common Microsoft Office products such as Microsoft Word and Microsoft Excel. An understanding of commercial issues may be useful but is not essential. The book is written in such a way that it can be understood and implemented even without detailed technical SAP knowledge.

**Why is this book important?**

Data migrations can be performed using simple procedures, which also allows business departments to play a more active role in data migration projects. This book provides a number of methods for data migration that can also save time and money. We hope this book expands the role of business departments and eliminates the need to write and test separate programs for each data migration.

**Terminology Used**

The terms *data migration* and *data transfer* are used interchangeably in this book. A *data source* can be any legacy IT system, whether it is an initial data transfer or an SAP system that is already in production use, or whether the data set has to be adapted to organizational changes.

## Part I: Basic Principles

**Basic principles of data migration**

In the first part of this book, you'll learn everything you need to know about data migrations. **Chapter 1** provides an overview of various data migration strategies and approaches. Additionally, this chapter contains a brief excursion on the topic of historical data.

**Planning migration projects**

In **Chapter 2**, we'll show you how to structure your data migration projects using SEAMAP to avoid time and quality issues. Using real-world examples, we'll explain the different types of data migration projects, the stages of those projects, and the work packages involved in each phase. We also provide guidance on planning and effort estimation to make your data migration successful.

## Part II: Data Migration in SAP Solutions

**SAP S/4HANA and SAP S/4HANA Cloud**

The second part of our book is dedicated to the topic of migrating data in SAP solutions. In **Chapter 3**, we'll introduce you to the SAP S/4HANA migration cockpit data migration tool and the SAP S/4HANA migration object modeler, both of which were developed specifically for the migration of data to SAP S/4HANA. We explain step by step how a migration project can be carried out using different approaches and which other applications support you in data migration. We'll also show you how to create your own migration objects or extend existing migration objects. The chapter is rounded out with useful tips on information sources and performance aspects.

**SAP SuccessFactors**

**Chapter 4** addresses the technical options for data migrations relevant to the cloud-based HR management software—SAP SuccessFactors—and specifically for the SAP SuccessFactors Employee Central module, the HR management system of SAP SuccessFactors and thus the core of the suite. The corresponding challenges in this environment are addressed in this chapter. Insights into the procedure are provided, highlighting the standard SAP SuccessFactors procedure with comma-separated values (CSV) file imports and also outlining the options of migrating data from SAP ERP Human Capital Management (SAP ERP HCM) to SAP SuccessFactors Employee Central using SAP Infoporter as an example. Some tips to support a successful data migration project round out this chapter.

**SAP Ariba**

You'll learn about data integration with SAP Ariba procurement solutions in **Chapter 5**. Depending on the source system, different ways exist in which your data can be integrated into these cloud solutions.

**SAP Business ByDesign**

In **Chapter 6**, we show you how to migrate to the SAP Business ByDesign cloud solution. Through an example, we explain the central steps for this goal: *validation, value conversion, simulation*, and *import*. The options for correcting detected data errors are discussed in more detail. A separate section is dedicated to the migration of financial data and the cross-object posting forecast integrated there. Practical tips round out this chapter.

## Part III: Alternative Methods

In Part III, we review the classic data migration methods that are still available in the ABAP platform and also introduce you to more recent alternative methods.

**Batch input technique**

**Chapter 7** introduces you to one of the oldest methods of data transfer, called a *batch input*.

**IDocs and ALE**

IDocs and Application Link Enabling (ALE) distributions are SAP technologies that are used in the areas of data management and integration. We provide introductory information on this topic in **Chapter 8**. Specifically, we present the use of ALE distributions in a data transfer.

**SAP Legacy System Migration Workbench**

Although SAP Legacy System Migration Workbench is on the *simplification list* in SAP S/4HANA, it is still usable. You can use SAP Legacy System Migration Workbench when you reach the limits of the SAP S/4HANA migration cockpit and the SAP S/4HANA migration object modeler. Despite some limitations in SAP S/4HANA, SAP Legacy System Migration Workbench is still a quite flexible tool, which we'll introduce to you in **Chapter 9** through many useful application examples.

**Cloud Integration**

In **Chapter 10**, we'll show you how to prepare your existing XML-based documents for data migration using a file-to-file approach. Specifically, we'll introduce you to several action options in Cloud Integration (i.e., *message splitting, data merge*, and *message mapping*) and provide an overview of how you can implement these options and use them for your data migration. An introduction to the cloud connector and the connection of Secure File Transfer Protocol (SFTP) servers gives you the comprehensive knowledge to implement the file-to-file approach using Cloud Integration.

**Avoiding programming work**

In **Chapter 11**, you'll learn how to convert specific data types and data constellations without extensive programming work. Using practical examples, we'll provide guidance on how you can effectively prepare your dataset for migration using tools like Microsoft Excel. In this way, you learn at least one way to prepare your data for transfer to an SAP system using a process that doesn't require any programming.

**Related areas**

**Chapter 12** provides information on database migrations and system conversions.

In **Appendix A**, the terminology relevant for understanding the book is listed and explained in a glossary.

Appendix

The structure of this book allows you to pick out a chapter, depending on your interests, without having to read the preceding chapters. Nevertheless, with this work, we would like to offer you a comprehensive project manual for data migration, so that after reading the book, you can confidently plan, assess, and implement migration projects yourself. We hope our detailed case studies with numerous screenshots will help you achieve this goal.

Information boxes

The highlighted boxes in this book contain information that is good to know and useful but sometimes outside of the immediate context. To help you immediately identify the type of information contained in the boxes, we have assigned icons to each box:

**Tips**

The *tips* marked with this icon provide specific recommendations that can make your work easier.

**Notes**

Boxes marked with this icon contain information about *additional topics* or important content that you should note.

**Warnings**

This icon refers to *specifics* that you should consider. It also *warns* about frequent errors or problems that can occur.

**Examples**

*Examples*, which are highlighted with this icon, refer to real-life scenarios and illustrate the functions described.

We hope this book will help you migrate your data to your SAP systems and enable you to take full advantage of SAP solutions. Now, we hope you enjoy reading and digesting all that you're about to learn!

**Frank Finkbohner**, **Martina Höft**, **Michael Roth**, **Jonas Kinold**, **Wolfgang Kuchelmeister**, and **Lukas Widera**

PART I

# Basic Principles

# Chapter 1
# Basic Principles of Data Migration

*You cannot successfully plan, execute, and complete your data migration project until you have dealt with all aspects of this complex topic at an early stage and have acquired the relevant knowledge about it. In this chapter, we'll lay that foundation of knowledge.*

According to the Cambridge Dictionary, *migration* (from the Latin *migrare* = to wander, to move away) stands for "the process of beginning a new computer system, or of moving information from one system to another" (see *http://s-prs.co/v581701*). A *data migration*, then, is the "moving" of data from one system to another. The data can migrate from an old system to a new system, from a local database to cloud-based storage, or just from one database to another. The motivations behind a data migration are most often to migrate to a newer or more powerful system, to improve data integrity, to consolidate data, or to enable a better analysis of data. This chapter will introduce the basic principles of data migrations. For this purpose, we'll explain the different types of implementation approaches and migration strategies available. In general, a foundation concept to remember is that the implementation approach of an SAP system (Section 1.1) determines the best migration strategy (Section 1.2).

## 1.1 Implementation Approaches

Generally speaking, several different approaches are available when implementing an SAP system. The most important approaches are the following:

- Big bang implementations
- Moving into an existing system
- Phased rollouts
- System optimizations or landscape optimizations

In the following sections, we describe these implementation approaches in more detail, starting with the big bang implementation.

### 1.1.1 Big Bang/Initial Load

A *big bang approach* is, as the name suggests, similar to "The Big Bang." The system is built within a short period of time (during the *cutover window*) with all the necessary functionalities for productive operation and then put into operation at a fixed point in time (for *go-live*). Poetically speaking, you could reference the Bible: "And God said, 'Let there be light,' and there was light" (Genesis 1:3). When ideally everything has been implemented cleanly, you can just flip the switch, so to speak, and the system starts running. Existing predecessor systems that are no longer necessary, called *deprecated systems* or *legacy systems*, are "decommissioned" at the same time. Interfaces to these legacy systems are shut down, and new interfaces to the new system are put into operation. In terms of migration, therefore, all the data required for the go-live must be loaded into the target system from the legacy systems within the cutover window. This activity is also referred to as the *initial load*. The pros and cons of the big bang approach are listed in Table 1.1.

| Pro | Con |
|---|---|
| All data is available in the new system. You can take full advantage of the new software right away. | If the project or go-live fails, it will have a greater impact on your business. An interruption of business operations can then not be excluded. |
| Since you no longer have any legacy systems, you save yourself the trouble of setting up temporary interfaces and help solutions. | Due to the many tasks, you tie up many of your own resources or you have to buy in additional resources (consultants, etc.). |
| | The amount of data to be loaded is often quite large and requires a detailed migration and go-live planning that does not allow for delays. |

**Table 1.1** Pros and Cons of a Big Bang Solution

### 1.1.2 Moving Into an Existing System

In general, moving into an existing system means that an already live SAP production system is enhanced with new functionalities or data. A typical example in this case is a company acquisition, for example. Moving into an existing system (the *target system*) is usually a singular event. In this context, data is transferred from other systems (the *source systems*), and under certain circumstances, new functionalities are introduced at the same time. The target system has usually been productive and filled with data for some time. A data migration to such a system is therefore more complex than to an empty system. As a result of changes to the existing data in the target system, you must exercise care to avoid data inconsistencies.

### 1.1.3 Phased Rollout

A *phased rollout*, sometimes also referred to as a *phased implementation*, is a combination of a big bang approach (its associated initial load) and a move into an existing system. The difference to moving into an existing system is that here, multiple implementations are planned one after the other or in parallel. This type of implementation is often encountered in a restructuring of IT landscapes. Typical examples include company mergers, the changeover to a central system landscape, or the consolidation and relocation of all IT systems to a cloud landscape. The most common case, however, is the rollout of a central SAP system (on-premise or cloud) to various subsidiaries or international affiliates. Such projects are usually planned over a period of several years. Due to the initial load and the multiple introductions of other systems/functionalities into an existing production system, great demands are placed on the project duration and quality of the data migration. Phased rollouts usually result in synergy effects over the years that reduce costs and increase quality. Table 1.2 lists some pros and cons of phased implementations.

| Pro | Con |
|---|---|
| If the project or go-live fails, only part of the business is affected, which mitigates the risk. | You often need many temporary interfaces to your systems, which you want to replace gradually. |
| You can expect to use fewer resources over time because, apart from the first rollout (the initial load), all subsequent projects run in a similar way as before, so the learning curve is shallow, and the synergy effects are high. | Because of the systems that have not been replaced, your system landscape becomes more complex and accordingly requires more complex and costly planning. |
| By not having to load all the data at once, the reduced amount of data involves easier planning of production start. Smaller cutover windows are possible than with a big bang. | An SAP system is a fully integrated system. The fact that not all parts of the company use the same system at the same time means that it takes longer for you to reap the full benefits of the new system, namely, until you have fully completed the rollout. |

**Table 1.2** Pros and Cons of a Phased Rollout

| Pro | Con |
|---|---|
| The complexity of the project is lower due to its division into different phases. | Migrating additional data into a running production system is complex and requires extensive testing. |
| Managing changes is less complex; you and your company have time to adapt to the changes. | |
| From phase to phase, you have the opportunity to learn from your mistakes and avoid them in the new phases. | |

**Table 1.2** Pros and Cons of a Phased Rollout (Cont.)

Phased rollouts must be planned rather well. Figure 1.1 shows an example of such a rollout. Notice how more time is spent on the configuration and implementation of the first rollout (the initial load) than on subsequent rollouts. Subsequent rollouts can be operated in parallel and usually require less time for configuration and implementation by leveraging the configuration template of the first rollout. Depending on whether you use test automation software or tools, you can also greatly reduce testing times and efforts. However, a data migration should always be fully tested as the data is moved into a production system.

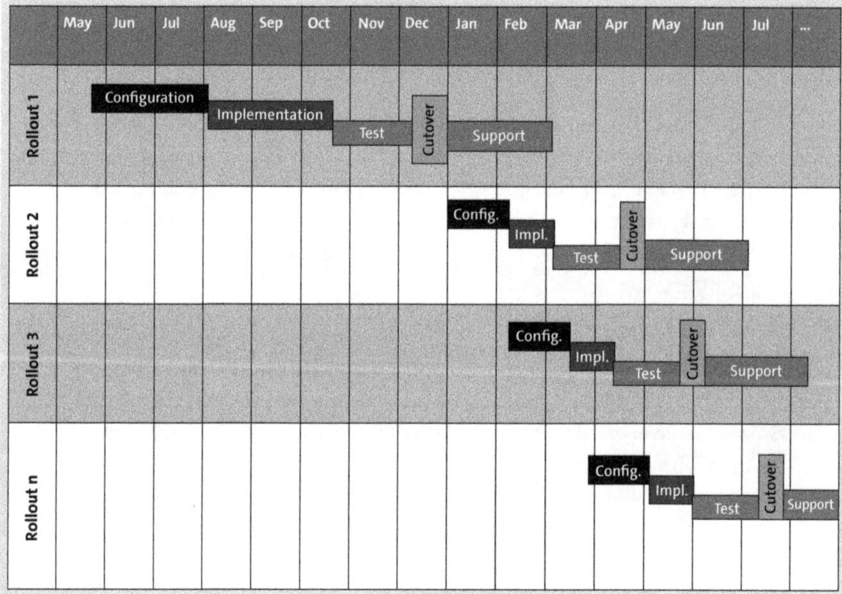


**Figure 1.1** Example of a Phased Rollout

### 1.1.4 System Optimization

A *system optimization*, also referred to as *landscape optimization* or *system landscape optimization (SLO)*, is often applied when parts of a company are sold, production is outsourced, or a company merger takes place. If processes or organizational structures of an SAP production system are swapped out to another SAP system or newly integrated into a system, special techniques and services are required, such as the transfer of all data of a certain company code to another system or the merging of different company codes into one system. Such cases are quite complex and often subject to financial audits. This kind of project is therefore offered as a paid service by specialist providers and is usually carried out directly at the table level. These transformations can't actually be done on your own. SAP offers such services, for example, through the Data Management and Landscape Transformation (DMLT) group.

## 1.2 Migration Strategies and Migration Scenarios

In general, you can choose from two main strategies: *greenfield* and *brownfield*. In addition, a third strategy is called *selective data transition*. In the following sections, we'll describe all three strategies, starting with the greenfield strategy.

### 1.2.1 Greenfield Migrations/New Implementations

*Greenfield*, a term used in the construction industry, refers to a new development on a previously undeveloped site. In terms of software, "greenfield" represents a new implementation. In the field of data migration, we understand the task as a migration to a new system that is to be completely installed and configured from scratch.

**New beginnings**

Greenfield means that you can start completely from scratch without any existing restrictions. This strategy is often used when space has become too limited in an existing environment or when an existing system can no longer be expanded or adapted to new conditions. In that case, *business process reengineering* and its accompanying simplification of processes are the first choices. The reasons for this recommendation are manifold and can include, for example, that you want to return to a standard configuration because of the following reasons:

- Your existing system is no longer maintainable or upgradeable.
- You want to save costs in the long term for one or more reasons:
  - The cost of maintenance of all modifications is increasing more and more.

- The cost of customization after an upgrade is increasing more and more.
- No one in your company is familiar with the existing modifications and applications anymore.

**Standard or not?** Many projects start by following an industry standard as closely as possible in terms of business processes and then programming their own applications or modifying standard applications as little as possible. Unfortunately, this approach is all too often abandoned. People often shy away from the effort in reorganizing their business processes after analyzing the chaos. But consulting companies sell preconfigured solutions that involve a lot of in-house development and thus are departures from the standard. Turning to the standard has many advantages and saves money in the long run. Many companies that had heavily modified their systems back in SAP R/3 later had come to realize (painfully) that these systems could not be upgraded to SAP ERP or only with rather complex and cost-intensive projects at high cost.

Figure 1.2 shows a schematic representation of the greenfield strategy for a new implementation of an SAP S/4HANA system. In this case, the data from an SAP ERP system or a non-SAP system is migrated to an SAP S/4HANA product. The target system contains the configuration delivered by SAP and often must but sometimes can optionally be adapted more or less to your specific requirements, depending on the product. Data is then loaded into the target system using the defined data migration tools.

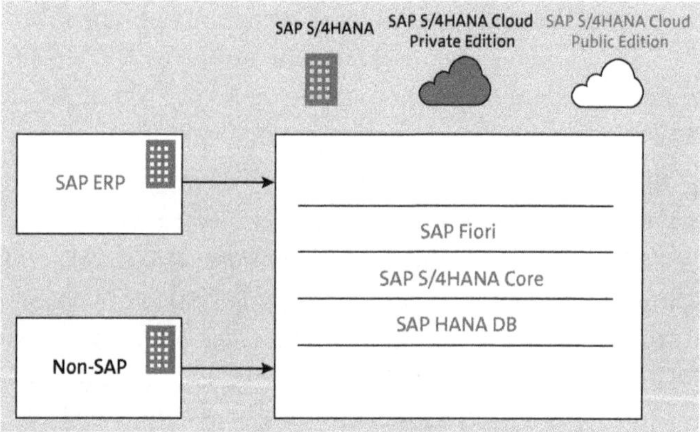


**Figure 1.2** New Implementation of SAP S/4HANA

**SAP S/4HANA Cloud** SAP S/4HANA Cloud Public Edition projects are always new implementations due to technical restrictions. Because of the limited configuration options, as a customer, you must deal with adapting your processes to the standard. Cloud projects can therefore usually be implemented and planned more quickly than on-premise projects, for example. Clients and consultancies that embrace the cloud concept of constrained configuration and adapt their processes and templates accordingly will be rewarded by their projects being completed within their planned timeframes. Think of the sandwich or pizza store around the corner: You're standing in front of this counter, and the friendly person behind the counter asks you which sandwich or pizza you want to buy. If you take the standard offer, you'll be done quickly. However, if you have special requests, such as extra cheese, onions (but only red onions!), everything takes longer, and the line behind you gets longer too. If you accept the standard that also tastes good and is satisfying, everything is easier, faster, and less error prone. As a customer, if you and/or your consulting firm fail to adopt the necessary *cloud mindset*, so if you try to adapt this system to your requirements, instead of taking what is already available and bringing your processes back to standard, then everything takes longer and costs time and money. If you're unlucky, the project will fail in the end.

**On-premise system**

Greenfield projects in on-premise systems often fail to meet the project scope in terms of budget and time. As mentioned earlier, issues arise from the manifold process configuration options available and the resulting coordination processes, additional developments, and costs.

**Get rid of the pounds**

A greenfield implementation means, however, daring to make a fresh start and getting rid of old burdens (your old data stocks) through archiving and data cleansing in the source system, for example. The result will be a more powerful and streamlined target system with increased storage capacities.

### 1.2.2 Brownfield Migrations/System Conversions

The concept of a *brownfield migration* doesn't necessarily mean the dried-up patch of earth you might see in your backyard in the summer. Rather, a brownfield migration is the core renovation of an existing building, the expansion in an existing building, or a new building in an already developed industrial area with existing buildings. In contrast to the greenfield approach, the restrictions governing a brownfield migration are higher. In terms of sustainability, however, a brownfield migration should not be underestimated either because the data remains in the system, and you'll worry less about data migration. However, this approach also means that you must continue to include the old data burden in your considerations. This data burden often contains, for example, incorrect master data (e.g., duplicates, incorrect addresses, etc.) and should be cleaned up in any case (e.g., through archiving, data cleansing, etc.).

System conversion

The typical SAP data migration scenario of the brownfield approach is a *system conversion*, that is, the conversion of an existing SAP ERP system to a new SAP S/4HANA system while retaining the existing configuration and all the data. In the diagram shown in Figure 1.3, notice how a system conversion is only possible to products based on SAP S/4HANA, specifically the on-premise version. A system conversion to the SAP S/4HANA Cloud Public Edition is technically impossible. One advantage of a system conversion is that the conversion is performed within the same system. For instance, an SAP Business Suite system can be converted into an SAP S/4HANA system in multiple steps (see Chapter 12, Section 12.2), which is not a normal upgrade, because the system is completely transformed.

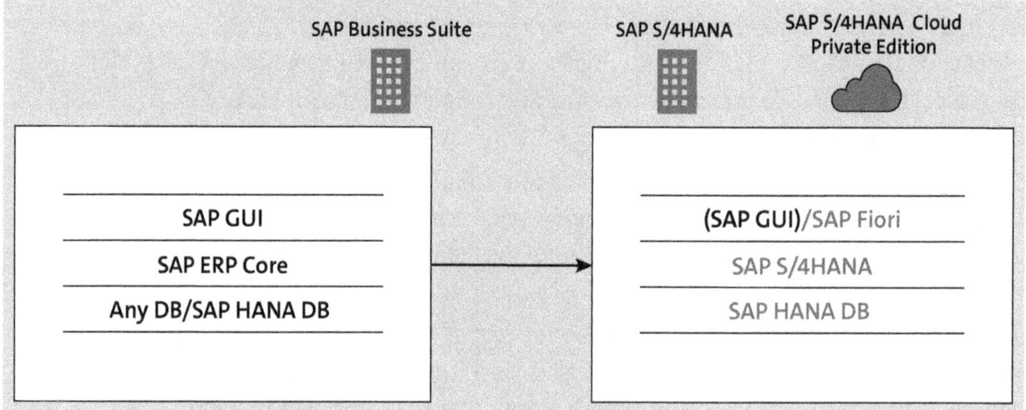


**Figure 1.3** System Conversion to SAP S/4HANA

For an overview of a system conversion, see Chapter 12, Section 12.2.

### 1.2.3 Selective Data Transitions

Unlike the previous two terms, the *selective data transition* concept has no origin in building construction. A selective data transition is used by various companies in relation to their products and solutions to allow data to be migrated from an existing system to a new system while the old system remains in operation. A selective data transition always requires thorough planning and preparation to ensure that data integrity is maintained and that no data is lost.

Minimal interruption of business operations

One advantage of this type of data migration is its ability to minimize business disruption and maintain data integrity. Users can continue to access the data they need while the migration is taking place. It also allows both the new environment and the old environment to run in parallel, and the new system can be tested before it finally goes into production.

**Parallel operation migration**

A selective data transition can be regarded as a *parallel operation migration* to express the concept of migrating data from an existing system to a new system while the old system remains in operation.

**Selective data transitions**

When SAP refers to a selective data transition in connection with SAP S/4HANA, the focus here is on the selective character of the data transition. Data from one or more SAP ERP systems is transferred to the SAP S/4HANA target system. The target system of a selective data transition must first be created either via a new implementation or via a system conversion. The data is then selected, converted to the SAP S/4HANA data model, and loaded into the target system. Figure 1.4 shows some examples of selective data transition options:

- *System consolidation*, that is, the migration of functions and data from multiple source systems into an SAP S/4HANA target system
- *Migration of selected business units* of a system into an SAP S/4HANA system
- *Migration of selected application areas* into an SAP S/4HANA system

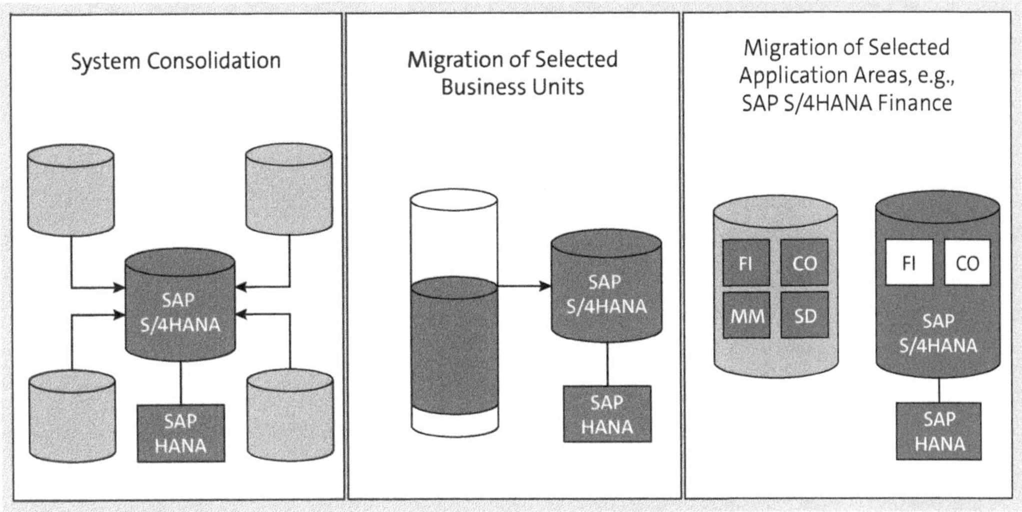


**Figure 1.4** Various Options of Selective Data Transition

To a limited extent, a selective data transition is also possible for non-SAP source systems via special *extract, transform, and load (ETL)* tools and solutions, such as SAP Data Services.

**Paid services and tools**

This type of data migration is highly complex and can only be performed by experts. In addition to SAP itself, however, other companies on the market offer the corresponding tools, consulting, and implementation services for a fee.

### 1.2.4 Comparison of the Strategies

**Greenfield migrations: new implementations**

Greenfield migrations are best suited for companies that want to adopt new technology requiring complete redesign from scratch or that want to completely renew their IT infrastructures. If you use a non-SAP system as your source system, the only migration path you have is a new implementation.

**Brownfield migrations: system conversions**

Brownfield migrations are best suited for companies that need to migrate their existing data and applications and that want to renew only parts of their IT infrastructure. This type of migration usually requires more time and effort because existing data and applications must be analyzed and adapted to the new system.

Selective data transitions are suitable for companies that want to modernize their IT infrastructures without compromising on the availability of their applications and their data. This type of migration is especially useful for companies that want to update their IT systems without experiencing downtimes. Companies that want to expand their IT infrastructure can have more capacity without sacrificing performance through a selective data transition.

## 1.3 A Digression on Historical Data

A frequently asked question concerns the migration of historical data. Many customers want to transfer their complete data inventory, including all the historical data from their legacy systems, to the new target system. But what is *historical data* anyway? At SAP, this term is understood to mean the following kinds of data:

- *Transaction data* that pertains to a completed or long-running process. This data includes, for example, completed or partially delivered orders, financial documents, or partially produced production orders.
- *Master data* that is marked for deletion or already archived or legacy data that has not been touched for years, such as products that have not been ordered for years but are still in the warehouse.

**Time period**

In addition, the time period during which your historical data can cover must not be forgotten. A huge difference exists whether you're talking about all the data in a system or "just" the data from the last two fiscal years.

**What goes where?**

Depending on your data migration scenario, several ways exist to migrate (or not migrate) historical data:

- *New implementation*: Migration of historical data, especially historical transaction data, is not possible in this approach for the most part. The possibility here is to load historical balances and material consumption values, for example.
- *System conversion*: In this case, as far as possible, all historical data (master data and transaction data) can be converted using standard SAP tools.
- *Selective data transition*: In this case, it depends on the service or tool offered whether and which historical data can be transferred as part of a project.

Master data

As you may know from your own recent experience or your company's experience, some moving boxes that get packed once will never be unpacked, but are retained every time you move. The big question you should be asking yourself now is, "Do I really need these moving boxes anymore, or can I just dispose of them?" In data migration projects, master data is analogous to moving boxes. In this context, you should really ask yourself what master data you really still need and what you can delete or clean up in your source systems.

[!]

**Historical Master Data and Dependent Transaction Data**

When you cleanse master data, you should also look at the dependencies between master data and transaction data. Products that are marked for deletion, for example, may still be in your data inventory. Therefore, you must either write off the stock or transfer the product marked for deletion. As part of a new implementation, you would then need to remove the deletion flag before migrating the master data and dependent transaction data and set it again after the migration. Similarly, certain statuses in some master data do not allow the creation of a transaction date for this master data. The status would block the migration process.

Transaction data

As mentioned earlier, the migration of historical data, especially historical transaction data, is largely not possible in a new implementation. Why is that? Let's consider some answers:

- In most countries, closed financial documents may not be posted again from an accounting perspective. So, if you have posted a financial document in your source system, when you post it again in the target system, it will go through a completely new process (Customizing) of the new system. Therefore, the document will most likely end up looking different from before (e.g., different tax rules) than in the source system and will also generate different follow-on documents. This result is usually rejected by a financial auditor. What you can migrate to an SAP S/4HANA system, however, are historical balances. These balances have no document reference and are written directly into the corresponding database table.

- Completed or partially posted documents, such as purchase orders or production orders, with their many subsequent documents and dependencies, are simply too complex to be cleanly extracted, converted and loaded as part of a new implementation. Due to this extremely high complexity, there are also no interfaces or application programming interfaces (APIs) from SAP for data migrations involving such documents or complete document flows. The only option is to transfer open documents or, from a financial point of view, open items. A common principle for partially open purchase orders is to cancel the open remaining quantities in the source system and then create new open purchase orders in the target system.

But why can I transfer these documents as part of a selective data transition or system conversion?

- As the name of the system conversion suggests, the data remains in the system, and only organizational units, such as company codes, may be converted to the new organizational units, or the data may be copied from one table to another as part of a specified migration schema if the data model has changed.
- In the case of a selective data transition, this ability depends on the service offered as to whether or not closed documents are supported. However, in this scenario, too, the data is usually extracted directly from the source systems and copied to the relevant tables in the SAP S/4HANA target system.

In both scenarios, the data is written directly to the tables, and thus the configuration of the business process is not run through. This approach is thus virtually the only way to migrate financial documents in such a way that they are accepted by most financial auditors.

New implementation

What alternatives do you have for a new implementation in this scenario?

We recommend considering the following options:

- **Keep the system running**
  - You can run your legacy systems in parallel with the new system until all open documents are closed (purchase orders, etc.). Temporary interfaces to the new system may need to be implemented in this case.
  - If you are required to uphold certain legal retention periods (which is why you or an auditor will require access to the historical data), you can also keep your legacy systems running or executable and boot up

as needed until the retention period expires. After that requirement passes, shut down the system and scrap it.

- You can also look for reputable service providers that have systems ready with the latest backup of their legacy system that you can access remotely. Such systems are then only booted up when required, and only fixed costs for provisioning and variable costs for access are incurred.

- **Business warehouse**
If the data is required for evaluations, you can swap out the data in aggregated form, for example, to a business warehouse system. If, for example, you need the sales per product per quarter, you don't need to store complete documents in this case, only the relevant aggregated data, thus saving space and money.
- **Side-by-side scenario**
In this case, the relevant data is simply swapped out. This data can include, for example, customer-owned tables in the same SAP S/4HANA system or tables in another database where they can then be queried by means of custom reports or queries.

[«]

**Extracting the Data Before Shutdown**

SAP offers a few tools and sample implementations for extracting your data before the final shutdown of a system, such as the following:

- *SAP Information Lifecycle Management (SAP ILM)*
  - With this tool, you can transfer the data to a standalone SAP ILM retention warehouse system (RW system) as part of the decommissioning process. Attachments to the documents can also be stored in this system.
  - Learn more at SAP Help Portal at this link: *http://s-prs.co/v581702*.
- *Data Retention Tool (DART)* excluding attachments
  - DART helps you meet the legal requirements for data retention for tax audit purposes. You can extract fiscally relevant data into sequential files, keep them, and view them again with other tools. DART is used, for example, in Germany for its Principles for Data Access and Verifiability of Digital Documents, which have been valid in Germany since January 1, 2002.
  - Learn more at SAP Help Portal at this link: *http://s-prs.co/v581703*.
- Example side-by-side scenarios:
  - The SAP Extensibility Explorer for SAP S/4HANA Cloud (*http://s-prs.co/v581704*) provides sample scenarios of enhancement options for your SAP S/4HANA Cloud system. One such sample extension is

"Migrated Sales Order Data." The extension describes how you can create a side-by-side application to perform search, sort, and filter operations on historical sales order data that you migrated from a legacy system.

- For the "Migrated Sales Order Data" example, refer to *http://s-prs.co/v581705*.
- For the setup guide, refer to *http://s-prs.co/v581706*.

## 1.4 Summary

In this chapter, we provided an overview of the basics of data migrations. You're now familiar with the various available implementation approaches, which range from a big bang to a system optimization. You now know the different migration strategies, including a greenfield approach, a brownfield approach, and a selective data migration. With our addendum into historical data, you should now also know better about this specific type of data and how you can deal with it in your projects.

# Chapter 2
# Planning Data Migration Projects



*Structured migration projects are more successful than unstructured ones. For this reason, you should clarify the necessary work packages and resource requirements for your upcoming data migration subproject early on.*

In Chapter 1, we introduced you to the fundamentals of data migrations; in this chapter, we'll now look at how you can structure a data migration project. Section 2.1 describes the phases of a data migration project, then we'll explain how they fit into the phases of an implementation project in Section 2.2. In Section 2.3, you'll then learn which tasks occur within a data migration phase and how you should divide these tasks into work packages. In the final section of this chapter, we'll provide tips for data migration object identification and effort estimation.

## 2.1 Phases of a Data Migration Project

As described in Chapter 1, different types of data migration projects exist, but their phases are usually the same.

You can use different process models and methods for SAP implementation projects. The current process model or framework for implementing SAP software is SAP Activate, which contains roadmaps developed by SAP for the implementation of your products and is constantly adapted and further developed. In this section, we explain which project phases exist in a data migration project and how they are integrated into the standard project phases of an SAP S/4HANA implementation project.

[«]

**Roadmap Viewer in SAP Activate**

For more information about SAP Activate, you should visit the Roadmap Viewer web page at *https://go.support.sap.com/roadmapviewer.*

SAP Activate is divided into six phases:

1. Discover
2. Prepare

3. Explore
4. Realize
5. Deploy
6. Run

**Seven data migration phases**

The best way to set up a data migration project is to run it as a subproject. The tasks of a data migration project differ from those of a product launch using SAP Activate. These tasks are distributed across seven data migration phases:

1. Data analysis
2. Data cleansing
3. Mapping
4. Implementation
5. Tests
   - Functional tests
   - Productive load test (PLT)
6. Validation
7. Productive load

Figure 2.1 shows a representation of the individual phases.

**Necessary overlap of the data migration phases**

Unlike SAP Activate phases, data migration phases are not strictly sequential but instead are iterative and overlapping. Thus, findings from a downstream phase intentionally affect the preceding phases. This is necessary so you react flexibly if errors arise. In the following sections, we'll go into more detail about the individual data migration phases.

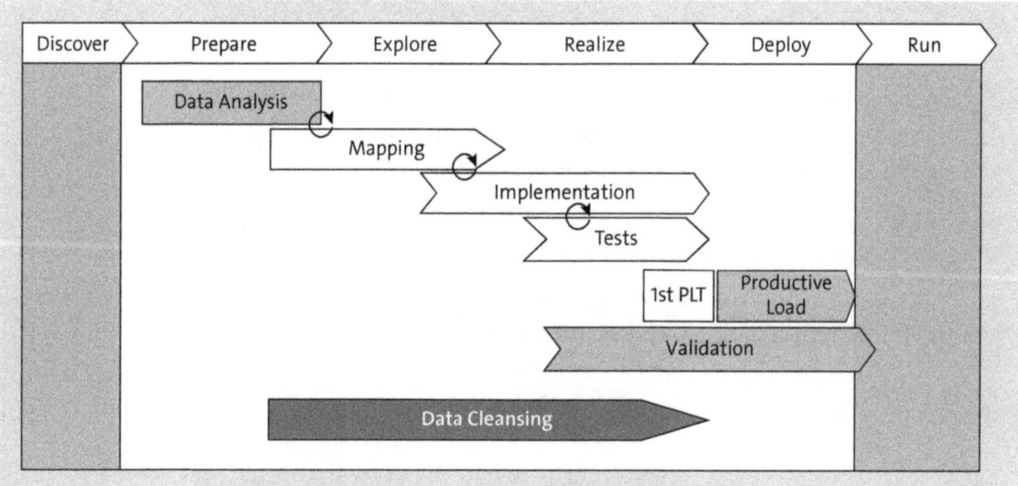


**Figure 2.1** Data Migration Phases in SAP Activate

### 2.1.1 Data Analysis

The analysis phase is usually started within the Prepare phase, but it does not necessarily have to start with this phase. You can just as well start the analysis halfway through the Prepare phase because, by then, the processes to be implemented are known, and some of them have often already been fully described. This approach makes it easier for the data migration analysis to identify the business objects, data migration objects, and source systems.

[«]

**Business Objects versus Data Migration Objects**

A *business object* is an individual data object that is needed to model a business process, such as a material, a customer, or a purchase order.

A *data migration object* (or simply a *migration object*) is usually a business object or a part of one. A migration object is used to migrate a business object. Business objects sometimes must be divided into smaller units for technical reasons, perhaps having different data sources or data migration interfaces. As a result, one business object may require more than one migration object. One example is the business object product, which is divided into multiple migration objects in SAP S/4HANA.

[«]

**Master Data versus Transaction Data**

*Master data* is data that is stable and used as the basis for business objects. Master data changes only slightly or not at all over a specific period. Known master data objects include product, customer, supplier, bank, and bill of materials (BOM).

*Transaction data* is data that is volatile and subject to continuous change. Common transaction data objects include stock levels, account data, all types of orders, and documents.

We recommend that you start the data analysis process as early as possible as it lays the foundation for all other activities.

### 2.1.2 Data Cleansing

Data cleansing is one of the most underestimated activities within a data migration project. If possible, data cleansing should always be performed in the source system. Under certain circumstances, however, it can also be done in the respective data transfer tool. Systematic errors in your source data can sometimes be converted to clean load data using conversion rules.

**Garbage in, garbage out**

You should also use insights from machine validations to clean up your data in the source system. The cleaner the data in your source systems, the cleaner the data in your target systems.

Data cleansing should start as early as possible in your migration project. Ideally, you already have defined processes in your source systems that permanently check the data for errors and clean them up. You should closely involve these processes, or the teams tasked with them, in the data migration project. Likewise, you must identify existing data cleansing rules and match them with the conversion rules. The findings on the quality of the source data from the initial data transfer tests are important for further data cleansing processes.

The data cleansing phase should be completed, at the latest, with the last productive load test. This is the only way to ensure that the findings of the last test can be adopted one-to-one at go-live.

### 2.1.3 Mapping

Once the data migration objects have been identified, the mapping phase can start. In the mapping phase, the structures and fields of your source system data to be migrated are mapped to the structures and fields of the data migration interfaces and programs of the target system. Since this task is usually done on paper, this step is often referred to as *paper mapping*. The mapping phase often overlaps with the analysis phase. However, you should not wait for the analysis to complete; you should start mapping early. A good time for mapping is when the first master data objects have been identified and the configuration for maintaining these objects is largely complete.

**Conversion rules** Configuration documents (formerly also referred to as *business blueprints*) usually already contain hints for conversion rules. The mapping phase can extend into the testing phase. Often, insights emerge from loading the initial test data that have a strong influence on the mapping and conversion rules. We therefore advise you start migration activities as early as possible so that sufficient time can be scheduled for testing the data migration and the post-processing resulting from these tests.

### 2.1.4 Implementation

Implementation, including your first small functional tests of the data extraction and data migration programs, can be started quite early in the realize phase of the project. The best time to download the first test data is when most of your conversion rules have been established so that the initial load can be pulled from the source systems.

### 2.1.5 Testing the Data Migration

Testing is the be-all and end-all of any data migration, and you should start the first functional tests as early as possible. You can't run enough tests. The more complex a data transfer is and the more target systems, source systems, data migration objects, and conversion rules involved, the more important performing tests becomes.

**The epitome: testing!**

There is no substitute for testing—except more testing! Test the entire data migration process and cutover at least once before go-live. However, our experience shows that one productive load test alone is often insufficient. We recommend at least two productive load tests (load tests under go-live conditions and in systems that technically correspond to the productive system) for a positive effect on data quality and adherence to scheduling.

[+]

**Testing Reduces Costs**

We can reduce this discussion into a simple formula: The *higher* the *number* of tests and the *higher* the *quality* of the loaded data, the *lower* the *rework* in the target system shortly before or after go-live and the *lower* the *overall costs.*

**Developing a test strategy**

For testing, especially for multiple tests of a data migration, an early planned data migration testing strategy is required. You must carefully plan the systems, timing, and sequence of testing and clean up or recover your test systems. Define an appropriate backup-restore concept for your test systems and determine the duration required for a backup and restore. Taking what are called *database snapshots* is often a matter of choice. In this context, you can create snapshots of a database at specified times and can quickly reset it.

### 2.1.6 Validation

**Data validation as a quality guarantor**

Validating the loaded data is essential for the quality of your processes in the target system. Only high-quality master data and transaction data guarantee the clean operation of your processes. In some industries, such as the pharmaceutical industry, such validations are even mandatory and must be documented. Without an appropriate validation of all data in the target system, for example, you cannot receive an *Food and Drug Administration (FDA)* certificate, and without this certification, you won't be allowed to sell goods on the US market. Thus, don't neglect the validation of your data.

The best time to start defining your validation rules is when determining the conversion rules. A machine-based validation process allows you, for example, to use the results and any identified erroneous source data for the data cleansing process.

**Validation before or after data migration?**

You should consider early on how you can validate your data later, during integration testing and go-live.

Two approaches are available for validating your data during a data migration:

- **Validation before the data migration**
  In this case, the source data is checked for syntactic correctness (e.g., length and data type) and semantic correctness (e.g., meaningful values with reference to the business process configuration). Conversions that are made subsequently will be ignored.
- **Validation after the data migration**
  In this context, you check the number and completeness of data records in the target system by asking yourself a few questions: Have all values been converted correctly? Have the corresponding fields and structures been filled correctly? Can the business processes, based on the loaded data, be executed without errors?

A combination of both approaches is frequently used.

### 2.1.7 Productive Load

The productive load phase is the most critical one and forms the basis for the later go-live decision. If you have done everything correctly in the previous phases, tested sufficiently, and planned your cutover cleanly and above all in detail, you can start this phase with a clear conscience. However, often small, inconspicuous things can mess up your cutover.

Let's consider a few examples from our previous work as data migration consultants:

- **Works councils, trade unions, working time legislation, and authorities**
  Works councils, trade unions or the trade supervisory office should be involved in planning your implementation project. Night work, weekend work and overtime must be arranged with or registered with the appropriate groups and offices. These processes require lead times and cannot be rushed. Otherwise, you may find that none or only a few employees are available for the final inventory at the go-live weekend.

Public holidays and vacations

- **Public holidays, time zones, and calendar weeks**
  Every country has specific periods where you can expect reduced productivity or the absence of project members, as well as of their subcontractors and other companies. In Germany, you can often write off May and June as "useless" months for your project. Depending on the state in Germany, the months of May and June have the highest density of public holidays, bridging days and vacation days (e.g., May Day, Ascension Day, Corpus Christi, and Pentecost). In France, the whole summer is a wash, as is the US between Thanksgiving and Christmas or China around

Chinese New Year and Golden Week. Thus, if possible, you should not schedule any elaborate work, tests, or go-lives within or shortly after those periods.

**Time zones and daylight saving time**

If you plan a data migration across time zones with multiple teams, you should make sure that the different time zones are also included in your planning. Starting a data migration, for example, at 10 a.m. CET, assumes that a team member in the San Francisco data center has also been scheduled at the dead of night and is available to provide the data. Pay attention to the different handling of daylight-saving time and standard time and their different changeover dates in different countries.

**Calendar weeks**

Another curiosity is the various ways to calculate the beginning of the first calendar week of a year. In Europe, calculation according to International Organization for Standardization (ISO) 8601 is common. Here, the first calendar week is the week that contains at least 4 days in the new year. Other countries, such as the US, calculate the first calendar week as the week which includes January 1. As a result, there may even be a calendar week 54 in the US, which is not even possible with DIN 1355/ISO 8601. Depending on the calculation method, the calendar weeks within a year may therefore differ from country to country. Confirming deadlines based on calendar weeks can be rather dangerous if you need to adhere to your schedule. For example, 2022 was a year in which calendar weeks differed between the US and ISO-8601 countries.

- **Communication of downtime**
  Communicate downtime with your suppliers and customers so that trucks aren't parked in front of your facility during that downtime, waiting to be loaded or unloaded.
- **Check and adjust automatic system processes**
  Interfaces to and from existing source systems or workflows in the target system should be disabled during the data migration process. In this way, you avoid falsified data, a system breakdown due to bulk generation of workflows, or the flooding of inboxes of the people in charge of workflow entries. Another problem that is frequently underestimated involves unplanned automatic backups. A backup process suddenly starting in a source system during the extraction of data its or in the target system during the loading of data can delay your schedule by more than 12 hours. In other words, if you notice only the next morning that the overnight data extraction or data transfer was too slow or was canceled, you must start all over again.
- **Upgrades and hotfixes in cloud environments**
  In cloud environments, such as SAP S/4HANA Cloud Public Edition, upgrades and hotfixes are deployed to the respective systems at specified

time slots (in what is called a *blue-green deployment*). Your data migration activities such as extracting data, loading data, etc. should be scheduled so that they do not conflict with downtime. Although these dates and time slots should be known to every customer, unfortunately a common problem is that customers schedule their activities in parallel with these time slots. In SAP S/4HANA Cloud Public Edition, you should definitely observe the following SAP Notes:

- SAP Note 3024158: Blue-Green Deployment for Hotfix Collection for RISE with SAP S/4HANA Cloud (for production systems)
- SAP Note 3042314: Hotfix Collection Deployment for RISE with SAP S/4HANA Cloud (for non-production systems)

## 2.2 Work Packages in Data Migration Packages

In this section, we'll describe the work packages involved in a data migration project. The sequence of packages and their dependencies on one another are also discussed.

Basically, you should always ask yourself the same questions before each project:

- What are the necessary tasks?
- When do these tasks occur?
- In which phase are the tasks included?
- In which order are the tasks processed?
- Who can do these tasks?
- How much does it cost?

When implementing a particular software, the tasks are often always the same and can therefore be easily incorporated into a process model or framework, such as SAP Activate. Looking at data migration projects, some tasks are recurring and thus can be structured and divided into work packages.

SEAMAP

The structure we describe for effectively dividing migration projects into work packages (*19/2 SEAMAP*) has been developed over the years and is used to plan and estimate the effort for required for data migration projects. This template for planning can be customized for each project. In 2022, another package was added to the structure, changing its name from *19/1 SEAMAP* to *19/2 SEAMAP*. Figure 2.2 shows an overview of the work packages of this structure. For reasons of clarity, work package 20 (tool selection) is not shown. The package relationships related to tool selection are shown later in Figure 2.4.

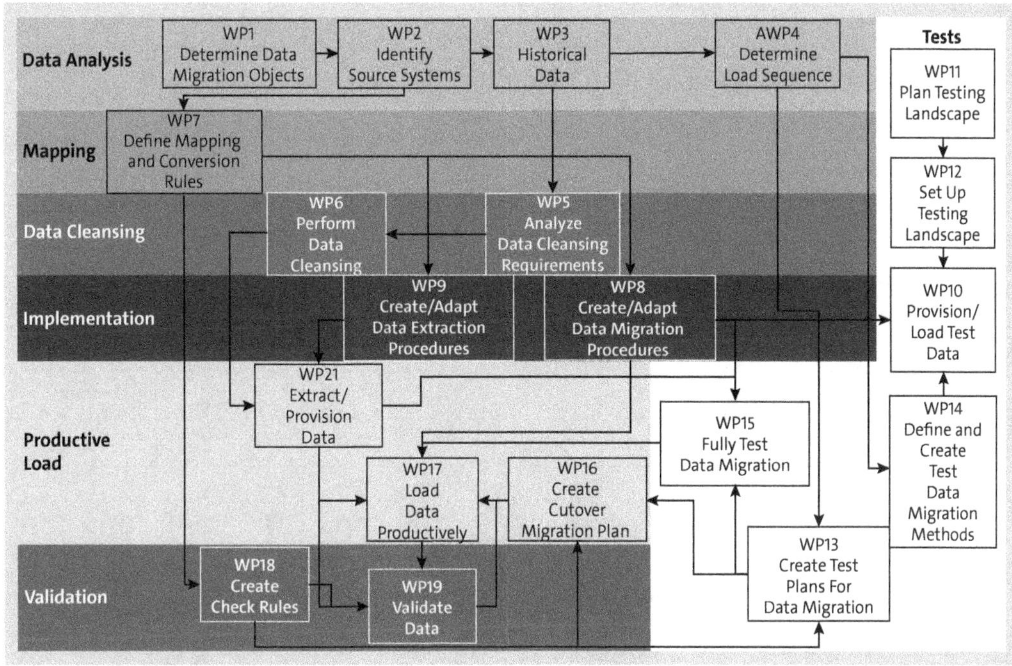


**Figure 2.2** Overview of the 19/2 Work Packages of a Data Migration Project (Source: 19/2 V7 SEAMAP, © Frank Finkbohner)

**19 full and 2 optional packages**

Basically, each migration project can be divided into approximately 19 work packages. However, there are two work packages (WPs) that are more optional in nature. In contrast to all other work packages, WP14 (Define and create test data migration methods) is not necessarily a classic data migration work package. It is only rarely used in parallel with data migration. For more information, see Section 2.2.6. The provisioning of the data is normally included in WP17 (Load data productively). However, in larger projects, companies or organizations, the data is usually provided by different teams than being loaded. To better support planning in this context, there is now an additional work package 21 (Extract/provision data). For details, see Section 2.2.7. That's why we talk about 19/2 work packages here. However, due to the size, you can download a complete overview of all packages and their dependencies from the book's bonus page (at *https://www.sap-press.com/5817*).

### 2.2.1 Work Packages Related to Data Analysis

Figure 2.3 provides an overview of the *data analysis* work packages.

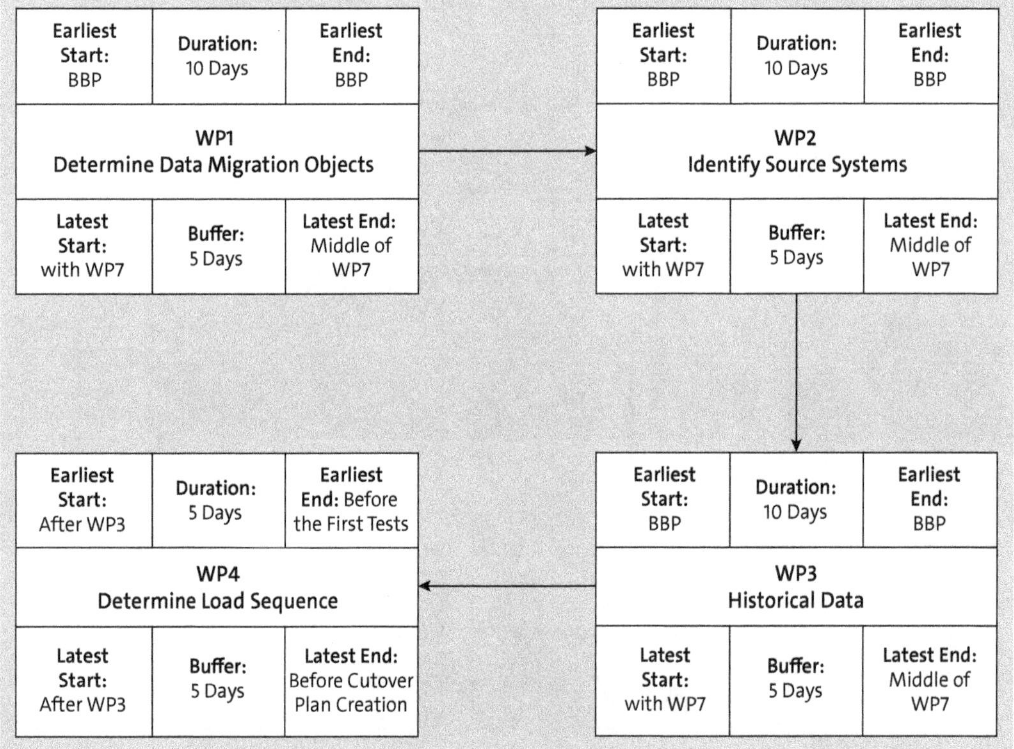


**Figure 2.3** Work Packages for Data Analysis

### Work Package 1: Identify Data Migration Objects

**Identify data migration objects**

The first work package of a data migration project serves to identify the necessary data migration objects. Here we distinguish, as already described in Section 2.1.1, between business objects and data migration objects. A data migration object is any separately loadable sub-unit of a business object that cannot be loaded in its entirety.

**Customer Master**

The customer master is a business object in terms of the business process. For data migration, however, it can consist of multiple migration objects if, for example, it is not technically possible to load the data as an object.

**Start time**

The earliest possible start time for this work package is the start of the prepare phase in SAP Activate. However, it is often better to start when the prepare phase is halfway through because the business objects required for these processes only emerge once they have been completed to a certain degree. Maintaining a list of business objects in parallel during the prepare phase is a good idea and requires little effort, while simplifying the work of your data migration subproject. Based on this list, the data migration subproject can then identify and describe the business objects. We'll explain what basic information should be collected in such a list in Section 2.3.1.

Identifying data migration objects is a process that does not end with the prepare phase. Gaps in the configuration of business processes are often found during the mapping or even the implementation phase of a data migration. You may thus need new data migration objects that no one thought of at first and that also have an impact on the existing configuration.

**Missing Configuration for Units of Measure**

In a large data migration project, the configuration may not have provided alternative units of measure for material masters. Alternative units of measure are additional units of measure that can exist in parallel to the base unit of measure of a material or product. Stocks are posted in base units of measure.

However, during the investigation of the datasets of the individual country systems, it might become apparent, for example, that several hundred materials in different country systems had different units of measure. Country A has stored material XY in "rolls," country B in meters, and country C in inches. Since country A was the main system, that country's system inventories would have been posted in terms of rolls during the data transfer. In other words, 400 meters of stock in Country B have become 400 rolls, and 2,000 inches of stock in Country C have become 2,000 rolls. Massive problems and equally massive costs would arise if not been detected in time. As a consequence, alternative units of measure were included in the configuration and manually maintained in a Microsoft Excel file for these materials. This file was then considered as an additional file for alternative units of measure when loading the material master data.

**Capturing data manually or automatically?**

Furthermore, you should determine whether the data should be migrated automatically or manually. In the following circumstances, migrating data manually may be easier and less expensive:

- The number of records to be migrated is small.
- The effort to manually enter the data into the system is smaller than the effort to create and test the programs. The hourly wage of an assistant employed for this purpose is often many times lower than the hourly wage of a programmer. Since employees need to be trained in the processes and data entry in the new SAP systems, you can also have such data entered as part of an employee training course, for example. Doing so saves development costs, and at the same time, employees become more familiar with the system's data entry processes.

**Effort Calculation**

Entering a data record in the target system using an app or transaction takes one minute. It then takes one person 600 minutes or 10 hours to manually enter 600 data records. Four people do the same job in 150 minutes (2.5 hours).

You also need to consider the number of tests required for this data and the complexity and error rate of creating it manually. A repeated manual creation may then be more cost intensive than an automated process.

**Timeframes**

In general, however, the identification of data migration objects should be completed by the middle of the mapping phase. You should not need more than 10 days (net) to identify the data migration objects.

#### Work Package 2: Identify Source Systems

**Identifying source and target systems**

Once the processes and data migration objects are largely known, you can start identifying your source systems. As with data migration objects, the earliest possible time to start is the beginning of the prepare phase in SAP Activate. However, you should have completed this work package by the middle of your mapping phase at the latest.

**High-quality data sources**

Identifying the data sources may seem simple at first glance, but on closer inspection, it can prove to be a complex task—especially if you have generated a highly complex infrastructure with many specialized databases and interfaces between these databases over the years. The "zoo" of different systems that has accumulated over the years often needs to be sorted out first. In such a case, a vitally important step is to clarify at an early stage—based on the data streams—from which systems you're going to take your data. You should find out where the most reliable, high-quality, and up-to-date sources are for your data to be adopted, but also take a close look at your target system. Having only one system to populate is usually the simplest case. If you need to fill in several different systems, you must be aware of this need early on.

**Timeframes**

It should be possible to identify your source systems within 10 days. In complex cases, such as the migration of two different systems with redundant data into one SAP system, you might encounter some difficulty to determine the leading system with the best data quality. In such cases, it may take a little longer.

#### Work Package 3: Historical Data

As described earlier in Section 2.1.4, a vitally important task is to clarify whether, and in what way, historical data should or may be transferred into the system.

In general, you can distinguish between the following types of historical data:

- **Master data no longer used**
  Master data that is no longer used includes, for example, data from suppliers that no longer exist, routings for finished products that are no longer produced, buildings as assets that have long since been sold, organizational data from staff positions that have been outsourced, etc. Migrating this type of master data into one system to another does not make sense.
- **Completed transaction data**
  This data includes delivered and paid orders, invoices, account activities, etc. The desire to migrate this data to the new system is often expressed. However, the migration of this data is not easy since in most cases there are no interfaces for the transfer of such data at all and it involves many risks.

**Obsolete master data**

Obsolete master data should only be transferred to a new system if absolutely necessary. The term *obsolete* already indicates that this data is no longer up to date. It only costs storage space and often causes problems when you try to load it. For more details on this topic, see Section 2.1.1 and Section 2.1.3.

**Migrating Human Resources (HR) Master Data**

In a large data migration project for HR master data, the organizational levels and their respective assignments to the HR master records of the last 10 years needed to be transferred. Taking over the complete organizational structure of a company from the last 10 years, with all its changes, involves a great deal of work. When the question came up as to why this was necessary, reference was made to quarterly statistics for the Board of Directors. These statistics list the number of employees per organizational level once every 3 months. Since the customer also uses a *business warehouse system* (BW system), it was agreed to transfer all previously created statistics to this BW system and only the current organizational levels. All changes to the current organizational structure were transferred to this BW system so that the Executive Board can now access current and old statistics at any time. This procedure massively simplified the data transfer, saved on costs, and ultimately achieved the same result in the end.

**Prohibiting the migration of completed documents**

As mentioned earlier in Chapter 1, Section 1.3, completed transaction data should always be viewed critically. For example, from an accounting point of view, in some countries, it is forbidden to transfer closed financial documents from another system. This rule exists because you would actually post a previously posted document for a second time, which is against accounting rules. Only with certain paid tools and services from various providers is this allowed and possible under certain circumstances. These tools write the data directly to the database tables, which technically speaking, is not a posting of the document. Before accepting closed transaction data, you should consult with your tax and financial auditors about your legal situation.

Often, the transfer of completed transaction data is simply too complex due to having many dependent subsequent documents. Thus, usually, no built-in data migration programs from SAP exist for the transfer of completed transaction data. So, before you insist on adopting historical data, you should understand the risks and the effort and consider alternative solutions.

**SAP S/4HANA**

For SAP S/4HANA and SAP S/4HANA Cloud, a brief guide for the migration of financial accounting master data and transactional data called *Migration of Financial Accounting Master Data and Transactional Data* is available in English at SAP Help Portal via the following link: *http://s-prs.co/v581707*.

**Timeframes**

For the earliest possible start time and the latest end, the same applies in this context as for the work packages previously described. Due to the many restrictions, it is usually possible to complete this work package within 10 days.

### Work Package 4: Determine the Load Sequence

Once the data migration objects and source systems are known, and a decision has been made about loading historical data, you can start planning the sequence of loading data, i.e., the load sequence. Since some data migration objects depend on other data migration objects, it is best to complete this planning process to a large extent prior to the first tests. For example, you cannot load sales documents without material master and customer master data.

**Determining the Load Sequence in SAP S/4HANA**

In SAP S/4HANA and SAP S/4HANA Cloud, the predecessor objects of any given data migration object are listed in the online documentation for each data migration object. In the SAP S/4HANA migration cockpit, you can also view these objects in the object overview of the project settings.

Planning the load sequence is the basis for subsequent *cutover planning*. The latest end date for this work package should therefore be set before the cutover planning, that is, before work package 16 (create cutover migration plan).

### 2.2.2 Selecting Tools

After defining the data migration objects and source systems, the selection of the right tools for your data migration also begins. The choice of tool is so important for a data migration project. Choosing the wrong tools at the start of a project can result in high additional costs for the project, especially in training, maintenance, and licenses.

#### Work Package 20: Tool Selection

Selecting tools

The selection of tools is made in this work package. Basically, you can distinguish among the following types of tools during a data migration:

- Tools for data cleansing
- Data extraction tools
- Tools for loading data
- Data validation tools

Figure 2.4 shows an overview of this work package and its dependencies.

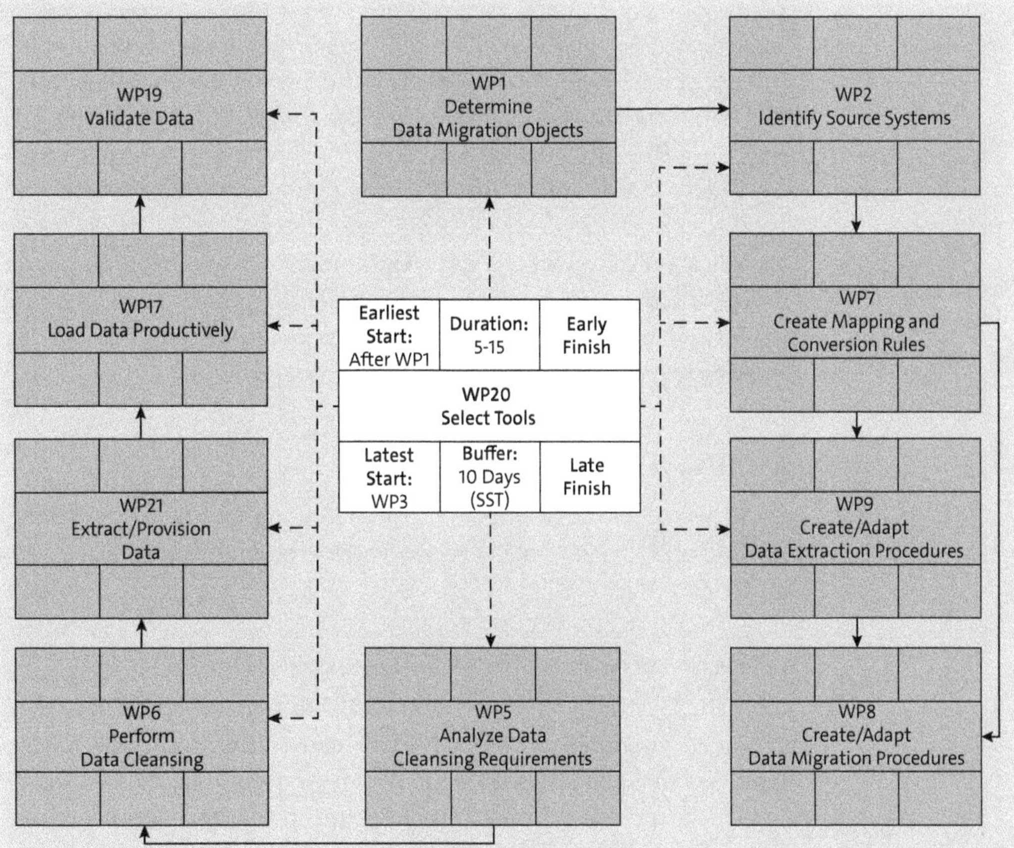


**Figure 2.4** Work Packages for Selecting Tools

ETL tools

A wide variety of tools exist on the market, some of which are called *extract, transform, and load (ETL) tools*, which can cover all aspects of data migration.

Time-consuming tendering procedures

Some tools built into SAP products, such as SAP Legacy System Migration Workbench (Transaction LSMW; see Chapter 9) or the migration cockpit (see Chapter 3), cover only some data migration requirements. Others, such as the ETL tool called *SAP Data Services*, cover a considerably broader range of data migration requirements. When choosing your tools, in addition to compatibility with your source and target systems, you must find and secure buy-in from your relevant experts for these tools.

**Long Selection Process for Tools in the Public Sector**

Also note that, in government agencies, the purchase of a software tool often must be "put out to tender" on a national or even international basis. In one such authority, for example, the acquisition of a data transfer tool was put out to tender across Europe. After a very long tendering phase (> 6 months), a number of tools were shortlisted. In the end, a particular tool was chosen, and a lot of money was spent on it. In addition, consultants were hired for this purpose, and some internal employees were trained on it. But since this tool was not common on the market, something happened that always happens in such a case: No migration consultants or specialists were found who had knowledge of the tool. There was also no budget for training external staff. In the end, this bidding process delayed the project by months, and it was only used for a very small area.

Procedure

Thus, you should always get a sufficient number of comparable quotes for the different tools. Ask the vendor for references and also check freelancer portals, IT job market forums, and project exchanges to see if you can find specialists for specific tools in your domestic market.

### 2.2.3 Data Cleansing

Data cleansing is, as already mentioned, an important component of a data migration project. Figure 2.5 provides an overview of these work packages and their dependencies.

#### Work Package 5: Analyze Data Cleansing Requirements

Once the bulk of the data migration objects and source systems have been determined, you can begin to analyze the data cleansing needs. As a rule, this process should not exceed 5 to 10 days per system. Analyze your source data to make sure it is accurate, complete, and up-to-date. Correct and complete any erroneous data and delete obsolete data and duplicates from the system.

2

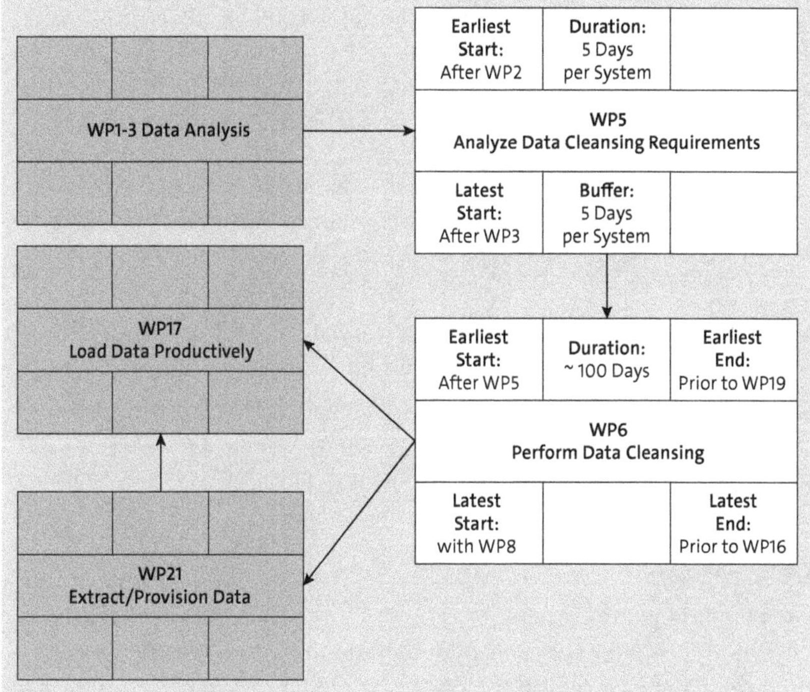


**Figure 2.5** Work Packages for Data Cleansing

### Work Package 6: Perform Data Cleansing

The data cleansing process may take several days or even months to complete. However, the data cleansing itself should be completed at the latest before the start of the productive data loading (final data extraction) of the data into the production system. The best time is before the last productive load test. Then, this data can be used for the dress rehearsal for productive loading. For better data quality, you should set up data cleansing as a permanent process that you run until the final data migration.

**Archiving**

When performing data transfers to existing systems, you may need to delete existing data in the target system. However, data in SAP systems cannot be deleted so easily. In such a case, you should familiarize yourself with the archiving of SAP data records.

**Deleting Existing Material Master Records Using Archiving**

In a multistage data migration project, we needed to load material master records into an existing production system. In previous data migrations, however, many material master records had been loaded in advance based on rules that subsequently proved to be incorrect. The easiest thing to do seemed to be delete these materials from the existing system. However, materials for which transaction data still existed in the production system could not be touched. After thorough examination, we decided the best solution was to archive these materials with an SAP archiving object for the material master, since the SAP archiving program identifies material masters with existing transaction data. Using a small, additionally developed program in the corresponding *Business Add-In (BAdI)*, we could determine and collect exactly those material masters for which there was no transaction data or other references in the SAP system. In the second step, exactly these material masters were then archived and deleted from the system.

Archiving data can also be very useful if you want to delete incorrectly loaded records of business objects from the system during data migration tests. For this purpose, depending on the archiving object, it may be necessary to set appropriate deletion indicators for the existing business objects. These indicators can then be used to specify the business objects that can be archived.

### 2.2.4 Mapping

The mapping phase is the right time to think about the appropriate data migration technique for each data migration object. Defining the conversion and mapping rules is best done based on the data structure of the data migration object offered by the selected data migration technique.

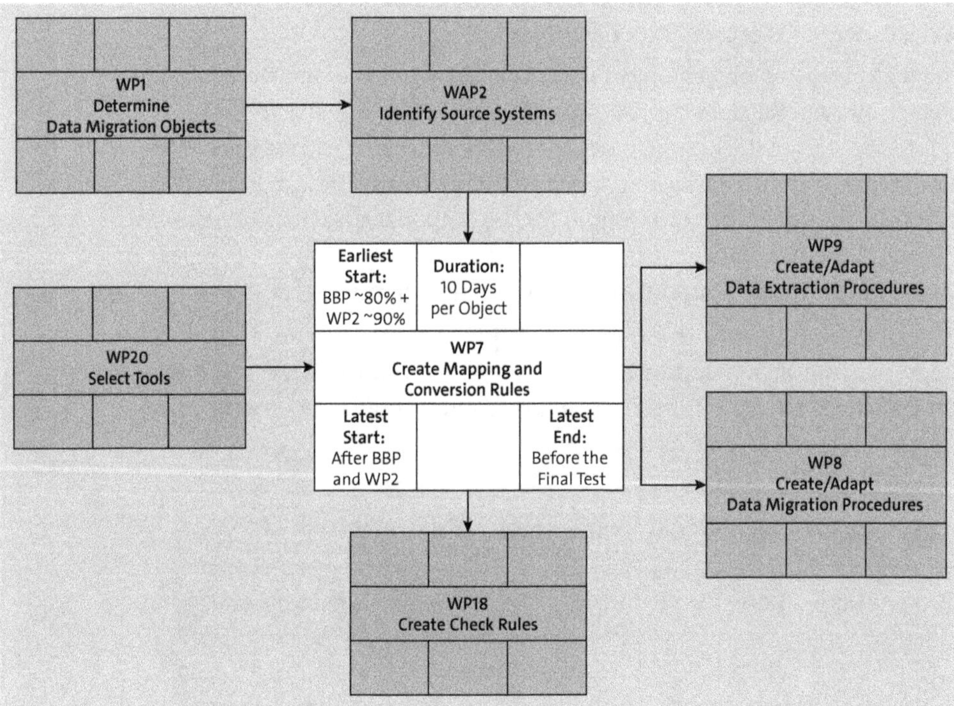


**Figure 2.6** Work Packages for Mapping

IDoc or batch input



For example, the more organizational structures you need to map, the more suitable a data migration tool such as the migration cockpit or a tool that uses IDocs (see Chapter 8) is for data migration. In general, you should always check the delivered migration objects of an existing data migration solution first before using alternative techniques. Batch input (see Chapter 7) should not be ruled out as a data migration technique from the outset. There are definitely data migration scenarios where batch input can be used to solve a specific problem for which there is no standard solution. Figure 2.6 shows an overview of the work packages for mapping and their dependencies.

[Ex]

**Map Product/Material Master in SAP S/4HANA**

The product master is best migrated via the migration cockpit and its migration object **Product**. A migration via IDoc is also possible, but it must be checked whether the data structures required for SAP S/4HANA are comprehensively mapped in IDoc type MATMAS06. Using the migration cockpit and IDoc, several organizational levels (company code, plants, sales organizations, storage bins, etc.) can be transferred in one load process. The corresponding IDoc can be highly parallelized for this purpose by using dialog processes. Batch input programs must be recorded, implemented, maintained, and tested specifically for each material type and view. This drives upload time and implementation and testing costs tremendously. New customer-specific views or fields can also be integrated relatively easily in migration objects of the migration cockpit using the SAP S/4HANA migration object modeler or IDoc.

### Work Package 7: Create Mapping and Conversion Rules

The earliest possible time to create the conversion rules is when the configuration of about 80% is complete because then most of the conversion rules are known. However, you must have started at the latest when all data migration objects have been determined; but you should be finished with them before the final tests.

Paper mapping

You should use existing migration templates to map individual data migration objects. This so-called *paper mapping* can be performed, for example, with a corresponding Microsoft Excel file. Talk with the appropriate team members and IT staff through the mapping of the source structures and source fields to the target structures and target fields and enter it there. You should also define appropriate rules for converting source values to target values.

[+]

**Use the Object Overview**

If you use SAP Legacy System Migration Workbench, you can use, for example, the *object overview* described in Chapter 9, Section 9.2.18. Clicking the **Objects as table** button lists the selected target structure of the data migration object and any existing field mapping in a table, which can then be easily exported to a comma-separated values (CSV) file, imported into Microsoft Excel, and used for paper mapping.

For the migration objects of the migration cockpit, the *field list* of the migration template file is suitable as it contains all necessary information of the target structures. Copy this field list into another Microsoft Excel document and extend it with columns for source structures and source fields, and your paper mapping Microsoft Excel is ready!

Persons involved

In mapping meetings, we recommend you "bring to the table" both employees with technical knowledge regarding the source system and employees with technical knowledge regarding both current and future business processes. Be sure to invite the right people. Never invite too many people to such meetings and, if possible, limit the meeting to one business item. You can thus reduce the risk of getting bogged down and going from the smallest to the largest. Nowhere is the saying "too many cooks spoil the broth" more true than when creating a mapping.

[»]

**Talk to the Right People**

In an HR data migration project, details for a specific data migration object of an HR infotype were to be discussed with a person in charge. During the conversation, we quickly realized that this person was a key person who had been virtually uninvolved throughout the implementation project. The result of this meeting was that additional necessary infotypes were identified that had not previously been considered at all in the configuration and therefore not in the data migration. So, the project suffered a few weeks of delay to complete the new tasks for configuration and data migration, and the go-live date had to be postponed by a month.

During the configuration process, changes occasionally occur in the business process that affect the mapping and conversion rules. For this reason, you should be in constant contact with the relevant subprojects and the responsible people in order to identify such rule changes at an early stage. Dig deeper if rules are not delivered. In some cases, it can take weeks or months to establish certain conversion rules.

**Time Management in the Project**

In a data migration project, we needed more than 10 months for a specialist department to agree on the exact classification of purchasing groups for certain material masters and their associated conversion rules. The rules were then delivered a few days before the cutover and could not be tested until just before the final productive load. Make sure that you involve all the necessary people and groups in the planning in a binding manner and that the supplying specialist departments are included in the data migration processes.

Creating the mapping and conversion rules is a process with a high communication content that should not be outsourced *offshore*. The individual parties involved in this process should speak a common business process language and, if possible, the native language of those who are responsible in order to avoid any misunderstandings.

**Delays Due to Insufficient Language Skills**

A company in the manufacturing industry had outsourced parts of the software development and also the data migration process to service providers with developers outside the German-speaking region. English was chosen as the project language. The German staff did have English skills, but they tended to be at a primary school level. Their knowledge was sufficient to order a pizza on vacation, but not to prepare an appropriate performance description in English. The language barrier in their English-language meetings was accordingly high. Miscommunication often led to misunderstandings when working with service providers, which resulted in rather long development times because everyone involved was talking past each other. Do not assume that all your project members or your customers and suppliers have sufficient skills to speak business English fluently.

**Checking conversion rules**

With some data constellations, problems may only occur after mapping and initial testing, which then flow back into a new analysis of the object and result in modified mappings. Conversion rules that were clearly defined during the analysis phase may turn out to be incorrect during initial mappings or implementation tests.

**Address Data**

Address data is often maintained in contravention of existing regulations. Legacy systems often have too few or no fields to capture mobile numbers, email addresses, or web addresses, for example. The responsible persons then often use free fields to store this data. Many of these systems also allow this capability in the absence of context checking. Initial tests then

> reveal that email addresses, for example, were maintained as a third address line, or that phone numbers did not adhere to any particular maintenance schema and were sometimes maintained with or without country codes. Sometimes, it also turns out that the processors maintain their own databases or Microsoft Excel spreadsheets, and the system's data was never really up to date. Such findings then led to new data migration objects, mappings, and also data cleansing activities.

**Testing and validation rules**

Also, when defining the conversion rules, you should think about the later validation of the data and record your findings.

Additional testing and validation rules must always be defined and agreed to by the customer. The customer alone is responsible for the subsequent verification of the data in the target system. Only they can decide whether a data record is correct or incorrect.

### 2.2.5 Implementation

Once you have defined your initial conversion rules, you can start implementing the data extraction and data migration procedures. A guideline here is that approximately 90% to 95% of the conversion rules for a data migration object are completed. In some cases, this value may be lower. On average, you should expect about 10 days per object to implement the data extraction and data migration procedures. For some objects you may need only one day, and for other, complex objects, however, the value of 10 days can be exceeded by far. On average, however, ten days is a good estimate. You should use predelivered data migration objects, such as those of the migration cockpit, which you can use without customization, because then you only need to include times for the extraction routines. However, you should be finished with the programs before the last productive load tests.

**Customizing affects the implementation**

Depending on the migration technique, it is important that the configuration of the individual processes is completed as far as possible. This is a necessary requirement especially for data migration processes via data migration programs that use batch input. If the customizing of the processes, especially the screen control, is not completed, errors may occur during the first functional tests. In the case of transaction records, the records or any programs generated from them must be continuously adapted to the modified screen control. IDocs and migration objects of the migration cockpit are usually somewhat less sensitive with regard to screen control, since the data is usually stored here using internal consistency, check and posting modules. However, there are also IDocs that store data using the batch input technique. For these IDocs, screen control customizing must also be completed before you can test without errors. However, unlike transaction records, you can start the implementation here at any time. Figure 2.7 shows the work packages for the implementation process.

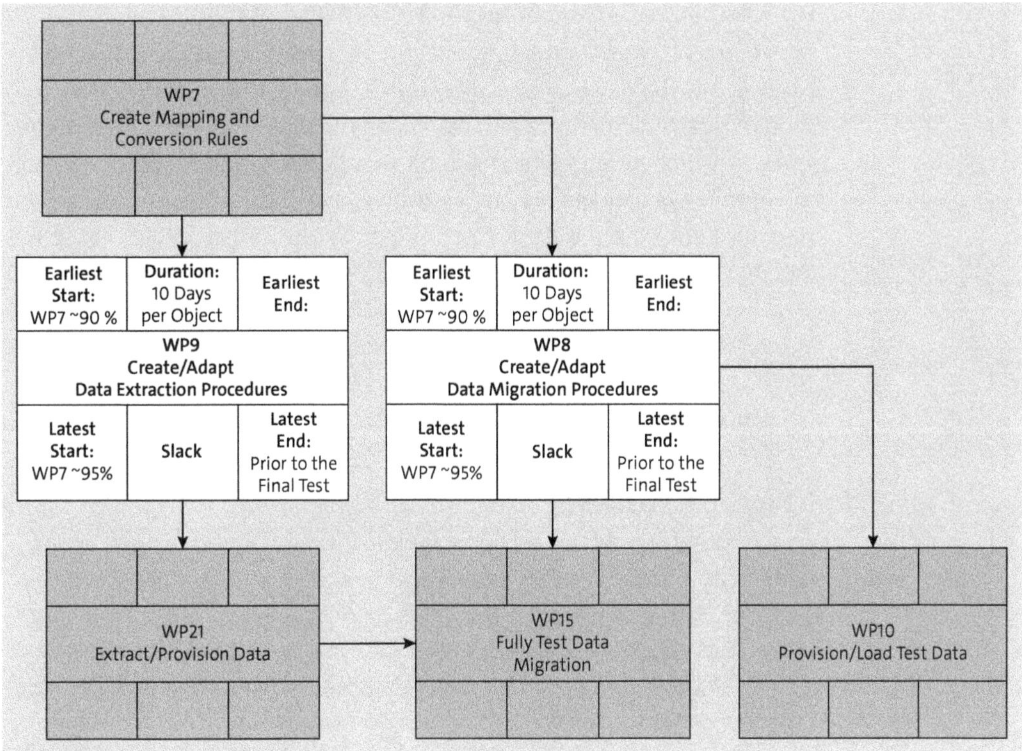


**Figure 2.7** Work Packages for the Implementation Process

**One-way programs**

Normally, data migration or data extraction programs are so-called *one-way programs*. This means that they are usually used only once and then never again for data migration. Only in the case of phased rollouts are these programs used over a longer period of time and possibly also adapted.

**Loading via permanent interfaces**

The use of a permanent interface for loading data must always be analyzed critically. It can be useful if the target system is a subordinated system in a system network and is therefore permanently filled with data via an interface. However, when such a system is initially filled, a very large number of data records are transported through this interface in a very short time. The large number of data records then often exceeds the normal load of these interfaces many times over. This usually leads to an overload of the input queues of the interfaces and to delays. So, if you plan to implement the initial filling of a system via a permanent interface, you should consider all the risks that could jeopardize your production startup. Under certain circumstances, it may then be useful to carry out the initial filling via separate data migration programs of work package 14 (Define and create test data migration methods). These can then also be used to provide test data for the functional and integration tests of the business processes.

### Work Package 8: Create/Adapt Data Migration Procedures

Depending on the data migration technique and data migration tool, work package 8 is the package that involves the greatest or smallest effort. You should start implementing the conversion rules early on, namely, once they are 90% complete. If possible, use built-in validations and checks of the tools to improve the data quality and identify erroneous data records or conversion rules. For example, if you can program in ABAP and use SAP Legacy System Migration Workbench, you can search for existing check function modules and use them in field mapping. The correctness of the target value of a value mapping of a migration object of the migration cockpit is usually checked against a stored domain or target table. There are only a few exceptions in which, for technical reasons, such a test cannot be performed.

**Use Check Modules**

Every SAP system offers a wealth of internally used function modules and methods that can be used, for example, to check zip codes, bank accounts, etc. for plausibility. You should use these function modules or methods in your field mapping and use them to output in a log erroneous data records that have been found. You then submit this log to the appropriate department or data cleansing team who can then correct the incorrect records in the source system. Check if the corresponding rules and configurations of the checks (such as zip codes) are up to date.

**Continuous adaptation**

Always assume that, especially at the beginning, your source data is often not of the quality that your mapping and conversion rules will work properly right away. For this reason, if possible, you should actively check the data for errors during the data conversion. Through this task and by permanently checking and adapting conversion rules to your existing data records, you can successively increase the data quality of your target data.

### Work Package 9: Create/Adapt Data Extraction Procedures

For the creation of the extraction programs, you must work closely with the maintainers of your source systems. Often, these people are the only ones who have sufficient the experience and knowledge to extract this data successfully. Particularly when dealing with poorly documented, self-programmed customer systems, a necessary step is often to transfer this work package to the appropriate support staff and to integrate them closely into the project.

### 2.2.6 Testing

The work packages of the testing phase, along with the work packages of data validation, are the crucial packages in a data migration project to ensure sufficient data quality. Figure 2.8 shows the different work packages for testing and their dependencies.

**Simultaneous testing and mapping**

The findings of the initial functional tests can be used to improve the mapping. Sometimes, the analysis of a data migration object can only be completed when the first source data is mapped, and the corresponding conversion rules are thus subjected to an initial check.

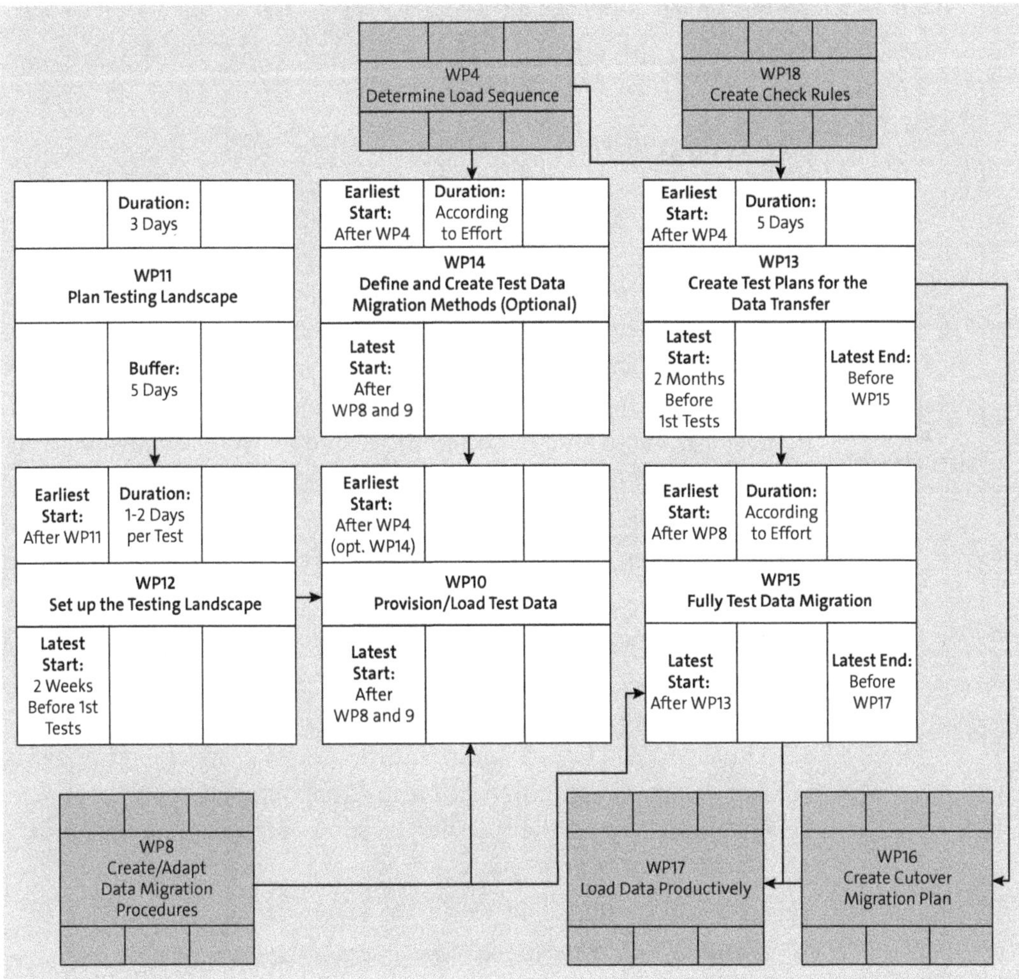


**Figure 2.8** Work Packages for the Testing Process

**Dealing with Bad Legacy Data**

In an international HR master data migration project, the rules for transferring employee home addresses had proved unusable after the initial tests. Part of the data in the source system was so outdated and so poorly maintained that an address transfer would not have made any sense (zip codes had the wrong format, country information was missing, etc.). The data quality would have been possible only with extensive manual changes subsequently. Our customer therefore decided not to adopt this data. After the go-live, each employee had to enter their private address in the system using an employee self-service (ESS) scenario. This updated the database and greatly reduced the cost of post-maintenance activities.

### Work Package 10: Provide/Load Test Data

Work package 10 (provide/load test data) covers all the activities for providing data for the functional tests based on the data migration programs, interfaces, or also using other methods. This work package differs from work package 15 (fully test data migration) in that there are no test plans here yet and the data will be functionally loaded and tested after a data migration program is completed.

**Recording the duration of migration steps**

Now is also the right time to record the duration of the individual migration steps. You should start recording this time data for the individual loading steps as early as possible:

1. *Duration of the extraction* of the data in the source systems
2. *Duration of any adaptation of the data* or transfer to a special migration template file
3. *Duration of the provision* of the data on the corresponding load servers (sending files of more than 500MB over a poor network or internet connection can take some time)
4. *Duration of reading the data into the load tool* (A difference exists whether a very large file is read directly from the application server, via a network drive, or via the operating system frontend, or whether the data is read directly from the source tables.)
5. *Duration of converting* the data in the tool
6. *Duration of validating the conversion results* (first pre-validations)
7. *Duration of loading the data* in the target system
8. *Duration of data validation* before and/or after loading

The sooner you have an overall picture of these data migration times, the better you can plan your go-live and testing. Simply recording the load times is often not sufficient because load times are often only 15% of the total migration time, starting with the extraction of the data and ending with the final data validation. Most of the time—about 75%—should be spent on data validation before and after loading the data.

### Work Package 11: Plan the Testing Landscape

Custom migration systems

An important step for any data migration project is that you have either your own migration system or at least your own migration tenant in the development system and the quality test system.

[«]

**Rolling Test Systems**

In a large international logistics project, a separate migration system was set up in parallel with the Q-system. The individual migration tests served to set up the system for the *integration tests*. Once the migration tests were completed, the system was used for the *integration test* and the old integration test system was deleted and used for additional migration tests. During the integration test, the loaded data was intensively tested against the implemented processes. Any errors found or changes requested were immediately sent to the migration team and incorporated into the new data transfer tests.

These rolling test systems ensured that a fresh system with data based on the latest findings of the preceding integration tests was available at the start of each new integration test.

System snapshots

You should also take advantage of the option to use *snapshot backups* during migration testing.

Each time you load a larger object, you should create a *snapshot* of your system because, if you find any errors after the load process, this way you can easily reset the system. Snapshots consume a lot of storage space, but they save a lot of time and effort compared to a conventional backup and restore, which usually take hours and often have to be requested days or weeks in advance in a cloud system.

[«]

**More Information on SAP HANA Backup Types**

You can find more information on SAP HANA backup types and data snapshots in SAP Help Portal by navigating to **SAP HANA Administration Guide for SAP HANA Platform • Availability and Scalability • SAP HANA Database Backup and Recovery • SAP HANA Backup Types** or visit *http://s-prs.co/v581708*.

The **Data Snapshots** subitem also provides a comparison of backups and data snapshots: **Comparison of Data Backups and Data Snapshots**.

### Work Package 12: Set Up the Testing Landscape

Don't skimp on the wrong end when setting up the testing landscape.

**Testing environment close to production**

Make sure that you test in a testing environment that is either identical to that of the future target system or at least technically very close to it. The more the testing environment corresponds to the target environment, the more accurate the measured values and thus the more accurate your planning will be for the go-live. Even the smallest deviations can lead to widely differing results—but you'll only find this out if you allow enough time for testing.

[!]

> **Effects of Changes**
>
> Even one additional application server in a planned system network can cause the overall system to run more slowly than without the additional server if there's is an incorrect constellation, incorrect configuration, or errors in the operating system. Also, changing the internal memory has a significantly higher impact on load times than increasing the number of processors.

### Work Package 13: Create a Test Plan for the Data Migration

Once you have determined the order of loading the data migration objects and the first data migration programs have been created, now is the time to create the test plans for the data transfer. The findings of these test plans will then be seamlessly incorporated into the cutover migration planning.

The durations of the migration steps, recorded in work package 10 (provide and load test data), is necessary for creating the test plans. It forms the basis of these test plans. The test plans are a prerequisite for testing the entire data migration (work package 15).

### Work Package 14: Define and Create Test Data Migration Method (Optional)

**Provisioning data**

In extremely large implementation projects that are planned over a long period of time, an often necessary step is to provision data for individual tests. For example, if the target system is subordinate in terms of its data (i.e., is a non-leading system of a system network), permanent interfaces are often used to load the data into this system. These interfaces are then also used to load the data productively. The development of these interfaces, for instance, Application Link Enabling (ALE), is often rather complex and time consuming. For testing the business processes, test data is therefore provided via special data loading programs. If you use the same transfer techniques for these data loading programs as the permanent interface, the knowledge gained from this data loading process can be used to optimize the interface.



In case of capacity issues with the permanent interface, we advise using these test data loaders for the initial load of the target system.

This work package is optional and rarely found in implementation projects.

#### Work Package 15: Fully Test Data Migration

**Once is good; twice is better**

For the best data quality and cutover timing, you should fully (holistically) test the data migration as early as possible and as often as possible. Testing can be most useful when performed in the runup to an *integration test*. An example of this scenario is described in work package 11 (plan test landscape). The special case of this testing is the *productive load test*, which often takes place as part of what is often called a *user acceptance test*.

[«]

> **Productive Load Tests**
>
> As mentioned earlier, the testing of productive loads should be done completely at least twice in a testing environment that resembles the production environment. This task is the only way to detect and eliminate bottlenecks (whether technical, time related, or personnel related) at an early stage.
>
> Start the tests as early as possible. The best time to for starting testing is when the data migration programs and data transfer test plans have been completed.

### 2.2.7 Productive Load

The productive load phase is the most critical phase of the entire project, often characterized by fixed deadlines and tight timeframes. In this phase, we see how well you've done your homework. The result is mercilessly analyzed by means of data validation. Figure 2.9 shows an overview of the work packages for productive loads.

#### Work Package 16: Create Cutover Migration Plan

Before the data gets loaded productively, a *cutover migration plan* must be created. The best way to start the initial migration planning activities is to schedule the loading sequence of the migration objects. However, you should start, at the latest, when you have created the test plans for the data migration. The test plans are the basis of the migration planning. As a rule, the migration plan is completed after the final testing of the entire data migration. The total time to create this plan should not exceed 5 days. In the case of *phased rollouts*, the duration can of course be longer due to the number of planned go-lives. Due to the learning curve, migration plans for further rollouts should not take as long as for the first initial data migration.

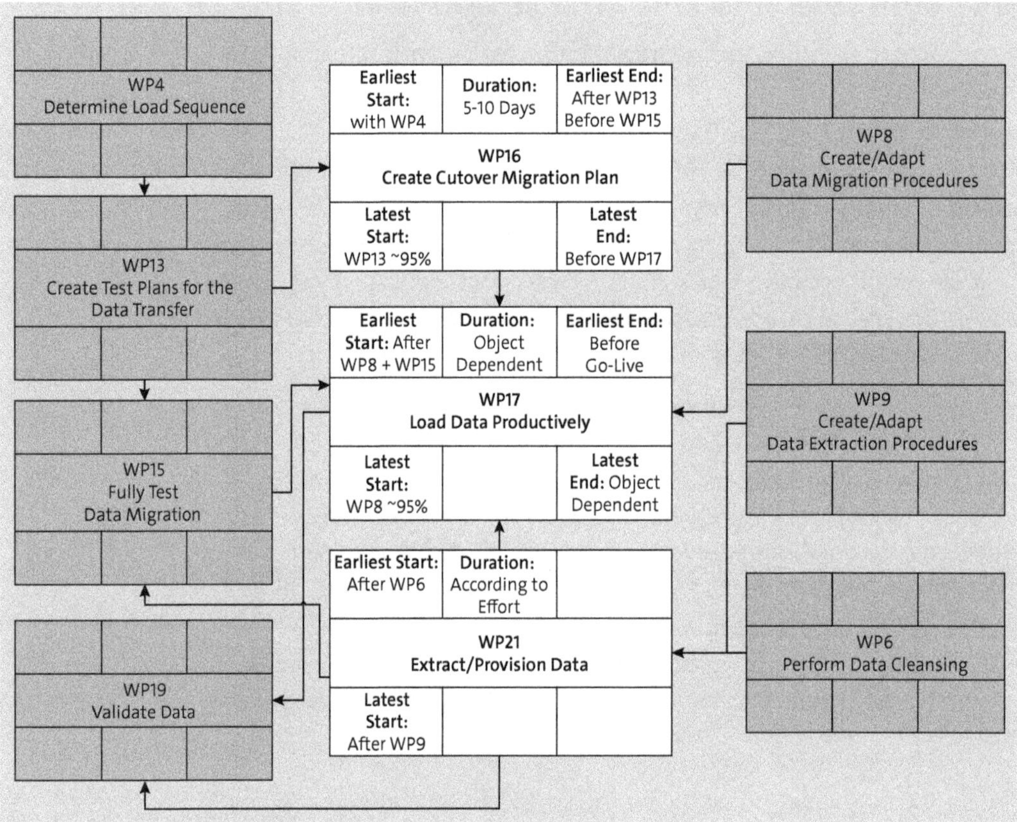


**Figure 2.9** Work Packages for Productive Load

**Narrow time slots** If you only have a narrow time slot for the initial data migration, you should categorize your individual data migration objects by importance for the go-live time. Some business objects of the source system are stable and are rarely changed or added to. These business objects can therefore be loaded into the target system days or weeks before the go-live. The prerequisite at this point is, of course, that the target system is supplied with the necessary customizing. When you load data into an existing production system, a certain risk is involved that you should evaluate upfront.

Some data migration objects do not necessarily need to be present in the system at go-live. Such data migration objects can also be loaded after the go-live date. However, you should also be aware in this case, for example, of the risks that arise from an increased system load at load time and the fact that you're loading data into a production system here.

#### Work Package 21: Extract/Provision Data

As mentioned at the beginning, you can either schedule the extraction and staging of data within work package 17 (load data productively) or swap it out to a separate package. In many large organizations, source system data is usually the responsibility of other teams. Thus, you need to schedule the staff of these other teams and their tasks into your test and cutover plans. This approach is the only way to ensure that the data is made available on time. From many years of practical experience, we know how cumbersome and time consuming it can be to communicate with teams that are not part of the implementation project and are not the responsibility of the project management. If the extraction and provisioning of data is done by multiple teams, individual sub-packages may still be involved.

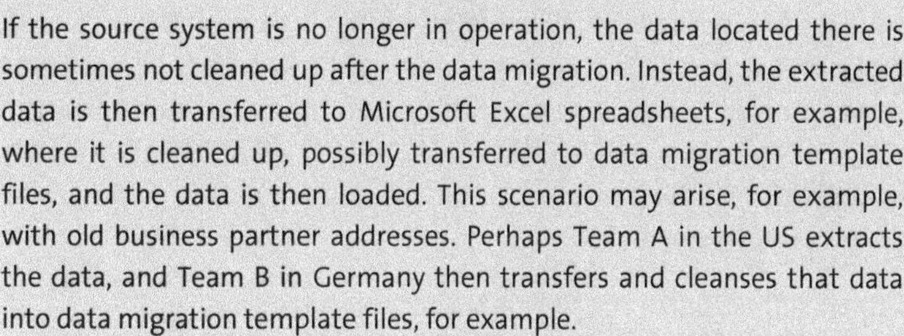

**Extraction and Provisioning Through Multiple Teams**

If the source system is no longer in operation, the data located there is sometimes not cleaned up after the data migration. Instead, the extracted data is then transferred to Microsoft Excel spreadsheets, for example, where it is cleaned up, possibly transferred to data migration template files, and the data is then loaded. This scenario may arise, for example, with old business partner addresses. Perhaps Team A in the US extracts the data, and Team B in Germany then transfers and cleanses that data into data migration template files, for example.

The extraction and provisioning of data is usually done after work packages 9 (create/adapt data extraction procedures) and 6 (perform data cleansing). It is a precursor package for work package 15 (fully test data migration) and for work package 17 (load data productively). Depending on whether you validate the data before loading or after loading, it can also be a precursor for work package 19 (validate data).

#### Work Package 17: Load Data Productively

The productive loading of data is one of the last work packages of a data migration project. The requirements for this work package include the following:

- The completion of all data migration procedures
- A successful test of the entire data migration
- The completion of the cutover plan including the final migration load plan.

As a rule, the steps of the migration load plan determine all other steps of the cutover plan.

### 2.2.8 Validation

Figure 2.10 shows an overview of the work packages related to validation. Two ways exist to validate the migration data at this point:

- After extracting or provisioning the data (i.e., before the actual data migration)
- After the data migration

**Validating data before loading**

In the first case, all data is checked for integrity and correctness directly after the extraction or after its provisioning, sometimes even before the conversion step. Only then will the data be loaded.

**Validating data after loading**

In the second case, all data is validated according to predefined rules after loading an object. This approach has some advantages because, in this case, you can also perform entire processes (automatically) using testing tools. In principle, it usually makes more sense to take the second approach and check each business object after the migration. Not until the data of this business object has been released can you proceed with the migration of dependent objects. While this process is rather time consuming, it ensures that your data meets quality standards. In particular, if you have highly complex conversion rules, a validation of the final data is preferable to a validation of the source data.

**Scheduling the Validation Process in a Customer Project**

In a large implementation project in the pharmaceutical industry, the validation of data was carried out in accordance with FDA regulations. Due to the limited time slot for data transfer and go-live, a 24/7 project plan was developed that included working around the clock. One hour before the calculated and scheduled completion of the migration of an object, the responsible employees were called to the company for validation. A separate team was responsible for adhering to the plan and coordinating the project staff. This team had worked three overlapping 9-hour shifts to verify the execution of the plan and to track load times. For this purpose, many tests and especially the documentation of the duration of the migration steps and the validation of the individual objects were very important. The distribution of tasks to individual employees in compliance with the applicable working time laws would not have been possible without detailed testing and the recording of times. The intensive testing process was rewarded with a smooth go-live and an extremely low, negligible number of corrections and rework.

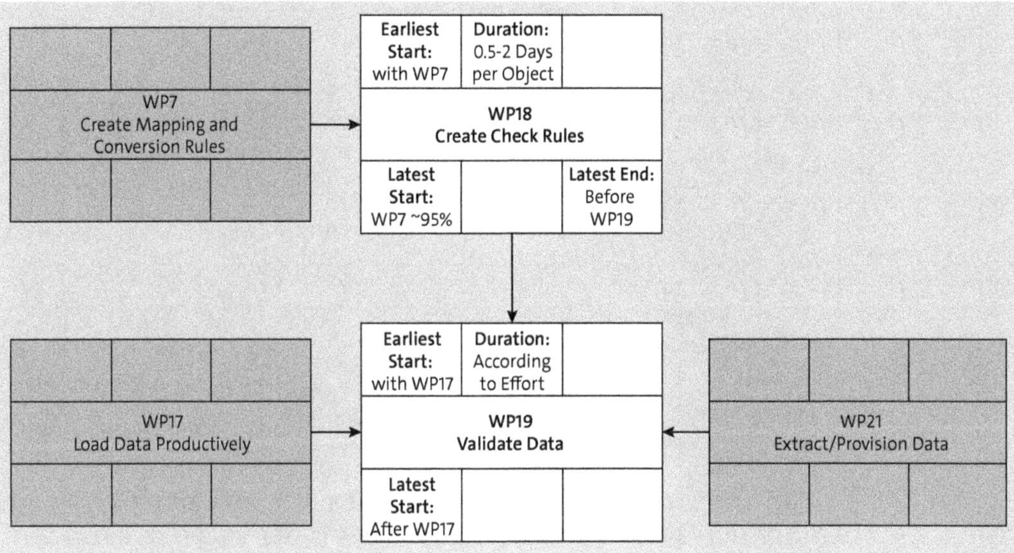


**Figure 2.10** Work Packages for Validation

### Work Package 18: Create Check Rules

The creation of the check rules must always be done in collaboration with the customer/client. The prerequisite for these check rules is the completion of the corresponding conversion rules and mapping.

Syntax checks

Some tools or procedures for data migration, such as SAP Data Services, provide functions for validating the loaded data.

[Ex]

#### Syntax Checks

These checks are often hard-coded routines of the data migration tool or API used for loading. Nevertheless, it never hurts to check the loaded data itself for technical correctness according to certain syntax rules, for instance, by asking yourself the following questions:

- Are the data types correct?
- Are the field lengths correct?
- Are the values in the date fields plausible?
- Have all mandatory fields been filled?
- Are the formats of numerical values, among other things, correct?

Semantics checks

In addition to syntax checks, semantics checks are also possible.

**Semantics Checks**

A semantics check evaluates the data for the correctness of its content but cannot usually be implemented in advance. This check is heavily dependent on the context of the data and varies from project to project and from object to object. Possible requirements may include the following:

- Each customer from a specific sales district must be assigned to a specific sales group.
- The material masters of specific product groups are assigned to different purchasing groups.
- Material masters that exceed or fall below certain values (e.g., in volume, weight, etc.) must be assigned to special storage bins.

Such checks must be created individually for each project. They can be partially automated by defining search queries (SAP Query). If you want these checks also used during ongoing operations, for instance, to continuously test and improve data quality, you should consider using relevant tools, such as *SAP Information Steward*. This tool allows you to define the rules and thresholds against which you can check and classify data records.

The steps for checking compliance with these rules in the system must be recorded in writing in a comprehensible manner.

#### Work Package 19: Validate Data

As mentioned at the beginning, the validation of your data can be performed before or after loading the data. If the validation is performed after loading an object, the successful check of this object should be considered as a prerequisite for the further loading of dependent objects. This approach improves data quality and avoids possible problems when you start loading subsequent objects. It is important that the check be carried out by the customer or the client and that the results be documented.

## 2.3 Planning and Effort Estimation

Now that you have learned about the process and work packages for data migration, we'll show you how to determine the objects for migration and estimate the duration of the overall process. For this purpose, we'll provide some formulas and figures from the evaluation of our previous migration projects.

### 2.3.1 Determining Data Migration Objects

When you determine the data migration objects, you should record the following information for each object:

- The SAP target system
- The migration object and a description of the object
- The object type (i.e., master data, transaction data, Customizing data, other data)
- The estimated complexity of the migration object (1 = less complex to 10 = very complex)
- Dependencies on the other data migration objects
- The number of data records to be imported
- The source system(s)
- Important contact persons
- The planned data transfer method
- Associated business processes

The best tactic is to save this information in a spreadsheet, such as Microsoft Excel, as shown in Figure 2.11, and add any other object or status information that you need.

| | | | | | | | | | Source systems | | | | | Region 1 | Region 2 |
|---|---|---|---|---|---|---|---|---|---|---|---|---|---|---|---|
| No. | Area | Object No. | Migration Object | Object Sub No. | Description | Comment | MD= Master Data CU= Customizing TD= Transaction Data | Data Migration Y (yes) N (no) M (manual) | System 1 | System 2 | System 3 | System 4 | Contact Person (SAP) for Validation and Mapping | Contact Person (Customer) for Validation and Mapping | |
| 1 | SD | 001 | **Customer** | 001 | **Customer from Spain** | 2016-09-11 Mr SAP1: Ask Ms Doe to get further information | MD | Y | X | | X | | Mr SAP1 | Mr Murphy | Ms Killroy |
| 2 | SD | 001 | **Customer** | 002 | **Customer from Mexico** | 2016-09-27 Email from Ms Killroy towards sales groups | MD | Y | X | | X | | Ms SAP2 | Mr Doe Ms Killroy | N.N. |
| 3 | MM | 002 | **Supplier** | 001 | **European Suppliers** | 2016-09-28 Herr Murphy: only eurpean suppliers are migrated | MD | Y | | | X | | Mr SAP1 | Mr Doey | N.N. |
| 4 | MM / SD | 003 | **Product** | 001 | **Raw material** | 2016-09-28 Ms SAP2: and Mr Doe: only RAW materials of System 1 | MD | Y | X | | | | Ms SAP2 Mr SAP1 | Mr Doe Ms Killroy | N.N. |
| 5 | MM / WM / SD / PP | 003 | **Product** | 002 | **Produced** | 2016-10-08 Mr Doey: clarify MRP settings | MD | N | X | X | | | Mr SAP1 | Mr Doe Mr Doey | N.N. |

**Figure 2.11** Collecting Data Migration Objects in a Microsoft Excel Spreadsheet

### 2.3.2 Effort Estimation

**Effort per project phase**

In well-planned implementation projects, the consulting effort for the migration portions (person-days for external consultants) should be between 7% and 10% of the total consulting effort. However, in poorly prepared or planned projects, this consulting effort may well exceed 30% of the project's consulting effort. These figures are based on internal analyses of migration projects carried out by SAP. You can use these numbers as a rough guide; however, they may vary from project to project:

- **Prepare phase**
  The effort for the migration project is rather low during the prepare phase; you should plan between 2% and 5% of the total migration effort.
- **Explore and realize phases**
  In these two phases of the project, the migration effort is between 50% and 70% of the total migration effort if you need to implement data migration programs, such as SAP Legacy System Migration Workbench. However, if you use migration tools with ready-made migration objects (such as SAP S/4HANA migration cockpit), you can greatly reduce the effort here. The prerequisite is that you implement your configuration close to the standard. However, if you modify the standard processes, adjustments to the delivered migration objects may be necessary, which will increase your effort again. For testing and test support alone, you can expect about 30%–40% of your total migration effort.
- **Deploy phase**
  In the production preparation phase, the effort after the overall data transfer tests drops to about 10% of the total migration effort—provided you have done your tasks in the prepare phase and no serious errors in the configuration occurred during the tests.
- **Run phase**
  For the productive data transfer and the support after the go-live, you should plan about 5% to a maximum of 10% of the total migration effort. The better the data transfer has been tested and the smoother the cutover has been, the less effort must be planned for the post-go-live support for the data migration. You can never be completely rule out that, for example, incorrect or outdated data has been migrated by mistake or that a value mapping has been set incorrectly and thus data must be corrected.

#### Effort Per Work Package

Once the necessary work packages of the migration project have been determined, a Microsoft Excel spreadsheet can be used to estimate the project effort. The effort for some work packages, for example, can be determined as flat. For some phases, such as the data analysis phase, a simple but highly effective formula can be used to estimate the effort. The following formulas have been estimated and verified in various migration projects over the years but, of course, are only a rough guide and may vary from project to project.

**Effort Estimation Formula for Data Analysis**

You should estimate the complexity of the corresponding migration object on a scale from 1 (low complexity) to 10 (very high complexity) and divide this value by 2.

Some criteria to assess the complexity of a migration object include the following:

- Number of data sources
- Number of fields to be mapped
- Number of conversions
- Are they simple 1:1 conversions, or is additional coding necessary?
- Number of data records
- Quality of source data and associated recoding to correct errors automatically

The resulting number is the approximate number of person-days you'll need for work package 1 (determine data migration objects).

**Effort Estimation for Mapping**

For work package 7 (create mapping and conversion rules), you can generally use the same formula as for the effort estimation for the data migration objects. However, if you have multiple technical data migration objects for a business object, the synergy effects and the learning curve when determining the conversion rules reduce the effort for the individual objects.

If, for example, the customer master is loaded from different source systems using an IDoc, the effort required for mapping and the conversion rules for the first data migration object can be determined using the formula. The effort for the further objects can be assessed at only half to a quarter of the effort for the first object.

**Implementation effort**

On the other hand, no sufficient formula can be given for the implementation of the data migration programs. In this case, the number of fields to be mapped and the number of conversion rules to be implemented are decisive. The experience of the person implementing the programs is also a factor that should not be underestimated. While a highly experienced *migrator* (a person involved in the data migration of a business object) may need only one day for the implementation and testing of a given data migration object, for example, a less experienced migrator may need several days. If you exclusively use the delivered migration objects for, say, the migration cockpit, no extra implementation work is required.

The effort for the work packages of testing and data loading depends on the migration duration and can only be calculated more precisely after the first tests. To estimate such figures, it helps to draw on empirical values from other projects.

**Validation effort** Validation times depend on the quantity and complexity of the validation rules and can therefore not be calculated using a formula.

### Estimating Effort Using Microsoft Excel

The worksheet of the Microsoft Excel file we describe next has an info area and an effort area. You can enter flat effort values and the effort for each migration object in this file.

#### *Info Area*

The info area contains the following columns, as described in Table 2.1 and shown in Figure 2.12:

❶ Number (**No.**) for numbering migration objects consecutively.

❷ **Area** to sort objects by area, for example, FIN, LOG, etc.

❸ Columns 3 and 4 list migration objects (**MObj**), in English and another language, in this case, German. These two languages often prove useful since most of the literature (e.g., blogs, documentation, books, etc.) on business processes, business objects, and migration objects is written in English. Finding information about the corresponding object is much easier if you also know its English name.

❹ Another optional column (**Type**) helps you to differentiate between master data and transaction data.

❺ In the next column (**In scope**), you define whether the migration object is relevant or not. In a list of more than 150 migration objects, only a minor portion is relevant to your project. In a phased rollout, it may also be possible that migration objects are not relevant for all phases and are then identified differently in various lists.

❻ In the **Tool/Method** column, you define the transfer method. Do you want to migrate the data manually, or do you use a specific tool? If so, which one?

❼ Optionally, you can also list the number of data records to be migrated in the **Data Volume** column. This is approach often useful for estimating the load times for testing and productive loading.

| No. | Area | MObj | MObj | Type | In Scope | Tool/ Method | Data Volume |
|---|---|---|---|---|---|---|---|
| | Flat | | | | X | | |
| 1 | SD | Customer | Kunde | MD | X | MC/File | 12,000 |
| 2 | MM | Supplier | Lieferant | MD | X | MC/File | 250 |
| 3 | LO | Product | Produkt | MD | X | MC/File | 32,000 |

**Table 2.1** Columns of the Information Area A of the Efforts Table

| | A | B | C | D | E | F | G | H | I | J | K | L | M | N | O | P |
|---|---|---|---|---|---|---|---|---|---|---|---|---|---|---|---|---|
| 1 | | | | | | | | | | | | | | | | |
| 2 | | | | | | | | | Data Analysis | | | | Tools | Data Cleansing | | Mappi |
| 3 | No. | Area | Migration Object english | Migrationsobjekt deutsch | Type | in scope | Tool / Method | Data Volume | WP1 flat | WP2 flat | WP3 | WP4 flat | WP20 | WP5 flat per system | WP6 | WP |
| 4 | | Flat-rate number of | | | | X | | | 5 | 5 | 5 | 2 | 15 | 5 | | |
| 5 | 1 | SD | Customer | Kunde | MD | X | MC / File | 12.000 | | | | | | | 1,000 | 2,00 |
| 6 | 2 | MM | Supplier | Lieferant | MD | X | MC / File | 250 | | | | | | | 1,000 | 2,00 |
| 7 | 3 | LO | Product | Produkt | MD | X | MC / File | 32.000 | | | | | | | 1,000 | 2,00 |
| 8 | ❶ | ❷ | ❸ | | ❹ | ❺ | ❻ | ❼ | 5,000 | 5,000 | 5,000 | 2,000 | 15,000 | 5,000 | 3,000 | 6,00 |
| 9 | | | | | | | | | | | | | | | | |

**Figure 2.12** Information Area and General Efforts in Microsoft Excel

### *Efforts Area*

In the efforts area, you enter the estimated effort in days. This area consists of a few column areas, as shown in Figure 2.13:

❽ The individual phases and work packages from Section 2.2

❾ A column for the totals

Table 2.2 and Figure 2.13 show an example of the efforts area with general and object-related efforts for the **Data Analysis** and **Data Validation** phases.

| Data Analysis | | | | ... | Data Validation | | |
|---|---|---|---|---|---|---|---|
| WP1 | WP2 | WP3 | WP4 | ... | WP18 | WP19 | Total |
| 5 | 5 | 5 | 2 | ... | | | 49.50 |
| | | | | ... | 0.5 | 0.5 | 7.75 |
| | | | | ... | 0.5 | 0.25 | 6.625 |
| | | | | ... | 1 | 0.25 | 9.25 |

**Table 2.2** Columns of the Efforts Area B of the Efforts Table

| P | Q | R | S | T | U | V | W | X | Y | Z | AA | AB | AC | AD | AE |
|---|---|---|---|---|---|---|---|---|---|---|---|---|---|---|---|
| Company | | | | | | | | | | | | | | | |
| Mapping | Implementation | | Tests | | | | | | Productive Load | | | Data Validation | | Column | Row |
| WP7 Create Mapping and Conversion rules | AP8 Create/adopt data migration procedures (incl. Test and documentation) | WP9 Create/adopt data extraction procedures (incl. Test and documentation) | WP14 Define and create test data load procedures **flat** | WP10 Provide test data **flat** | WP11 Plan test landscape **flat** | WP12 Implement test landscape **flat** | WP13 Create test plan for date migration **flat** | WP15 Test complete data migration | WP16 Create cut-over data migration plan **flat** | WP21 Extract / Provide date (optional) **flat** | WP17 Perform productive data migration | WP18 Define validation rules | WP19 Validate data | Sum (Number of Days) | |
| | 0,0 | | 0,0 | | 2,0 | 1,0 | 4,0 | | 5,0 | 0,5 | | | | 49,50 | 49,50 |
| 2,000 | | 2,000 | | 0,500 | | | | 1,000 | | | 0,250 | 0,500 | 0,500 | | 7,750 |
| 2,000 | | 2,000 | | 0,250 | | | | 0,500 | | | 0,125 | 0,500 | 0,250 | | 6,625 |
| 2,000 | | 3,000 | | 0,500 | | | | 1,000 | | | 0,500 | 1,000 | 0,250 | | 9,250 |
| 6,000 | 0,000 | 7,000 | 0,000 | 1,250 | 2,000 | 1,000 | 4,000 | 2,500 | 5,000 | 0,500 | 0,875 | 2,000 | 1,000 | 73,13 | 73,13 |

**Figure 2.13** Efforts Area in Microsoft Excel

### *General Effort*

**Estimated effort** In the cells of the row for general effort, you enter the estimated effort in person-days that arises independently of a migration object. In the SEAMAP structure, the efforts can be specified as flat for the following work packages:

- **WP1: Determine data migration objects**
  In this case, you can estimate 2 to 10 person-days depending on the number of processes to be implemented and the source systems.
- **WP2: Identify source systems**
  The more complex and heterogenous your system and data retention landscape is, the longer it takes to find the appropriate source systems. Also consider unofficial sources of your processors here. In many areas, they have extra data on customers in additional, unofficial Microsoft Excel files, card boxes, or databases, which are often more up to date than the official databases. Here, too, you should plan 2 to 10 person-days.
- **WP3: Historical data**
  You should not underestimate the effort for this work package. You should calculate 10 person-days here.
- **WP4: Determine load sequence**
  It is usually no big effort to determine the load sequence. Loading the data also requires data extraction and data provisioning. If the systems are distributed across continents, you must know the load sequence so that the data gets provisioned at the right time. Allow for 1 to 2 person-days here.
- **WP5: Analyze data cleansing requirements**
  More cleansing requirements often arise after the initial tests. To analyze the data cleansing requirement, you should calculate 5 to 10 days for each system.
- **WP11: Plan testing landscape**
  In a cloud environment, the test landscape is predefined. You have either a two-system landscape with a quality assurance (QA) system and a production (P) system, or a classic three-system landscape with a development system, a QA system, and a P system. The effort is low, and usually

only involves preceding source systems. In an SAP S/4HANA-based on-premise or hosted system, you can allow for 2 days.

- **WP12: Set up the testing landscape**
  This case is similar to the one for planning the testing landscape. However, the effort must be calculated for each test performed. After all, resetting or implementing a system anew may involve several days. Schedule 1 to 2 person-days for each test.
- **WP13: Create test plans for the data transfer**
  Test data is usually transferred using the same tools and programs that are used for the productive data transfer. You usually adapt the test plan after each test. For this reason, you should enter 2 to 5 person-days for the entire effort here.
- **WP14: Define and create test data migration methods**
  If you use a service now, such as SAP S/4HANA Cloud, test data refresh, which provides you with the test data on a scheduled basis, then you should enter the corresponding provision runtime here as a flat rate.
- **WP16: Create cutover migration plan**
  Basically, the cutover migration plan is part of the entire cutover plan. It results from the test plan for data migration (WP13) and the effort times that were estimated for the individual work packages, data provisioning (WP21), loading of data (WP17), and data validation (WP19). Creating the cut-over migration plan usually takes 3 to 5 person-days. In complex projects, you may need up to 10 days.
- **WP21: Extract/provision data**
  The time required to extract or provision the data is usually less than the time required to load the data. As a result, you can also use flat rates here after the first-time measurements and do not necessarily have to break the figure down to the objects.
- **WP20: Select tools**
  Do not underestimate the effort required to select the right tools. In a cloud environment, the transfer tools are usually predefined. The tools and methods for data analysis, data cleansing, and data extraction of your source systems can be versatile, and your selection depends on the offering and possibly existing licenses. In the public sector, software procurements often have to be made through public tenders. Here, the effort can be several weeks, which must be considered for each customer specifically.

Normally, you must allow for 30 to 60 person-days just for flat rates. Table 2.3 shows a sample calculation for flat rates.

| WP1 | WP2 | WP3 | WP4 | WP20 | WP5 | WP11 | WP12 | WP13 | WP16 | WP21 | Total |
|---|---|---|---|---|---|---|---|---|---|---|---|
| 5 | 5 | 5 | 2 | 15 | 5 | 2 | 1 | 4 | 5 | 0.5 | 5 |

**Table 2.3** Example Flat Rates for Individual Work Packages

#### *Object-Related Efforts*

Effort of each migration object

All other work packages have efforts that arise for each migration object. Table 2.4 shows a sample list of the relative effort of work packages for each migration object in an SAP S/4HANA Cloud-based system.

| Object | WP6 | WP7 | WP9 | WP10 | WP15 | WP17 | WP18 | WP19 | Total |
|---|---|---|---|---|---|---|---|---|---|
| Customer | 1 | 2 | 2 | 0.5 | 1 | 0.25 | 0.5 | 0.5 | 7.75 |
| Supplier | 1 | 2 | 2 | 0.25 | 0.5 | 0.125 | 0.5 | 0.25 | 6.625 |
| Product | 1 | 2 | 3 | 0.5 | 1 | 0.5 | 1 | 0.25 | 9.25 |
| Total | 3 | 6 | 7 | 1.25 | 2.5 | 0.875 | 2 | 1 | 23.625 |

**Table 2.4** Example Object-Related Effort in Person-Days

[»]

**Downloading a Microsoft Excel File**

You can find the Microsoft Excel files described in this chapter in the materials available for this book at *www.sap-press.com/5817*.

## 2.4 Summary

In this chapter, you learned about the phases of a migration project and how they fit into the phases of an implementation project. We have shown you how to structure your migration project in work packages and what dependencies there are between the individual work packages. Using this structure, it should now be easier for you to understand the complexity of a migration project and to better estimate the efforts. The aids to estimating efforts presented in the preceding section are, as noted, only a rough guide, and there is no guarantee that these figures will apply to your project. Via the bonus pages, you can download sample files for object identification and effort estimation in Microsoft Excel format, which you can use in your project and adapt to your needs.

PART II

# Data Migration in SAP Solutions

# Chapter 3
# Data Migration to SAP S/4HANA



*This chapter presents the SAP S/4HANA migration cockpit. This tool for migrating data is one of the most important tools of a new implementation in SAP S/4HANA and SAP S/4HANA Cloud.*

Which different migration approaches does the *SAP S/4HANA migration cockpit* (often referred to as *migration cockpit*) support? How does the migration cockpit work? Where can you obtain further information on data migrations and on supported migration objects? This chapter provides answers to these questions and more. Section 3.1 describes the various data migration approaches of the migration cockpit and illustrates their underlying concepts. Section 3.2 describes how you can migrate data using the approach **Migrate Data Using Staging Tables** (i.e., staging tables) approach using files. Section 3.3 describes how you can migrate data using the **Migrate Data Directly from SAP System** (i.e., direct transfer) approach. Section 3.4 provides a basic overview of the modeling environment in the *SAP S/4HANA migration cockpit*. Section 3.5, Section 3.6, and Section 3.7 address the specifics of the different data migration approaches.

[«]

**SAP S/4HANA versus SAP S/4HANA Cloud**

In this chapter, "SAP S/4HANA Cloud" refers to SAP S/4HANA Cloud Public Edition, which is technically based on the *SAPSCORE* product version. SAP S/4HANA refers to SAP S/4HANA products (on-premise or SAP S/4HANA Cloud Private Edition) that are technically based on the *S4CORE* product version.

## 3.1 Overview of the SAP S/4HANA Migration Cockpit

**Limited usability of SAP ERP tools**

Migrating to SAP S/4HANA Cloud differs from migrating to the SAP S/4HANA on-premise because the known migration tools, such as SAP Legacy System Migration Workbench or SAP Data Services (rapid data migration), cannot be used at all or only to a limited extent. Pure backend tools like SAP Legacy System Migration Workbench cannot be called due to missing backend access. In case of external tools, such as SAP Data Services, the number of cloud-enabled interfaces that can be called from external sources to migrate data is limited. The migration cockpit and its two migration approaches are therefore the first and partially only data migration tool available, particularly in the cloud.

[»]

**Familiarize Yourself with Data Transfer Processes**

Before you start the data migration, you should familiarize yourself with the phases and activities of a data migration project in general. Unfortunately, we frequently observe that both customers and consultants underestimate the complexity of a data transfer. Particularly with regard to the cloud, a popular misconception has circulated that everything works at the touch of a button. We therefore recommend you first read Chapter 2, internalize the tips provided there, and incorporate them into your project planning.

### 3.1.1 Overview of the Migration Approaches of the SAP S/4HANA Migration Cockpit

**From a pure file approach to a file/staging approach**

The transfer data using files migration approach was delivered together with the old Web Dynpro-based migration cockpit to SAPS/4HANA Cloud 1511. Many of you know the old migration cockpit by its transaction name *Transaction LTMC*. Since SAP S/4HANA 1610, this migration cockpit has also been provided for the on-premise version of SAP S/4HANA (S4CORE). **Migrate data using staging tables** (staging) has been supported since SAP S/4HANA release 1709 FPS02 or SAP S/4HANA Cloud 1905. Because both approaches use the same migration cockpit, they are also referred to as migrate data using files and staging tables (file/staging).

**One migration cockpit for all approaches**

With SAP S/4HANA 1909, the new migration approach of **Migrate data directly from SAP system** (direct transfer) was added for the SAP S/4HANA on-premise versions. The migration cockpit for this approach is an SAP Fiori app, and it is launched via the SAP Fiori launchpad. With SAP S/4HANA Cloud 2008 and SAP S/4HANA 2020, the migration cockpit of the two migration approaches—**Migrate data using files and staging tables** and **Migrate data directly from SAP system**—was combined into one SAP Fiori app: the Migrate Your Data app. Transaction LTMC is thus obsolete and can no longer be used as of SAP S/4HANA 2021.

[«]

3

**Note**

The migration objects delivered are identical for the two migration cockpits. Due to varying data retention limits, however, the migration projects cannot be exchanged between the two migration cockpits. Thus, the projects that you had created in the old migration cockpit cannot be opened in the new migration cockpit, and vice versa.

The old migration cockpit will be switched to read-only mode as of SAP S/4HANA Cloud 2105 or SAP S/4HANA 2021. You can still call the app/transaction, but you can no longer run any data migrations.

**Files as containers for populating staging tables**

At the same time, the technical data retention of the two approaches, *file* and *staging*, was combined in the migration approach **Migrate Data Using Staging Tables** (staging tables). Previously, the data retention of the data to be processed was physically separated between files (staging area) and staging tables; now, data is stored uniformly in staging tables. Thus, a file is only a container for uploading the data to be processed to the staging tables. Previously, a separate migration project was required for the files or staging tables to upload data from files or staging tables. They can now be combined in one migration project. In theory, there is no limit for the data volume to be loaded. The maximum file size of a *migration template (template)* is limited due to technical reasons. But the data to be loaded is located in the same staging tables after upload and no longer in cluster tables that are separated by files.

**Comparing the migration approaches**

Table 3.1 compares the two migration approaches of the migration cockpit for SAP S/4HANA Cloud 2008 and SAP S/4HANA 2020, unless another release is indicated.

| Data Transfer... | ... Using Staging Tables | ... Directly from SAP System |
|---|---|---|
| Availability | SAP S/4HANA Cloud staging via SAP Business Technology Platform (SAP BTP) and database as a service (DBaaS) as of 1905; SAP S/4HANA staging as of 1709 FPS02 | As of SAP S/4HANA 1909 |
| User interface (UI) | SAP Fiori app (as of SAP S/4HANA Cloud 2008/2020) | SAP Fiori app |

**Table 3.1** SAP S/4HANA Migration Cockpit: Comparison of Migration Approaches

| Data Transfer... | ... Using Staging Tables | ... Directly from SAP System |
|---|---|---|
| Supported migration methods | Compatible function modules<br>ABAP RESTful application programming model business objects (core data services [CDS] behavior definition) | |
| Source systems | Every SAP and non-SAP source system | ABAP-based SAP source systems (Section 3.3) |
| Data provision | Staging tables populated via Microsoft Excel 2003 XML templates; comma-separated values (CSV) files; or extract, transform, and load (ETL) tools | Data is selected and extracted directly in the SAP system |
| Data cleansing | In the source system, CSV file, or in the XML template | In the source system |
| Data protection | The files used must be retained securely. | Access to the source system must be limited. |

**Table 3.1** SAP S/4HANA Migration Cockpit: Comparison of Migration Approaches (Cont.)

### 3.1.2 Available Migration Objects

With every release, the number of supported migration objects and their functions change. However, these changes always depend on the function of the released data migration interfaces. Moreover, the number of available migration objects differs depending on the migration approach and the target product.

[!]

**Important Information for SAP S/4HANA 2022**

If you install a new system with SAP S/4HANA 2022 FPS0/FPS1 or upgrade your existing system to SAP S/4HANA FPS0/FPS1, be sure to follow SAP Note 3291483. Due to an issue during the creation of the SAP S/4HANA 2022 FPS0 delivery, some migration objects and their components have not been delivered consistently. For this reason, it is mandatory to follow the instructions in the aforementioned SAP Note and import the corresponding changes before you start loading data. The issue has been resolved with SAP S/4HANA 2022 FPS2.



**Direct transfer approach**

The *direct transfer approach* is distinguished by scenarios and separate migration objects are delivered for every scenario. In other words, the number of available migration objects varies depending on the source system.

**Staging approach**

In the *staging approach*, the number of available migration objects depends on the target product and the release. In SAP S/4HANA Cloud Public Edition, new migration objects are delivered with a new release or within a *continuous feature delivery (CFD)*. To minimize the impact on customers, enhancements to existing objects are currently only delivered with a new release. For SAP S/4HANA and systems and products that are based on it, adaptations are provided only for the main release of a new product version of SAP S/4HANA. For technical reasons, it is currently not possible to deliver migration objects of a new release to lower product release levels (i.e., you cannot *downport*). Moreover, a migration object is only provided for SAP S/4HANA if SAP Best Practices is available for a migration object and if the new functionality is available, not only for the cloud version, but also for the on-premise version. These limits explain the varying number of migration objects for cloud and on-premise. Table 3.2 shows the number of available migration objects for each product version at the time of this writing.

**Current object list on SAP Help Portal**

The migration objects that are available for release can be found on SAP Help Portal through the following quick access URLs for the individual products:

- For SAP S/4HANA Cloud: *https://help.sap.com/S4_CE_MO*
- For SAP S/4HANA: *https://help.sap.com/S4_OP_MO*

| Staging | Direct Transfer |
|---|---|
| SAP S/4HANA Cloud 2402: 250+<br>SAP S/4HANA 2023: 220+ | SAP S/4HANA Cloud: 19<br>SAP S/4HANA 2023 FPS0:<br>▪ SAP ERP: 145+<br>▪ SAP Apparel and Footwear: 73<br>▪ SAP Extended Warehouse Management (SAP EWM): 29<br>▪ SAP S/4HANA Supply Chain for extended service parts planning: 7<br>▪ SAP Customer Relationship Management (SAP CRM): 1 |

**Table 3.2** Number of Available Migration Objects for Each Migration Approach

[»]

**Migration Objects of the Staging Approach with the Addition "Restricted"**

The names of migration objects of the staging approach can have additions. For migration objects that have such additions, you should read the migration object documentation carefully. Then, consider the following additions we used earlier:

- **(restricted)**
  If a migration object does not cover all fields and structures or functions of all relevant business processes, the addition "restricted" is used.
- **(deprecated)**
  If a migration object is newly built, the migration object used so far is indicated with the addition "deprecated." Obsolete objects are then still delivered together with the new migration object for a release in SAP S/4HANA Cloud and usually removed in the subsequent release. In SAP S/4HANA, they are removed with the next product version, in some cases, already with the same product version as the new object.

Changes on SAP Help Portal

You can find a detailed overview of the changes of a migration object and the staging approach on the following SAP Help Portal pages:

- For SAP S/4HANA Cloud: *http://s-prs.co/v581727*
- For SAP S/4HANA: *http://s-prs.co/v581728*

## 3.2 Migrating Data Using Staging Tables

With the new data retention of the new migration cockpit for SAP S/4HANA Cloud 2008 and SAP S/4HANA 2020, data retention was completely switched to storage in staging tables.

### 3.2.1 Technical Realization

One staging table per data structure

One spreadsheet of a migration template file corresponds to one staging table. As a result, the data of every structure of a migration object is stored in separate staging tables. When you populate these staging tables directly, in contrast to the migration template files, you avoid splitting the data into several files and uploading these files to the migration cockpit.

Migration project

From a technical perspective, in the **Migrate data using staging tables** approach, migration objects are part of a migration project in the *migration workbench*, which you can use via the migration cockpit. The migration workbench is not released for direct use by SAP customers. Migration projects, as delivered with their migration objects and rules, are templates only.

Migration projects are copied to a customer-specific project (Z project) when you create a customer project in the migration cockpit. This copy is then used in the migration cockpit to generate a function group that is deployed in the migration cockpit.

### SAP S/4HANA Cloud

Staging tables only via database as a service

In the new SAP Fiori cockpit, the staging tables of the local SAP S/4HANA database schema in SAP S/4HANA Cloud—in contrast to SAP S/4HANA—can only be written as containers via the file. The staging tables of the remote SAP HANA database schema, on the other hand, must be set up in an SAP HANA Cloud instance on SAP BTP. Then, the staging tables can be read and written from the outside. The connection to these staging tables is made via *communication scenario* SAP_COM_0678 (Migration Cockpit Integration). You can extract table content from any system and write it directly to the staging tables in an SAP HANA Cloud instance using additional tools. Communication scenario SAP_COM_0678 scenario establishes the database connection using *DBaaS* in SAP BTP.

Read the test script and setup instructions

You should read two important documents related to data migrations to SAP S/4HANA Cloud using remote SAP HANA database schemas:

- The *setup instructions* "Setting Up Data Migration to SAP S/4HANA from Staging (2Q2)" for setting up communication scenario SAP_COM_0678
- The "Data Migration to SAP S/4HANA from Staging (2Q2)" test script

You can find the documents in *SAP Best Practices Explorer*. Just search for the abbreviation "2Q2" or access the link *http://s-prs.co/v581729*.

Technical role for accessing SAP BTP

In addition to role SAP_BR_CONFIG_EXPERT_DATA_MIG for the migration cockpit (Configuration Expert - Data Migration/Configuration Expert - Data Migration) described in Section 3.2.2, you need role SAP_BR_CONF_EXPERT_BUS_NET_INT (Configuration Expert - Business Network Integration/Configuration Expert - Business Network Integration). This role is necessary so that you can connect to the staging tables in the remote SAP HANA database schema in the SAP HANA cloud instance on the SAP BTP from the migration cockpit.

### SAP S/4HANA

Local or remote database schema

Unlike SAP S/4HANA Cloud, the SAP HANA database for the on-premise version can be set up anywhere. Here you can either use the local SAP S/4HANA database schema of your SAP S/4HANA system or a remote SAP HANA database schema on a separate SAP HANA box.

Local schemas and data privacy

Because the data of the staging tables can be displayed in a local SAP S/4HANA database schema via Transaction SE16 (Data Browser) and in the remote SAP HANA database schema via Transaction SE16H (General Table

Display), you should restrict access to these tables by means of authorization object S_TABU_DIS for data security and data protection reasons.

**Entering the database connection in the positive list**

The database connection to a remote SAP HANA database schema must have been set up in advance. To be able to use the connection, it must be entered in the DMC_C_WL_DBCO_OP table—"Allowlist of database connection for on-premise scenario"—using Transaction SM30.

### 3.2.2 Assigning Roles for the Migration

**Technical roles**

To use the migration cockpit via the SAP Fiori launchpad, a business role must be assigned to the user performing the migration (migration user). You can create this technical role from the delivered technical template role SAP_BR_CONFIG_EXPERT_DATA_MIG (Configuration Expert - Data Migration). For SAP S/4HANA, you must assign the user a role that was created from the backend template role SAP_CA_DMC_MC_USER (SAP S/4HANA Migration—Data Transfer).

**Application roles**

In addition, you'll require further authorizations and roles to import or validate the individual migration objects. These authorizations vary depending on the migration application programming interface (API) executed. In SAP S/4HANA Cloud and SAP S/4HANA, predefined business role templates are provided for specific business cases. You can find the application role template for each migration object for SAP S/4HANA Cloud in the documentation of the migration object or on the SAP Help Portal page at *https://help.sap.com/S4_CE_MO*. For SAP S/4HANA, the documentation of the migration object can be found at *http://help.sap.com/S4_OP_MO*.

### 3.2.3 Initial Screen of the Migration Cockpit

**Calling the SAP S/4HANA migration cockpit**

You can find the migration cockpit in the **Data Migration** area ❶ of the SAP Fiori launchpad. Then, click the **Migrate Your Data - Migration Cockpit** tile ❷, as shown in Figure 3.1. This step takes you to the migration cockpit. Using **Migrate Your Data - Migration Cockpit (Deprecated)** ❸, you could call the old migration cockpit, which is only used to display old migration projects.

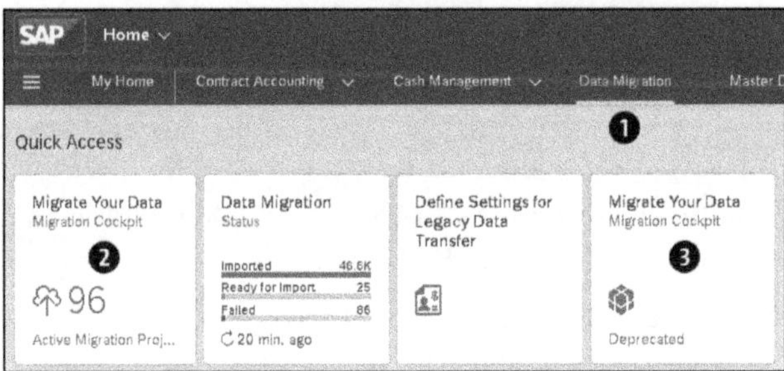


**Figure 3.1** Data Migration Group Tiles in SAP S/4HANA Cloud 2302

### 3.2.4 Data Migration Process

The seven process steps

The data migration process of the migration cockpit consists of seven consecutive steps. Most of these steps have an action assigned in the migration cockpit. This action is listed in parentheses in the following list:

1. **Selecting migration objects**
   The migration cockpit automatically generates the staging tables for every migration object selected.
2. **Load data to the staging tables (upload file)**
   See also Section 3.2.11.
3. **Prepare staging tables (prepare)**
   See Section 3.2.13.
4. **Process mapping tasks (mapping task)**
   See Section 3.2.14.
5. **Simulate data migration (simulate)**
   See Section 3.2.16.
6. **Migrate data (migrate)**
   See Section 3.2.17.
7. **Validate data (migration results)**
   See Section 3.2.18.

Before you select a migration object, you must first create a migration project or edit an existing one.

### 3.2.5 Creating a Migration Project

Creating staging projects

Click the **Create** button and select the migration approach for which the migration project should be created from the dropdown menu shown in Figure 3.2. You can choose between the following two approaches:

- **Migrate Data Directly from SAP System**
- **Migrate Data Using Staging Tables**

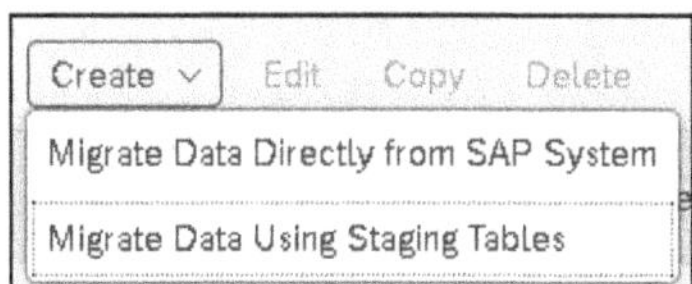


**Figure 3.2** Selection Box for the Creation of Staging Tables

We'll focus on the staging table approach **Migrate Data Using Staging Tables** in this section. The **Migrate Data Directly from SAP System** approach is described in Section 3.3. At this point, select **Migrate Data Using Staging Tables** from the **Create** dropdown list.

### Step 1: Enter General Data

General data for the migration project

The **New Migration Project** dialog box is displayed. In the **General Data** section, enter a name for the migration project in the **Name** field ❶, as shown in Figure 3.3.

The system automatically determines the **Mass Transfer ID** ❷. If you click the **Display Next Available** button ❸, the system generates a new ID if the automatically generated ID does not meet your requirements. The **Mass Transfer ID** must start with "M," "N," or "O."

Select a database schema

In the **Database Connection** area ❹, enter the location of the *database schema* in which the staging tables are to be created. You can choose either the *local SAP S/4HANA database schema* of the SAP HANA database of your SAP S/4HANA system or a what's called a *remote database schema* in an external SAP HANA database. You have these two options:

- **Local SAP S/4HANA Database Schema** ❺
- **Remote SAP HANA Database Schema** ❻

Select a database connection

If you want to transfer data from files only, you would usually select **Local SAP S/4HANA Database Schema** ❺. If you want to use the remote **SAP S/4HANA Database Schema** ❻, you must select the corresponding schema in the selection box ❼. (For more information, see Section 3.2.1.) Figure 3.3 shows the completed step 1 for the general data of an SAP S/4HANA cloud system.

Once you have specified the general data for your project, click **Step 2**.

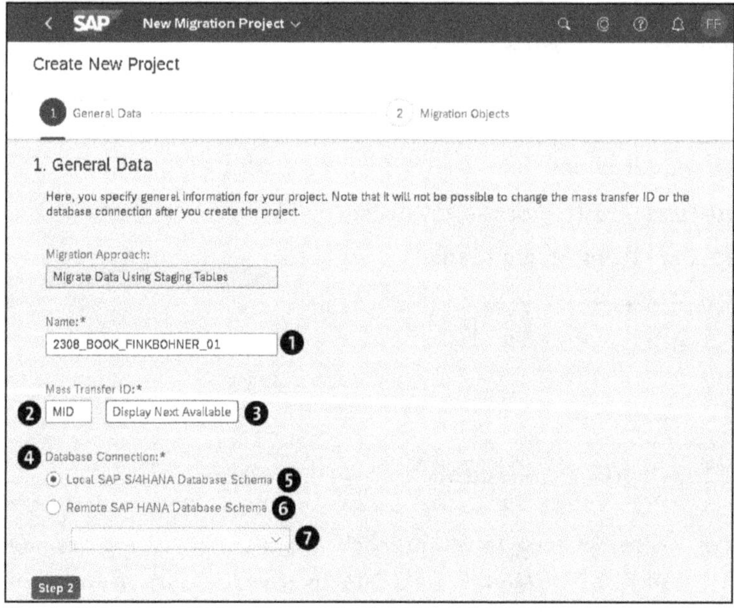


**Figure 3.3** Creating a Migration Project, Step 1: Enter General Data (SAP S/4HANA Cloud)

### Step 2: Assign Development Package in SAP S/4HANA

**Specifying the development package**

In an SAP S/4HANA system, step 2 prompts you to specify a development package (development class). Only with the help of this development package can you transport migration projects to another system. The option to export migration projects via files no longer exists Since SAP S/4HANA 2021. If you do not plan to transport your project, as shown in Figure 3.4, select **Local Package ($TMP)** ❶ under **Development Package**. If you want to transport a project to another system, select the **Other Package** option ❷ and then select the appropriate development package using the input help ❸. When you're done, click **Step 3**.

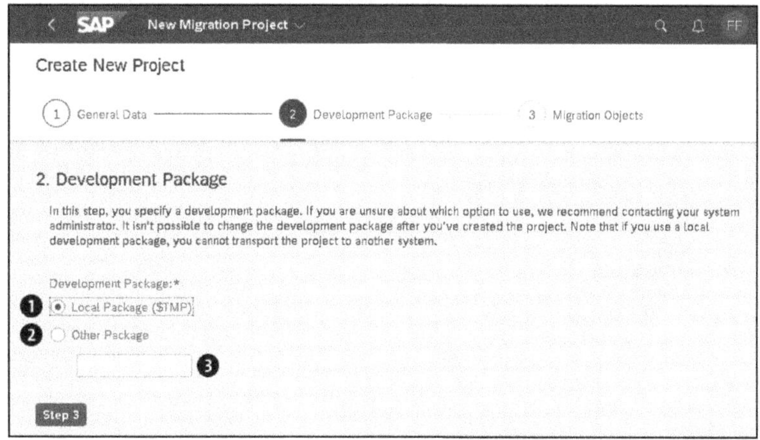


**Figure 3.4** Creating a Migration Project, Step 2: Enter Development Package (SAP S/4HANA)

[!]

**Restrictions When Creating Migration Projects**

Migration projects cannot be created in SAP S/4HANA systems that are marked for production or locked against repository changes (QA systems). In this case, you need to create a project in another system and then transport this project through your system landscape into the target system. Note that value mappings and other configurations are not included in the transport. To transfer value mappings, you can export and import them as described in Section 3.2.14. A migration project assigned to a local package can later be assigned to a transportable development package via a report delivered with SAP Note 3119168. A project is then transported using the SAP S/4HANA migration object modeler, as described in Section 3.5. For more details, see Section 3.5.9.

### Steps 2/3: Assign Migration Objects

**Select available migration objects**

Depending on whether you're in an SAP S/4HANA or SAP S/4HANA Cloud system, you're now either in **Step 3** or in **Step 2** (Migration Objects). You can select the objects that you want to use in your project in the **Migration Objects** area, from the list under **Available Migration Objects.** For our purpose, select the migration objects from the list and click the [>] icon to insert them in the **Selected Migration Objects** list. You can remove objects by clicking the [<] icon. The **Predecessor** column displays the number of predecessor objects of the migration object. If you click this number, the system displays the predecessor objects in a list. Additionally, you can call the **Documentation** for each object. In Figure 3.5 (SAP S/4HANA Cloud), we selected two objects from the list of available migration objects for our project.

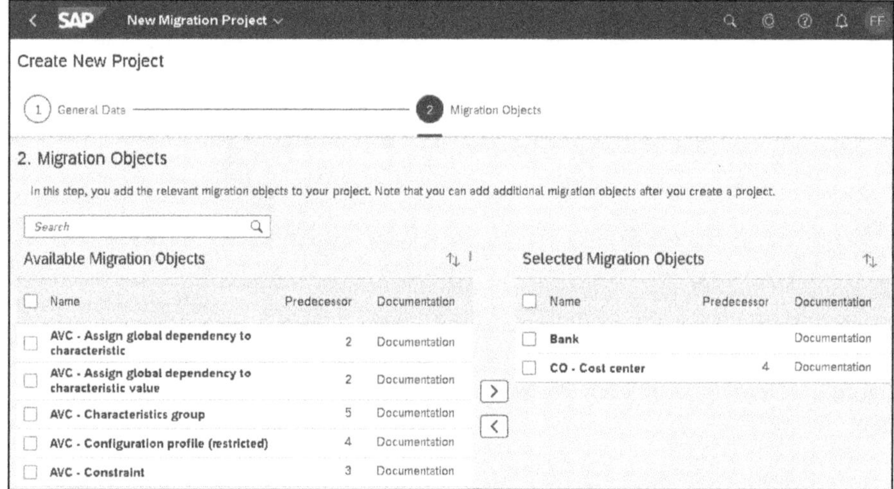


**Figure 3.5** Creating a Migration Project, Step 2: Assign Migration Objects

**Assigning objects using the "Check" button**

By clicking the **Review** button, you can assign objects to the migration project. If one of the migration objects has at least one predecessor object, the **Add Objects** dialog box appears. Now, select the predecessor objects via **Select Objects**, which you then add to the project using the relevant button. If you do not want to add a predecessor object, click **Do not add.**

**Creating a project**

The next screen provides a summary of all project data, as shown in Figure 3.6. If you notice that you missed something or that you made a mistake, you can click the **Edit** button to return to previous dialog boxes and correct your entries. You create your project by clicking the **Create Project** button. As a result, the migration objects are copied from the standard SAP-delivered project to your own project. This step can take several minutes depending on the number and size of your selected migration objects.

**Removing objects only via the SAP S/4HANA migration object modeler**

Removing a migration object from a migration project later is only possible using the *SAP S/4HANA migration object modeler*, which is described in Section 3.4.

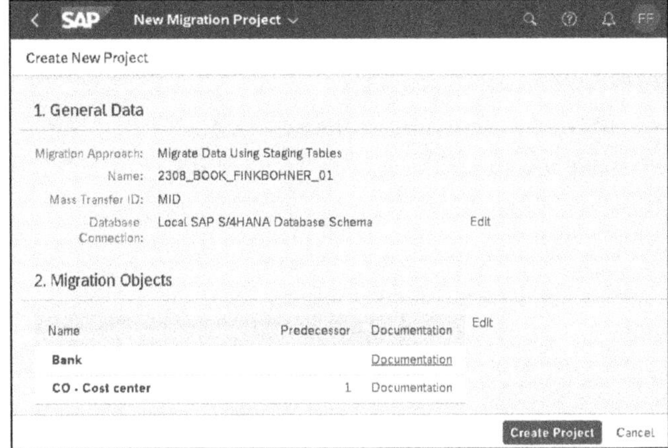


**Figure 3.6** Creating a Migration Project, Step 3: Create New Migration Project

Once the prepare project activity is completed, your migration project will appear in the list of migration projects, as shown in Figure 3.7.

> **Controlling Activities**
>
> Almost all activities within the migration cockpit are executed via background jobs. You can track their activities using the **Monitoring** function (see also Section 3.2.8). The creation of a project appears there as the **Prepare Project** activity.

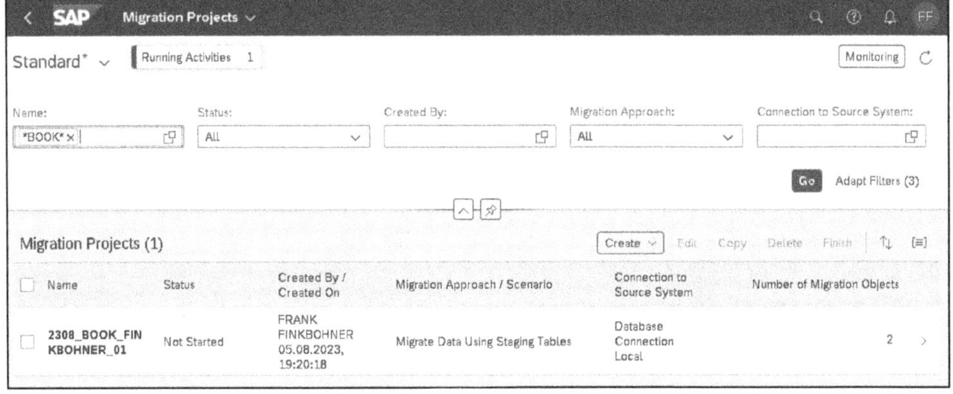


**Figure 3.7** Migration Project Created

### 3.2.6 Editing a Migration Project

In this section, we'll discuss the steps for editing a migration project, beginning with subsequent changes to product data and walking through the migration objects area, CSV file options, and the data retention period.

#### General Information

**Subsequent changes to project data**

Subsequent changes to the project name or the addition of further migration objects can be made via the **Edit** dialog box. For this task, select the project to be edited and choose **Edit**.

#### Migration Objects

**Assigning additional migration objects**

In the **Migration Objects** area, you can see all available migration objects, and you can add objects that have the **Not used in project** status by selecting the **Used in Project** checkbox. The status of this object changes to **To be added**.

#### Comma-Separated Values (CSV) File Options

In SAP S/4HANA Cloud and since SAP S/4HANA 2022, you can migrate your data using CSV files in addition to XML files.

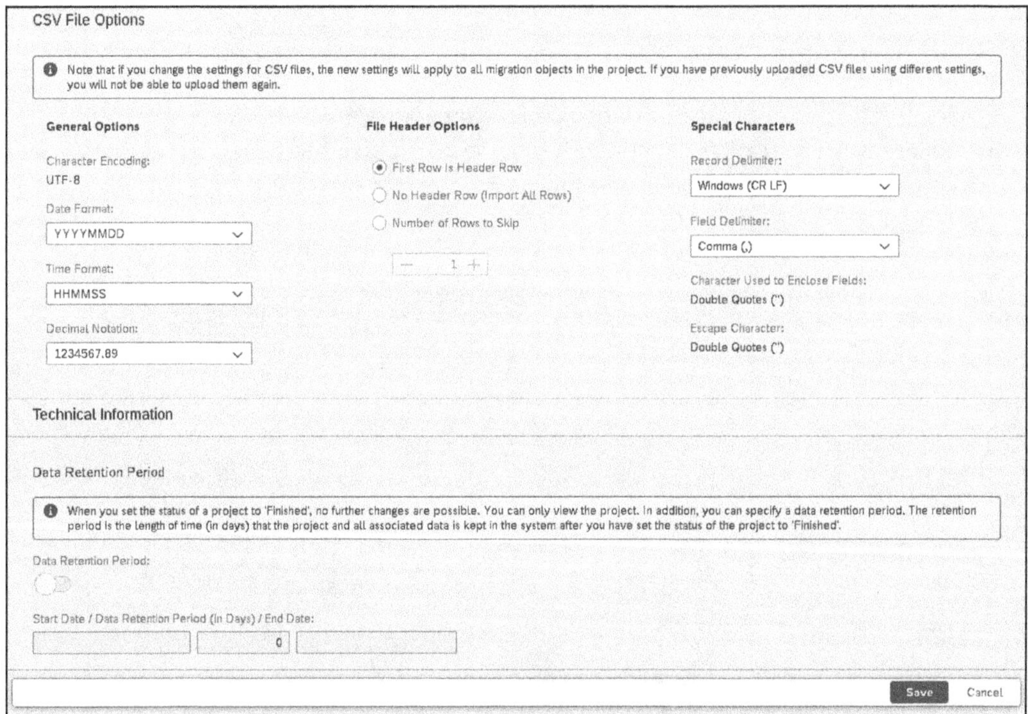


**Figure 3.8** Project Settings: CSV File Options and Technical Information

Via **CSV File Options**, you can maintain the technical settings for the CSV files, as shown in Figure 3.8.

### Technical Information

**Data retention period**

In the **Technical Information** area, you can find the settings for the **Data Retention Period**. You cannot edit this period until you have set the project's status to **Finished**.

## 3.2.7 Calling the Migration Project

**Overview of migration objects**

Now, select a migration project and start it by clicking the [>] icon (**Next**) on the right. The system displays an overview of the migration objects of the selected migration project, as shown in Figure 3.9.

**Header area**

The header area of the project overview provides various information and functions (from left to right):

- **Migration Project** column: Name of the migration project.
- **Running Activities** button: For example, if you add a migration object, you must copy it to the project. You can monitor this process via this dynamic button.
- **Monitoring** button: Monitoring running and finished activities (Section 3.2.8).
- **Mapping Tasks** button: Maintenance of fixed values, control parameters, and value mappings that are cross-object (Section 3.2.14).
- **Job Management** button: Maintenance of the maximum number of background jobs that can be used in the migration project (Section 3.2.15).
- **Settings** button: Overview of the migration objects, the history of activities, CSV file options, and technical information for support.
- **Finish Project** button: Click this button to finish a migration project.

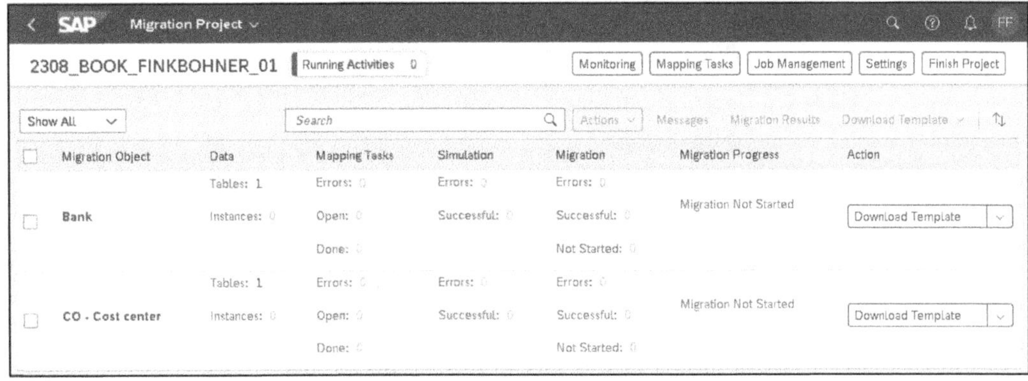


**Figure 3.9** Overview of Migration Project

Object list

The columns of the object list contain the following information for each migration object:

- **Migration Object**: Name of the migration object and number of predecessor objects, if any.
- **Data**: The number of staging tables of the object and the instances uploaded. An instance is a unique data record of the header structure. For complex migration objects with more than their structure, such as **CO - Cost Center**, one instance corresponds to exactly one cost center.
- **Mapping Tasks**: Number of incorrect, open, and completed value mappings.
- **Simulation**: Number of erroneous and successful data records of the last simulation run.
- **Migration**: Number of non-faulty, successfully migrated, or started data records.
- **Migration Progress**: In this column, the numbers of the **Migration** column are displayed as a percentage through a progress bar. If a migration has not been started yet, the status **Migration Not Started** will be displayed.
- **Action**: From this dropdown list, you can select the action to be executed. In the new migration cockpit, you can select your actions flexibly. The migration cockpit suggests the appropriate next action in each case, but you can also choose another action.

Filtering the object list

The list of the migration objects can be filtered using the **Show All** selection field on the left. Possible values include **Show All**, **Errors**, **Not Started**, and **Success**. Moreover, you can use the **Search** field to search for specific migration objects.

### Detail Screen of Migration Object

Information in the upper half

The migration object name is available as a hyperlink. If you click on the name, the system displays detailed information about the migration object. The upper half shows the following information:

- **Documentation**: Click this hyperlink to access the documentation in the SAP Help Portal.
- **Views**: The **Views** function provides the current view of the migration object. The **Edit** option lets you adapt the view depending on the system and the availability of alternative views as long as the migration object does not contain any data yet (Section 3.2.6).
- **Table Structure**: This function displays the structures' hierarchies, which is important to understand along with the relationships of the individual structures and their key interrelationships.

**Functions in the detail screen**

You can call the following functions:

- **Check for updates**: This function checks if updates are available for the migration object. This check is usually done automatically when certain actions are executed, but in certain cases, you can also trigger this check manually. An update is available in the various product versions in the following circumstances:
  - In SAP S/4HANA Cloud after a release upgrade or possibly after implementing a hotfix
  - In SAP S/4HANA after an upgrade to a new product version or after implementing a *transport-based correction instruction (TCI)*
- **Update**: If an update is available, it can be brought up to date via this function.
- **Check for Custom Fields**: You can define custom fields for a business object using *in-app extensibility* in SAP S/4HANA. If the migration object API has been registered for in-app extensions, custom fields will be automatically included in the migration object template. To find out whether a migration object supports custom fields, you should refer to the documentation or the overview of available migration objects in SAP Help Portal.
- **Mapping Tasks**: You use this function to maintain fixed values, control parameters, and value mappings of the migration object.
- **Instances**: This function shows all uploaded data records of the header structure and their migration statuses. For more information, refer to Section 3.2.12.

Figure 3.10 shows the upper half of the detail screen.

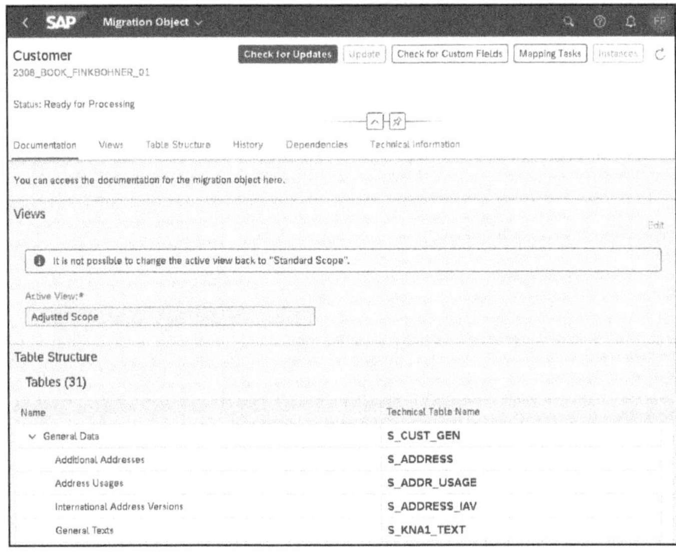


**Figure 3.10** Migration Object Detail Screen (Upper Half)

**Information in the lower half**

The lower half provides the following information:

- **History**: The history lists all completed actions (activities) of the objects including the data about the user who started this action. Click **Show Messages** to display the messages of an action (see also Section 3.2.21).
- **Dependencies**: This section displays migration objects that can be predecessor objects for the current migration object. The display contains the fields **Migration Object** (name of the migration object) and **Type** (of the object), such as **Predecessor**. An important fact is that not every object listed here must be migrated in advance. Whether you do often depends on the constellation of the data to be migrated. The **Bank** migration object is a predecessor of the **Customer** object. In the US, however, in contrast to Europe, the **Bank** migration object is not a predecessor because in the US invoices are still paid by check and thus the bank data of a customer is not available.
- **Technical Information**: This section lists the **Object ID**, **Technical Name** of the object, and the **Template ID**. These three pieces of information are important for support.

Figure 3.11 shows the lower half of the detail screen.

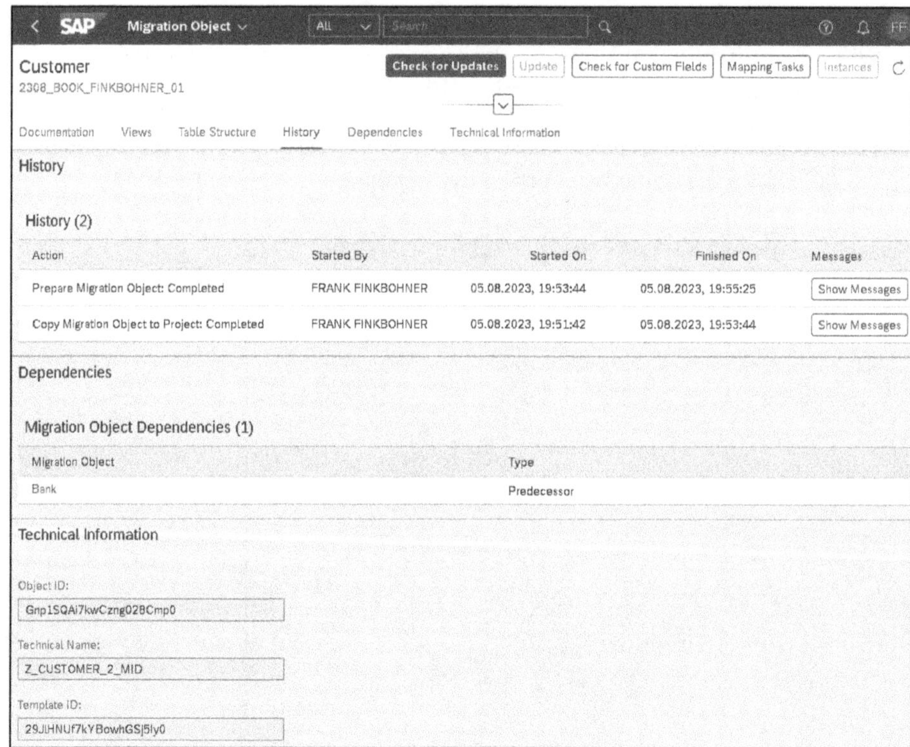


**Figure 3.11** Migration Object Detail Screen (Lower Half)

#### Migration Object Documentation

The migration object documentation is available at SAP Help Portal. For example, you can view the documentation and download it as a PDF file without logging on to the system.

### 3.2.8 Monitoring the Activities

**Migration object documentation**

Almost every activity/action is scheduled as a background job, as mentioned earlier. The easiest way to monitor the activities is to use the **Monitoring** function, as shown in Figure 3.12 ❶.

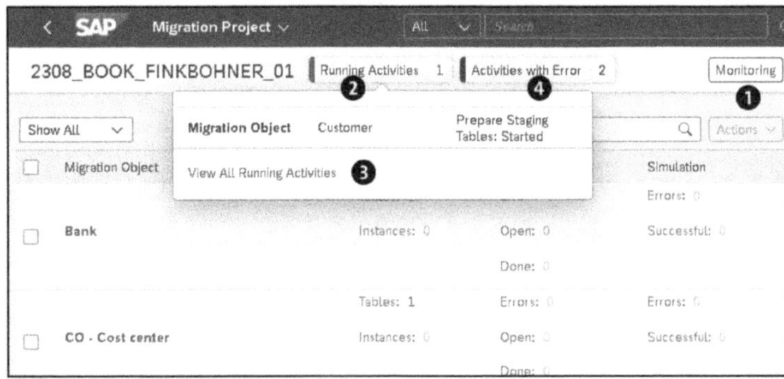


**Figure 3.12** Monitoring Activities

#### Running Activities

Additionally, you can get information about the *generic tag* **Running Activities** ❷, where you can see the number of all running activities. Via **View All Running Activities** ❸, you can access the **Monitoring** function. If an activity encountered an error, another generic tag, **Activities with Error** ❹ is displayed to the right of **Running Activities**. Via **View All Failed Activities**, you can open a monitoring view with preset filters for the **Completed with Errors** and **Failed** statuses.

#### Canceling an Activity

In SAP S/4HANA Cloud and since SAP S/4HANA 2021, you can also cancel scheduled or started simulation or migration activities via **Monitoring**. For this task, in the **Options** column of the activity you want to cancel, click the **Cancel** button. A small dialog box will then appear asking if you really want to cancel the activity. You can confirm this action by clicking the **Cancel current activity** button, which is sometimes hidden behind a meatball icon ⋯.

#### Message Overview

In the **Options** column, you can click the **Display Messages** button to display all messages/messages related to this activity. Depending on the activity, a menu button also displays in the **Options** column with other options you can execute, such as **Download File** or **Show File**.

> **Hide Faulty Jobs**
>
> You can use the **Set Status to "Error Fixed"** function in the monitoring view to mark activities with errors as fixed. As a result, the generic tag **Activities with error** is no longer displayed, and the functions in the header bar do not disappear behind the meatball icon [...].

### 3.2.9 Downloading Migration Templates

New file type CSV

You download migration template files to your local computer in the migration project overview using the **Download Template** option. Since SAP S/4HANA Cloud 2208 and SAP S/4HANA 2022, a new way is available to populate the staging tables. In addition to the previous XML file, you can now also use CSV files to populate your staging tables.

#### Downloading Migration Template Files

Downloading one or more templates

You can download migration templates in several ways. One way is to select **Download Template** in the **Action** column of the migration object, which directly downloads the migration template. You can select the format, **XML file** or **CSV files**.

> **CSV Files Always as ZIP Files**
>
> CSV files are always downloaded as a ZIP file regardless of whether you have one or more structures (staging tables).

However, you can also select one or more migration objects and choose the **Download Template** function. The templates of the selected migration objects are downloaded in a ZIP file. You must unpack the migration template files to your local computer. When selecting only one migration object, the corresponding template is downloaded directly as an XML file.

#### XML Files

The **XML File** migration template uses the file format **Microsoft Excel XML Spreadsheet 2003**. You can open and populate it using Microsoft Excel 2003 or better or any other spreadsheet program that can process this format, for example, *OpenOffice Calc*. Figure 3.13 shows an example of such a migration template.

| | 1 | 2 | 3 | 4 | 5 | 6 |
|---|---|---|---|---|---|---|
| 1 | **Source Data for Migration Object: Customer** | | | | | |
| 2 | Version SAPSCORE 130 - Adjusted Scope - 15.06.2023 © Copyright SAP SE All rights reserved | | | | | |
| 3 | | | | | | |
| 4 | S_CUST_GEN | | | | | |
| 5 | KUNNR | BU_GROUP | KTOKD | NAMORG1 | NAMORG2 | NAMORG3 |
| 6 | ETE;80;0;C;80;0 | ETE;80;0;C;80;0 | ETE;80;0;C;80;0 | ETE;40;0;C;40;0 | ETE;40;0;C;40;0 | ETE;40;0;C;40;0 |
| 7 | **Key** | **BP Group** | **Account Group** | **Customer Organization** | | |
| 8 | Customer Number* Key fields are mandatory and must always be filled with a unique key. If there is internal numbering, you can provide the key of your source system or any other unique key. The key is technically mandatory for the tool to internally connect the records of subordinated sheets with the records of the superordinate sheets. Type: Text Length: 80 | BP Grouping* Assign each business partner to a grouping when you create the business partner. The grouping determines the number range and the account group (customer, one-time customer, supplier, one-time supplier). You cannot change the assignment afterwards. SAP delivers standard groupings for the internal and the external number assignment. BU_GROUP: BP01 NRRNG: 01 From No: 0000000001 To Number: 0000999999 Ext: X BU_GROUP: BP02 NRRNG: 02 From No: 0001000000 To Number: 0009999999 Ext: BU_GROUP: BP03 NRRNG: 03 From No: 0010000000 To | Customer Account Group The Account Group serves as a feature for classification of customer master records. As a standard, the Account Group will automatically be determined by the BP Grouping value. If flexible grouping is activated, you can provide a different Account Group using the values defined in customizing. Type: Text Length: 80 | Name If you want to create an organization, enter the first name of your account organization. All fields for Account Person must be empty. Type: Text Length: 40 | Name 2 Second name field for business partners in the Organization category. All fields for Account Person must be empty. Type: Text Length: 40 | Name 3 Third name field for business partners in the Organization category. All fields for Account Person must be empty. Type: Text Length: 40 |
| 9 | MIG_USC001 | Ext. Alpha (BPAB) | | Euclidus University | | |
| 10 | MIG_USC002 | Ext. Alpha (BPAB) | | World Banks | | |
| 11 | MIG_USC003 | Ext. Alpha (BPAB) | | 22nd Century Fox | | |
| 12 | MIG_USC004 | Ext. Alpha (BPAB) | | Abarcombie & Finch | | |
| 13 | MIG_USC005 | Ext. Alpha (BPAB) | | Americans Apparels | | |

Introduction | Field List | General Data | Data Controller | Additional Addresses | Address Us...

**Figure 3.13** Migration Template of the "Customer" Migration Object

### Structure of the XML Migration Template

**Spreadsheets of the migration template**

The migration template usually includes several spreadsheets. For more information, see the **Introduction** worksheet.

[!]

**XML File Structure Changes Result in Errors with Downloaded Templates**

For compatibility reasons, you should only use templates that you have downloaded from the latest release. You may not be able to upload templates from other releases. The migration cockpit checks if the internal data structure of the XML file to be uploaded corresponds to the data structure of the current release. If there are deviations in the structure, for example, another field sequence, new fields, or new structures, the file is usually not accepted. Because many migration objects in new versions are adapted to business process changes, this mechanism ensures data integrity.

**Key fields**

The key fields in the individual spreadsheets define the relationship between the header structures and substructures of an object. As a result, a key record must exist in the spreadsheet of the superior structure for each substructure record. Otherwise, the migration cockpit displays an error message, and the record is not imported.

[»]

**Mandatory Structures and Mandatory Fields**

The migration template contains what are called *mandatory structures*. These structures (spreadsheets) must contain records and must not be empty. All other structures are optional and don't need to be populated. For example, in the **Purchase Order** object, the header structure and the item structure are mandatory. In the **Customer** object, only the top-level structure (**General Data**) is mandatory. Mandatory structures are marked on the **Field List** worksheet with the addition "mandatory," for example, **General Data (mandatory)**.

However, you don't need to fill in "mandatory" fields and key fields in optional spreadsheets if you don't want to use these structures. For example, if you don't want to migrate long texts in an object, you don't have to make any entries in this spreadsheet. Leave this spreadsheet blank.

### The Most Common Problems with XML Files

Because the templates are XML files with a specific format and an internal structure, you should consider the following:

- Use only the format of the downloaded XML file; other Microsoft Excel file formats will not work.
- Do not change the structure of the XML file. Custom XML templates often cause data loading errors.

**Always read the documentation**

To avoid errors, always read the **Introduction** spreadsheet in a migration template and the relevant documentation for specific migration objects. Important notes are also available in the product assistance for the migration cockpit at SAP Help Portal.

**Important Knowledge Base Articles and SAP Notes**

Several important SAP Help Portal pages, SAP Notes, and SAP Knowledge Base articles (KBA) can help you better populate your migration templates. You should download and read them and their attachments:

- **"Data Migration Template Samples for SAP S/4HANA Cloud"**
  This SAP Help Portal page provides migration templates with sample data populated for selected objects and a description of the sample data: *http://s-prs.co/v581730*.
- **KBA 2683477, "SAP S/4HANA Cloud Data Migration content: Why fields on apps are not available in the content?"**
  This KBA describes a phenomenon when you see fields in the maintenance app that do not exist in the migration template: *http://s-prs.co/v581731*.

- **KBA 2692715, "Migration Cockpit: How to Correctly Fill Data into XML Template"**
  Examples show how manipulated inputs (incorrect copying, format changes) can have an effect: *http://s-prs.co/v581732*.
- **KBA 2728194, "Decimal Numbers Are Truncated during Upload - Warning Message"**
  In this KBA, you'll learn how to correctly deal with the problem of storing floating point numbers according to IEEE 754 in Microsoft XML. A value that is displayed in Microsoft Excel as "0.57" is often stored as "0.57999999999996" in the XML file. Since the field type definition often provides only 2 or 3 decimal places, when uploading this value to the migration cockpit, a warning appears that the value is too long and will be truncated; see *http://s-prs.co/v581733*.

### Comma-Separated Values Files

One CSV file per structure

An XML file contains all structures of a data migration object. So, in the best case, you can load all the staging tables of a migration object with one XML file. However, by means of CSV files, you always fill only one structure (i.e., one staging table). If the migration object, such as **Customer**, consists of multiple structures, you must populate and upload a separate CSV file for each structure used.

Storing the files in folders

If you select **Download CSV Files** via the **Download Template** action, you'll always get a ZIP file with a folder containing all the CSV files of the migration object. The prefix of the ZIP file and folder name is *Source data for* followed by the language-dependent name of the migration object. Thus, the file for the **Customer** migration object is named *Source data for Customer.zip*, and the folder, *Source data for Customer*.

ZIP file for multiple migration objects

If you have selected multiple migration objects, then the ZIP file will not be downloaded directly. The action is swapped to a background job, and you must then download the ZIP file via the **Monitoring** view. This ZIP file contains one folder per migration object, which contains the respective CSV files. The name of the ZIP file is composed of the prefix *Source data* and the project name.

Prescribed file names

A CSV file contains the technical structure name such as *S_ADDRESS* followed by *#FreeText.csv*. You must not change the names of the structure. However, for the *#FreeText* placeholder, you can insert text of your own choice.

[»]

**Downloading XML Files for Help**

A CSV file handles only the technical structure and field names and, unlike the XML file, contains no structure or field descriptions and no tooltips. If you use CSV files, you should always download the corresponding XML file as well. As described in Section 3.2.8, when we covered the structure of the XML migration template, you should make the hidden columns and rows with the technical information visible. This visibility makes it easier for you to populate the files correctly.

#### Frequently Asked Questions and Answers about Migration Templates

Why are some fields longer?

One of the most common questions is why many fields have 80 digits in the migration template, although the SAP field is shorter. The whole aspect of different field lengths is vividly described in KBA 2563055: *http://s-prs.co/v581734*.

Units of measure and currencies in ISO code

Another frequently asked question is why it is necessary to map units of measure (UoMs) and currencies to an International Organization for Standardization (ISO) code as a target value for the SAP S/4HANA system. For a detailed answer, see KBA 2907822, "SAP S/4HANA migration cockpit - Unit of Measure Handling" at *http://s-prs.co/v581735*, and KBA 2661986, "Migration Cockpit: Currency key issues regarding ISO code" at *http://s-prs.co/v581736*.

### 3.2.10 Filling Migration Templates or Staging Tables Directly

Upper file size limit: 100MB

The maximum size of a file (whether XML, CSV, or ZIP file) that can be uploaded via an *HTTP request* to SAP S/4HANA Cloud is technically restricted to 100MB. For details, see KBA 2719524 (*http://s-prs.co/v581737*).

If you use a ZIP file, the total of packed file sizes of all the migration files that are contained in the ZIP file must not exceed 160MB.

[+]

**Splitting XML Files Easily**

On the GitHub development platform, you can find sample code for a command-line utility that helps you split larger XML files into smaller XML files. This code is called the *SAP S/4HANA Migration Cockpit XML File Splitter* and is publicly available at *SAP-samples/s4hana-mc-xml-file-splitter* at *http://s-prs.co/v581738*.

Mandatory fields

Each migration object contains mandatory fields. The API specifies these as mandatory and must have a value (mandatory fields). Furthermore, however, there may be other fields that can either be configured as mandatory fields or that become mandatory fields depending on other field values. Such fields cannot technically be designated as mandatory in the migration template files. If such mandatory fields have not been filled, this will be displayed as a message in the **Simulate** action.

### Filling CSV Files

**Structure of the CSV file**

A CSV file of the migration template consists of a header row followed by the data rows. The header contains the technical field names. Mandatory fields are marked with an asterisk (*), and key fields are marked with the letter k (for "key"). Depending on the CSV file options you set up earlier in the project settings, the fields and data are separated from each other by the selected field separator, such as the header for the S_ADDR_USAGE structure of the **Customer** migration object:

```
KUNNR(k/*),BU_ADEXT(k),ADR_KIND(k/*),XDFADU
```

[«]

**Additional Information**

For more information on using CSV files in SAP S/4HANA Cloud, see KBA 3210687 at *http://s-prs.co/v581739*. The KBA includes, among other things, a link to a user guide that describes the whole topic in its entirety.

For information on using CSV files in SAP S/4HANA 2022, see KBA 3296020 via the following link: *http://s-prs.co/v581740*.

We also recommend the blog "SAP S/4HANA migration cockpit - Transferring Data from CSV Template Files to Staging Tables" by Iliana Olvera via the following link: *http://s-prs.co/v581741*.

An especially useful video can be accessed via SAP Help Portal or via this direct link: *http://s-prs.co/v581742*.

### Populating Staging Tables Directly

**Tools for populating**

Your staging tables can be located in a local SAP S/4HANA database schema or a remote SAP HANA database schema. You populate them using external tools, for example, *SAP Data Services*, *SAP HANA smart data integration*, *SAP HANA Studio*, or other tools with that you can access the staging tables of the local or remote database schema. You cannot directly populate the staging tables from the migration cockpit. An important source of information on how to populate this data can be found in KBA 2733253, which can be accessed at *http://s-prs.co/v581743*.

**Run consistency checks**

After the staging tables have been populated, the process continues like before, when we uploaded a file in Section 3.2.11. The only difference is that the data is now uploaded directly to the staging tables and not from a file. Therefore, an imperative step is that you run a consistency check afterwards. You can perform this task, for example, via the **Migration Object Tables** view via the table overview (described in more detail in Section 3.2.11). However, you'll also be prompted to perform a consistency check later when you perform the **Prepare** action for a migration object. We strongly recommend that you enable this consistency check if you have populated the staging tables directly.

### 3.2.11 Loading a File to Staging Tables

**"Migration Object Tables" screen, in three parts**

Once you have populated the migration file, import this file to the migration cockpit by choosing **Upload File**. Now, the **Migration Object Tables** view opens, which is divided into the three overviews: **Tables** ❶, **Upload Files** ❷, and **Correction Files** ❸, as shown in Figure 3.14. We describe the **Correction Files** overview in Section 3.2.19.

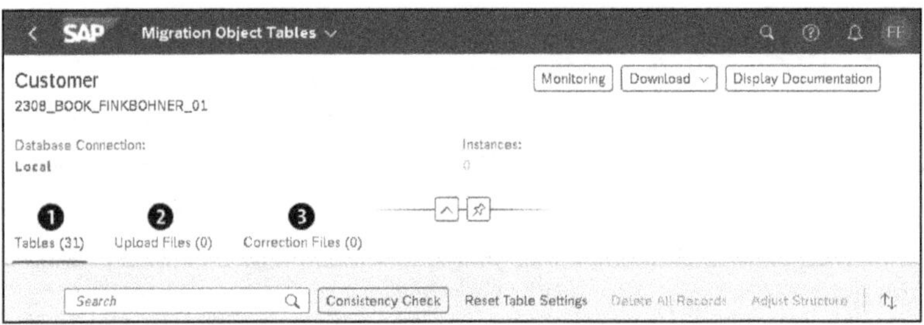


**Figure 3.14** Migration Object Table View

From the **Migration Project** view, navigate to the **Migration Object Tables** view by clicking the number to the right of **Tables** in the desired migration object in the **Data** column.

#### Upload Files Overview

**Uploading files**

In the **Upload Files** overview, select **Upload** to upload your migration template file. However, you can also drag your files to the designated area. The name of the file is unique. You cannot upload several files with the same file names. Subsequently, the file is validated. Once the upload is complete, the name of the file is displayed in the list, and you can view the validation status. After a successful validation, the data gets transferred to the staging tables. You can monitor both activities—**Validate File** and **Transfer Data to Staging Tables**—via **Monitoring**. If the file does not contain any errors, the status is **Data Successfully Transferred to Staging Tables**. You can click the **Show Messages** next to the status display to show messages on validation.

The header area indicates the number of instances that you have uploaded. An instance is a unique data record of the header structure, for example, a unique customer number. In IT terms, this is a closed unit. If an instance has already been loaded to the staging tables, you can no longer load it via another file.

Instances = number of data records of the header structure

The list of the source files indicates the **Status** of the file, its **File Name**, **File Size**, the name of the user who uploaded the file (**Uploaded by**), and the date when the upload took place (**Uploaded on**).

Information on the source file

The status of a file can have the following values:

- **Upload**: The file is being uploaded, and a counter appears showing what percentage of the file has already been uploaded. Then, the **Upload completed** status displays very briefly before the automatic validation gets triggered.
- **Validation scheduled**: The file is currently being validated.
- **Validation completed with warning**: The file contains errors and cannot be uploaded. Check the messages, and, according to the warning, check the file. Often, a field value is too long, or you're using an old migration template file, and structures and/or fields are missing.
- **Check failed**: The file contains critical errors and cannot be uploaded. This error is often a missing value for a required field (DMC_MD_INTERPRETER 045) or, for example, that key values on a substructure do not find a corresponding record with the corresponding key at the higher level in the hierarchy (/LTB/MC 329).
- **Transfer to Staging Tables**: The transfer to the staging tables is currently taking place but is not complete yet.
- **Data successfully transferred to staging tables**: The file was transferred to the staging tables.

You can use the following **Actions** on these files:

Executable actions

- **Validate Data**
- **Transfer Data to Staging Tables**: You usually use this action if you want to take over the data after a warning message anyway or possibly in the context of uploading CSV files, see the next subsection.
- **Delete Files**: If you delete a file, the data is not deleted from the staging tables. You can also remove a file using the ⊗ icon (**Delete**).
- **Download Files**: With this option, you can download the files as XML files.

Figure 3.15 shows an example of the successful upload of a file.

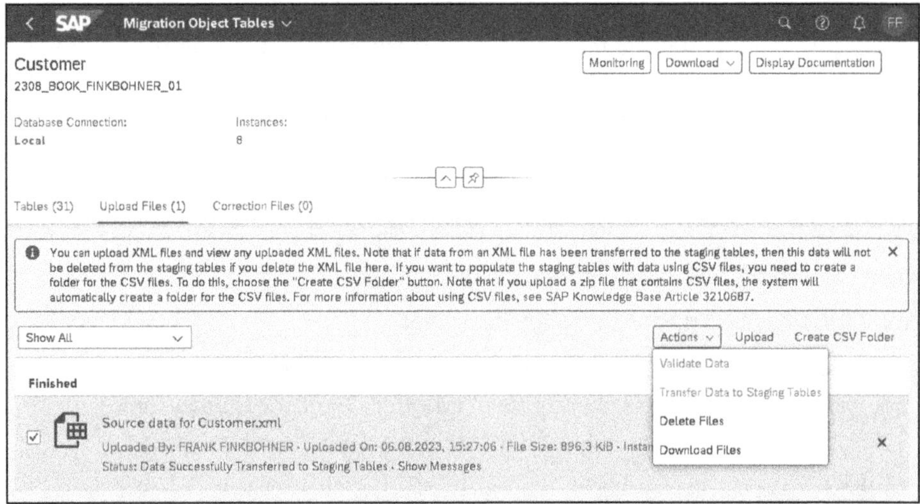


**Figure 3.15** Migration Object Tables View: Uploaded Files Overview

You can also upload multiple files via a ZIP file.

**Checking the data after upload**

Inconsistent data structures are not stored at all, for example, by storing files in staging tables. The entire internal XML structure can be corrupted if you copy data with the wrong function within Microsoft Excel and thus cause changes to the data types and manual manipulation of the XML file. You should therefore always check the data in the migration cockpit after uploading. You should especially pay attention to the following points:

- **Record shifts**
  Are all values in the correct fields?
- **Data types**
  Are date fields populated correctly? Is the date information still valid?
- **Leading zeros**
  Are leading zeros still present?
- **Numerical values**
  Are numerical values (numbers and their decimal places) correct?
- **Data records not transferred due to missing key relationships**
  Have all records from the file been uploaded?
- **Truncated values**
  Were any values truncated? Since the cells in Microsoft Excel have no length limit, it may happen that you enter a value, for example, that is truncated when you fill it into the database table. You'll then get a warning when filling the staging tables.

**Displaying the contents of files**

For this task, click on the file name in the **Upload Files** view. The content view of the file is displayed (**Migration Object File Content**). The file's detailed information and the project name are displayed in the header area (which can be hidden), followed by the *Icon Tab Bar* with tabs to the relevant structures and the number of uploaded records in parentheses after each structure's name. The content area shows the data records for each structure. You can search for data records on this screen, put the records in the desired order using the button (**Sort**), or filter them by clicking the icon (**Settings**). Figure 3.16 shows you the content view using the sample file for the **Customer** migration object.

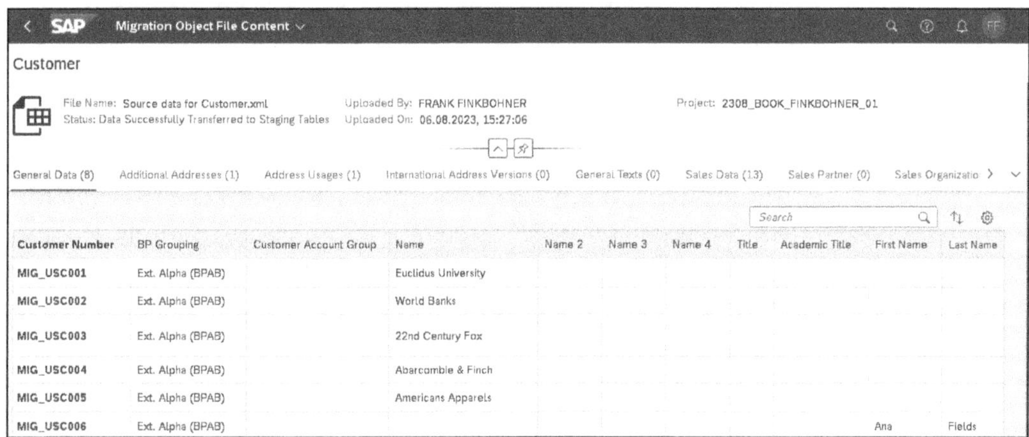


**Figure 3.16** Content of a Loaded Migration Object Files for the "Customer" Object

### Uploading CSV Files

Uploading CSV files is also described in detail in the User Guide mentioned earlier in the previous section; for this reason, we won't go further in this section. Access KBA 3210687 at *http://s-prs.co/v581744*.

[!]

**Where Is My File?**

When you upload CSV files, you may see that the file has been uploaded, and then the file suddenly disappears. In that case, you should look in **Monitoring**; usually, the validation has been scheduled and not finished yet. If the validation was finished without errors, then the file appears again; otherwise, it remains hidden. Then, check the messages in the **Validate File** activity in **Monitoring**, correct the errors in the CSV file, and reupload it.

### Tables Overview

**Structures and staging tables**

The table overview provides an overview of the structures and their staging tables. The header contains the **Database Connection** information and the number of uploaded **Instances**. Generally, you can jump to the **Monitoring** view here or call the migration object documentation via **Display Documentation**. Using the **Download** menu button, you can download either the **Metadata** or the migration template files (XML or CSV format) via **Template**. The metadata file is a text file that you can also display as a CSV file (separated by semicolons) in Microsoft Excel. It basically contains the technical object name, structure information, the staging table and the data you know from the field list of the migration template file, extended by the name of the staging table and the tool tip for the fields. Every structure is assigned to a separate staging table. You can execute the following actions on these tables:

- **Consistency check**
  If you have data directly at database level in the staging tables, you can perform the necessary consistency check here. It checks whether the corresponding data records exist on the respective structure above for all data records of a lower structure. If you load data via files into the staging tables, this check happens automatically when you upload the files in the validation activity.
- **Reset table settings**
  Under certain circumstances, sometimes, you need to unlock your staging tables. This problem may arise, for example, if a job was canceled manually. With this option, you reset the locks so that you can work with the data again. The number of erroneous and simulated instances is thus set to 0. The number of migrated data records is not adapted.
- **Delete all records**
  This action allows you to delete all data records for selected structures regardless of their status. You should use this option with care because improper use can result in errors and inconsistencies.
- **Adjust structure**
  This option enables you to update the structure of the selected table. In this case, the table status is **Adjustment Required**. This action may be necessary, for example, for tables in a remote SAP HANA database schema or after the deletion of a staging table on the database. No data is lost for existing tables.
- ⇅ **(Sort)**
  Sorts the list

**Information of a table**

Each row of the table contains the following information:

- **Name**: Name of the structure.
- **Technical name**: Technical name of the structure.

- **Technical ID**: Name of the staging tables in which the data is stored.
- **Rows in Staging Table**: Number of data records of the respective staging table.
- **Table Status**: The table status includes the values **OK** and **Adjustment Required**. With this status, it indicates whether adjustments to the table structures must be made using the **Adjust Structure** option.

Figure 3.17 shows an example of a table overview.

Displaying the table contents

If you click on the number in the **Rows in Staging Table** column, you navigate to the overview screen, **Migration Object Table Content**. All data records of the staging table of the selected structure are displayed here, as shown in Figure 3.18. You can display the documentation via the **Display Documentation** button. The overview shows the name of the staging table (**Technical ID**) and the table status.

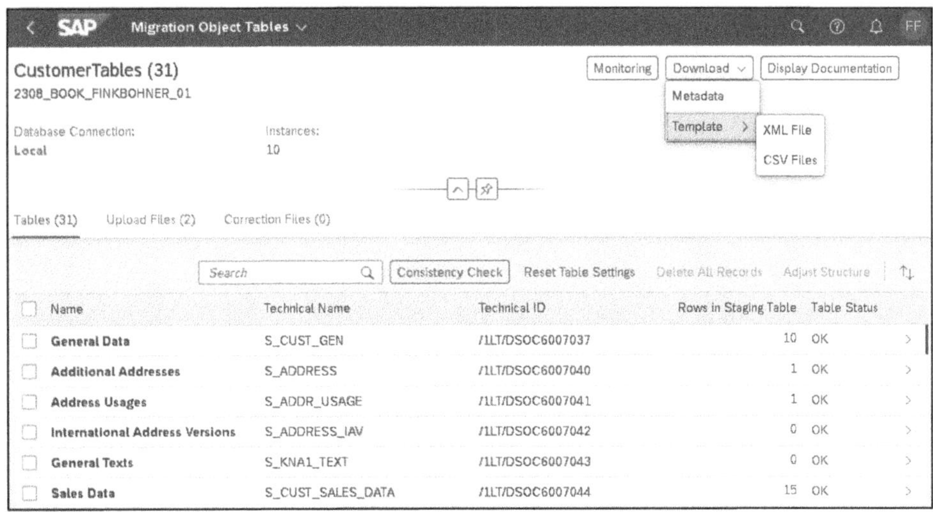


**Figure 3.17** Migration Object Tables View: Overview of the Tables

The row of the table list contains the following information, as shown in Figure 3.17:

- **Name:** Name of the structure
- **Technical Name**: Technical name of the structure
- **Technical ID**: Name of the staging table
- **Rows in Staging Tables**: Number of data records in the respective staging table
- **Table Status**

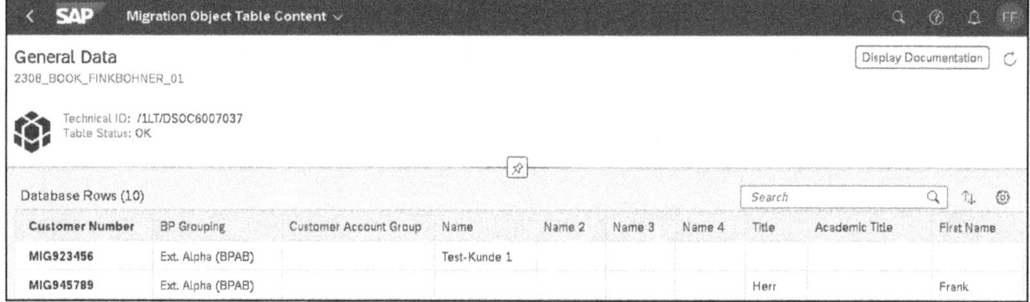


**Figure 3.18** Migration Object Table Content View: The Content of a Structure

**Displaying the table structure**

In the table overview, select the name of the structure or click the [>] icon (**Next**) to navigate to the **Migration Object Tables, Structure** view. This view has a flexible column layout and is divided into two parts. On the left, you can view the list of structures, and on the right, you can view the metadata of the selected structure. The overview contains the following information in the header area:

The list provides information on the structure, which you know from the field list of the migration template file:

- **Name**: Name of the field and tool tip about the field
- **Technical name**: Technical name of the field
- **Key field**: Determines if it is a key field.
- **Mandatory**: These fields must have values when you populate the tables with data.
- **Data type**: Data type of the SAP HANA database
- **Length**: maximum length of the field
- **Decimals**: Number of decimal places for numerical fields
- **Group**: Fieldes are classified in groups to ensure a better overview. In the maintenance app, related fields are usually grouped in so-called areas. The group names are based on these area names.
- **Not Null**: You can technically define a field of a database as a *not null* field, which means that this field must have a value. Key fields, mandatory fields, numerical fields, and date fields are usually defined as **Not Null** fields. These fields cannot be left empty. When populating them, you must transfer a defined initial value if you do not want to provide any value.

Figure 3.19 shows an example of a **Migration Object Tables, Structure** view.

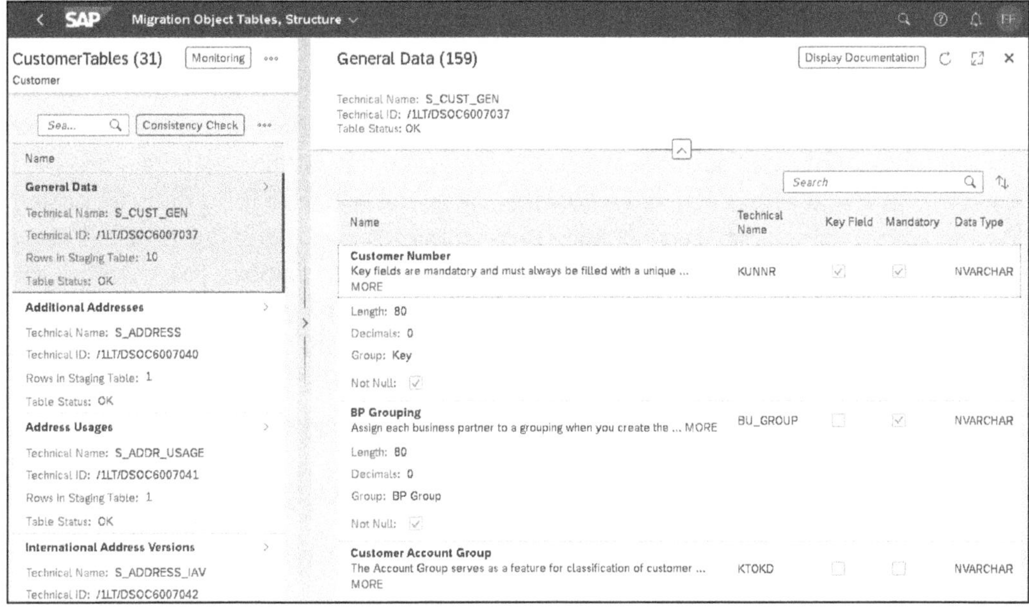


**Figure 3.19** Migration Object Tables, Structure View: Metadata of the Structure

### 3.2.12 Using Migration Object Instances

Once you have loaded data into the staging tables, you'll notice that the **Instances** function is selectable in the Migration Object detail screen of the **Migration Object** view. This function takes you to the **Migration Object Instances** view. Alternatively, you can also navigate there using the object list of the migration project overview, as shown in Figure 3.20. In this case, you click either the number of instances in the **Data** column, which are hyperlinks, or one of the numbers in the **Simulation** or **Migration** columns.

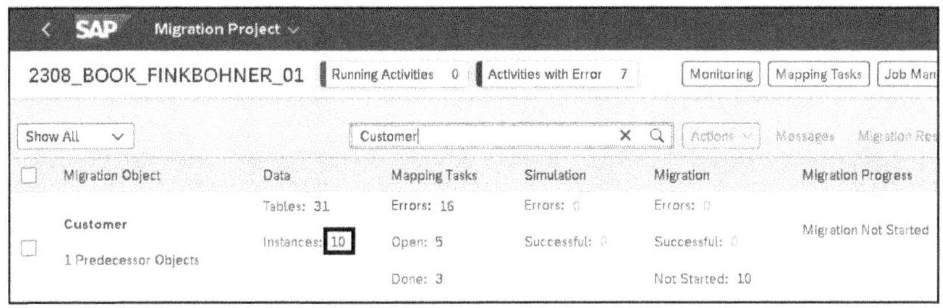


**Figure 3.20** Migration Project Overview and Instances

**Migration control center**

Used skillfully, you can use the **Migration Object Instances** view as a control center for your entire migration. The advantage of this view is that you can select specific instances to perform actions on them. Especially for testing and subsequent analysis, this view offers just about everything your heart desires.

**Filtering the view of instances by status**

Depending on how you called the list, the **Migration Object Instances** view displays either all instances or only those instances that correspond to the status selected, as shown in Figure 3.21. The list shows the values of the key fields and the first fields of the instance. The list can be filtered using the fields **Action/ Status** ❶, **Source** ❷, and **Key**, as well as other fields ❸ of the migration object and using **Adapt Filters** ❹ and the **Go** button ❺. Additional fields and their values can be displayed in a separate window using the **Display All** hyperlink ❻ of the **Additional Fields** column. You can also see the **Source** of the instance (e.g., the file name); the last **Action** on the instance (e.g., **Initial**, **Simulated**, or **Migrated**); and the **Status** of the instance (e.g., **Initial**, **Success**, or **Error**).

**Downloading a summary**

Using the **Download All** ❼ function, you can download a summary of all displayed instances with selected fields that you can choose in a dialog box. The file can then be downloaded using the **Download Instance Records** activity in **Monitoring**.

**Customizing the list view**

Furthermore, you can sort the list by clicking the [⇅] icon, group it using the [≡] icon ❽, and configure the displayed columns using the [⚙] icon.

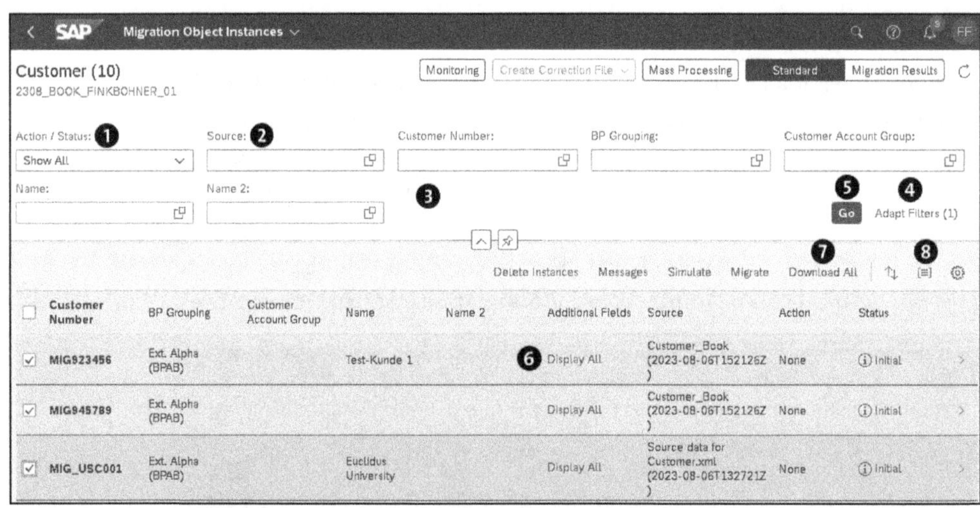


**Figure 3.21** Displaying Migration Object Instances

**Viewing an individual instance**

When you click on a data record, the **Migration Object Instance** view is displayed, as shown in Figure 3.22. The overview is divided into two areas: On the left, you're provided with information on the instance you've selected, and you can choose the table (structure) to be displayed. The values of the selected staging table are then shown on the right.

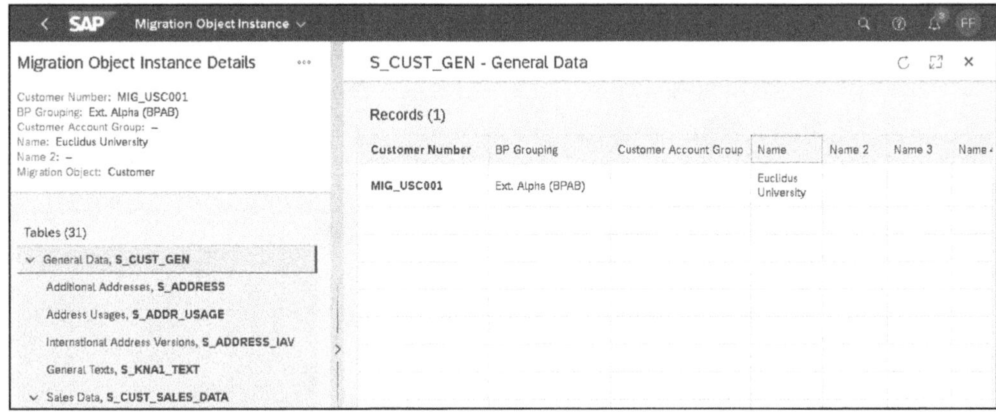


**Figure 3.22** "Migration Object Instance" View: Customer General Data

**Functions for migration object instances**

In the **Migration Object Instance** view, you can perform the following functions on the selected instances:

- **Delete Instances**: All instances selected are deleted.
- **Messages**: The messages for each instance selected are displayed.
- **Simulate**: Selected instances are simulated.
- **Migrate**: Selected instances are migrated.

> **Impact of Value Mapping on the Display of Instances**
>
> The **Migration Object Instance** view shows the values of the data records as you loaded them to the staging tables. When you start the **Simulation** action, the data of the fields to which a mapping task is assigned are converted to the confirmed target value. In other words, after the simulation is complete, the system displays the converted data, which is then used for migration.

Regardless of the selected instances, you can perform the following functions:

- **Create Correction File** (Section 3.2.19)
- **Mass Processing** on migration object instances (Section 3.2.20)

### Deleting Instances

**Migrated instances cannot be deleted**

The **Delete Instances** function deletes the selected instances from the database. You can only delete instances that have not been migrated successfully. After you have selected the function, a text window is displayed informing you about this deletion. You can either run the function by selecting **Delete** or exit by selecting **Cancel**.

#### Displaying Messages per Instance

The **Messages** function displays all messages of the selected instances, as for example shown in Figure 3.23. Select the instance on the left, and the messages for this instance are displayed on the right.

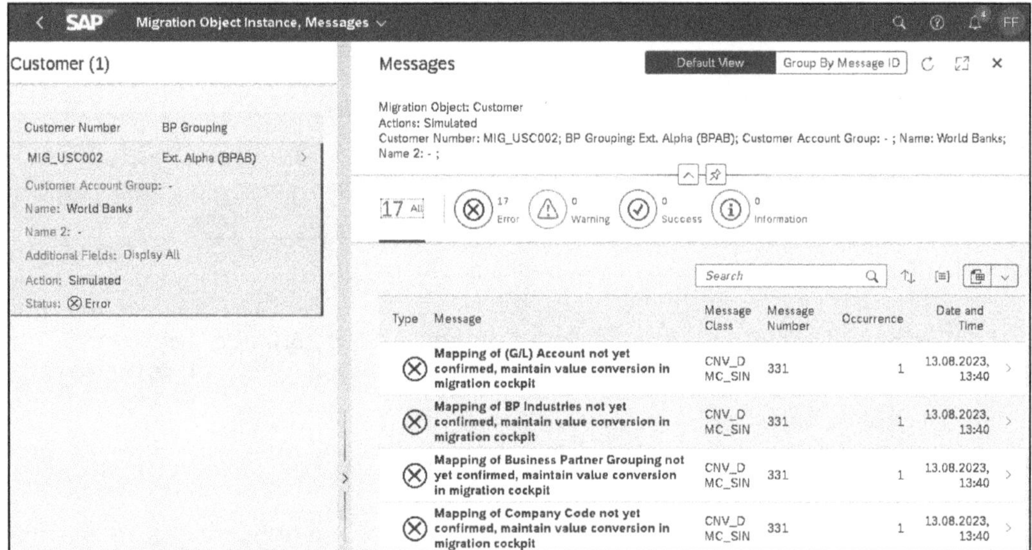


**Figure 3.23** "Migration Object Instance, Messages" View

### 3.2.13 Prepare

**Preparing staging tables**

Once you have uploaded the data, navigate back to the **Migration Project** view to perform the **Prepare** action on the migration object. If the menu button displays the **Mapping Tasks** action, you should perform that action first (Section 3.2.14). The corresponding background job is scheduled and displayed under **Monitoring** and **Running Activities**. The **Prepare** action is comparable to the validation step, which you probably know from the old migration cockpit. This action reads the source data from the staging tables and writes it to the defined *value mappings* (mapping tasks).

**Consistency checks**

Now, as shown in Figure 3.24, you can either call the **Prepare** action ❶ for each migration object individually via the menu button or select several migration objects at once ❷ and execute **Actions** ❸ via the general menu button. Next, the **Prepare Staging Tables** dialog box appears. If you have populated your staging tables directly using a tool at the database level, then you must turn on the **Consistency Check**. If you have populated them using files, then click the **Prepare Staging Tables** button ❺ and leave the **Consistency Check** ❹ turned off. In this case, a consistency check is unnecessary because one was already performed as part of the validation process when the files were uploaded. Doing another check will then only cost time and, if run again at this point, may even result in false positive errors for less than a handful of migration objects. This problem arises when the data type definition of a key is not consistent in the dependent structures. While the issue has since been resolved for SAP S/4HANA Cloud, it can still occur in SAP S/4HANA.

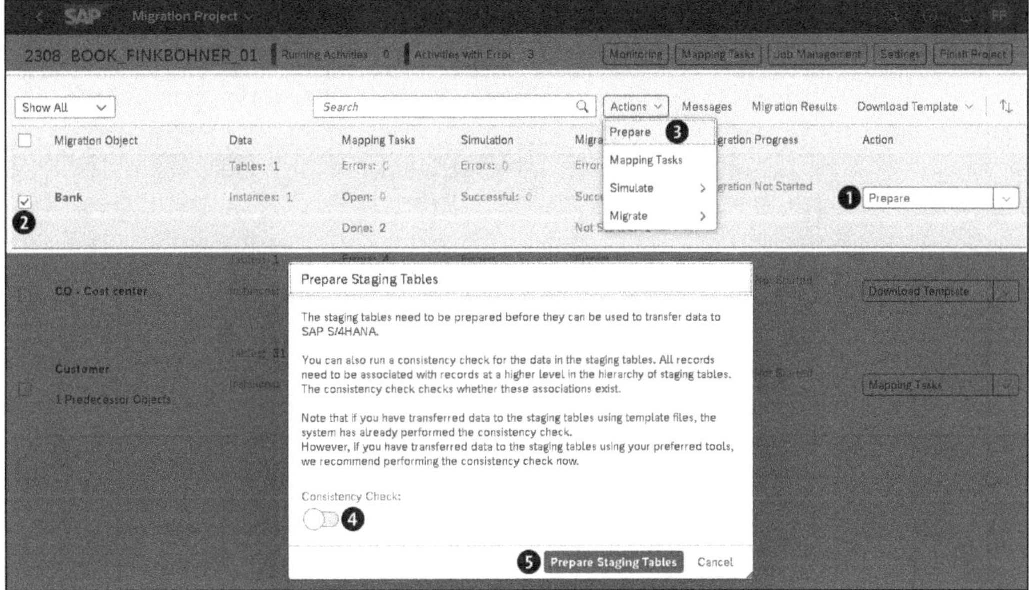


**Figure 3.24** Preparing Migration Objects

### 3.2.14 Maintenance of Mapping Values

**Editing mapping tasks**

The next action after preparing the data and validating it is processing the mapping tasks. The great advantage of using the migration cockpit for a data migration is its ability to apply a value mapping across multiple objects. As a result, you only need to maintain source values of a mapping task, such as "Mapping of Cost Center Type," once and all other objects will use this mapping. This option has particular advantages if most of your data is to be migrated from a source system.

[«]

**Performing Mapping Tasks before Preparing**

For some migration objects, such as the **Customer** migration object, **Mapping Tasks** are already present immediately after the migration object has been added, and these tasks must be performed before **Prepare.** Most of the time, these objects are control parameters that influence the generation of the value mappings during the **Prepare** action and are therefore mandatory to execute before **Prepare** to avoid errors and problems and to avoid needing to call the **Prepare** action again.

You can edit either all mapping tasks of the project or edit only the mapping tasks of a migration object. For the first option, you select the **Mapping Tasks** function in the upper area of the **Migration Project** view. For mapping tasks per object, you select the **Mapping Tasks** action directly for the migration object in the **Migration Project** view of the respective object. As shown in Figure 3.25, the number of mapping tasks and the number of tasks in the **To be Confirmed**, **Confirmed**, or **Error** status are located in the header area. The list of mapping tasks shows the name of the **Mapping Task**, the **Status**, the **Type** of the mapping task, and the number of **Values** in this mapping task.

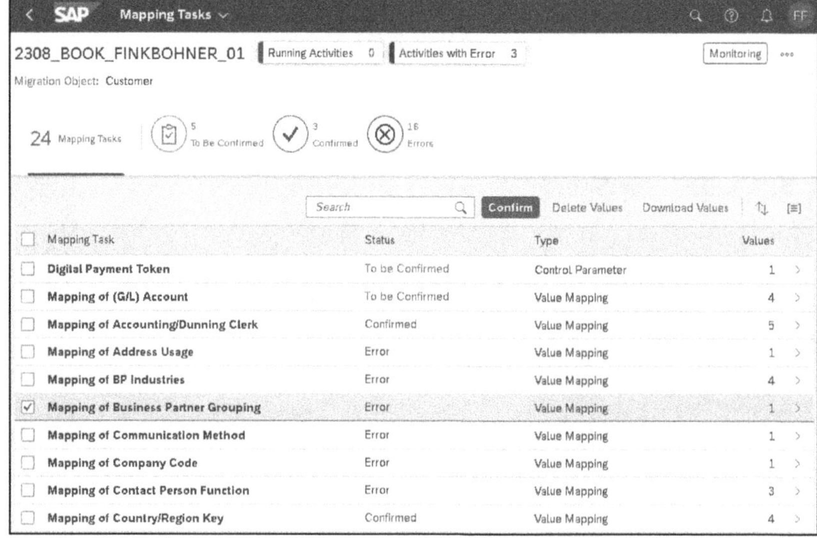


**Figure 3.25** Mapping Tasks of the Project

[»]

### What Does the "Error" Status Mean?

Mapping tasks are in the **Error** status if field values of a source system are longer than the field definition of the target structure, for example, if plants in your source system are longer than the 4 digits of a plant of an SAP system. You usually get a DMC_RT_MSG 206 warning for such values, such as **MAP_REGIO: Due to different lengths or invalid value, Indiana (IN) set to initial value** during the **Prepare** action. If the field values are less than or equal to the field definition, the source values are copied directly to the target values. However, if they are longer, only the source value is written, and the target value remains empty. You must then manually assign this in the value mapping of the mapping tasks. This behavior avoids inconsistencies due to the truncation of target values. Especially with object IDs, such as customer numbers that are too long, it could otherwise become an issue if several source customer numbers pointed to the same target customer number.

**Types of mapping tasks**

A distinction is made between the following types in the **Type** column:

- **Control Parameter**: You use control parameters to control the behavior of specific migration objects during data migration. You can specify here, for example, whether the customer ID is to be used for integration with external systems via the *Unified Key Mapping Service (UKMS)*.
- **Fix Value**: In this case, you define fixed values. A typical fixed value is the value of the controlling area, which is usually specified as "A000" in SAP S/4HANA Cloud.
- **Value Mapping**: In this case, you map fixed values of the source to the target values of the SAP S/4HANA system.

**Bulk confirmation of mapping values**

If you have maintained the correct target data already in your source data, you can simply select all mapping tasks to be confirmed and confirm them in bulk using the **Confirm** function.

**Checking the data after bulk confirmation**

During confirmation, a small information box is displayed on the screen (called a *message toast*). When this message disappears, the bulk confirmation is complete. Mapping tasks that could not be confirmed are still in the **To be Confirmed** or **Error** status.

### Downloading and Uploading Mapping Values

**Editing the mapping templates file**

Beside **Monitoring**, you can **Download Mapping Templates**. In the selection window, you select the corresponding mapping tasks and load them via **Download** into a packed ZIP file, which you can then download via **Monitoring**, as shown in Figure 3.26.

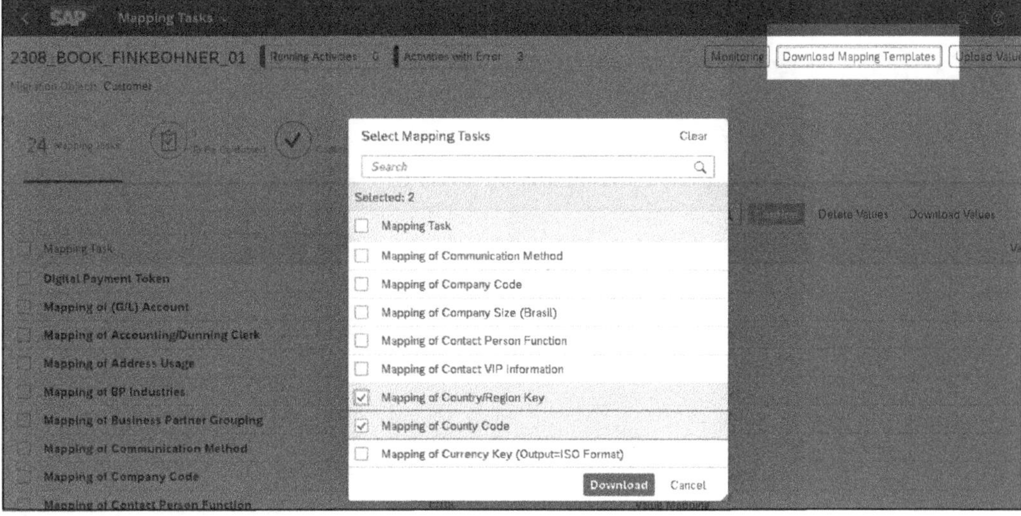


**Figure 3.26** Download Mapping Templates

You can also download selected mapping tasks including their values, to a file, via **Download Values**. You then navigate to monitoring.

Downloading the file

The **Download Mapping Values** activity is displayed. You can select from two options:

- **Show Messages**: This option displays the message overview screen, described previously in Section 3.2.8, which contains the respective messages for the activity.
- **Download File**: These values are downloaded as a ZIP file.

In addition, you can use **Upload Values** to reimport value mappings of XML files you have downloaded. Figure 3.27 shows the monitoring view with the two selection options for the **Download Mapping Values** activity.

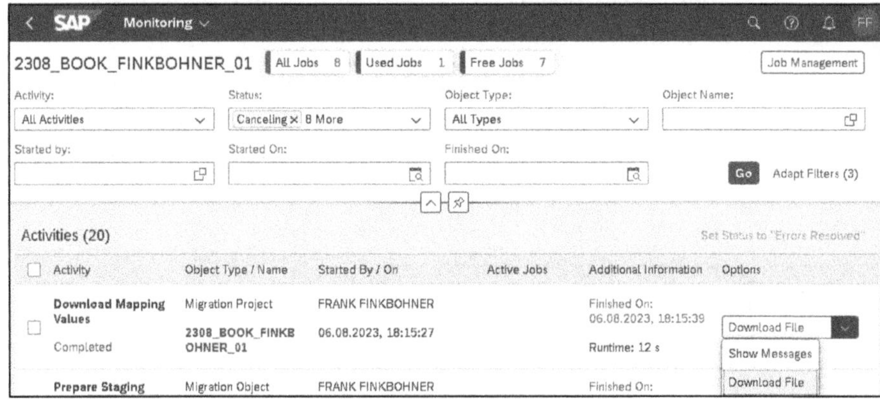


**Figure 3.27** Monitoring View: Download Mapping Values

Once you have maintained the XML files, click **Upload Values** to upload them back to the system using the **Upload Mapping Values** view.

### Deleting Values

Delete values

In SAP S/4HANA Cloud and as of SAP S/4HANA 2022 FPS01, it is possible to delete entire mapping tasks. For this task, select the mapping tasks to be deleted and click **Delete Values**. A dialog box appears, which displays the following text:

> *All mapping values for the selected mapping tasks are deleted. Mapping values are needed for migration simulation and data migration. When mapping values are deleted, errors or data inconsistencies may occur, especially if the mapping values are used by more than one migration object.*
>
> *Note that selected mapping tasks are no longer displayed on the Mapping Tasks screen. When you re-upload mapping values for a mapping task, the mapping task is also displayed again on the Mapping Tasks screen.*

*Delete all mapping values for the selected mapping tasks?*

If you're sure that you want to delete the mapping tasks, confirm your decision by clicking **Delete**.



### Maintaining Individual Values of a Mapping Task

Click the [>] icon (**Next**) of a mapping task to maintain it. The view that is displayed, shown in Figure 3.28, has a flexible column layout again.

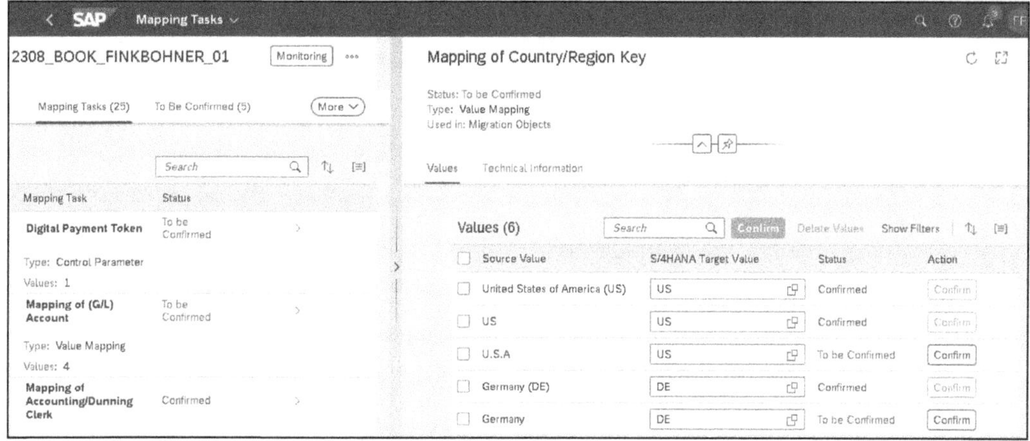


**Figure 3.28** Maintaining Mapping Values

The left shows the previous list of mapping tasks, and the right shows the mapping task values. You can select individual values and **Confirm** or **Delete** these values.

**Adapting the view**

If you have a large number of mapping values, you can **Show Filters** and use the [⇅] icon (**Sort**) and the [[≡]] icon (**Group**) to customize the display of the **Mapping Tasks** view.

[!]

**New Open Mapping Tasks after Simulation**

For some migration objects, you may notice that, after successfully confirming mapping tasks and starting the **Simulate** action, new mapping tasks will suddenly appear again. This occurrence is not a bug in the migration cockpit, but a result of the interdependence of some mapping tasks. Cost centers, for example, have a reference to the controlling area, but many objects do not have a controlling area in their transfer structure. In such a case, the required controlling area is determined via the plant or company code. To determine the cost center value, you must first confirm the value mapping for the plant or company code in the initial **Prepare** action, since the mapping tasks for the cost centers may not have been populated during the initial **Prepare** action. First confirm the open mapping tasks and then perform a second preparation action. With the plant or company code mapping confirmed, the controlling area can now be determined, and the mapping task for cost centers, assigned.

[»]

**Additional Information**

Additional information on mapping tasks can be found in the following KBAs:

- KBA 2976393, "Using Mapping Values in Different Projects": *http://s-prs.co/v581745*
- KBA 3003068, "Inserting and Deleting Entries for Mapping Tasks": *http://s-prs.co/v581746*

### 3.2.15 Job Management

**Parallelizing the transfer**

Because the new migration cockpit is built on the principle of staging tables, parallelization no longer occurs through the number of files. For this purpose, in job management, you can increase or decrease the number for **Maximum Number of Jobs for Project** ❶ and **Jobs Used for Migration Objects** ❷. For this task, click on the **Edit** ❸ button in the **Job Management** view and use the plus and minus icons (❹/❺) to adjust the desired number of jobs per project and per migration object, as shown in Figure 3.29. Save your selections by clicking the **Save** button ❻.

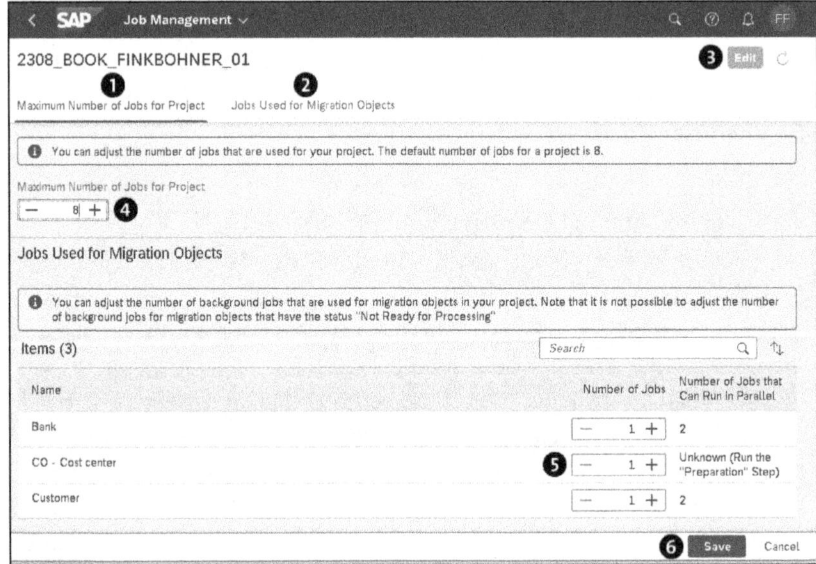


**Figure 3.29** Job Management

The number you set **Maximum Number of Jobs for Project** depends on the number of jobs available in the system. Each migration object occupies at least two jobs: one for the controller or calculation job and *n* more for the load jobs (e.g., simulate, migrate).

**Eight parallel jobs at maximum**



[«]

**Maximum Number of Background Processes in SAP S/4HANA Cloud**

In SAP S/4HANA Cloud Public Edition, the maximum number of background processes (BP) is limited. The migration cockpit can use a maximum of 80% of these background processes. At the time of this writing (January 2023), the maximum number of background processes was the following:

- QA system (1 application server): 10 BPs, maximum 8 usable
- PRD system (2 application servers): 20 BPs, maximum 16 usable

Contact your system administrator to obtain more information on the number of available background processes for your SAP S/4HANA on-premise system. It is not useful to specify more jobs than are available. Note also that some background jobs are usually already occupied with standard jobs. Using multiple parallel migration projects may have been useful before SAP S/4HANA 1809. Since SAP S/4HANA 1809, however, it is rather counterproductive and provides no performance gain. You may instead cause problems due to an overflowing job queue.

For more information, see also KBA 3034738, available at *http://s-prs.co/v581747*.

### 3.2.16 Simulating the Migration

**Start simulation**

After you have confirmed all mapping tasks and distributed the jobs in job management, the next step is to simulate the data migration. For this purpose, in the **Migration Project** view, select the **Simulate** action for the respective migration object. The **New Simulation** dialog box appears where you can choose between the following simulation options, some of which are self-explanatory, and start the simulation via **Start simulation**:

- **Random 10%** of the existing migration object instances are simulated.
- **Random 500** migration object instances are simulated.
- **All instances** are simulated.
- **Custom Selection**

  This option is where you jump off to mass processing for simulation. A three-step selection process is used to select the instances to be simulated. We describe the general structure of the **Mass Processing of Migration Instances** view in Section 3.2.20.

However, you can also select one or more objects and select the **Simulate** option from **Actions** dropdown list, with the following two options, as shown in Figure 3.30:

- **All Instances**
- **All Instances with Errors**

For each selected migration object, the **Validate Data** activity is scheduled as a background job, the status of which you can check under **Running Activities** or under **Monitoring**. If simulation fails, you can see this directly by clicking the **Activities with Error** button.

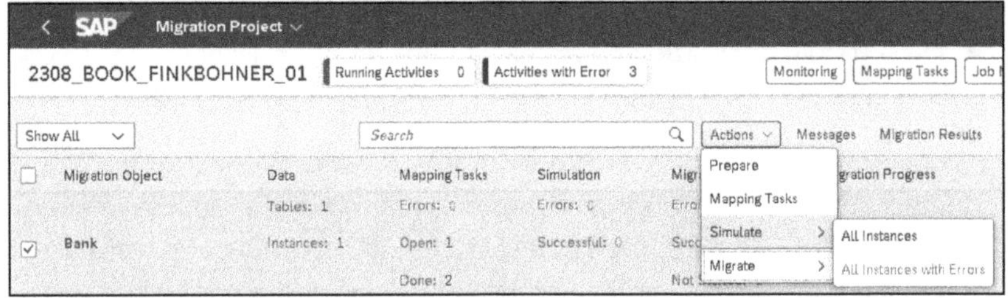


**Figure 3.30** Simulating Migration Objects

Figure 3.31 shows an example with errors arising during simulation. The migration object is marked with a red bar on the left to emphasize its error state.

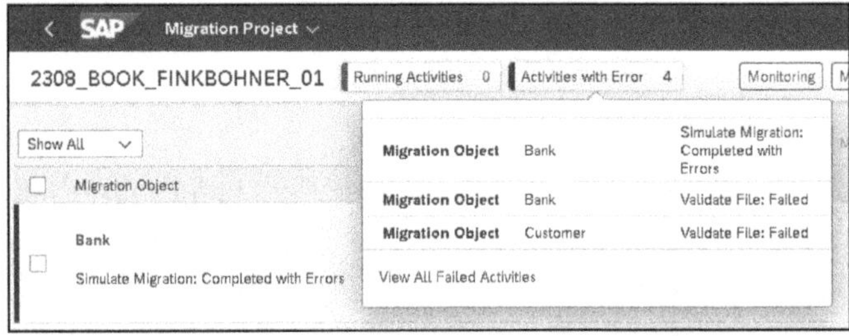


**Figure 3.31** "Activities with Error" Button after Simulation

**Checking errors in monitoring**

You can display the errors using the **View All Failed Activities** option or via **Monitoring**. For this purpose, in the **Monitoring** view, select the **Show Messages** option in the **Simulate Data** activity for the simulated migration object.

**Long text to help solve the problem**

If you click on a message, you navigate to the **Message** detail view. The detail view shows the **Long Description** for a message, which provides further information on the message and on troubleshooting (in case of errors). Under **Technical Information**, you can find the **Message Class** and **Message**

**Number** fields, which are important for the support team and for finding solutions on the internet.

**Message Classes Tell You a Lot about the Origin of an Error Message**

The message class enables you to find out, for example, if the error was output by the migration cockpit, the migration content, or the API. Message classes of the migration cockpit usually contain `DMC_`. A common message class of the migration content is `CNV_DMC_SIN`. All other message classes are usually used by APIs for data migration. You can thus better classify the problem and assign it to the relevant support component.

**Simulate Single Data Records**

Before simulating a very large number of data records, you should first start the simulation for a few selected records. You can use the **Migration Object Instances** view to select specific data records for simulation. Access the **Migration Object Instances** view by clicking on the **Migration Project** view and the number of instances in the **Data** column of the migration object. In the **Migration Object Instances** view, you can use selection fields, such as **Action/Status** ❶, in the headers to narrow down the instances displayed. Then, select the records to be simulated ❷ and click **Simulate** ❸. You can check the status of the activity via **Monitoring** ❹, as shown in Figure 3.32.

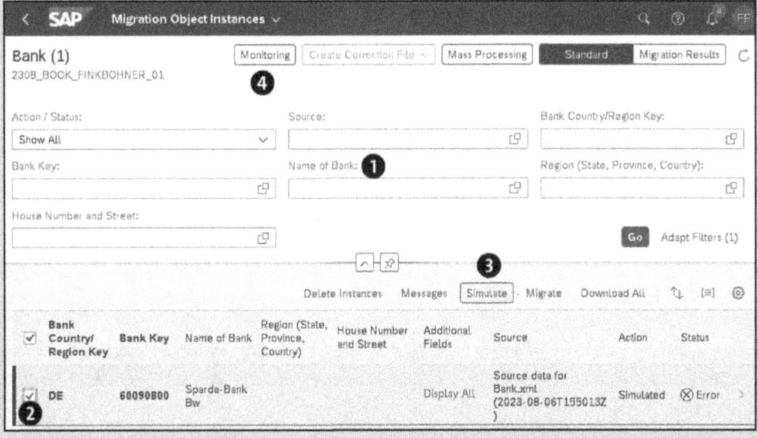


**Figure 3.32** Simulating Individual Migration Object Instances

For more information on the **Migration Object Instances** view, see Section 3.2.12.

**Migration after successful simulation**

Once the problem is solved, you can restart the simulation. After successful simulation, the migration object's **Action** column button is set to **Migrate**, as shown in Figure 3.33.

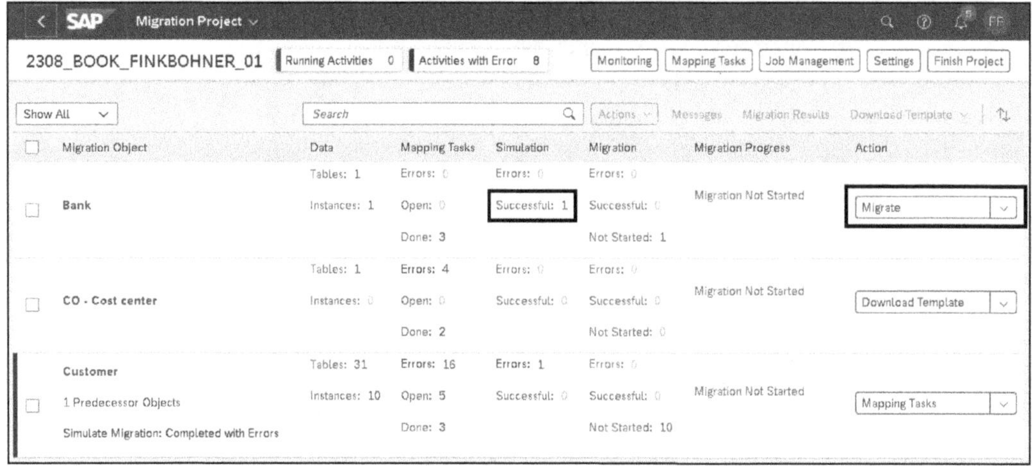

**Figure 3.33** Simulation Results

### 3.2.17 Migrating Data

**Start migration**

You can now migrate your data. For this task, you run the **Migrate** action for the migration object. Similar to **Simulate**, the **New Migration** dialog box appears where you can select your desired migration option (Section 3.2.16).

The migration object's status is then set to **Migrate Data: Started**. In monitoring, you can monitor the activity's progress. If the data migration is successful, the status is set to **Completed**.

**Checking the messages after the migration**

After each data migration, you should view the messages of the migration run in the **Messages** view via **Show Messages**.

In the **Migration Project** view, if the migration is successful, the migration object is marked with a green vertical bar on the left, and you'll see a progress bar in the **Migration Progress** column, as shown in Figure 3.34.

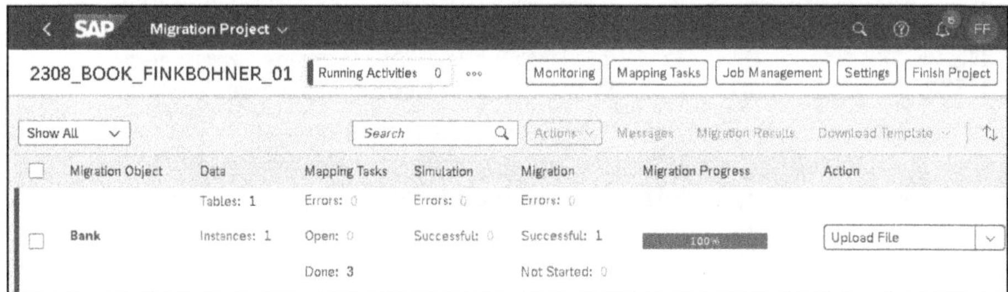

**Figure 3.34** Successful Migration in the Migration Project View

### 3.2.18 Migration Results

In SAP S/4HANA Cloud and as of SAP S/4HANA 2021, the new **Migration Results** function is available. You can call this function from within the **Migration Project** view. For this task, select a migration object ❶ and choose the **Migration Results** ❷ function. As an alternative, you can switch to the **Migration Object Instances** view via **Instances** ❸. Figure 3.35 shows the different ways to call the migration results.

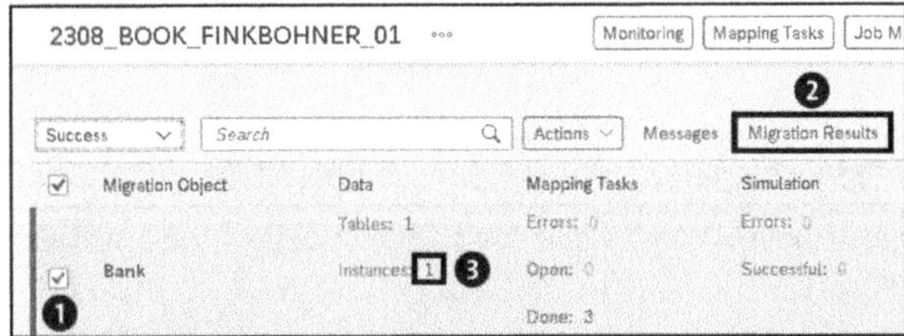


**Figure 3.35** Calling the Migration Results

You can access the **Migration Object Instances** view via either path. In the header, you'll see the segmented **Standard/Migration Results** button. This button is automatically set to **Migration Results** when you switch to it by clicking the **Migration Results** button. In the **Instances** field, you must then switch from **Standard** to **Migration Results.**

The **Migration Results** option shows a new column named **Result of Step "Migration"**. For successfully migrated instances with **Migrated** in the **Action** column and **Success** in the **Status** column, the key parameters of the migrated instance are shown. These key parameters are hyperlinked. Clicking a hyperlink calls the validation app stored for the migration object, as shown in Figure 3.36.

**Figure 3.36** Migration Results with Links to the Validation App

After selecting the key ❶, the **Display Additional Information** link appears. Click on **Open in New Window** ❷, as shown in Figure 3.37.

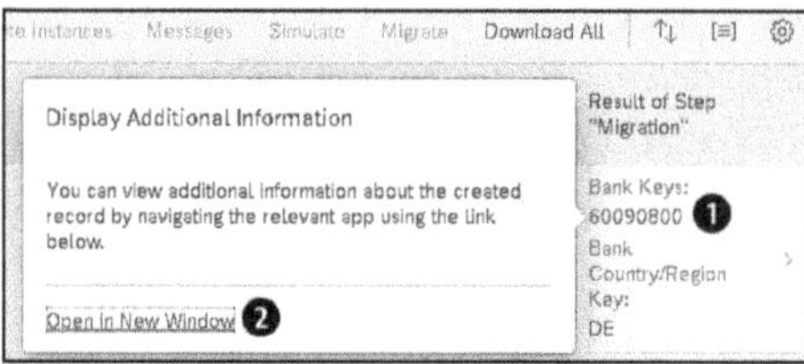


**Figure 3.37** Opening the Validation App in a New Window

To find out which app ❶ is assigned to a migration object for the purpose of validating the data, you should refer to the documentation of a migration object, as shown in Figure 3.38. The information is stored in the **Post-Processing** section. The area also contains the user role template ❷, from which your system administrator can take the corresponding user catalog role, which is then assigned to a migration role for your user. However, you can also use the Data Migration Status app in SAP S/4HANA Cloud to validate the data. For more information, see *http://s-prs.co/v581748*.

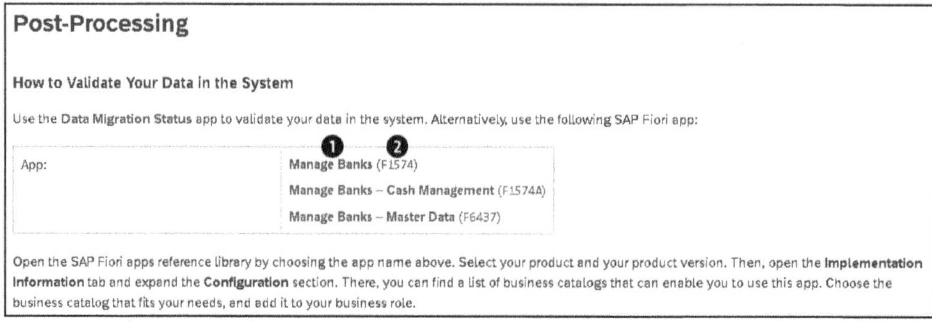


**Figure 3.38** Documentation: Post-Processing Area

**What to Do If the Link Does Not Work?**

If you're not authorized to use the validation app, then the hyperlink will not work, and you won't be able to access the app at all.

However, for some migration objects, jumping to an app is not offered for technical reasons. Perhaps the app is based on a transaction or is an SAP Fiori app that does not allow parameters. Perhaps there is more than one app for validation. Since the Migration Results app and the Data Migration Status app use the same configuration tables, you can get the list of migration objects for which no direct jump is supported from KBA 2681413. Then, follow the instructions in the **Post-Processing** section of the migration object documentation.

### 3.2.19 Creating a Correction File

Migration status "Completed with error"

If errors occurred despite previous simulation, the **Migrate Data** activity is displayed with the **Migrate Data: Completed with Errors** status, and the object is marked with a red bar on the left. You can now use the **Create Correction File** action ❶ to download an XML file or download CSV files containing the incorrect data records. For this purpose, select the source(s) in the **Select Source** dialog box that appears ❷ and start generating the file by clicking **Create Correction File** ❸, as shown in Figure 3.39.

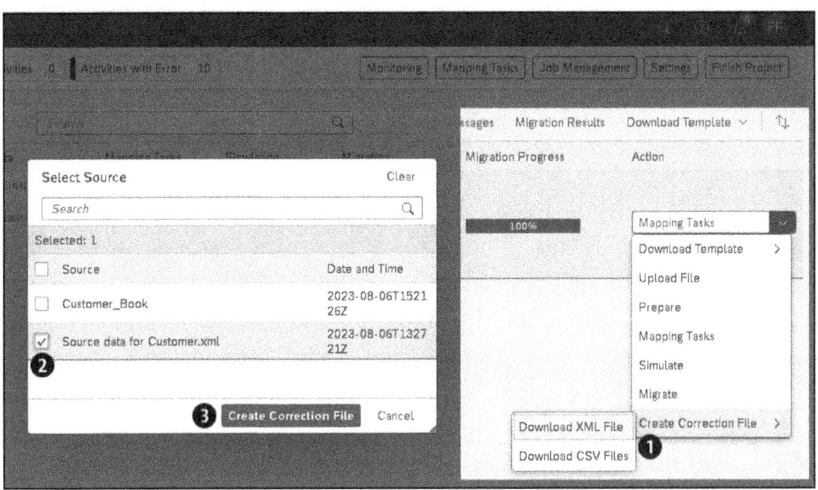


**Figure 3.39** Selecting the Source for the Correction File

Downloading file(s)

Then, switch to **Monitoring** and select the **Create Correction File** activity and under **Options**, select **Download**.

In addition, all created correction files also appear in the **Migration Object Tables** view in the **Correction Files** ❶ area. On this view, you can filter these correction files via the **All/XML/CSV** button ❷. You can then download or delete selected correction files again through the **Download** ❸ or **Delete** ❹ functions, respectively. For a correction file, you can make a selection in the **Options** column ❺ to download the files or view the messages generated during creation, as shown in Figure 3.40.

Packed files

When you download a file, the system generates a ZIP archive that contains the correction file. The name of the file is composed according to the schema *Correction file for<Source>(<Date>).xml*. An example would be *Correction file for EN_Customer_Sample_01_2208 (2023-01-01T151037Z).zip*.

You can now edit, upload, prepare, simulate, and finally re-migrate this file or these files. Before you upload the corrected data records from a revised correction file, you must delete the associated erroneous instances.

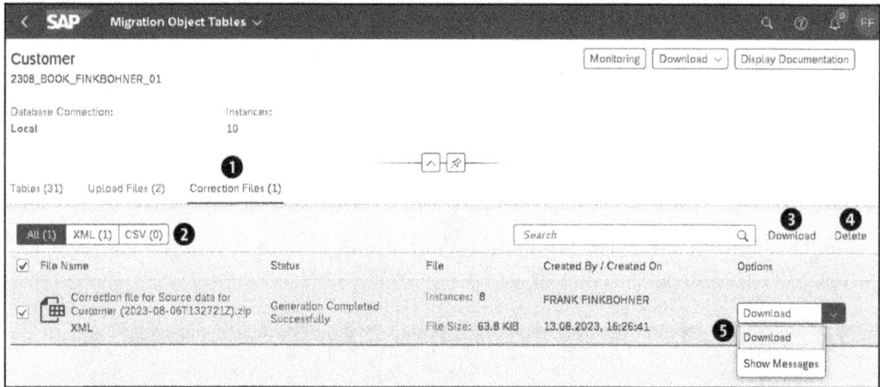


Figure 3.40 Overview of the Correction Files

### 3.2.20 Mass Processing via Free Selection on Instances

In this section, we explain the new mass processing feature implemented in the migration cockpit as of SAP S/4HANA 2020 FPS1. This mass processing feature can be called either from the **Migration Project** view via the **Simulate** and **Migrate** actions or from the **Migration Object Instances** view.

**Simulate and migrate**

When you click on the **Simulate** or **Migrate** action, the menu previously described in Section 3.2.16 and Section 3.2.17 is displayed. Both menus contain the selection criterion **User-specific selection** ❶❷, as shown in Figure 3.41.

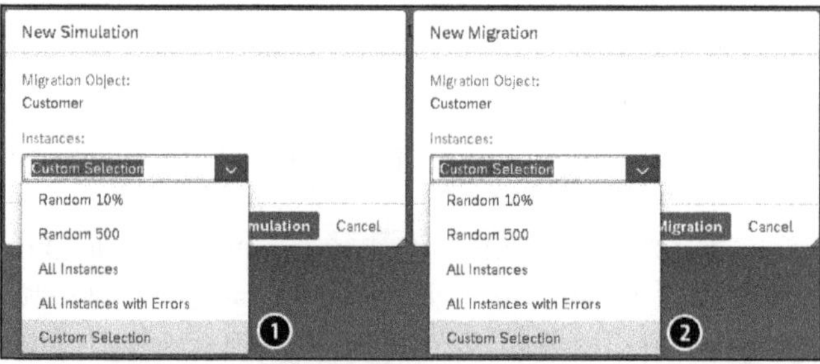


Figure 3.41 Simulate and Migrate: Selection Criteria

If you select **Custom Selection** in either menu, the **Mass Processing of Migration Object Instances** view opens. As shown in Figure 3.42, you'll see a *guided procedure* with three steps.

In the first step, the action of the mass processing activity is displayed. If the **Simulate** or **Migrate** mass processing was called via the **Migration Project** view for a single object, the selection is preset, and only **Step 2** can be clicked. As shown in Figure 3.42, the function was called from the **Migration Object Instances** view. For this reason, you can choose between the **Simu-**

**late**, **Migrate**, and **Delete Instances** options. In our example, we've chosen the **Simulate** option. Click **Step 2** to go to the next step (**Selection**).

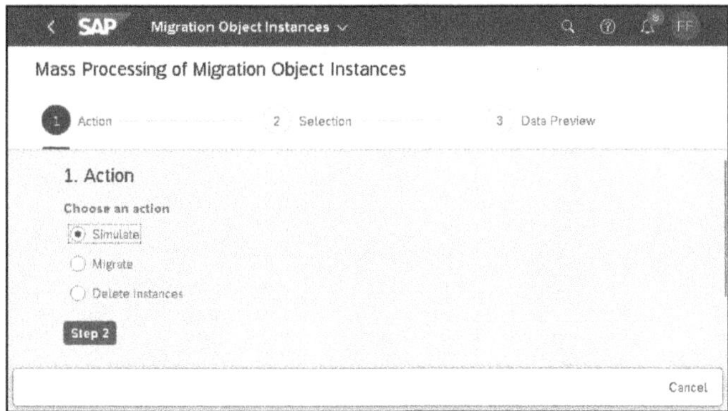


**Figure 3.42** Step 1: Mass Processing Action

In the **Selection** step, you can enter values for the first seven fields of the main structure and the **Action/Status** field of the instances as selection criteria to select only those instances for which you want to perform mass processing. You can choose between the following actions or statuses in the **Action/Status** field:

- **General**
  - **Show All**
  - **Unprocessed**
  - **Migration Not Started**
- **Simulation**
  - **All Simulated Instances**
  - **Simulated Successfully**
  - **Instances with Simulation Errors**
- **Migration**
  - **Instances with Migration Errors**

In our example, for the **Customer** object shown in Figure 3.43, we have entered the value "A*" in the **Name** field and selected the **Show All** in the **Action/Status** dropdown list.

To start the **Data Preview**, click **Step 3**. You'll then see a list of instances that match the selection criteria. You can make changes to the selection criteria to refine the selection of instances for mass processing by either selecting the step 2 (**Selection**) again, as shown in Figure 3.43, or simply by scroll up. Figure 3.44 shows step 3 (**Data Preview**). Click **Review** to prepare the mass processing step.

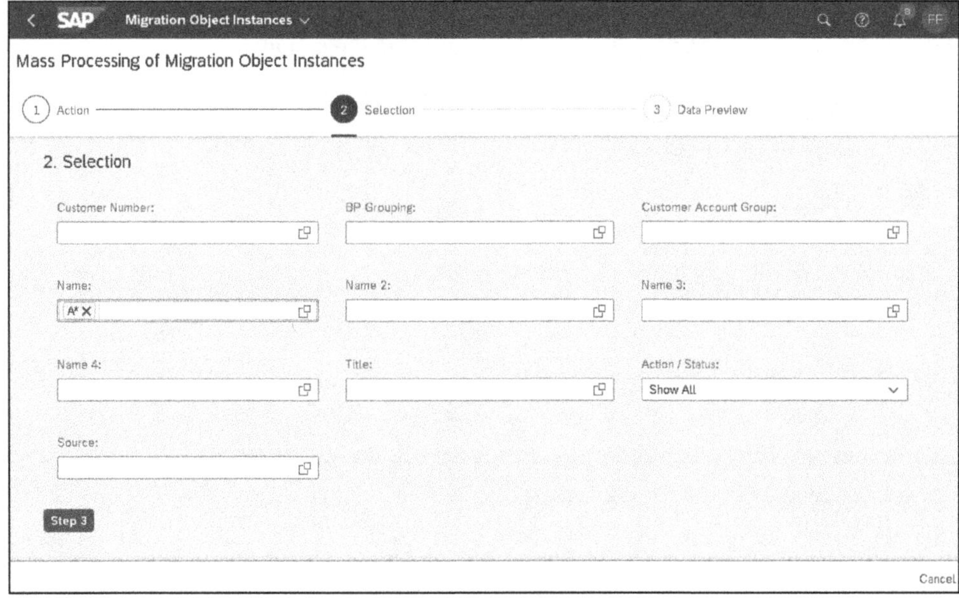


**Figure 3.43** Step 2: Select Mass Processing

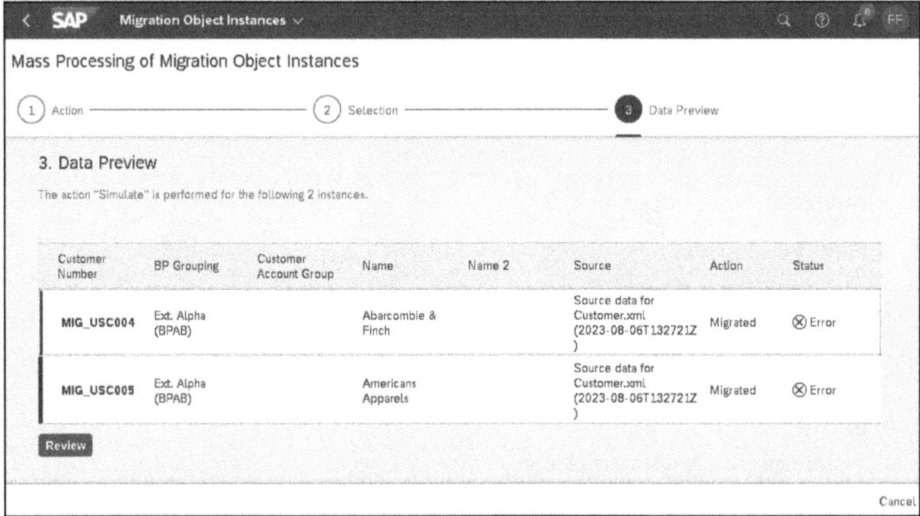


**Figure 3.44** Step 3: Mass Processing Data Preview

Figure 3.45 shows a summary of the action, the selection, and the number of affected instances of the data preview. When you click **Start**, the mass processing will start. This background activity can be tracked via Monitoring.

With this new mass processing feature, you can now easily select a specific data record or group of records to run the simulation or migration, or you can selectively delete instances from your staging tables.

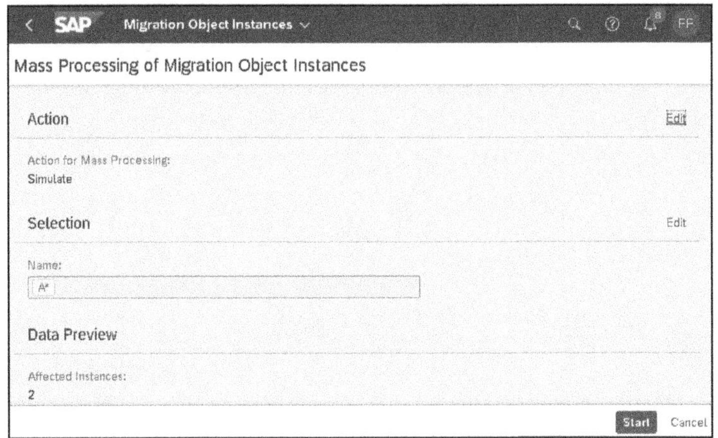


**Figure 3.45** Summary of Mass Processing Before Execution

## 3.2.21 Messages

In this section, we'll discuss how to download messages, as well as viewing, filtering, and downloading migration object instance messages.

### Downloading Messages

**Downloading messages via the history**

If you want to download messages, for example, the errors of a migration object's specific action, first, open the **History** of the migration object's actions in the **Migration Object** view, as shown in Figure 3.46.

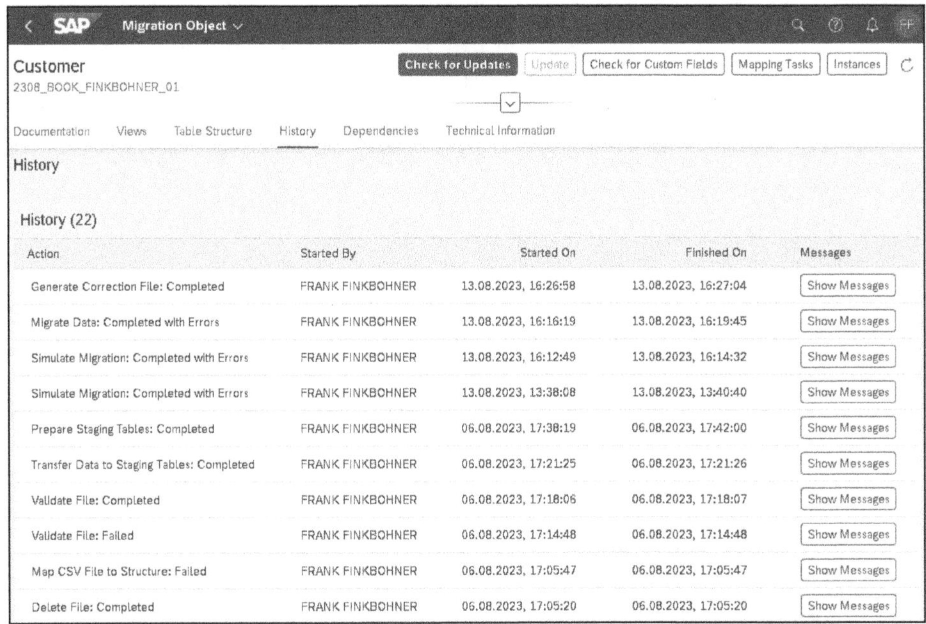


**Figure 3.46** History of a Migration Object

**Grouping provides information on systematic problems**

Then, click the **Show Messages** button for the corresponding action. The known overview of the migration messages is displayed. For a better overview, you can, for example, group the messages here via **Group By Message ID** ❶ and display only the errors ❷. As a result, you receive a summary of the individual error groups. Systematic errors, for example, a missing configuration, can be detected quickly here. In the example of Figure 3.47, the corresponding button was used to first select the errors only and then the grouping.

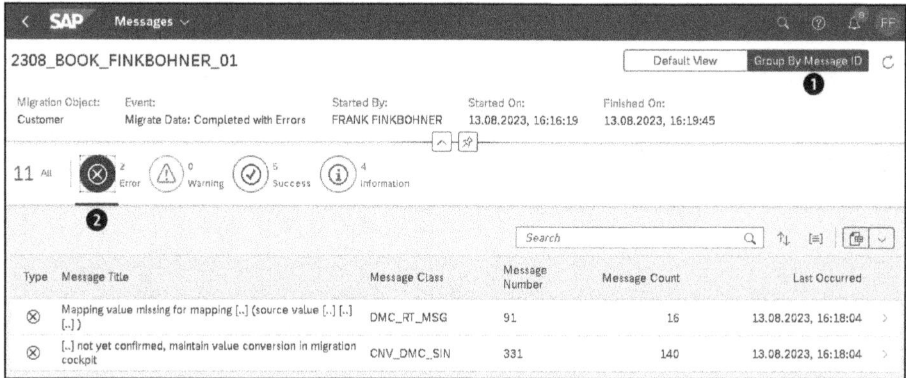


**Figure 3.47** Migration Messages View: Grouped by Error

**Exporting messages**

If you select the number below **Message Count**, the **Migration Messages** view displays the messages and their details, as shown in Figure 3.48.

**Downloading messages**

If you click the [export icon] icon, you can download the selected messages to a Microsoft Excel file (*Messages.xlsx*). You can then forward this file, for example, to your key users or user departments for troubleshooting.

**Figure 3.48** Message Details

## Migration Object Instances and Messages

For a more complex overview of the messages of selected migration objects ❶, use the **Messages** function ❷ of the **Migration Project** view, as shown in Figure 3.49.

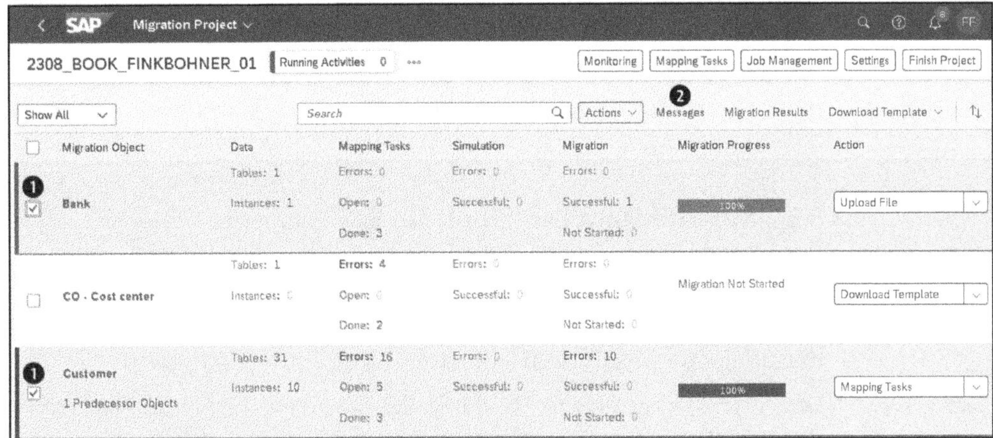


**Figure 3.49** Calling Messages for Migration Objects

In the **Migration Object Instances, Messages** view shown in Figure 3.50, you must select the respective object in the header ❶. In the pane on the left, you filter the messages using **Recent Actions** ❷ and **Message Type** ❸. In the pane on the right, you'll then see the **Messages** ❹ in a list. The list shows the **Type**, **Message**, **Action**, **Message Class**, **Message Number**, and the number of **Instances**. You can sort, filter, and group the messages as usual ❼ and hide the filter area on the left (**Hide Filters** ❻).

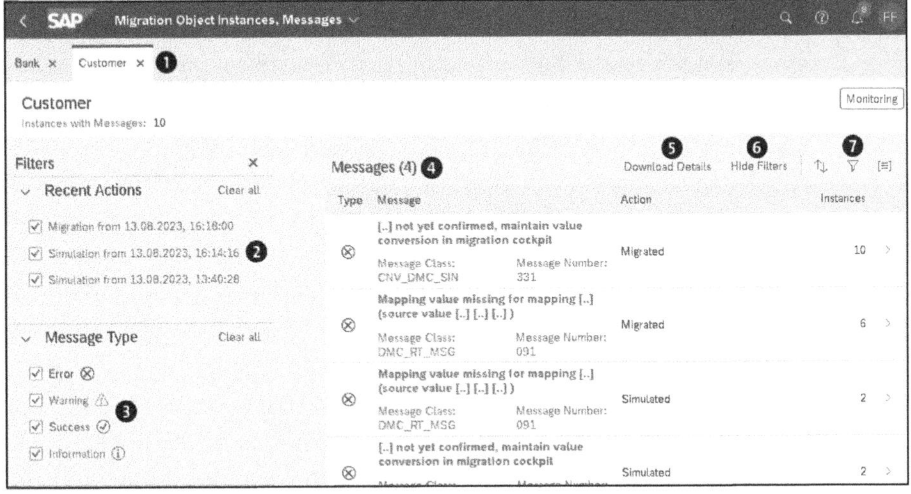


**Figure 3.50** The "Migration Object Instances, Messages" View

**Downloading details**

Select the **Download Details** ❺ function to download information about the messages and the generating instances. For this task, in the **Select Column** dialog box that appears, select the fields (e.g., key fields, message fields, non-key fields) to be included in the exported list.

You can then download this file via **Monitoring**. This file is a ZIP file named *Messages for <migration object>_<date>.zip*, for example, *Messages for customer (2023-01-02T183002Z).zip*. This ZIP file contains a TXT file of the same name that contains selected information in tab-separated format.

## 3.3 Migrate Data Directly from SAP System (Direct Transfer)

As of SAP S/4HANA 1909, another option is available in the migration cockpit: migrating data directly from SAP system (*direct transfer*, for short).

**Connecting systems via RFC**

Direct transfer means that the data is selected from the SAP source system using a *remote function call (RFC)* connection and stored in the migration cockpit, which is physically located in the SAP S/4HANA target system. As in the *staging approach*, the **Mapping**, **Simulation**, and **Migration** functions are available for further processing. Creating or posting data in the SAP S/4HANA target system is done via *SAP standard APIs*.

**Predefined selection**

This approach has one decisive benefit: The selection from the source system is already predefined—based on organizational units and migration objects. It can be adapted to customer requirements in the *SAP S/4HANA migration object modeler*. We provide more details in Section 3.7.

This option is available for migrations with SAP S/4HANA as the target. Migration of data from SAP S/4HANA systems to other SAP S/4HANA systems is not supported. SAP S/4HANA Finance systems (formerly SAP Simple Finance) are classified as SAP S/4HANA systems.

The **Migrate Data Directly from SAP System** approach has also been released for SAP S/4HANA Cloud Public Edition, release 2402. However, the direct transfer is not described further in this book. For more information, refer to the following link: *http://s-prs.co/v581748*.

### 3.3.1 Differentiation and Areas of Use

The migration cockpit has been developed for the initial provisioning of data to an SAP S/4HANA system. It is intended for use in new implementations of SAP S/4HANA. The direct transfer approach is *not* suited to keep data synchronized between systems, map a permanent data exchange between systems, or build an integration scenario. For these requirements, you're provided with other SAP products, for example, SAP Master Data Governance, SAP Product Lifecycle Management (SAP PLM), SAP Landscape Transformation Replication Server, SAP Data Intelligence, and Central Finance.

**Non-Unicode source systems**

The migration cockpit can also process data from a non-Unicode system. In the direct transfer approach, the two systems are connected via RFC. When importing data from the source system to the migration cockpit, the data is automatically adapted to the SAP S/4HANA data model.

### 3.3.2 Technical Requirements and System Preparations

Let's now look at the technical conditions for using direct transfers and the necessary preparations on the system side.

#### Technical Requirements

The following technical requirements apply, depending on the source system:

- **SAP ERP to SAP S/4HANA**: The lowest supported release is SAP ERP 6.0 (SAP NetWeaver 7.0).
- **SAP Apparel and Footwear to SAP S/4HANA**: The only supported release is SAP Apparel and Footwear 6.0.
- **SAP EWM to EWM in SAP S/4HANA, based on SAP S/4HANA**: The lowest supported release is SAP EWM 5.1 (SAP NetWeaver 7.0). In addition to decentralized EWM, a migration to embedded EWM will also be supported from SAP S/4HANA 2022 FPS0.
- **SAP CRM to SAP S/4HANA service**: The only supported release is SAP CRM 7.0 (scenario available as of SAP S/4HANA 1909 FPS1).
- **SAP Advanced Planning and Optimization SPP to SAP S/4HANA Supply Chain for extended service parts planning**: SAP Supply Chain Management (SAP SCM)7.14 is the lowest supported SAP SCM release (scenario available from SAP S/4HANA 2021 FPS0).

As mentioned earlier, the SAP S/4HANA migration cockpit does not support the migration of data from SAP S/4HANA systems to other SAP S/4HANA systems.

#### System Preparations

The following preparations must be made before you can start with the migration cockpit approach for direct data transfer:

- The SAP S/4HANA installation guide describes the steps with regard to the migration cockpit. You can find the guide you need depending on the release:
  - For SAP S/4HANA 1909: *http://s-prs.co/v581749*
  - For SAP S/4HANA 2020: *http://s-prs.co/v581750*
  - For SAP S/4HANA 2021: *http://s-prs.co/v581751*
  - For SAP S/4HANA 2022: *http://s-prs.co/v581752*

  For SAP S/4HANA releases, import the `DMIS (Data Migration and Integration Server)` or higher add-on specified in Table 3.3 into the source system. The add-on is included in the SAP S/4HANA license. For more info, see SAP Note 2973957 via the following link: *http://s-prs.co/v581753*.

| Target System | Add-On in the Source System |
|---|---|
| SAP S/4HANA 2019 | `DMIS_2011_1 SP17` or higher |
| SAP S/4HANA 2020 | `DMIS_2011_1 SP19` or higher |
| SAP S/4HANA 2021 | `DMIS_2011_1 SP21` or higher |
| SAP S/4HANA 2022 | `DMIS_2011_1_SP23` or higher |

**Table 3.3** Valid DMIS Add-Ons for Source Systems

- Define an RFC connection from the source system to the SAP S/4HANA target system.
- Set up SAP Fiori in the target system. For more information, refer to the installation guides mentioned earlier and KBA 3039598, available at *http://s-prs.co/v581754*.
- Implement *all* SAP Notes for the migration cockpit that apply to your release. Central SAP Notes include the following:
  - Composite SAP Note 2780378 (Composite SAP Note for SAP S/4HANA Migration Cockpit [all releases])
  - Composite SAP Note 2747566 (SAP S/4HANA Migration Cockpit: Composite Note for direct data transfer from the SAP system)

[»]

**SAP Note Analyzer**

Before you create a migration project in your SAP S/4HANA system, use the note analyzer to check whether all the necessary notes (`DMIS` add-on and corresponding components in SAP S/4HANA) have been implemented in your systems. For more information, see SAP Note 3016862 and the SAP Community blog post by Prithvi Spandana "How to implement the latest corrections released for Migration cockpit," available at *http://s-prs.co/v581755*.



- Run report `CNV_PE_MC_LOAD_OBJECTS` in your target SAP S/4HANA system. This report loads the migration objects delivered by SAP from client 000 into the client in which you're working. Note that, when you update your SAP S/4HANA system to a new release or install a new feature or support package, you must run the program again. For more information, see SAP Note 2870546.
- Set up the necessary authorizations (see Section 3.3.4).
- Create the customizing for the applications affected in the target system. The relevant customizing must already be available in the target system before you can start with the first migration tests. You require, for example, organizational units like company codes, plants, and sales areas. You also require a document type for the migration, counter-posting accounts for the migration, open posting periods, material types, etc.

### 3.3.3 Migration Objects for SAP ERP Systems and Other SAP Source Systems

Selection options

In addition to the technical platform, the default delivery for this approach also includes predefined selection options. Depending on the source system (referred to as a *scenario* in the system), you would first choose the organizational unit from which the selection is to be made. Within the chosen organizational unit, you select migration objects for which you select the relevant data records from the source system.

Table 3.4 provides an overview of scenarios (source systems), including their predefined selection options.

| Source System | Predefined Selection Criterion |
|---|---|
| SAP ERP | Company code |
| SAP EWM | Warehouse number |
| SAP CRM | Sales organization |
| SAP Apparel and Footwear | Company code |
| SAP eSPP | Plant |

**Table 3.4** Delivered Scenarios and Predefined Selection Criteria

**Documentation for All Migration Objects and All Versions**

A complete list of migration objects for all migration approaches is available on SAP Help Portal at *http://help.sap.com/S4_OP_MO*. You can filter this list by release and migration approach. Each migration object is described in detail. You also learn about the scope, availability, useful transactions, etc.

**Information on Planning**

You can find current information with regard to planning on the SAP S/4HANA migration cockpit roadmap, which is included in the standard presentation on the migration cockpit: *http://s-prs.co/v581756*.

### 3.3.4 Assigning Roles for the Migration

Technical roles

For the direct transfer approach, the migration user requires specific roles both in the source system and in the SAP S/4HANA target system, as shown in Table 3.5.

| Tool | Role in the Source System | Role in the SAP S/4HANA System | Role in the SAP Fiori Frontend System |
|---|---|---|---|
| SAP S/4HANA migration cockpit | – | SAP_CA_DMC_MC_USER | SAP_BR_CONFIG_EXPERT_DATA_MIG |
| Migration object modeler | – | SAP_CA_DMC_MC_DEVELOPER | – |
| RFC user | ■ SAP S/4HANA 2020 or lower: SAP_DMIS_OBT_REMOTE Communication type<br>■ As of SAP S/4HANA 2021 or higher: SAP_DMIS_MC_DT_REMOTE Communication type | – | – |

**Table 3.5** Technical Roles for Direct Transfer

In addition, you'll require further authorizations and roles to select the data of the individual migration objects from the source system. In the SAP S/4HANA target system, you require application roles to display or validate the migrated data in the application. These authorizations vary depending on the migration API executed. In SAPS/4HANA, predefined business role templates are provided to you for specific business cases. You can find the relevant application role template for each migration object for SAP S/4HANA in the migration object documentation at *http://help.sap.com/S4_OP_MO*. If authorizations are missing, you can use Transaction SU53 (Authorization Trace). This transaction lists technical information if authorization checks have failed. You usually create roles with the corresponding authorizations within an authorization concept.

**Application roles**



### 3.3.5 Running a Project with Direct Data Transfer

Back in Section 3.2, we guided you through a complete project using the migration cockpit. This section now discusses in detail the functions and screens that are different in the direct transfer of data from an SAP system scenario.

#### Starting the App

As previously described, you start the migration cockpit for this approach via the **Migrate Your Data - Migration Cockpit** tile. You can find this app in the **Data Migration** area, as shown in Figure 3.1.

#### Creating a Project

First, create a project. If possible, you should create only *one* project for all objects that you want to migrate for your organizational units, for example, one project for all migration objects that you want to migrate for company code 0001.

[«]

> **Mappings Are Not Cross-Project!**
>
> Mappings are cross-migration objects as long as they are assigned to a project. Mappings are not cross-project, that is, if you create more than one project, for example, for a company code, the mappings must be downloaded from one project and uploaded to the other project.

Select the option **Migrate Data Directly from SAP System**. A wizard guides you through the necessary steps. You must populate the following fields in step 1 (**General Data**), as shown in Figure 3.51:

**Step 1: General Data**

- **Name**: Assign a unique name for the project in this field.
- **Scenario**: For example, SAP ERP to SAP S/4HANA. The scenario defines which organizational unit is provided for selection later on (for example, the company code) and which migration objects are available. Organizational units and objects are delivered by SAP and can be adapted to the modeling environment (Transaction LTMOM) if required. Particularly the adaptation of delivered objects and the creation of custom migration objects are relevant for many customers. You can find more information in Section 3.7.
- **Connection to Source System**: You can select this connection as soon as the system administrator has created the RFC connection between the source system and the SAP S/4HANA target system. Only one RFC can be assigned for one project. If you want to migrate data from several source systems to the SAP S/4HANA target system, you must create one project for each source system. These mappings (Section 3.3.5) must be kept in sync between the projects to ensure a consistent migration.

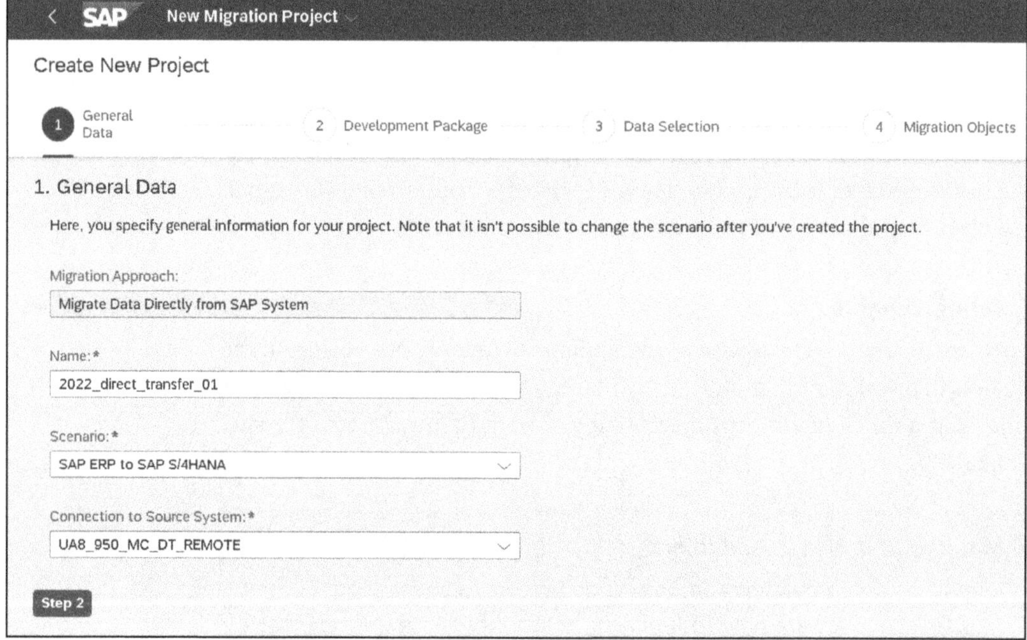


**Figure 3.51** Creating a Project: Defining General Data

**Step 2: Development Package**

When you click **Step 2**, you will navigate to the next step: **Development Package**, as shown in Figure 3.52. In this step, you must assign a development package when creating the project so you can transport the project later.

Note that you can no longer change the scenario or development package after the project has been created.

The transport comprises all project parts except for the mapping values. Mapping values are not transported. You can download and upload mapping values for individual, multiple, or all mapping tasks in the migration cockpit (**Mapping Tasks** screen) or in Transaction LTMOM. If you do not want to transport the project, do not specify a transportable development package, for example, $TMP. The transport itself is started in the transport system. For more information, see Section 3.7.7.

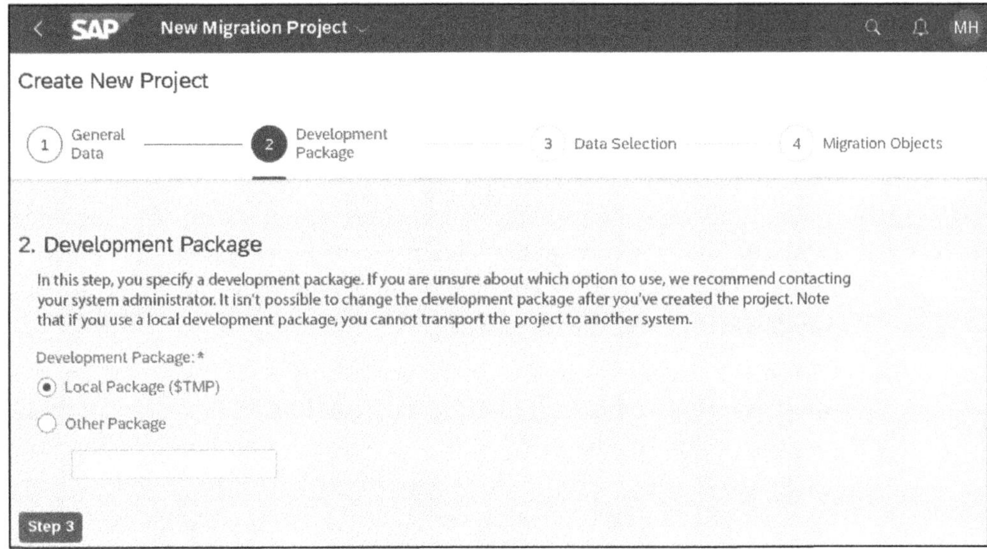


**Figure 3.52** Creating a Project: Maintaining a Development Project

[«]

**No Transport Connection in SAP S/4HANA 1909**

The transport connection is not available yet in SAP S/4HANA 1909. As a result, in this release, you must create projects manually in the production system. In SAP S/4HANA 1909, you can download and upload mapping values for individual, multiple, or all mapping tasks in Transaction LTMOM only.

Step 3: Data Selection

When you click on **Step 3**, you will navigate to the next step: **Data Selection**. Now, you can choose one or more organizational units from which data selection should be made, as shown in Figure 3.53. In our example, we're using the company code.

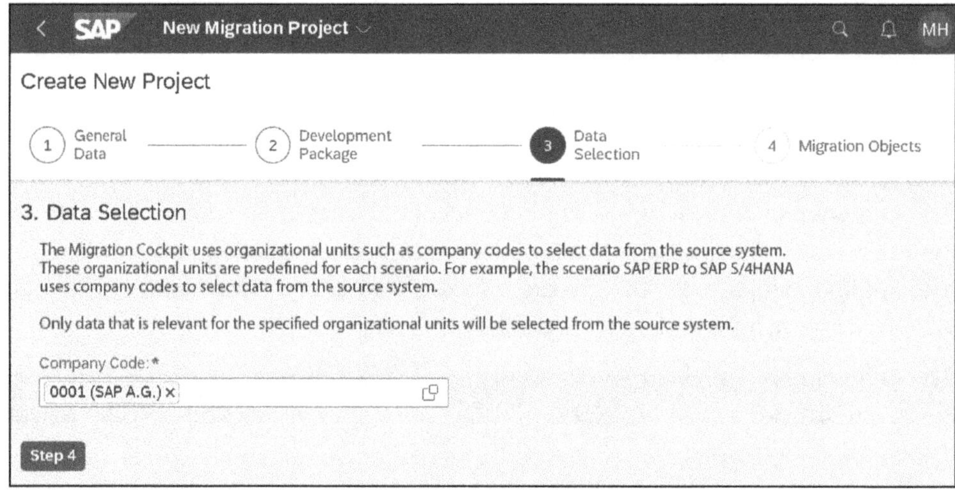


**Figure 3.53** Creating a Project: Storing the Data Selection

**Step 4: Migration Objects**

When you click **Step 4**, you will navigate to the next step: **Migration Objects**. In this step, migration objects are displayed that SAP provides for the scenario selected. Clicking on **Documentation** takes you to the object's documentation on SAP Help Portal.

The description contains an object description, the exact object scope, the Business Application Programming Interface (BAPI) that is used for creating or posting in the SAP S/4HANA system, transactions for checking migrated data, etc. If you click on the number in the **Predecessor** column, you can obtain a list of all predecessor objects for the present object. Select the migration objects you want to use in your project and move them to the **Selected Migration Object** area by clicking the arrow icon [>], as shown in Figure 3.54.

**Removing a migration object from the project**

You can add further objects to the selection at any time. Deleting migration objects from a project is currently only possible in Transaction LTMOM provided that no data has been migrated to the target system for this object. Additionally, you must delete the already existing (selected) instances for this migration object from the migration cockpit by calling report `cnv_pe_mc_del_items` in the SAP S/4HANA backend. Alternatively, you can use a workaround and rename the unwanted migration object in Transaction LTMOM, for example, in `OBSOLETE_Cost_Center`.

**Checking predecessor objects**

In the **Check** step, the system asks whether you want to add the predecessor objects to the project. A popup window displays the migration objects that are logical predecessors to the objects you selected, for example, for the **CO - Activity type** object, the **CO - Secondary cost element** migration object, and **CO - Activity type group**. It is not mandatory to add these migration objects because they may have reached the target system by other means or are not required in the individual case. No predecessors are added in the example shown here. You can now save the project by clicking the **Create Project** button.

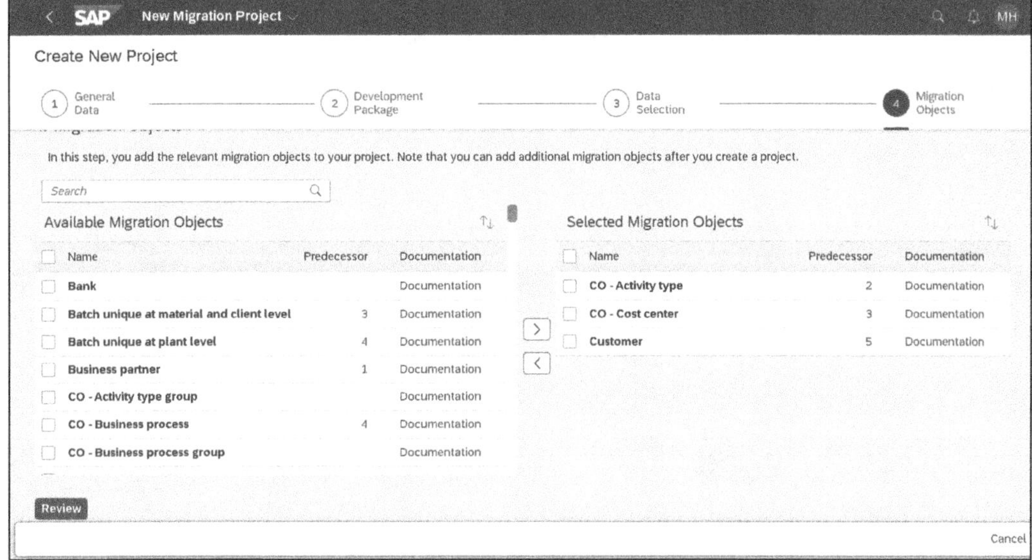


**Figure 3.54** Creating a Project: Selecting from the Available Migration Objects

**What to do in case of an error?**

If problems arise during the creation of a project, you'll be indicated via the **Migration Project** view, as shown in Figure 3.55. When you create the project, the migration objects are copied from a migration template to the present project, which can take some time.

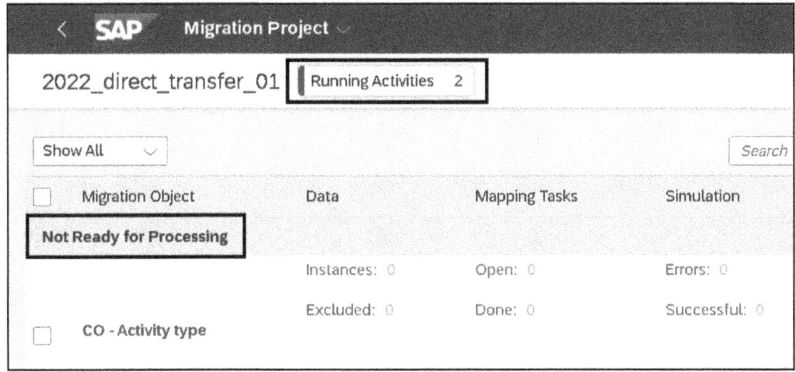


**Figure 3.55** Troubleshooting in the "Create Project" Step

As long as the objects are not ready for further processing, they are displayed as **Not Ready for Processing.** If the copy function is canceled, you can restart it. If other technical problems arise, the **Activities with Error** button appears in the upper area of the screen. You can click this button to analyze the problem, as shown in Figure 3.55.

**Additional settings for the project** If you click on **Settings** at the top right of the **Migration Project** screen, you can navigate to a screen with additional functions of the project, for example, **Data Retention Period**, **History**, etc.

**Job Management** The top right area also includes an option to navigate to the **Job Management**. For details on project settings and job management, see Section 3.2.7 and Section 3.2.15.

### Detail Screen of the Migration Object

The detail screen of the migration object is a central page. When you click on the name of the object, for example, **CO – Activity type**, you navigate to this page, which includes basic information on the object and its history, as shown in Figure 3.56.

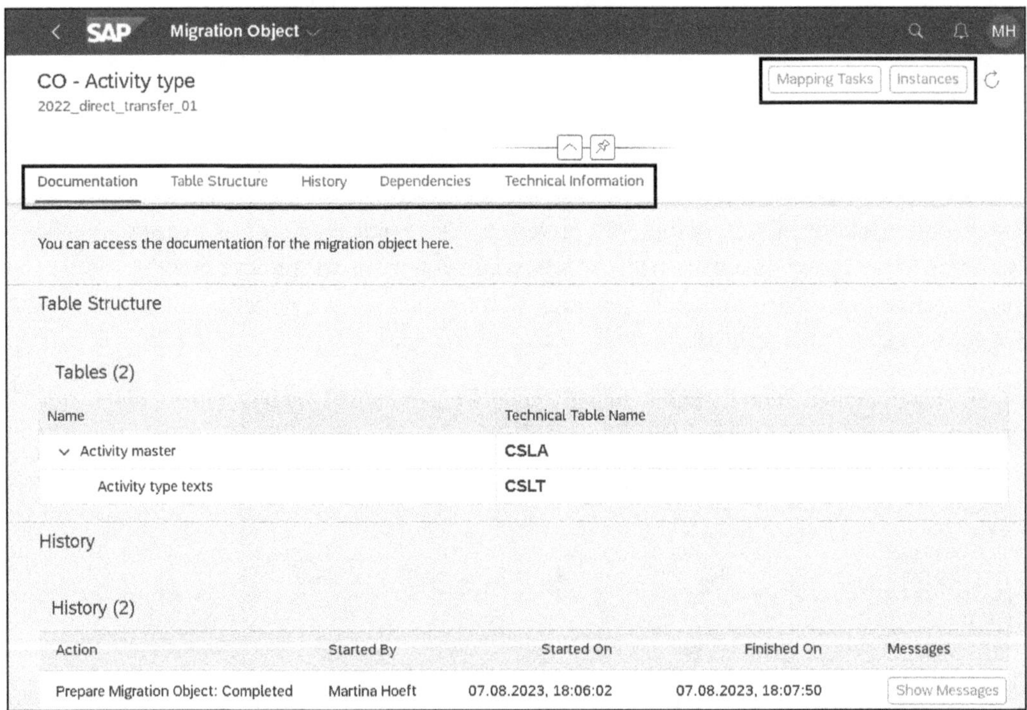


**Figure 3.56** Migration Object View: Detail Screen

**No check for updates** Compare this screen with the detail screen shown earlier in Figure 3.10. Notice that the information provided for a migration object is quite similar. In the top-right corner of this approach, you can find the **Mapping Tasks**

link and links to the individual **Instances** for this object. The current release does *not* provide any function for checking for migration object updates. If you want to use a content update for an object which was implemented via a TCI note, for example, you must then add the migration object to your project again. For this task, call Transaction LTMOM and select **Migration Object Modeler • Create Migration Object • From Template • Migrate Data Directly from SAP System.**

**No check for custom fields**

The **Migration Object** view currently does not provide any option for checking for custom fields in the SAP S/4HANA target system, as shown in Figure 3.56. Because the approach of direct transfer reads data from the SAP source system, we consider the check for field enhancements based on the source system. Section 3.7.5 provides more details on this topic.

### User Guidance in the SAP S/4HANA Migration Cockpit

**Flexible selectable actions**

The migration cockpit guides you through the migration process by suggesting the next step in the **Action** column. But it also offers full flexibility at the same time. For each object, you can select any step from the dropdown menu or execute an action from the header for all selected objects of the project, as shown in Figure 3.57.

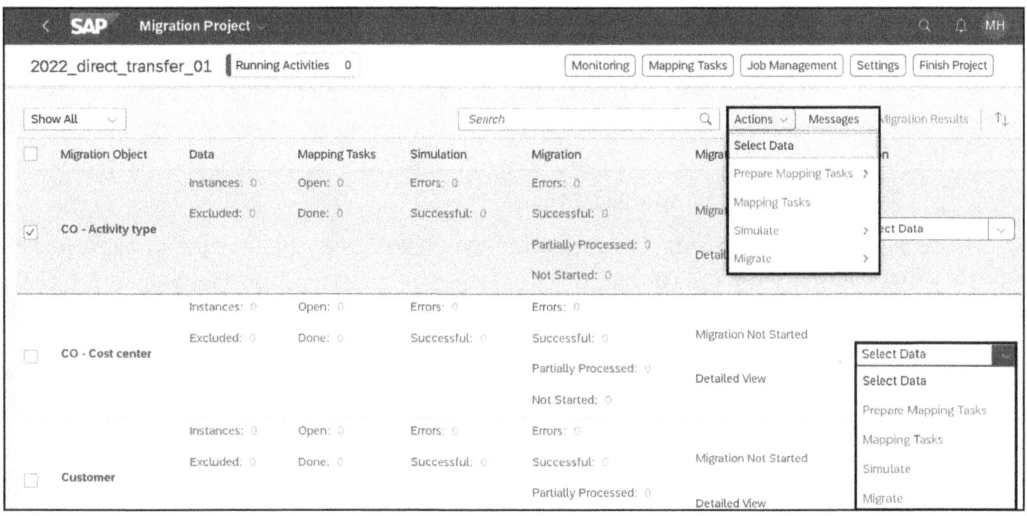


Figure 3.57 Migration Cockpit: Migration Project Screen

### Selecting Data from the SAP Source System

After you have defined the project scope, you can import the data from the source system.

[»]

**Modeling Environment Provides More Options!**

This section describes the selection options within the framework predefined by SAP. In Transaction LTMOM, you have more options to control the selection from the source system. There you can define additional selection criteria for individual or all migration objects within a project.

**Integrated organization derivation**

Based on the entered organizational units and selected migration objects, the migration cockpit specifies all relevant data records in the source. An implicit organization derivation ensures that all related data records are found.

**Instances**

The data model of the migration object defines which fields are imported from which tables and are assembled to one instance. The instances form the basis for subsequent posting or creation of data in the SAP S/4HANA system. The instances contain all necessary fields to correctly provide the respective BAPI of the application.

**Examples of organization derivation**

Let's consider an example: In your project, you defined the company code 0001 and selected the **CO – Activity type** migration object for migration. Activity types are not assigned to a company code, but to a controlling area (see the data model of the migration object and table CSLA). The implicit organization derivation states that table TKA02 is read in the source to find the controlling area that is assigned to the selected company code (for example, 001A). From tables CSLA and CSLT, the migration cockpit reads the fields that are required according to the definition of the migration object (for example, 001A) and generates an instance for each activity type from the controlling area 001A.

**Setting up parallelization for the selection**

Before you start the selection, you need to set the number of background jobs per project and per migration object under **Job Management** (parallelization). You can start the selection for individual, multiple, or all migration objects within a project, as shown in Figure 3.57.

On the **Migration Project** screen, the number of selected instances is updated automatically until the selection is complete. The number is displayed in the **Data** column, as shown in Figure 3.58.

In the **Monitoring** view, you can view the processing progress and the messages for all activities of the project (described earlier in Section 3.2.8).

**Selection repeatable?**

In principle, you can run the selection repeatedly or multiple times. Note, however, that, in this case, the entire data volume must be re-read in the source for the migration object concerned, which requires additional time. If you re-run the **Select Data** action, the dialog box shown in Figure 3.59 appears.

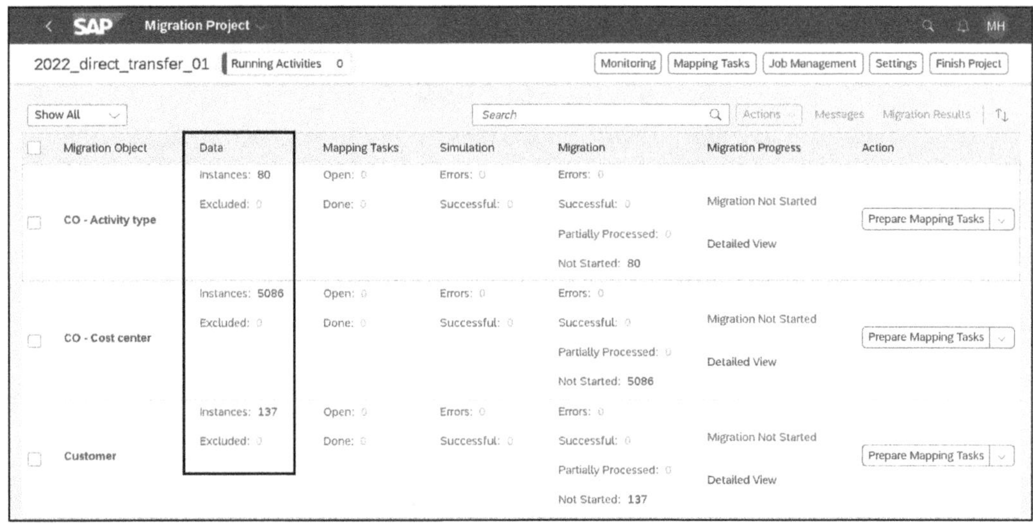


**Figure 3.58** Project View after Selection Has Been Made

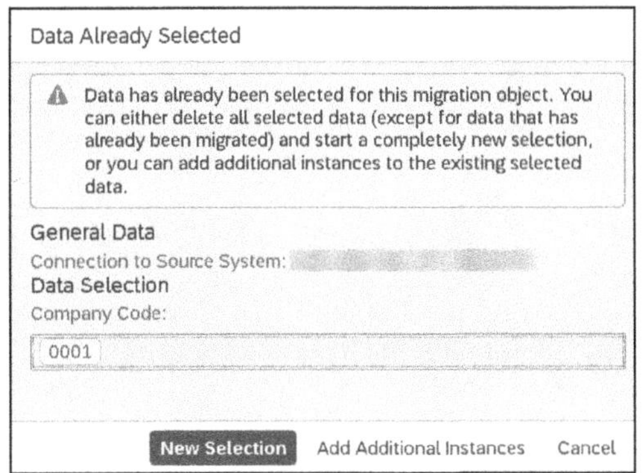


**Figure 3.59** Options for New Selection

If you expect *changed data records* in the source, you should click the **New Selection** button. If you know that no records have change, but you've only *newly added* data records in the source for this migration object, you can select **Add Additional Instances.**

If you only want to stop processing a limited number of instances, you may exclude individual instances in the **Migration Object Instances** view as an alternative.

**Manual Step Required for Data Record Updates in the Source in SAP S/4HANA 1909**

As of SAP S/4HANA 1909, you can repeat the selection process. If the migration cockpit finds new records in the source, they are added. But if you want to repeat the selection because data records have changed in the source, that is, the contents of the key fields are still the same, you must explicitly delete the previous selection in advance. For this purpose, log on to the SAP S/4HANA backend (SAP GUI) and start report CNV_PE_MC_DEL_ITEMS. With this report, you can delete instances based on various selection criteria. For more information, see SAP Note 2857598.

### Checking the Data Selected

**Set column view**

You're now on the **Migration Project** screen. If you click on the number of instances selected, the system takes you to the **Migration Object Instances** view, as shown in Figure 3.60.

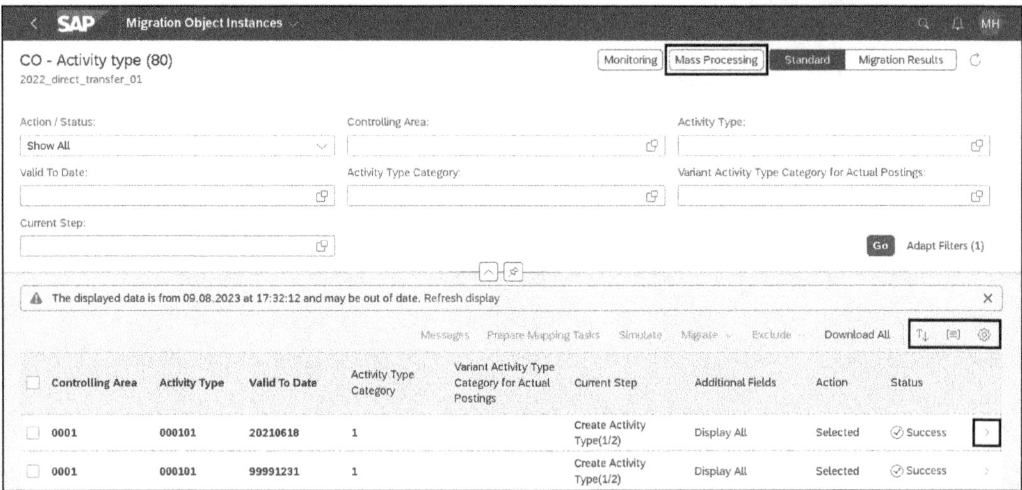


**Figure 3.60** "Migration Object Instances" View

Alternatively, you can navigate to the instance view from the **Migration Object** detail view. On the instance view, you can specify which columns should be displayed (via the cogwheel icon). You can search, sort, filter, and group objects in this list. You can display the entire data record for each instance (by clicking on the arrow icon in the row). This data record consists of fields that have been selected from the source for the migration object according to the data model. The data record usually contains fields from multiple source tables.

**Executing actions for individual records**

In the **Migration Object Instances** view, you can perform actions for the individual instances. Select the individual instances from the list and click the **Prepare Mapping Tasks, Simulate, Migrate**, and **Exclude/Undo Exclude** buttons to start the action to be performed.

As an alternative, you can use the **Mass Processing** function (top right) to filter records from the existing list according to specific criteria and the run the actions for this subset of records. This approach enables you to work specifically with certain records from the entire set of instances read, for example, with particularly critical records or records that represent specific business processes.

[«]

**No Cleansing Functionality**

The direct transfer approach does not provide any cleansing functionality, such as duplicate check, golden record, etc.

If data cleansing is necessary, this process must take place in the source system. Alternatively, you can set the selection in such a way that unwanted records are not read (Transaction LTMOM). In the migration cockpit, you can exclude records that have already been read from further processing either individually in the **Migration Object Instances** view or using the **Mass Processing** function.

### Preparing Mapping Tasks

Filling default values

SAP S/4HANA 2021 introduced the **Prepare Mapping Tasks** step. This step was previously part of the **Select Data** action. While the **Select Data** step selects the migration object instances, the **Prepare Mapping Tasks** step fills the mapping tasks with default values (source value = target value). First, this separation has the advantage of optimizing your downtimes because you can skip this activity if you upload the mapping values using Microsoft Excel. (More information on this topic in this next subsection on mapping values.) Second, this subdivision provides more transparency about the activities the system is currently performing.

### Mapping Values

Automatically generated mappings

Depending on the release status, the selection or the **Prepare Mapping Tasks** step generates mapping proposals. In each migration object, it is defined for which fields mappings are necessary or possible. During selection, a 1:1 suggestion is automatically generated for each value of such a field.

If fixed values, for example, the posting date or a counter-posting account for migration, are required for the migration object, a mapping task is automatically generated.

For the cost center migration object, for example, a total of 17 mapping tasks is displayed for the following fields and 13 additional mapping tasks:

- **Controlling area**
- **Company code**
- **Cost Center**
- **Cost center group**

Download/upload

The overview of mapping tasks provides links to the download/upload options so you can easily edit mappings outside the migration cockpit and exchange and synchronize mappings among various projects.

Keeping mappings in sync between projects

How to download and upload mapping values is described in detail in Section 3.2.14.

In the **Mapping Task** detail screen, you can check the mapping target values by clicking the **Check** button. Depending on the mapping task, this check may involve a pure check of the format or a check for the existence of a certain value. In other cases, such as for the posting date, further logical checks are performed, for example, whether the posting period for the company code concerned is open on this date. For performance reasons, checks for the direct transfer approach are not performed automatically when the values are confirmed but instead are performed explicitly via the **Check** button when desired. The check function is implemented for selected mapping tasks only from SAP S/4HANA 2020 FPS0 onwards.

[»]

**No Check of Mapping Target Values in the "Mapping Tasks" Step in SAP S/4HANA 1909**

In SAP S/4HANA 1909, no option exists for checking the mapping target values during the mapping maintenance. As a result, only during simulation can you determine if a mapping target value is correct in terms of form and content.

[»]

**No Central Storage or Integration with Other Tools Currently**

In SAP S/4HANA, no central repository exists for mappings or in integration with other tools, for example, SAP MDG. Keep your mappings synchronized using the download/upload function. Check the latest SAP Notes if any enhancements have been developed recently.

You can also delete mapping values in the *direct transfer* approach. For more information, see Section 3.2.14, especially our discussion on deleting values.

### Simulating the Migration

The next step is to simulate the migration. As a result, the migration cockpit applies the mapping rules to the instances and transfers the transformed instances to the BAPI of the migration object. The BAPI simulates the instances. For each instance, the system passes through all checks that are contained in the BAPI.

**Setting the parallelization**

Before you start the simulation, you must set the number of jobs per project and per migration object in the **Job Management** view.

**Records to be processed, flexible selection**

In the **Migration Project** view, you can start the simulation for several migration objects in parallel. Select them and click on **Actions** in the upper area of the screen. Choose whether you want to simulate **All Instances** or **All Instances with Errors** for these instances.

To simulate a single migration object, choose the **Simulate** option in the **Action** column. Now, you can decide if you want to start any 10%, any 500 instances, all instances for this object, or a custom selection (navigate to the **Mass Processing** function). Alternatively, you can first navigate to the **Migration Object Instances** view and select and simulate individual instances from the list.

### Analyzing and Cleansing Errors

**Three views of messages**

You'll receive various error messages after the simulation. Three different views of the messages are available. You can access a specific run for a migration object via **Monitoring** and then click the **Display Messages** button. This approach will take you to the **Messages** view. Click **Error** for an overview of error messages, as shown in Figure 3.61.

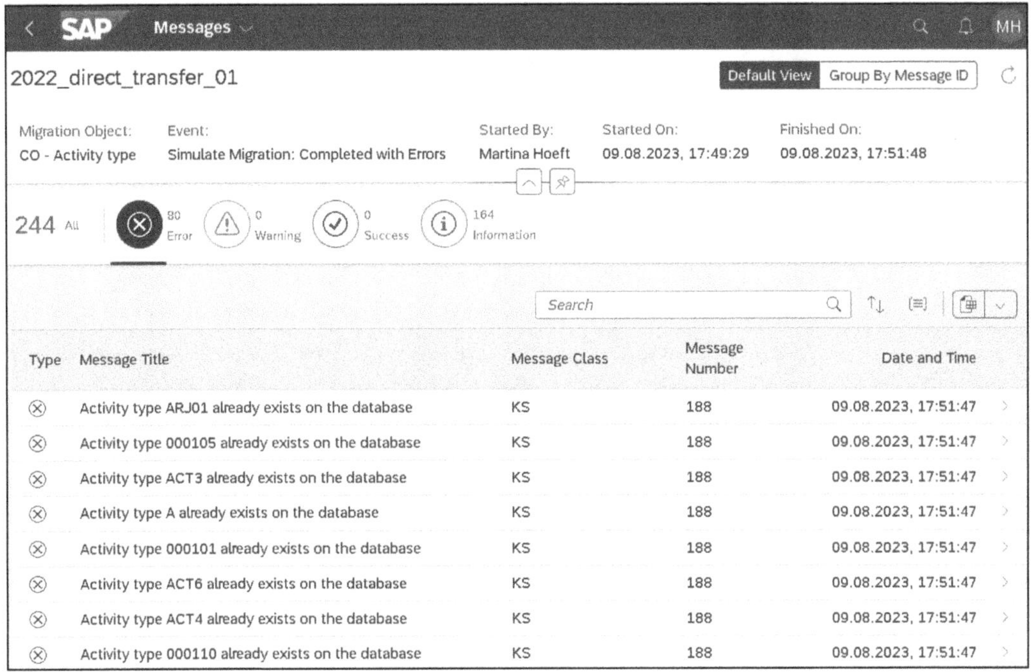


**Figure 3.61** Messages of a Simulation Run

You get a second view, for example, by clicking on one of the numbers of the project view to navigate to the **Migration Object Instances** view. All instances for this object are displayed if you click on the number after **Instances**. If you click on the number in the **Simulate** column (after **Error**), the system displays the **Migration Object Instance** list with the instances that were simulated with errors. You can change this view as required. In the instance list you now select the instances for which you want to analyze messages and click **Messages**. This opens a view that displays the messages per instance, as shown in Figure 3.62.

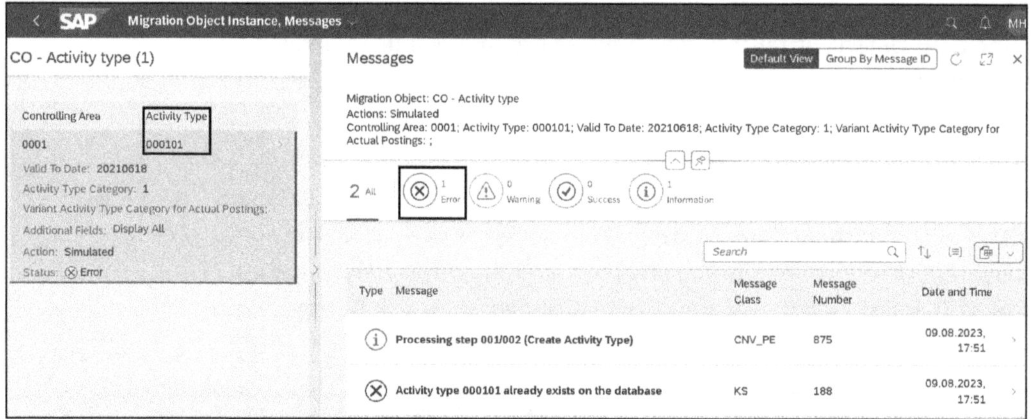


**Figure 3.62** Messages per Instance

**Displaying messages per instance**

Since SAP S/4HANA 2020 FPS01, you can display the affected instances per message in a third view. For this task, you must select the relevant migration objects in the **Project** view and click the **Messages** button. This step will take you to the **Migration Object Instances, Messages** view. As shown in Figure 3.63, for each selected migration object ❶, you can view its messages. Additional filtering options ❷ are also available. For example, you can filter by *action* (e.g., **Simulated**) or *message type* (e.g., **Error**). To access an overview of the affected instances per message, click on the number of instances ❸ of the message.

The **Migration Object Instances, Message Details** view, as shown in Figure 3.64, provides an overview of all instances affected by the message. Clicking on the key fields for migration object instances of an instance quickly takes you to the migration object instance details.

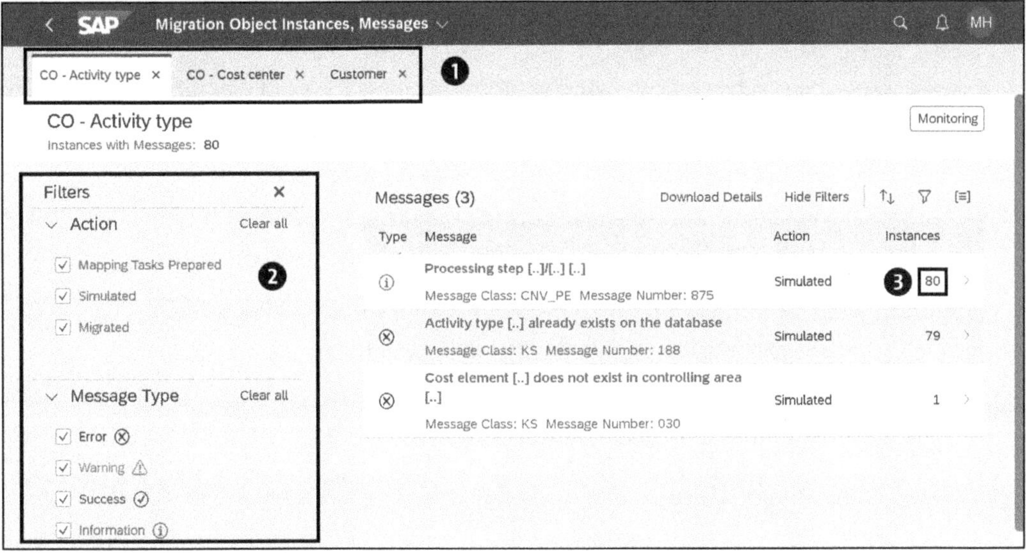


**Figure 3.63** Migration Instances, Messages

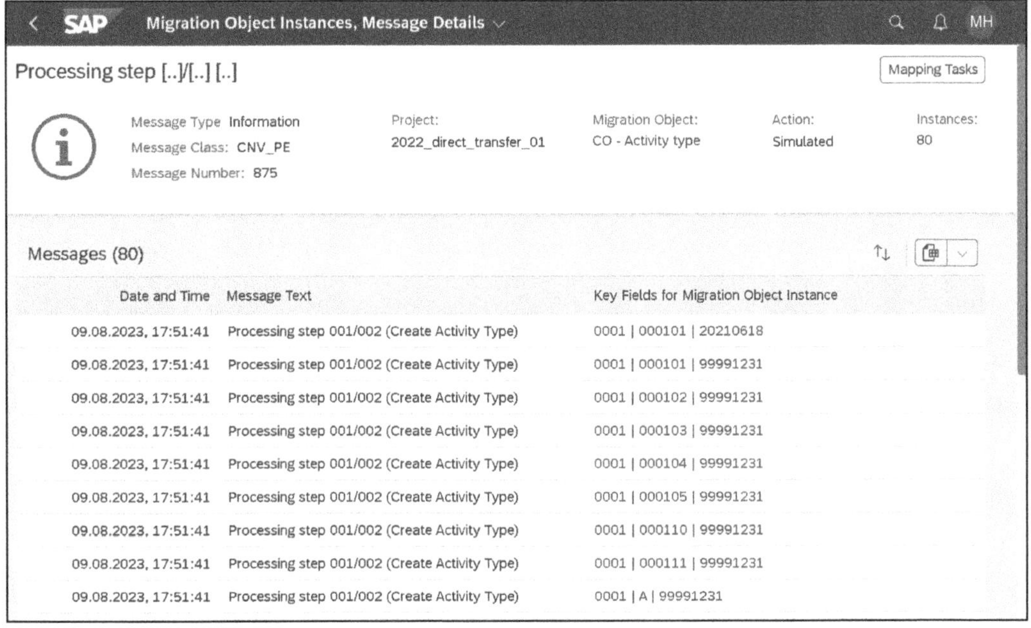


**Figure 3.64** Migration Object Instances, Message Details

You can sort, filter, and download these messages in all views.

Navigate to the detail screen of a message for further error analysis. The message ID, for example, KS 096, provides initial information about the application component concerned. You can use this abbreviation for internet research or for searching SAP Notes to obtain valuable tips for troubleshooting.

Analyzing errors

Basically, the following solution options are available for messages that originate from an SAP application:

- Change the data in the source and repeat selection
- Create or adapt customizing in the target system
- Create or adapt required master data in the target system
- Change mapping to an existing or suitable target value
- Exclude instance from further processing

**Additional Tools for Error Analysis**

In Transaction LTMOM (called in the backend), you can debug instances during simulation (Section 3.7.6).

If you want to delete all instances or specific instances for a migration object from the migration cockpit, call report `cnv_pe_mc_del_items` in the backend.

### Migrating

The migration cockpit provides the same functions for migration that are available for simulation. In addition, you can migrate only successfully simulated instances. From a technical perspective, the only difference is that the data is now created or posted in the SAP S/4HANA system.

### Checking the Migrated Data

Checking migration results

As of SAP S/4HANA 2021, you can display and review the migrated data using the **Migration Results** view. For this task, select a migration object in the **Migration Project** view and click the **Migration Results** button. As shown in Figure 3.65, depending on the migration object, you'll see one or more fields containing the migration results. For high data volumes, you can limit the resulting list of instances via filters.

As of SAP S/4HANA 2021 FPS01, you can download a list of the migrated data (both the instance and its migrated fields) via the **Download All** button. You even have the ability to customize this list according to your needs. As shown in our example activity type, you can decide whether to download only the key, result, and status fields or even more fields per instance.

Direct navigation to the relevant SAP Fiori app

In addition, you can navigate directly to the relevant SAP Fiori app for the stored migration object and display the data in this app. For this task, click on the icon in the **Result** column, as shown in Figure 3.65.

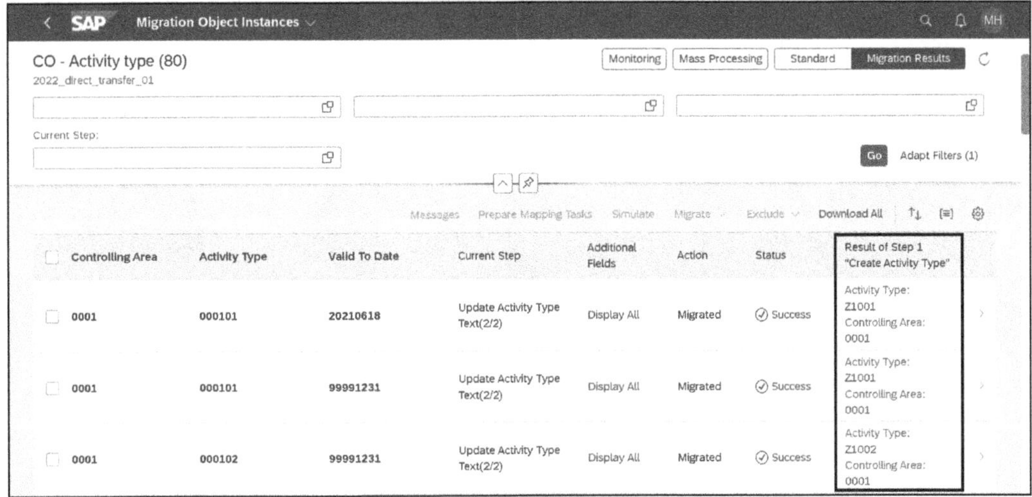


**Figure 3.65** Migration Results

Furthermore, you can check the data in the SAP S/4HANA target system by using the corresponding application app or transaction or by using Transaction SE16 to check the table contents. The migration object description provides information on how you can verify the result of the migration. Note that some prerequisites must be met for accessing apps, for instance, when user roles are necessary.

[+]

**Displaying a List with Selected Instances**

In the **Migration Results** view, you can display a list of all selected instances after data selection and even download this list. For this task, simply select the desired migration object in the **Project** view and click the **Migration Results** button. The **Download All** button allows you to download the list of selected instances.

### Detail View, Transfer Options, and Steps

**Detail view on the migration process**

The **Detailed View**, available as of SAP S/4HANA 2021, provides a more comprehensive overview of the migration process and enables you to control it in greater detail. You'll also see an overview of the relevant migration object transfer steps and options used to migrate different types of migration object instances to SAP S/4HANA. As shown in Figure 3.66, using the **Customer** migration object as an example, most migration objects have only one transfer option.

However, some migration objects, such as MM (Purchasing Scheduling Agreement), offer more than one transfer option. For more information on the detail view and transfer options, see "Transfer Options and Steps" at SAP Help Portal.

To access the **Detailed View**, click the **Detail View** button on the **Migration Project** view.

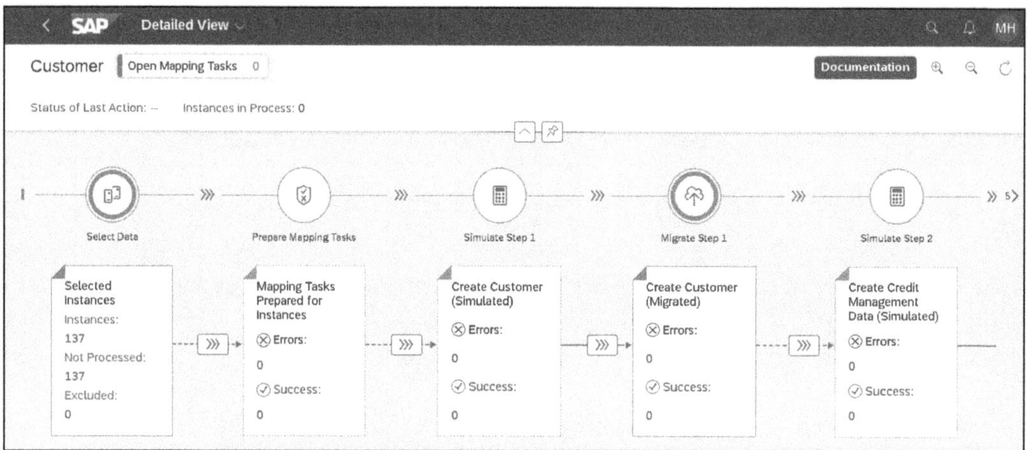


**Figure 3.66** Detailed View Transfer Options

### Copying an Existing Migration Project

**Mapping and fixed values are not copied**

As of SAP S/4HANA 2020 FPS01, you can create a new migration project that is a copy of an existing migration project. During creation, all migration objects, mapping task definitions, and settings (including user-defined changes) are copied. Mapping values or fixed values are not transferred to the new project.

To copy an existing project, select the project to be copied in the **Migration Projects** view and click the **Copy** button. Then, you can assign a name for your new project, specify a different connection to the source system if needed, and specify a different development package. If necessary, you can adjust the criteria you want to use to select the data. At this point, you cannot add additional migration objects to the project. However, you can add additional objects after the project has been completely created.

## 3.4 SAP S/4HANA Migration Object Modeler

Now that you understand the various migration approaches in the SAP S/4HANA migration cockpit for new implementations of SAP S/4HANA, as described in Section 3.2 and Section 3.3, we want to turn to your options for adapting the delivered content of the migration cockpit to meet your customer-specific circumstances. For this purpose, we'll use the modeling environment available in the migration cockpit.

**A powerful modeling environment**

The *SAP S/4HANA migration object modeler* is name of the modeling environment in the migration cockpit. Available in the on-premise version of SAP/4HANA, you can call it using Transaction LTMOM.

In SAP S/4HANA Cloud Public Edition, the *migration object modeler* is called via the SAP Fiori tile **Model Migration Objects - Migration Object Modeler**. However, this app is limited in scope and currently only available for modeling data migrations with files or staging tables. For more info, see Section 3.6.

The *SAP migration object modeler* is included in SAP S/4HANA Cloud Public Edition 2402 and is also available for the **Migrate Data Directly from SAP System** approach. We won't describe this option in more detail in this book, but for more information, see *http://s-prs.co/v581757*.

[«]

**Required User Role**

To use Transaction LTMOM in SAP S/4HANA, your user must be assigned role SAP_CA_DMC_MC_DEVELOPER.

The migration object modeler has become a powerful modeling environment since SAP S/4HANA 2020. New functions have been added with each release since its introduction back in SAP S/4HANA 1610. The current functionality is now comparable to SAP Legacy System Migration Workbench and even more comprehensive in some areas, such as field and structure mapping, cross-object usage of transfer rules, and thus by extension value mapping.

**Useful for all migration approaches**

The migration object modeler can be used for either migration approach available in the migration cockpit.

Before we discuss the functionality of the migration object modeler for each migration approach, we want to provide first a brief overview of its general structure. You'll learn how to select and search for existing migration objects, and we'll provide information about the migration APIs that are used. In Section 3.5 and section Section 3.6, you'll learn about the functionality of the migration object modeler when you use the migration cockpit to perform a data migration with files or staging tables. Section 3.7 describes the functionality of the migration object modeler with regard to data migrations using the direct transfer approach from an SAP system.

Which functions are available depends on the various approaches, which are discussed in more detail in the respective sections.

### 3.4.1 General Structure of the SAP S/4HANA Migration Object Modeler

**Entry point of migration object modeler**

As shown in Figure 3.67, the migration object modeler displays the object browser by default if you open it using Transaction LTMOM. You can then select whether you want to view a migration project with all its related migration objects or view an individual migration object directly.

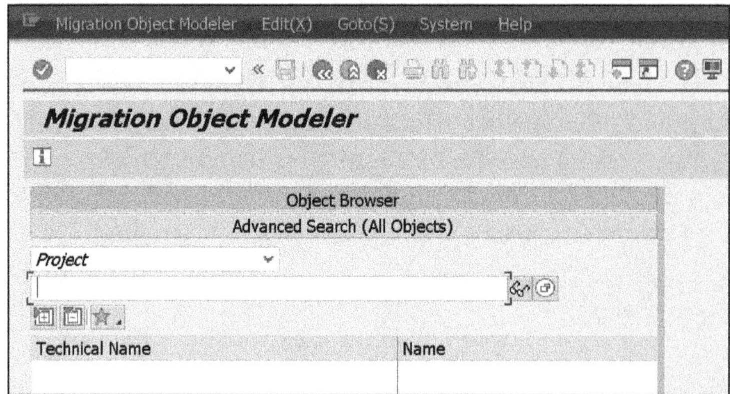


**Figure 3.67** Object Browser Selection in the SAP S/4HANA Migration Object Modeler

If you display an individual migration object, you can see a selection area including a navigation tree on the left and a workspace on the right, as shown in Figure 3.68.

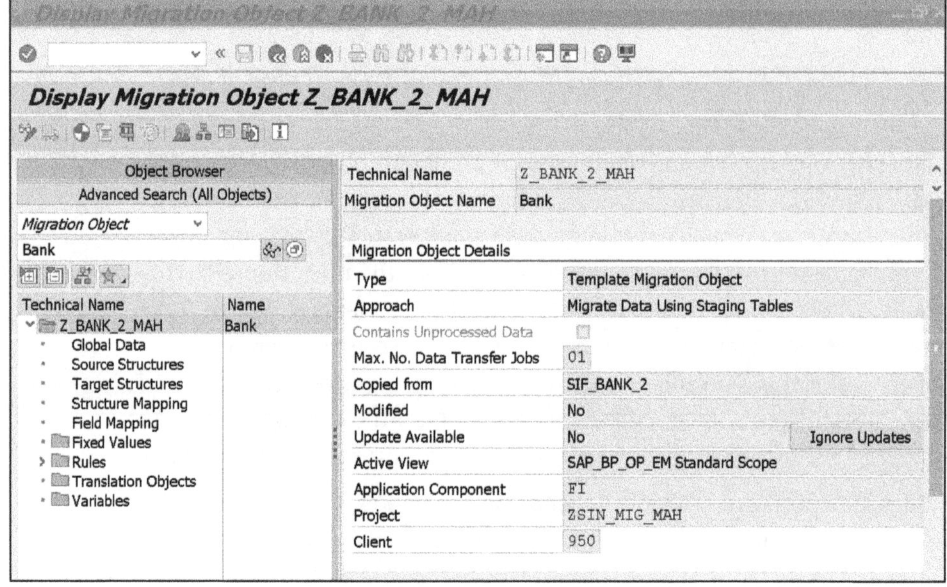


**Figure 3.68** Structure of the SAP S/4HANA Migration Object Modeler

> **SAP GUI Settings**
>
> Our SAP GUI screenshots are based on the Corbu theme. If you use other SAP settings, your screens may vary.

**Documentation**

You can click the icon (**Fullscreen On/Off**) to hide/display the selection area and thus display the workspace in full screen.

When you click the icon (**Information**), you call the documentation of the migration object modeler. Its initial screen is shown in Figure 3.69.

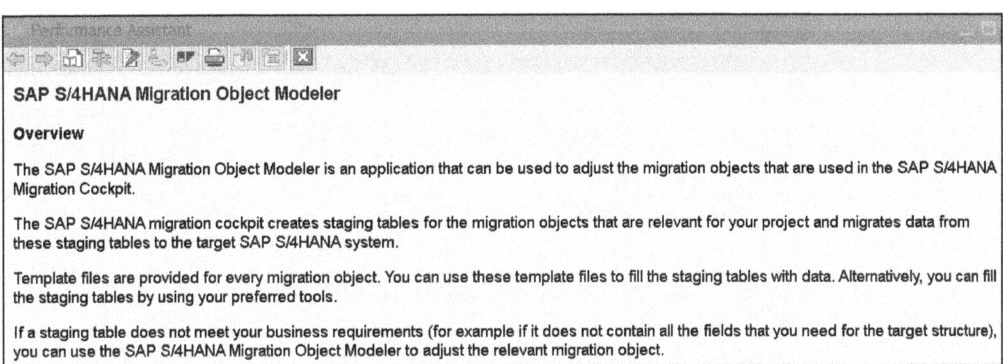


**Figure 3.69** Initial Screen of the Documentation for the SAP S/4HANA Migration Object Modeler

You can select the documentation of the various migration approaches depending on the approach you're currently working with. If you have already selected a project or migration object in the object browser, the system displays the documentation of the corresponding migration approach directly.

**Favorites list**

You can add projects and migration objects to your favorites list by selecting the relevant project or migration object and then clicking the icon (**Favorites**). For this task, select **Add**. Figure 3.70 shows the selection for the favorites function. In addition, you can edit your favorites list through this icon.

Migration Object Modeler
Object Browser
Advanced Search (All Objects)
Project
enter name or description
Technic
Add
Edit Favorites
Name

**Figure 3.70** Selection Options of the Favorites List

### 3.4.2 Selecting an Existing Migration Object

Selecting migration objects

If you want to select an existing migration object, click the icon on the right of the empty input field for the migration object, as shown in Figure 3.71.

The **Find Migration Object** search window appears. First, select the relevant migration approach in the **Approach** field, in our example, **Migrate Data Using Staging Tables**. In the **Project** field, enter the search term for the project that includes the migration object. Confirm your selections by clicking the icon (**Next**). Then, double-click the desired migration object or confirm by clicking (**Next**) again.

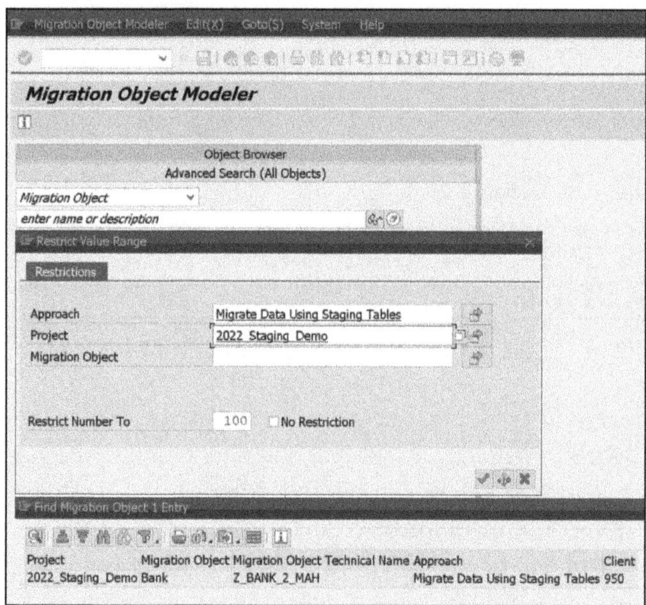


**Figure 3.71** Selecting an Object in the SAP S/4HANA Migration Object Modeler

Alternatively, you can also display an entire project and select the corresponding migration object from this list. The full range of the migration object modeler functions for this migration object is now available.

### 3.4.3 Advanced Search (All Objects)

Advanced search help

The second view of the migration object modeler is the **Advanced Search (All Objects)**. By means of the selection button in the left selection area, as shown in Figure 3.72, you can toggle between the **Object Browser** and the **Advanced Search (All Objects)** views.

In the advanced search view, you can search by **Project** and **Migration Objects**. Select the desired search screen, and then narrow that that search using the search fields in the search screen.

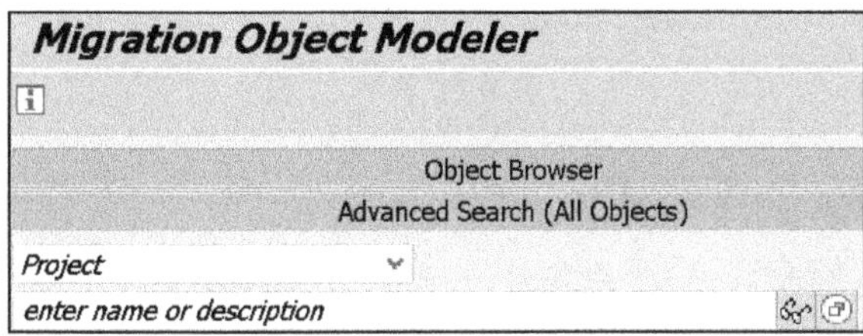


**Figure 3.72** Selection Button of the SAP S/4HANA Migration Object Modeler

The search also includes all user-created migration objects (**Search for User-Defined Migration Objects**) and migration objects that were created using a migration template (**Search for Migration Objects Created from Template**).

### 3.4.4 The Migration APIs Used

Standard APIs (both BAPIs and function modules) released by SAP for SAP S/4HANA can be used to transfer data via the migration cockpit.

**Function module documentation**

Some of these APIs, in particular BAPIs, include a what is called *function module documentation*. You can call this documentation via Transaction SE37 (*Function Builder*). The documentation often provides useful information on import structures and their fields.

**Standard interfaces**

In addition to using standard APIs, the system also supports the use of CDS behavior definitions since SAP S/4HANA 1909.

**Requirements on migration APIs**

To enable the use of a function module as a migration API, it must be compatible with the requirements of the migration cockpit. The requirements for a migration API are stored in the "SAP S/4HANA Migration Cockpit - Creating Your Own Function Modules in LTMOM" guide, linked in SAP Note 2590165.

## 3.5 Modeling for Data Migration with Staging Tables

In this section, we'll now describe how you can use the migration object modeler for these two approaches.

[+]

**More Information on the Staging Approach**

The following guides provide detailed information about using the migration object modeler with the staging table approach:

- SAP S/4HANA Migration Cockpit – Deep Dive LTMOM for File/Staging – OP2021 and higher: *http://s-prs.co/v581758*
- SAP S/4HANA Migration Cockpit – Deep Dive LTMOM for File/Staging – SAP S/4HANA 2020: *http://s-prs.co/v581759*

### 3.5.1 Displaying a Migration Object

After you have selected a project or a migration object, you can now view and edit it in the object browser.

Figure 3.73 shows the object browser for a selected migration object in display mode.

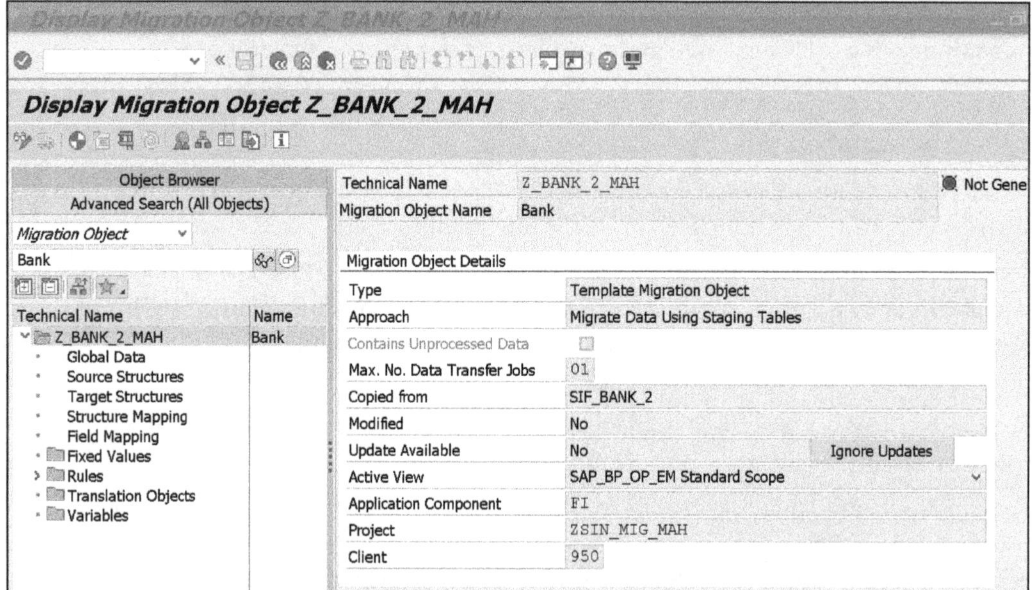


**Figure 3.73** Object Browser with Migration Object Selected

**Client Dependency**

Migration projects created in SAP S/4HANA 2020 or higher that use the **Migrate Data Using Staging Tables** migration approach can only be displayed and changed in the migration object modeler in the relevant client. This limit is not the case for older versions of the migration cockpit (Transaction LTMC).

Clicking he icon lets you toggle between display and change mode. Additionally, the **Migration Object** menu item offer more functions that you can partially also call via the application bar. Let's briefly describe these functions next.

**Navigation tree**

First, look at the navigation tree on the left. The navigation tree has a root node that represents either a migration object or a migration project. For example, if you select a migration object in the tree, you can then click the icon (**Go to higher level**) to navigate one level up to the migration project.

The navigation tree is structured in the following way:

- The top node is the migration project. Directly below this node, you can see its **Subprojects**. A 1:1 relationship exists between a project and a subproject; that is, exactly one subproject is assigned to one project.
- Below the subproject, you can find the migration objects and transfer rules of the migration subproject.

  Transfer rules can be created at the subproject or migration object level, and they are then available at the underlying levels. The transfer rules of the migration objects delivered by SAP have mainly been created at the subproject level and are therefore available for all underlying migration objects. Rules that are delivered at the project level are displayed for information purposes as **Read-Only**. These rules cannot be changed.
- A migration object has the following navigation nodes:
  - **Global Data**: In this node, you'll define global data types and variables.
  - **Source Structures**: In this node, you'll define the source structures.
  - **Target Structures**: In this node, you'll find the target structures of your migration API and define other important settings.
  - **Structure Mapping**: In this node, you'll link the source structures with the target structures.
  - **Field Mapping**: In this node, you'll assign the source fields to the target fields and link them with a transfer rule if required. You can also assign event-based rules to relevant processing times.
  - At the bottom, the system displays the rules of this level.

Figure 3.74 shows parts of the navigation tree for our sample projects.

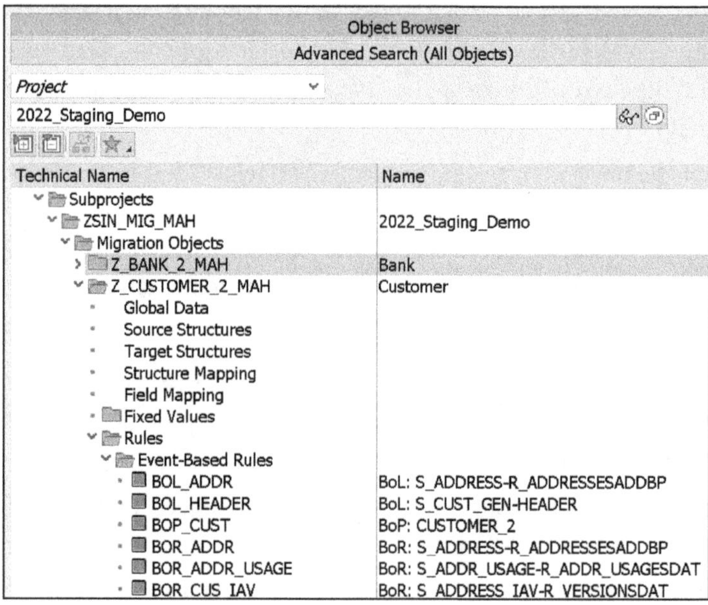


**Figure 3.74** Navigation Tree in the Object Browser

### 3.5.2 Transfer Rules

**Types of transfer rules**

The object browser distinguishes between four main types of transfer rules:

- **Fixed values**
  A fixed value is a constant value, for example, a fixed controlling area, to which you can assign multiple fields or rules. Changes to the fixed value automatically apply to all assignments. To assign a fixed value to a target field, you first create a rule, assign the fixed value to this rule, and then assign this rule to the target field. You cannot assign a fixed value directly to a target field.
- **Rules**
  - *Event-based rules*: Event-based rules comprise ABAP coding that is executed at predefined events, for example, at the start of a loop, the end of a record, after a loop, etc. By means of event-based rules, you can execute user-defined ABAP coding at specific events.
  - *Field-based rules*: Field-based rules can be based on translation objects, fixed values, or free ABAP coding and can be assigned to fields of a target structure. They can thus determine the target value either via translation objects or fixed values as well as via complex algorithms.
  - *Internal rules*: In contrast to event-based or field-based rules, internal rules are not assigned to events or fields, but they are called by these rules. Consider them something like locally defined, reusable subroutines.

- **Translation objects**
  Translation objects define how values from a specific field in the source system must be translated to the target values in the target system during data migration. Translation objects cannot be assigned directly to a target field. You must first assign the translation object to a field-based rule and then assign the rule to the corresponding target field.
- **Variables**
  In this case, you can define global variables that you populate, for example, in an event rule at the beginning and then read or change in other rules.

### 3.5.3 Generating and Displaying the Runtime Object

By clicking the (**Generate Runtime Objects**) icon, through this technical function, you can generate anew the local function group that is used by the migration cockpit. This newly generated function group is required after the source structures or fields have been mapped because the fields will not be processed otherwise. When you generate the runtime object, the system automatically synchronizes the structures of the migration object.

Display Generated Function Group

You can click the icon (**Display Generated Function Group**) to view the entire function group in the application bar. The ABAP Editor then displays the function group, as shown in Figure 3.75.

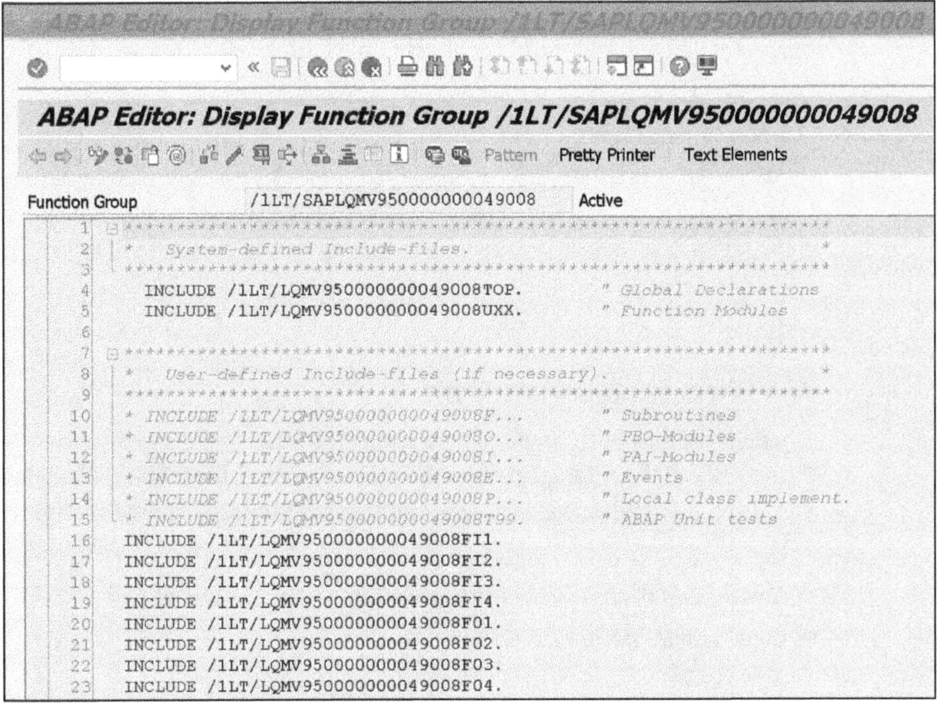


**Figure 3.75** Display Generated Function Group

### 3.5.4 Updating the Migration Object

**Update of a migration object**

After you have adapted a delivered migration object, sometimes, the system displays the message shown in Figure 3.76 when you open this object in the migration cockpit.

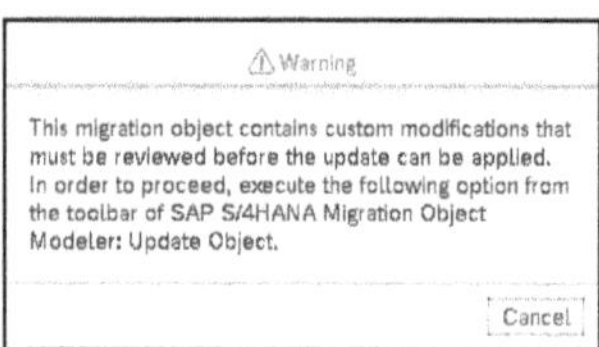


**Figure 3.76** Warning Message in the SAP S/4HANA Migration Cockpit

In this case, you must open the object in the migration object modeler. In the migration object's header data, the **Modified** and **Update Available** fields both have the value **Yes**, as shown in Figure 3.77.

**Figure 3.77** Available Update

A confirmation window is displayed if you now click the icon to run the update in change mode.

If you confirm this message, all customer-specific modifications to this object are deleted, and the object is set to the current state delivered by SAP.

[»]

**Ignoring Updates for This Object**

You can ignore the update for a migration object. As a result, no update takes place even if SAP modifies an object. To activate the update for the migration object again, you must delete the existing object and then create it anew from a template.

### 3.5.5 Creating a Migration Object

You have various options when creating a migration object in your project. When you call the migration object modeler, the **Migration Object Modeler** menu item provides the following editing functions, as shown in Figure 3.78:

- **Create Migration Object** either **From Template** or **User-Defined**
- **Copying a Migration Object**

For both options, you can select for which migration approach you want to perform this action: either for **Migrate Data Using Files/Staging Tables** or for **Migrate Data Directly from SAP System**.

You can also call these functions if you have already selected a project or migration object. In this case, the topmost menu entry changes to **Project** or **Migration Object**, and you can no longer choose between the various migration approaches because a migration approach always applies to the entire project.

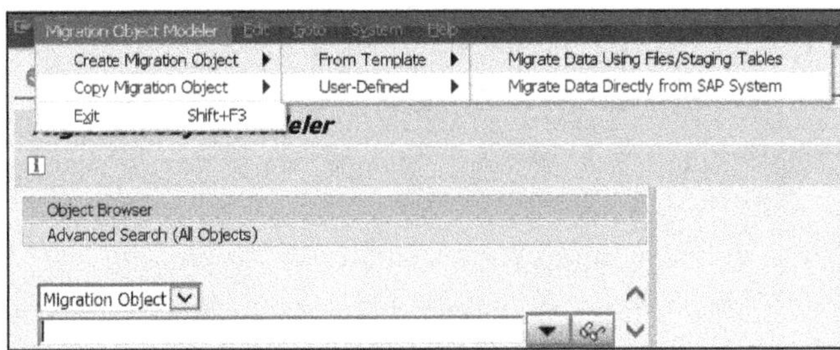


**Figure 3.78** "Migration Object Modeler" Menu Entry

For more information on creating and copying your own migration object for the staging approach, see the following user guides and click-through demos:

- Creating a New Migration Object (SAP S/4HANA 2020 or higher): *http://s-prs.co/v581760*
- SAP S/4HANA Migration Object Modeler: Creating a New Migration Object (SAP S/4HANA 1809 or higher): *http://s-prs.co/v581761*
- SAP S/4HANA Migration Object Modeler: Working on a Migration Object (SAP S/4HANA 2020): *http://s-prs.co/v581762*

  A one-time registration is required at *www.sap.com/account.html*.

[«]

**No Reference to a Template Object When Copying a Migration Object**

In contrast to creating a migration object from a delivered template, copying a migration object does not have any reference to the template object. The migration object is considered as a separate customer-specific migration object. You can adapt all fields (e.g., change the field length). All fields of the source structure are marked as **User-defined Field**. There is no update if the SAP system changes the migration object through an upgrade or correction.

You can copy migration objects both from projects that were created in another client and from older projects that were still created in Web Dynpro using Transaction LTMC.

[»]

**Copy Includes when Copying Migration Objects**

If you want to decouple the copied migration object completely from updates and corrections, you should also copy the existing and assigned includes in the **Target Structures** area to your customer namespace and reassign them.

### 3.5.6 Extending Existing Migration Objects

To learn how to extend existing migration objects, we recommend the click-through demos available at the following links:

- SAP S/4HANA Migration Object Modeler: Working on a Migration Object (SAP S/4HANA 2020): *http://s-prs.co/v581763*

  A one-time registration is required at *www.sap.com/account.html*.
- SAP S/4HANA Migration Object Modeler: Adding Additional Fields to a Migration Object (SAP S/4HANA 2020): *http://s-prs.co/v581764*

  A one-time registration is required at *www.sap.com/account.html*.

[»]

**Restrictions for Modified Migration Objects**

Before you adapt a delivered migration object, you must note that, if errors arise in adapted or self-developed migration objects, SAP only provides support for content-related problems that are based on the original template. SAP Note 2481235 provides more details.

### 3.5.7 Creating Custom Rules and Translation Objects

You can create your own rules and translation objects if the delivered rules do not fully cover your application case or if you require custom rules. Translation objects can never be mapped directly to a field but must be assigned to a field-based rule in advance. If you create a rule or translation object at the migration object level, you can only use it for the specific migration object. You should create the rule or translation object at the subproject level so that it can be used by all the migration objects in your project.

### 3.5.8 Simulating or Debugging Migration Objects

If you want to test how source fields are transferred to an SAP S/4HANA target system, you can simulate the import of data from staging tables. For this task, in the migration object modeler, select a migration object and switch to change mode. Click the icon (**Simulation**) to simulate the import of data from staging tables. This action takes you to the view shown in Figure 3.79.

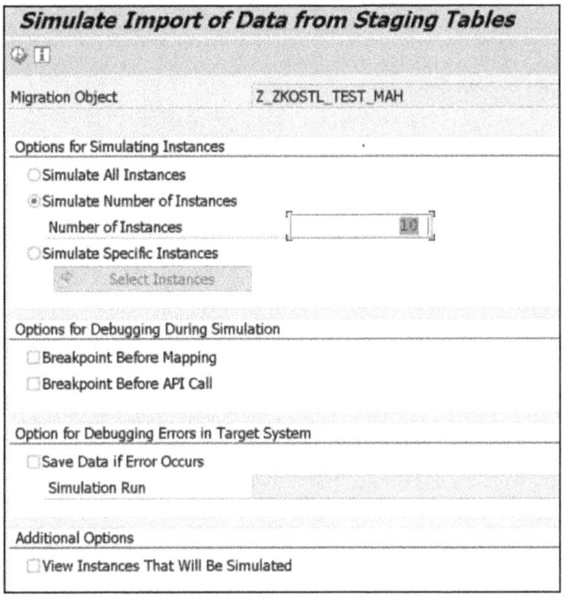

**Figure 3.79** Report for Simulating the Data Import

**Prerequisites for simulating the data import**

The following prerequisites must be met so that you can simulate a migration object:

- The migration object was generated successfully and is not locked by another user.
- A simulation parameter was defined in the target structure.
- There is no data in the staging tables.
- If the migration object was copied from a template, no update must exist.

You can call the documentation by clicking the icon (**Information**) in the simulation's selection screen.

You have the following options for selecting data for simulation:

- **Simulate All Instances**
- **Simulate Number of Instances**: The records are selected randomly.
- **Simulate Specific Instances**: Note that selection is not case sensitive.

For all records, a general rule applies that the system only simulates records with the processing statuses **Not processed** or **Error**. In addition, you have the option to automatically set breakpoints at specific moments, either prior to conversion of data (**Breakpoint Before Mapping**) or prior to posting data into the target system (**Breakpoint Before API Call**). Since SAP S/4HANA 1909, you can also display data that is simulated by selecting **View Records Before Mapping** or **View Records After Mapping** or both.

[»]

**Report for Testing and Debugging**

If you work with an older SAP S/4HANA release, you can use report DMC_MC_SIMULATE_IMPORT_STAGING to perform a simulation run for a staging migration object. Call the report using the ABAP Editor (Transaction SE38) or ABAP program execution (Transaction SA38). Then, select your migration object. You can also use the report for debugging your migration object. It is critical that you maintain all conversions (mappings) in the migration cockpit in advance.

The report is included in SAP Note 2665339.

If you want to simulate the import of files in an older release, you can use the report from SAP Note 2630183.

[+]

**Resetting the Transfer Status**

If you want to re-migrate records that have already been migrated for testing purposes, navigate to **Goto • Reset Transfer Status** in the migration object's selection menu to display all the records transferred for the migration object and reset the transfer status for these records.

### 3.5.9 Transferring Migration Projects from One System to Another

Transporting content

With SAP S/4HANA 2021, the standard SAP transport function is used to transport projects. For this task, you must have assigned a development package to your project when you created it (see "Step 2: Assigning a Development Package in SAP S/4HANA" in Section 3.2.5). To add a project to a transport request, you should first select the project to be transported and then click the button (**Add to transport request**). All associated migration objects, fixed values, rules, translation objects, and variables are automatically included in the transport request. Excluded from the transport are instances that have already been uploaded to the staging tables. Also, only the definition of a rule, fixed value, translation object, or variable is added to the transport request and transported. Values, such as mapping values, cannot be transported. Once you have imported your project into a new system, you must store a new database connection the first time you open the project.

For more information on transporting content, click the [i] icon (**Information**) and refer to the slide set "SAP S/4HANA Migration Cockpit - Migrate Data Using Staging Tables - Transport Concept" linked in SAP Note 3104790.

[+]

**Production Systems/Systems That Cannot Be Changed**

It is not possible to create or change projects in systems that cannot be changed (e.g., production systems). If you need to make such adaptations, you can adapt the project in a development system and transfer it again to the target system.

Using a project in another client

Additionally, you have the option to use a migration project in another client, as shown in Figure 3.80, by selecting **Goto • Use Project in Different Client** in the project view. Note that the project is not copied in this process but instead is only available in the newly specified client after this step. Because all mapping values are deleted, you should download them in advance so that you can upload them again to the new client. Note that you may have to revise the mapping values prior to the upload to ensure that no value pairs that originate from an internal numbering are uploaded.

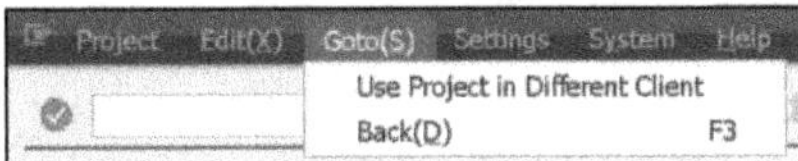


**Figure 3.80** Using a Project in a Different Client

[«]

**Using a Project in a Different Client**

The **Use Project in Different Client** function is relevant only for the migration cockpit for SAP S/4HANA 2020 or higher and only for the migration approach **Migrate Data Using Staging Tables** and not for older versions of the migration cockpit (Transaction LTMC).

## 3.6 Modeling for Data Migration Using Staging Tables in SAP S/4HANA Cloud Public Edition

Since SAP S/4HANA 2111, the migration object modeler in SAP S/4HANA Cloud Public Edition has been made available to all customers. You can call the modeler via the Model Migration Objects - Migration Object Modeler app. The migration object modeler in SAP S/4HANA Cloud Public Edition is limited in scope as currently only certain migration objects released by SAP can be extended with structures and fields. For more information, see SAP Note 2999428 (see *http://s-prs.co/v581765*) and in KBA 3216716 (see *http://s-prs.co/v581766*).

[»]

**Required User Role**

To use the migration object modeler in SAP S/4HANA Cloud Public Edition, your user must be assigned role SAP_BR_ADMINISTRATOR.

## 3.7 Modeling for Directly Transferring Data from SAP Systems

This section details the specifics of the modeling environment for directly transferring data from SAP systems.

The benefit of this approach is that it already provides some predefined selection logic for each migration object. It also contains a logic for deriving organizational units.

[»]

**Required User Roles**

Assign the roles described in Section 3.3.4 and additionally role SAP_CA_DMC_MC_DEVELOPER to enable the use of Transaction LTMOM. If you want to develop source code rules, additional authorizations are required; for example, see *http://s-prs.co/v581767*.

[»]

**Consider the Documentation!**

After you have called a project using Transaction LTMOM, the top bar shows an info button. This button provides the complete documentation for the modeling environment of the **Migrate Data Directly from SAP System** approach. If you're already in a step within the wizard or within the modeler, an additional info button is available that provides more information on the respective step.

[+]

**Detailed Information About the Modeler**

In the guide entitled "SAP S/4HANA Migration Cockpit - Deep Dive LTMOM for Direct Transfer," you can find highly detailed information about the migration object modeler: *http://s-prs.co/v581768*.

### 3.7.1 Basic Knowledge for Modeling

This section describes the basic terms for modeling. We'll also explain how the migration cockpit's *direct transfer* approach derives organizational units for selection.

### Organizational Units

To structure an SAP client, so-called *organizational units*, are defined in all SAP applications. The organizational units are assigned to one another so that the various SAP applications can process the data in an integrated way. In financial accounting, for example, the company code is an important unit; in controlling, the controlling area is an important unit, and in logistics, the plant is an important unit. Plants are assigned to company codes, company codes are assigned to controlling areas, etc.

When you migrate data, you must first decide from which organizational units you want to migrate data. Depending on the scenario, the appropriate organizational unit is predefined as an input parameter in the migration cockpit. For the "SAP ERP to SAP S/4HANA" scenario. this organizational unit is the company code. For the company codes selected, you'll then specify the migration objects for which data is to be migrated.

### Organization Derivation

**Integrated selection engine**

Integrated with the migration cockpit, a *selection engine* ensures that all data records are read that belong to the selected migration objects of this organizational unit. This applies even if the relevant organizational unit is not contained in the corresponding application table or if the relevant records must first be determined via child tables. So, the migration cockpit is provided with implicit complex selection logics.

### Filters

**View of the filter values via Transaction LTMOM**

In the modeling environment, the derivation is primarily controlled using so-called *filters*. A filter is a collection of values of the same data type that is stored centrally for the project. These values are available for all migration objects of a project. A filter can use, for example, controlling area, company code, plant, sales organization, purchasing organization, etc. Filters are defined by SAP, who also specify which filters are available for which scenarios, for example, for the "SAP ERP to SAP S/4HANA" scenario. With SAP S/4HANA 2022, you can display the derived values, which are called *filter values*, in Transaction LTMOM.

[«]

> **More Information about the Selection Process**
>
> The "Direct Transfer - Selection Guide" document stored in KBA 3249988 explains the selection process for the direct transfer of data based on numerous examples and detailed technical information (see *http://s-prs.co/v581769*).

**Selection options via BAdI**

Since SAP S/4HANA 2021, a BAdI is available for data selection. Using the BAdI, you can influence the data selection process by additional logic in the source system. If the filter logic provided in the SAP S/4HANA migration object modeler (Transaction LTMOM) or the simple selection criteria are not sufficient to meet your data selection requirements, you can implement ABAP code in the source system in the BAdI for the selection. The BAdI for the selection can be activated per project and per SAP ERP data table of the source system in the migration object modeler. For more information, see KBA 309292, available at *http://s-prs.co/v581770*.

[»]

**Customers are Responsible for Changes to the Standard Selection**

Once you change the standard selection that was predefined by SAP, it is your responsibility to ensure consistency of the selected data.

### 3.7.2 Creating a Migration Object (Copy)

As mentioned in Section 3.5.5, you have several different options when creating a migration object.

#### Creating a Migration Object from a Delivered Template

**Copying migration objects**

You have the option of copying the migration objects that are delivered by SAP. Call Transaction LTMOM and select your direct transfer project. Switch to change mode and follow the menu path **Migration Object Modeler • Create Migration Object • From Template • Migrate Data Directly from SAP System**. The system provides all migration objects which are available for the scenario that you assigned to your project, for example, "SAP ERP to SAP S/4HANA."

#### Creating Custom Migration Objects

For information on creating your own migration object for the direct transfer of data, see "SAP S/4HANA migration object modeler: Creating a New Migration Object," a video available at *http://s-prs.co/v581771*.

#### Copying a Migration Object

You can also copy migration objects that you have already created in your project. Note that you can copy objects only within an individual project. This limitation has technical reasons due to the structures generated. For this task, follow the menu path **Migration Object Modeler • Copy Migration Object • Migrate Data Directly from SAP System**.

### 3.7.3 Synchronizing Structures

Changed structures or objects

3

Migration objects refer to ABAP Dictionary objects, for example, tables and function modules. These ABAP Dictionary objects may change from time to time. If, for example, a new version or correction is implemented, this can lead to additional import fields for a function module. Table fields in the source system can be added or deleted. You can synchronize the structures in such cases. If a structure in a migration object does not correspond to the structure in the ABAP Dictionary, the system updates the structure in the migration object. Call Transaction LTMOM and double-click the migration object. Then select **Migration Object • Synchronize Structures**.

### 3.7.4 Creating Custom Rules

The central system documentation (info button after you have called Transaction LTMOM) provides comprehensive descriptions and examples on the topic of rules. You can also find useful information in the video "SAP S/4HANA Migration Object Modeler: Creating an Own Mapping Rule" (see *http://s-prs.co/v581772*). This video shows how to create your own mapping rule in SAP S/4HANA 2020 using the migration object modeler.

[«]

**Custom Rules in SAP S/4HANA 1909**

Creating custom source code rules is possible only with SAP S/4HANA 2020. In SAP S/4HANA 1909, you can already create custom value mapping rules. However, they are not yet provided with default values for the target values. This means for these self-created value mapping rules, you first download the Microsoft Excel template in Transaction LTMOM and populate it with source and target values. You then upload it using Transaction LTMOM. Custom MOVE rules and custom fixed value rules can also be created in SAP S/4HANA 1909.

### 3.7.5 Migrating Table Field Extensions (Z Fields)

Automatic detection of Z fields

The migration cockpit recognizes if the source system contains field extensions (custom fields or Z fields) for the SAP standard tables (only in the direct transfer approach). Call Transaction LTMOM and navigate to the **Source Tables** area within your migration object. The migration cockpit displays custom fields as source fields. Whether and how the custom field can be migrated to the target system depends on the structure of the BAPI that is used in the migration object.

There are three different concepts for function modules:

- Function modules that contain includes or append fields
- BAPI function modules with type 1 extension fields
- BAPI function modules with extension fields of type 2 with or without X fields

In all cases, the includes and append structures must be maintained by the customer (Transaction SE11). The user requires in-depth DDIC knowledge and an ABAP development authorization in the system.

### 3.7.6 Simulating and Debugging Migration Objects

Debugging only in change mode

You can simulate and debug individual or selected migration object instances also in the direct transfer approach. For this task, select a migration object in Transaction LTMOM and switch to change mode. The selection screen of Figure 3.81 opens if you select the button (**Simulation**).

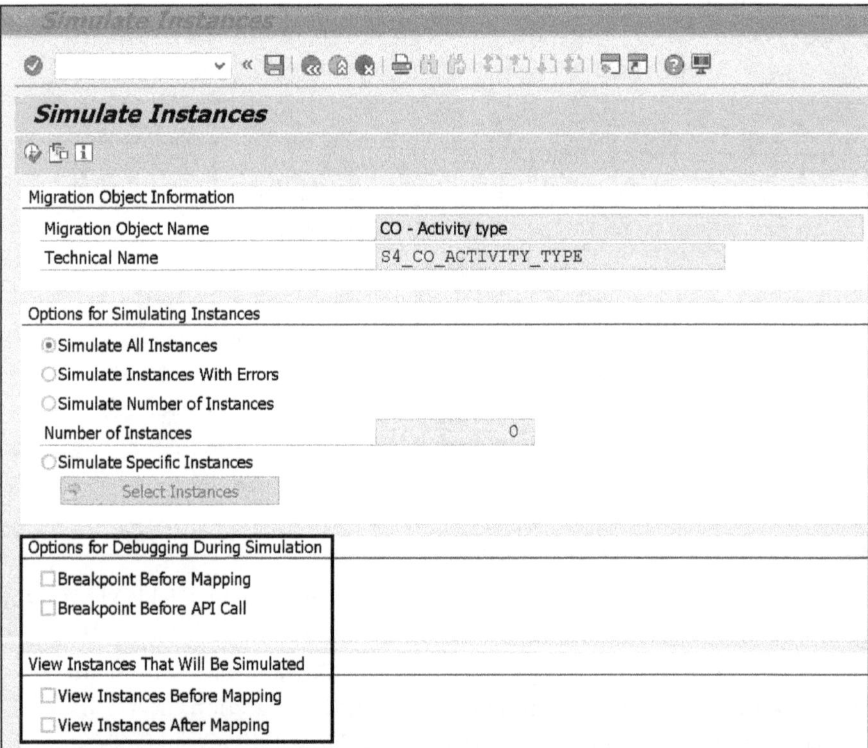


**Figure 3.81** Simulating and Debugging Instances

There are two points in time when you can set a breakpoint:

- **Breakpoint Before Mapping**: The program stops before the migration cockpit applies the mapping rules to the migration object instance. This way, you can track in which steps the system transforms individual values of the instance.
- **Breakpoint Before API Call**: The program stops before the migration cockpit transfers the transformed migration object instance and the transformed values to the API for simulation.

In the lower part of the screen, you can find two options that control whether the original migration object instance is displayed as it was imported from the source system or whether the converted instance is to be displayed.

### 3.7.7 Transporting Projects

**Import/export only via transport**

As described in Section 3.3.5, SAP transport connection is available as of SAP S/4HANA 2020 to transfer projects to other systems. When creating a project in the migration cockpit (SAP Fiori), you must already assign a transport request to this project.

In Transaction LTMOM, you can add projects, migration objects, and rules to a transport request. For this task, place the cursor on the corresponding element and click the button (**Add to Transport Request**).

[«]

**Transporting Rules**

Only the definition of a rule is added to transport requests. They do not contain any values, for example, mapping values. You can add them to the system via download/upload or maintain them directly in the migration cockpit.

[«]

**Transports in SAP S/4HANA 1909**

The transport connection is not available yet in Version SAP S/4HANA 1909. Projects must be transferred manually.

## 3.8 Additional Useful Sources of Information

You now know how to operate the migration cockpit and the migration object modeler for the various migration approaches. We have provided many useful examples and hints. Below you can still find important sources where you can find even more detailed information on these topics.

### 3.8.1 Landing Pages of Data Migration

In general, it is useful to deal with the options available and the sources of information prior to the data migration. Besides this book, the landing pages for data migration for SAP S/4HANA Cloud and SAP S/4HANA provide the best entry point. You can find them on the SAP Help Portal at the following web links:

- SAP S/4HANA Cloud: *https://help.sap.com/S4_CE_DM*
- SAP S/4HANA: *https://help.sap.com/S4_OP_DM*

### 3.8.2 SAP Communities

**SAP Community** You can use the *SAP Community* which can be accessed at *https://community.sap.com/*.

In general, and particularly for the SAP S/4HANA on-premise version, you should check out the migration cockpit page of the SAP Community at *http://s-prs.co/v581773*.

Here you can find a wide variety of information on all aspects of data migration in SAP S/4HANA.

### 3.8.3 Performance Aspects

Below we have listed what we consider to be the most important KBAs:

- 3034738, "Migration Cockpit (new): Parallelization of XML file loads - SAP ONE Support Launchpad"
- 3065607, "Performance tips & tricks for SAP S/4HANA Migration Cockpit: Migrate Data Using Staging Tables - SAP ONE Support Launchpad"
- 3066336, "Modifying Data Transfer Jobs to improve the data transfer performance of the SAP S/4HANA Migration Cockpit - Migrate Data Using Staging Tables"
- 3056885, "Performance tips & tricks, SAP S/4HANA Migration Cockpit: Transfer data directly from SAP system - SAP ONE Support Launchpad"

[»]

**Search for Current SAP Notes**

Always search for current notes on the performance topic. Besides corrections, there can also be KBAs or current consulting notes on this topic. Use the keywords "SAP S/4HANA Migration Cockpit" and "performance" for your searches.

### 3.8.4 Aligning the Time of Migration with Maintenance Periods

In SAP S/4HANA Cloud, you should not schedule a data transfer within a maintenance period, i.e., during the blue-green deployment phase. Information on blue-green deployment can be found at the following links:

- Blog by Akshay Sharma: "S/4HANA Cloud Systems Behavior During Blue-Green Deployment Uptime": *http://s-prs.co/v581774*
- SAP Note 3024158: *http://s-prs.co/v581775*
  - 3SL Maintenance Schedule: *http://s-prs.co/v581776*
  - 2SL Maintenance Schedule: *http://s-prs.co/v581777*

## 3.9 Summary

With this chapter, we showed you the SAP S/4HANA migration cockpit and its different approaches that are crucial for your data migration to SAP S/4HANA. You should now be able to migrate data into SAP S/4HANA. In addition, we covered the modeling environment for the SAP S/4HANA migration cockpit, which you can use to adapt and create new migration objects. We described the options available for adapting the delivered migration objects of the migration cockpit to your company-specific requirements. We first provided a brief overview of the migration cockpit structure, and then looked at the functionality of the migration object modeler when you perform the data migration in the migration cockpit using files or staging tables. Finally, we concluded with the functionality of the migration object modeler with regard to data migration directly from an SAP system.

# Chapter 4
# Data Migration to SAP SuccessFactors

*Every SAP SuccessFactors implementation requires the initial provision of data. An early understanding of these specific complexities and challenges will support your journey towards a successful data migration.*



In this chapter, we discuss in detail the strategies and considerations associated with the data migration of SAP SuccessFactors Employee Central (Employee Central for short). We present solution approaches for the migration of historical data specifically for the implementation of Employee Central in the initial phase.

When implementing this solution, there is always the requirement to initially load existing basic/organizational as well as employee data from one or more legacy systems into the Employee Central environment. Since Employee Central is a global enterprise application, you very likely have several different source systems that are relevant for this target environment. This diversity requires the definition of an appropriate strategy or, in some circumstances, different strategies that must be included for a successful migration to Employee Central.

The success of a data migration also depends on appropriate project decisions: Their drivers are, for example, the definition of the scope of a data history and also what budget is available for data migration. We discuss not only the functional and technical issues in the context of data migration but also address the potential process challenges.

Section 4.1 addresses the basic preliminary considerations associated with Employee Central data migrations. To better frame the topic of data migration for Employee Central, we turn to the foundational topics relevant to Employee Central in Section 4.2.

The two other sections describe standard tools that can be used when migrating SAP HR data from an SAP ERP Human Capital Management (SAP ERP HCM) system or from cost center information from FI/CO to SAP ERP. In Section 4.5, we then turn to the standard procedure for SAP SuccessFactors—the file import procedure. The final section provides some tips to support a successful data migration project in the Employee Central environment.

## 4.1 Basic Preliminary Considerations for Employee Central Data Migration

Chapter 2 addressed the various aspects of a data migration project. In this chapter, we highlight specific considerations for Employee Central.

In your Employee Central project, you should start preparing for data migration early on. Before you tackle the technical issues, it is important to review the business issues well and resolve any potential challenges.

**Aspects and areas of tension**

The following, concisely formulated points are a selection of aspects and also areas of tension that influence the approach to SAP SuccessFactors data migration:

- Local systems versus global system
- Project rollout approach—phased rollout versus big bang
- Data migration follows data model
- Data history—reasonable scope
- Tension between data migration and provision of data in integration

The aspects included in the list are detailed below.

### 4.1.1 Local Systems versus Global System

One point to keep in mind first for your consideration is that implementing Employee Central means that data from various local systems must be prepared for a global system so that it can be used to initially populate the Employee Central system.

**Drivers of a data migration**

As shown in Figure 4.1, several different local systems are affected by the migration in this constellation. Since each system offers different data structures and different data export options, the approach to data migration is driven by the need to find solutions for the different scenarios.

You can meet this challenge, for example, with the rollout approach described in the following section.

The requirement of having to migrate many different local systems makes it difficult to define or use a single toolset and approach for preparing and transferring data for data migration. A useful approach is to delegate the provision of data to local units. Appropriate information is provided to local units in the form of documents and workshops. At the global level, only quality validation and the actual import are then performed.

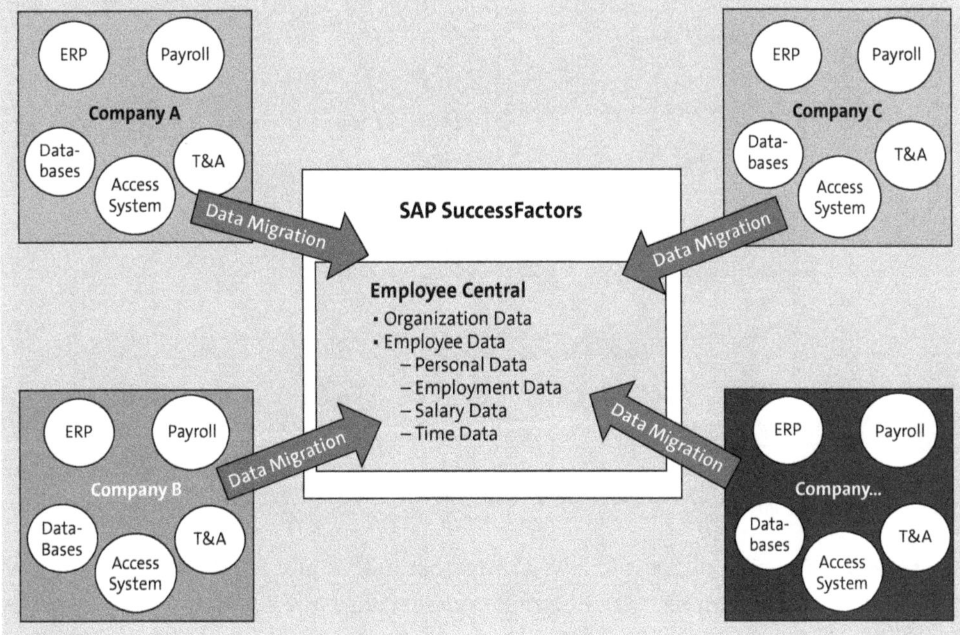


**Figure 4.1** Local Systems versus Global System

### 4.1.2 Project Rollout Approach: Phased Rollout versus Big Bang Approach

**Phased rollouts**

The reason for a phased rollout of an Employee Central project is usually that a large number of global enterprise companies are supposed to be merged. Project challenges that can be more easily mapped using this approach include varying availability in different time zones and the availability of project resources.

In the phased rollout you define individual phases of the rollout. In the first phase, usually a pilot phase, you define the so-called *global template*. The global template defines the core, which is equally valid for all corporate companies. This core essentially comprises the organizational structure, the field definitions and the processes to be supported globally. For the first phase you must also define a number of companies where a corresponding localization of the data model will be performed.

**Emerging process split**

The depth of localization depends on the definition of the process split created in this constellation. *Process split* means that with the introduction of Employee Central, parts of the process are mapped in Employee Central because Employee Central is the leading system here, and then further process steps occur in the local system. Figure 4.2 illustrates this approach.

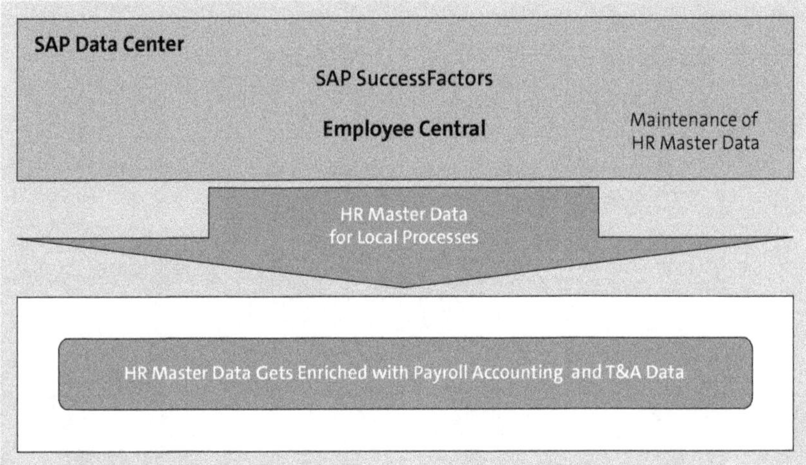


**Figure 4.2** Overview of a Process Split

**Rollout phases**

The pilot phase is followed by further rollout phases. The number of these phases can be quite different and depends strongly on the size and especially on the complexity, for example, the number of companies in different countries and the number of employees per company. Appropriate data migration activities must be performed for each phase.

As addressed earlier, the phased rollout divides the employee population by relevant groupings, and based on those groupings, Employee Central is rolled out in individual phases.

The phased rollout has a shorter data migration period with respect to each phase and less impact on the overall employee population at go-live. However, the higher number of phases, usually defined by the number of countries, may increase the overall project timeframe.

This implementation approach reduces the resource burden within the individual project areas and the core team in the long term. Figure 4.3 shows how the individual phases are interconnected.

**Big bang approach**

In the big bang approach, all employees in the individual companies are included in the project scope from the beginning, and their requirements are analyzed and implemented.

At this point, the big bang, as shown in Figure 4.4, corresponds to the first phase of the phased approach as far as the global template is concerned and includes the localizations for all companies. The employee population may span multiple countries worldwide, and all are enrolled in the Employee Central system at the same time.

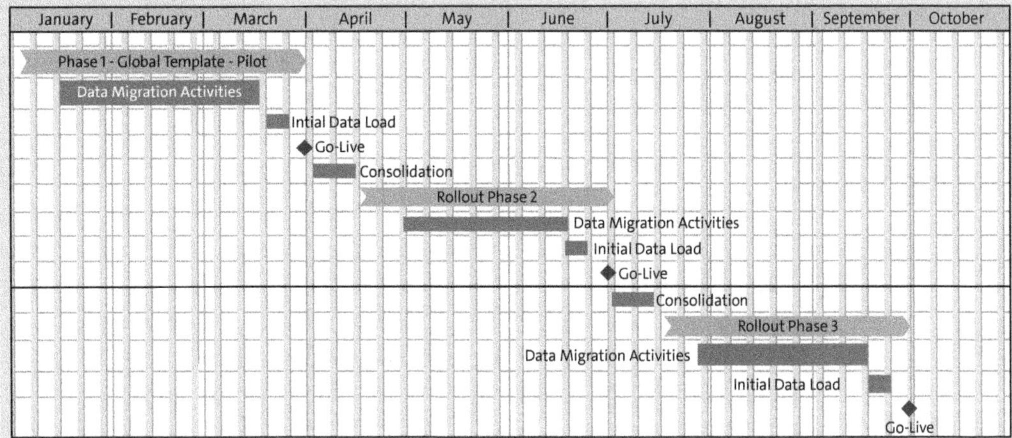


**Figure 4.3** Generic Process of a Phased Rollout

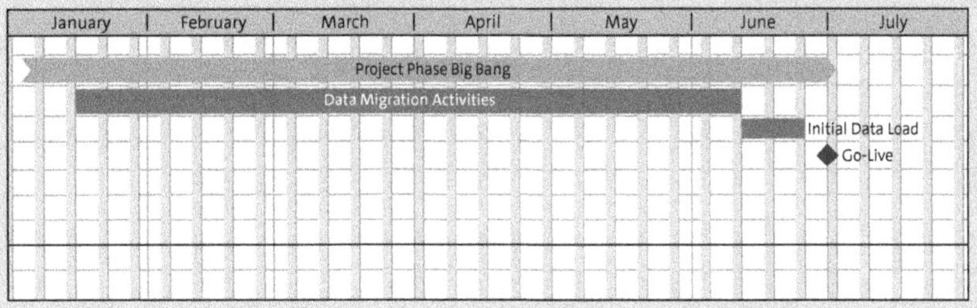


**Figure 4.4** Generic Flow of a Big Bang Approach

**Maximum resource requirement**

This approach has the advantage that the system can be put into operation for the entire workforce at the same time and supposedly shortens the project timeframe. However, the demands on the resources involved reaches capacity level for this period. More resources are needed for thorough testing and migration activities that are required for all legal entities at the same time.

### 4.1.3 Data Migration Follows the Data Model

**Employee Central data model**

The scope of and approach to data migration depends heavily on the definition of the Employee Central data model and the rollout approach. The dependency with regard to the rollout approaches was touched upon in Section 4.1.2.

As part of the implementation project, you need to define which objects and data entities are to be used. If you use a large number of entities or objects in Employee Central, the data migration effort increases as well. Each entity contains a number of global and/or local fields. The number and data type of these fields also affect the complexity of the migration.

Text fields are easier to handle than list fields or fields based on object information, such as department; the correct identification numbers or codes must be used here.

### 4.1.4 Data History: Reasonable Scope

During the initial data migration, there is often a requirement to transfer as much historical data as possible to the Employee Central system in order to have it completely available in the target system, for example, for reporting purposes, for better traceability, or because a legacy system is to be taken out of operation completely. Since Employee Central is a time-slice-based system that includes many objects and dependencies, it is necessary and useful to prepare—and if necessary simplify and harmonize—your historical data before importing it.

**Data for a cut-over point date**

We recommend importing the data on a cut-over date that is not far into the past. Historical data should only be loaded after a thorough check. We recommend you clean up historical data structures before importing so that the complexity is minimized according to your requirements. If you don't simplify your data, the import process can is be quite time consuming due to the many past dependencies that must be mapped.

**Migrating Employee Data History**

If you want to migrate employee data from several years in the past to the standard SAP SuccessFactors structures, you must consider the following points:

- The basic data valid at this historical point in time (e.g., cost centers, positions, organizational structures, and the like) must be available in the system.
- The employees who have performed management roles at this point and who may have already left the company must also be provided within the Employee Central minimum data scope.

However, this is often not applicable or necessary.

If a large amount of historical data must be loaded, we recommend you load the historical data into custom metadata framework (MDF) objects. This approach significantly simplifies the data model and its dependencies, which is sufficient for typical use cases (e.g., displaying important data, creating straightforward reports).

**Custom MDF objects**

You can then use these custom MDF objects for self-services and evaluate them in reports or pass them on to third-party systems. Figure 4.5 shows the eligible targets in SAP SuccessFactors.

[«]

**Historical Data for the Rehiring of Terminated Employees**

When migrating data to a new Employee Central environment, you must consider how the organization rehires a previously terminated employee. Depending on this strategy and the project specifications, you may lack a record in the new system for the terminated employee who is to be rehired. In this case, you must find a solution that considers whether or not a history exists for terminated employees.



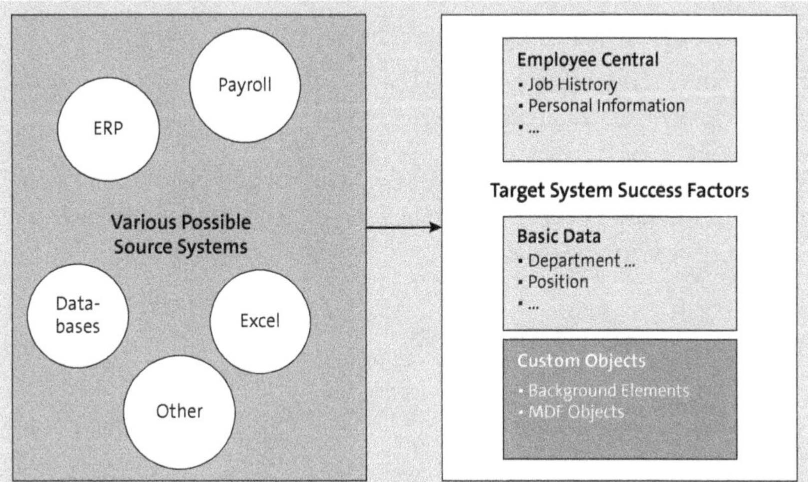


**Figure 4.5** Overview of Different Data Sources and Their Targets in SAP SuccessFactors

### 4.1.5 Tension between Data Migration and Provision of Data in Integration

Data usage: integration

One basic requirement that strongly influences the preparation for data migration is whether the data to be loaded into the Employee Central system will be maintained in a leading manner in Employee Central after go-live and whether the integration with the local system will be implemented for it.

Depending on the local system, the requirements can be rather restrictive. A prime example is integration with an existing SAP ERP HCM system. An imperative check is that the data transferred to Employee Central corresponds exactly to the data that exists at that time in the relevant SAP ERP HCM infotypes and time slices as of the specified cut-over date. If not the case, data changes will occur during the first integration run, which transfers the data as it exists in Employee Central. This description applies in the same way to both organizational and employee data.

## 4.2 Basic Principles of SAP SuccessFactors Employee Central

For further considerations regarding data migration, we think it useful to take a short digression into the data structures/objects of Employee Central as the target environment.

### 4.2.1 Employee Central Structures/Objects

Employee Central provides a number of options (entities/objects) for storing data.

**Main types for data storage**

At a high level, the following two main types of data are relevant:

- Basic and organizational data
- Entities for employee data

Figure 4.6 shows which Employee Central entities and objects can be relevant. This list varies depending on the requirements of your use cases and the process splits to be supported.

| Employee Central Employee Entities | Employee Central Organization Data |
|---|---|
| • Employment Information<br>• Job History<br>• Biographical Information<br>• Global Information<br>• Information on Email<br>• Information on Telephone Number<br>• Personal Information<br>• Job Relationships<br>• Addresses<br>• Passport Number Info<br>• International Assignment - Details<br>• Non-Recurring Salary Components<br>• Compensation Information<br>• Recurring Salary Components<br>• ... | • Legal Entity<br>• Business Area<br>• Division<br>• Department<br>• Position<br>• Location<br>• Cost Center<br>• ... |

**Figure 4.6** Relevant Employee Central Entities and Objects

Both main types are based on similar Employee Central data structures that support a similar technical approach to data migration. In the following sections, we describe these two main types in more detail.

#### Basic and Organizational Data

**Basic and organizational data**

Basic and organizational data refers to data that is stored in the SAP SuccessFactors system in order to be able to use it when maintaining employee data. Figure 4.6 shows a few of these objects listed in the **Employee Central Organization Data** section.

Most of this data is stored in MDF objects, which are *historicizing*, and thus, the data has a time reference and is stored in time slices with a start date and an end date. Basic and organizational data are partly based on each other. The position is a good example of this relationship because it is based on a large amount of other basic and organizational data.



**Position as an example**

Figure 4.7 shows a position in which various basic and organizational data is highlighted.

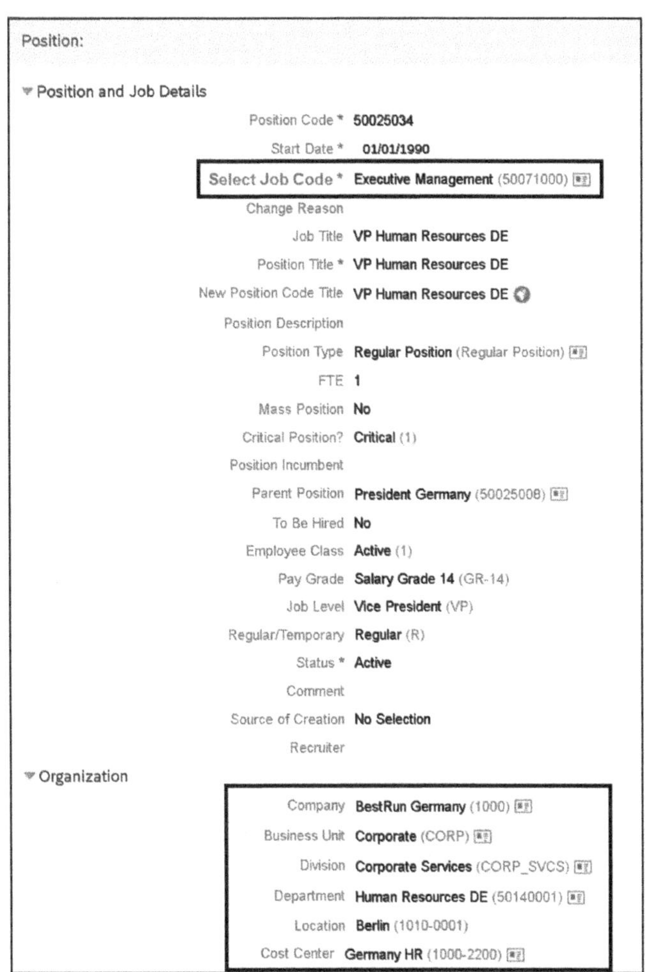


**Figure 4.7** Example of Using Other Basic and Organizational Data for a Position

The consecutive structure defines a sequence in which the data must be available in the target system during data migration. From structure, it is derived that the basic data must be loaded first, and this process in turn occurs in a specific sequence. For example, the data of the legal unit must be loaded before the locations, and they in turn must be loaded before the position data.

### Entities for Employee Data

**Entities for employee data**

The employee data entities are used to store, first, employee personal data and, second, their employment-related information. An exemplary list of Employee Central entities would include the following:

- Personal information
- Biographical information
- Global information
- Employment information
- Email information

**Figure 4.8** Position Data in the Employment Information

Time references

Some of these entities, such as the biographical information and the employment information, have no time reference. Other entities, such as personal information and job history, have a time reference. These are stored in time slices with a start and end date.



In the case of employee data, the basic and organizational data are used. Using the example shown in Figure 4.8, notice how the data from the position is used in the **Job Classification** entity.

Thus, now, we have a similar sequential structure that defines a sequence in which the data must be available in the target system during a data migration. The position must first be in the system so that it can be used when loading employee data and so that data can be transferred from it to the employee's data record.

### 4.2.2 Import Options in SAP SuccessFactors

Import options in SAP SuccessFactors

SAP SuccessFactors provides two ways to import the data:

- Using manual data imports via cascading CSV import files that reference each other.
- Using the OData application programming interface (API)

CSV file import

The method of importing data using CSV files is the standard approach for Employee Central. Whether you're performing a one-time migration, loading a number of new employees into the system, or updating various foundational objects: All these activities can be covered using the CSV file import function. To perform data imports, the system provides specific CSV import templates. The use of import templates is discussed in more detail later in this section, and the topic is revisited in Section 4.5.2.

Web Service OData API

Compared to CSV data imports, the Web Service OData API provides a more flexible and powerful way to perform data migrations. However, you require a higher level of technical knowledge, different skills, and additional technology and therefore also require more effort. An example of an additional technology is called *middleware*, on the basis of which the programming is created and then executed.

In Section 4.3 we address the use of a standard SAP tool called *Infoporter*. This tool can transfer the data to the Employee Central system via a web service.

[+]

**Cost–Benefit Analysis**

As part of determining the data migration strategy and the tools used in the process, you should perform a cost–benefit analysis and check whether you want to capture the data manually, import it via CSV file, or/and implement an API-based integration.

Let's dive a bit more deeply into CSV imports or CSV import templates. Import templates, whether they are used for basic and organizational data or for employee data, always have two headers and multiple columns. The top line of the template contains the technical field name, and the second line contains the description of the field. Note that these two lines must not be deleted. In addition, the values of the first line must not be changed. Figure 4.9 shows a template for importing department data.

Figure 4.10 shows an example CSV import template for employee data. Common to both figures is the double header, which contains identification information for various columns.

**Preparation for data migration**

The task of preparing for data migration is now to populate the individual columns with the relevant datasets.

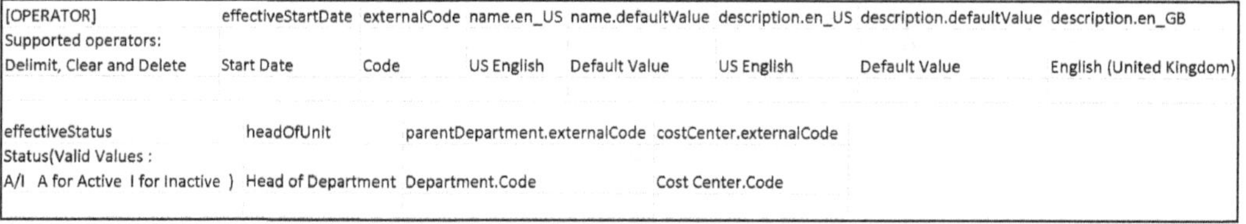

| [OPERATOR] | effectiveStartDate | externalCode | name.en_US | name.defaultValue | description.en_US | description.defaultValue | description.en_GB |
|---|---|---|---|---|---|---|---|
| Supported operators:<br>Delimit, Clear and Delete | Start Date | Code | US English | Default Value | US English | Default Value | English (United Kingdom) |

| effectiveStatus | headOfUnit | parentDepartment.externalCode | costCenter.externalCode |
|---|---|---|---|
| Status(Valid Values :<br>A/I A for Active I for Inactive ) | Head of Department | Department.Code | Cost Center.Code |

**Figure 4.9** CSV Import Template for the Foundation Object "Department"

| A | B | C | D | E |
|---|---|---|---|---|
| user-id | country-of-birth | date-of-birth | person-id-external | region-of-birth |
| User ID | Country Of Birth | Date Of Birth | Person Id | Region Of Birth |

**Figure 4.10** CSV Import Template for the Employee Entity "Biographical Information"

The way in which the data is provided can vary widely. The simplest approach is to use a tool such as Microsoft Excel or OpenOffice Calc to read data from source systems and further enrich it for import.

For larger companies/data migrations, a good idea is to use a database application in which the data can be aggregated and then also provided in the appropriate CSV import format.

One deployment challenge is always that some fields reference value lists/ basic data, and these data objects must be properly deployed in the CSV import file. As a result, a value mapping from legacy system values to Employee Central values might need to be performed.

Once the import files have been provided, they can be imported using the standard import tools. Different import tools exist here for the individual types. The complete procedure is described in more detail later in Section 4.5.

## 4.3 Infoporter for SAP ERP HCM Data

The standard tool or procedure for data imports to SAP SuccessFactors is the CSV file import. For data imports from SAP systems, standard tools exist for two use cases, which will be discussed first in further chapters; then, we'll deal with a standard import in more detail.



Infoporter

With the *Infoporter* tool, SAP offers a standard function for migrating data from SAP ERP HCM systems. Infoporter enables you to provide organizational data such as organizational units or positions, but also employee data from different SAP ERP HCM infotypes for the migration. This tool can be used via direct web service communication or by providing text-based CSV import files that are manually loaded into the SAP SuccessFactors system.

### 4.3.1 Comparison of Infoporter and Standard CSV Import Templates

Infoporter versus standard CSV import templates

Let's start with a comparison of the two approaches, that is, the use of SAP SuccessFactors standard CSV import templates and Infoporter. Table 4.1 lists their most important aspects.

| SAP SuccessFactors Import Templates | Infoporter |
|---|---|
| **Description** | |
| Standard import templates are provided for various SAP SuccessFactors modules. | Standard tool provided by SAP for migrating data from SAP ERP HCM to Employee Central. |
| **Benefits** | |
| ■ Import templates are automatically updated based on the configuration of the associated data objects.<br>■ Supports both full and incremental loads.<br>■ Monitoring of data load jobs within SAP SuccessFactors. | ■ Provides a configuration framework for managing field-value mappings.<br>■ Extensible framework to support custom fields and mapping logic.<br>■ Supports full and delta loads.<br>■ Supports the migration of web services.<br>■ Provides the option to test the data transfer in preview mode. |

**Table 4.1** Comparison of Infoporter and CSV Import Templates

| SAP SuccessFactors Import Templates | Infoporter |
|---|---|
| **Challenges** | |
| ▪ Data cleansing/transformation must be performed externally.<br>▪ The sequence of data loads must follow the instructions in the implementation manuals to avoid errors. | ▪ Data cleansing must be performed in SAP ERP HCM if required.<br>▪ Provisioning of data that is not available in SAP (this must be processed using the CSV standard)<br>▪ The mechanism for data validation after loading must be managed externally via reports. |
| **Use cases** | |
| ▪ Third-party tools are not included in the scope of application.<br>▪ Data conversion requirements are minimal or can be managed externally. | Supports only the migration of employee and organizational data from SAP ERP HCM to Employee Central |

**Table 4.1** Comparison of Infoporter and CSV Import Templates (Cont.)

The descriptions regarding the CSV approach refer to the end-to-end approach. As a result, the CSV files are not created using Infoporter. This approach will be considered later in Section 4.3.2.

### 4.3.2 Infoporter Integration: Architecture Components

**Business Integration Builder**

Infoporter is delivered as part of the PA_SE_IN integration add-on for SAP ERP HCM and Employee Central. The add-on integrates both systems for data replication. The definition of the necessary mapping takes place in what's called *Business Integration Builder (BIB)*. The result can be output as one or more CSV files, and the files can be uploaded to Employee Central manually or automatically.

Figure 4.11 shows a visualization of how individual architecture components interact.

Once the migration is complete, the setup can also be used for the regular replication of data via SAP Integration Suite through SAP Business Technology Platform (SAP BTP).

**Additional Resources**

For more information, including step-by-step instructions on how to perform these activities, see the "Migrating Data from SAP ERP to Employee Central" manual at *http://s-prs.co/v581709*.

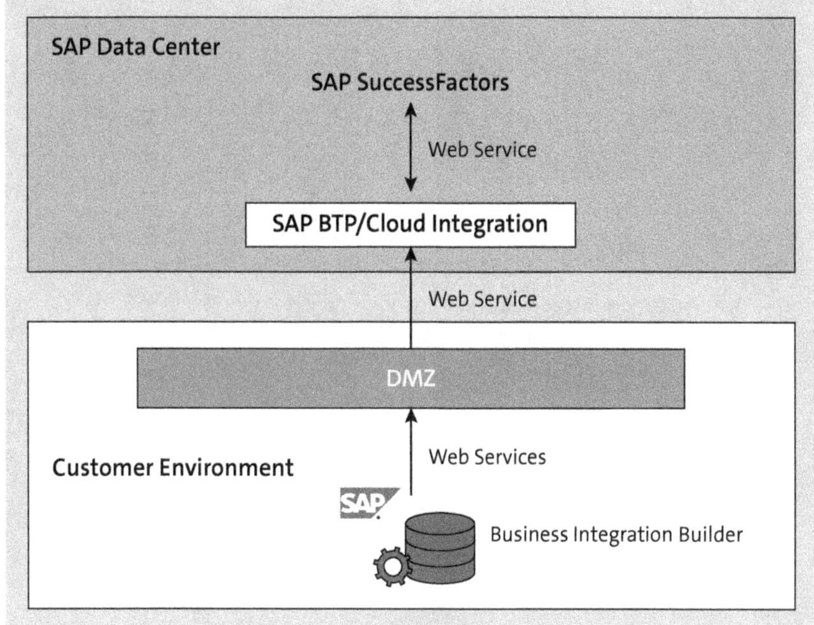


**Figure 4.11** Infoporter Integration of Architecture Components

### 4.3.3 Infoporter: Comparison of CSV and Web Service Options

As mentioned earlier, the decision as to whether to use a CSV file import or web services is an important one to make at the beginning of the project, as it leads to two different approaches.

**Impact on costs**

There is no blanket recommendation that applies to everything. Both options have cost implications. Once you have decided on a method, the relevant Employee Central entity metadata must be uploaded to BIB, and changing the technology means you must restart the configuration. You cannot easily convert CSV Infoporter templates to web service Infoporter templates, nor vice versa. To better assess whether Infoporter should be used in a CSV or web service configuration, see Table 4.2 for a comparison.

| CSV | Web Service |
|---|---|
| **Functional Scope** | |
| CSVs can only be used for migrating data from SAP to Employee Central. | In principle, the web service–based version can also be used for the replication from Employee Central to the SAP system as an option. |

**Table 4.2** Infoporter in Comparison to CSV or Web Service

| CSV | Web Service |
|---|---|
| **Scope of Changes and Timeframe for the Changeover** | |
| *Freeze period*: Period that starts with the export of data from the source environments and ends with the deployment in Employee Central. During this time, the data is frozen. | |
| The duration of a necessary freeze period is longer.<br>When using CSV, the data freeze window starts with the first extract from the SAP system, since no simple delta function exists. | Web services can be used to perform an initial data load in advance. Thus, most of the data has already been transferred. In a further step, only delta data is transferred.<br>This capability can be activated via the configuration of what's called the *template group*,<br>which makes it possible to start the data freeze later and shorten its duration. |
| **Configuration Effort** | |
| Using the CSV import requires the same configuration effort as deploying the web service, but there may be cases where it is necessary to manually enrich the CSV content for specific cases. This manual update must be performed each time you extract and reupload your data during testing. | Web services basically require a large portion of configuration work. Since no files are created, all assignments must be made in the Infoporter configuration. The advantages include consistency and automation. |
| Manual changes must be documented, followed, and repeated and still are prone to error (e.g., forgetting a step, performing steps in the same way). | |
| In some approaches, a combination of CSV import and web service approach is useful. One approach may be to load global data in an automated way using web services and then to process and enrich specific cases using CSV. | |

**Table 4.2** Infoporter in Comparison to CSV or Web Service (Cont.)

| CSV | Web Service |
|---|---|
| **Accuracy, Security, and Confidentiality** | |
| After extracting from SAP ERP HCM but before importing into Employee Central, changes can be made to the files. These changes may or may not be documented, and user errors may occur.<br>A copy of the data resides on file servers for a while and needs to be cleaned up. The deletion of this personal data is an important data protection issue. Manual splitting into smaller packages may be required. | All changes or assignments to the data are made in the configuration and transported to production via transports.<br>All data is transferred to Employee Central via encrypted HTTPS web services. No files are created in this one-step process.<br>Manual splitting is not required. No manual changes to the data are possible |
| **Error Analysis** | |
| No immediate import error messages are displayed when running the export from SAP ERP HCM.<br>The error is reported at the next step during and after the upload to the Employee Central system. This process is usually performed by other project staff.<br>The error can also be caused by format errors, local PC settings, splitting, or manual content changes and therefore can make error analysis more complex. | When running the web service call from SAP ERP HCM, you'll receive an immediate success/error message with details if an error occurs. |

**Table 4.2** Infoporter in Comparison to CSV or Web Service (Cont.)

[«]

**Summary Regarding Web Services or CSV**

Web services require more configuration effort, which is offset by less effort during the test and production runs of the data migration. If you need to integration Employee Central to SAP ERP HCM, these configurations can be reused.

CSV usage generally means less configuration work, especially if replication is not the main focus. At the same time, more migration effort is required, and a longer freeze period must be endured for the changeover.

### 4.3.4 Steps Involved in Data Migration Using Infoporter

Data migration steps

Data migration with Infoporter is divided into four main blocks, as shown in Figure 4.12:

❶ **Preparation**
In this step, the data required for the subsequent mapping is read.

❷ **Configuration**
The fields and values from SAP ERP HCM and Employee Central are mapped to each other and, if necessary, conversions are performed.

❸ **Extraction from SAP ERP HCM**
The data is extracted from SAP ERP HCM.

❹ **Import to Employee Central**
After appropriate processing, the data is transferred to Employee Central either via CSV or web service.

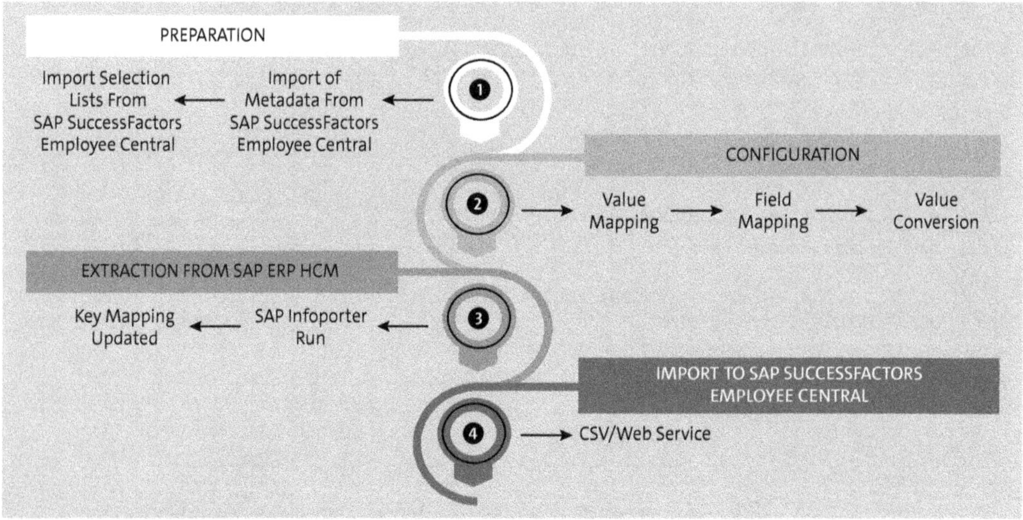


**Figure 4.12** Steps Involved in Data Migration Using Infoporter (Source: SAP)

## 4.4 Migrating SAP Cost Centers to Employee Central

Transferring cost centers to Employee Central

SAP provides a standard interface for transferring cost centers from an SAP ERP system to Employee Central. This interface can be used for a one-time transfer of information as well as for a cyclic update of the data. This interface is based on an add-on called `ODTFINCC`, which must be installed in the SAP source environment. The Cloud Integration package `ERPtoSuccessFactorsEmployeeCentralCostCenter` is required to transfer the data to Employee Central.

Figure 4.13 shows the interaction among the individual components.

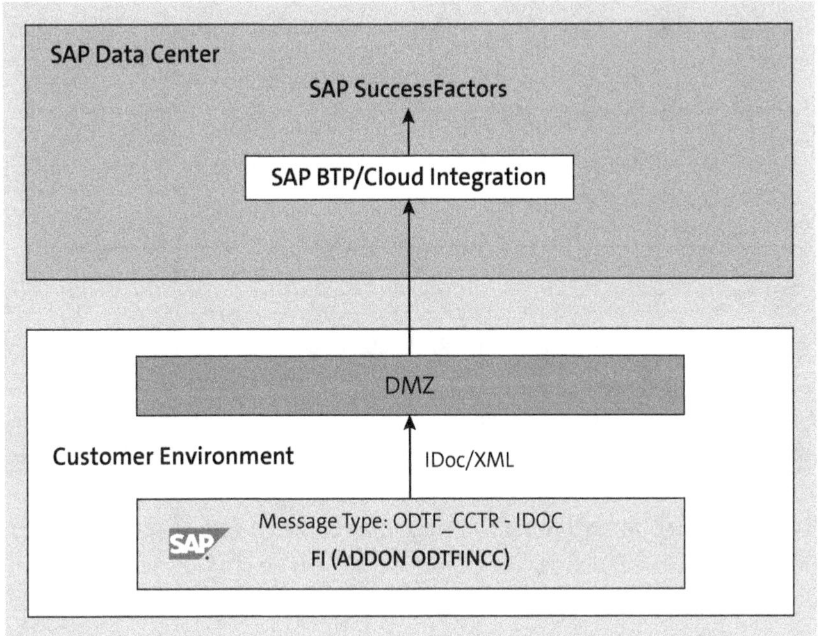


**Figure 4.13** Integration Architecture and Components of the Cost Center Migration

**Configuration of the architecture components**

The first part for the deployment includes the configuration of the architecture components as well as the installation of the add-on, the Cloud Integration Secure Sockets Layer (SSL) certificates in the SAP system and the Cloud Integration package. A technical user with the necessary permissions must be provided on the Employee Central side. On the SAP side, further configurations are required, namely, the following:

- Activate change pointer for message ODTF_CCTR (Transaction BD61)
- Create logical system for Employee Central (Transaction SALE)
- Maintain distribution model for message ODTF_CCTR (Transaction BD64)
- Edit RFC destination for Employee Central (Transaction SM59)
- Create logical port (XML/HTTP) for Employee Central (Transaction WE21)
- Create partner profile for Employee Central (Transaction WE20)

Once you have performed these tasks, a transfer of cost centers from SAP ERP to Employee Central can take place. For this purpose, report ODTF_REPL_CC must be executed via Transaction SE38.

SAP report

Figure 4.14 shows the SAP report, ODTF_REPL_CC. Inside the black frame, notice that it is also possible to output the result of this report to a CSV file. The file generated via this configuration can be used for uploads to Employee Central.

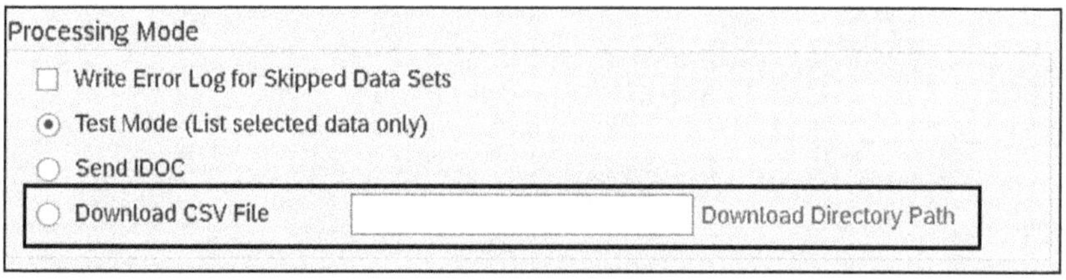

Processing Mode
- ☐ Write Error Log for Skipped Data Sets
- ◉ Test Mode (List selected data only)
- ○ Send IDOC
- ○ Download CSV File [ ] Download Directory Path

**Figure 4.14** Excerpt From the ODTF_REPL_CC Report

This integration process can also be configured to meet various data mapping and transformation requirements, such as mapping cost center codes and names from an SAP ERP system into the Employee Central data model. In addition to cost center data, the integration can also be used to transfer other related data, such as cost center managers and employees assigned to cost centers.

**Additional Resources**

Detailed step-by-step instructions for the SAP ERP cost center migration/integration deployment can be found at *http://s-prs.co/v581710*. You can find the SAP documentation at *http://s-prs.co/v581711*.

## 4.5 Importing Data into SAP SuccessFactors

CSV file import interface

SAP SuccessFactors provides a CSV file import interface that can be used as a standard method for data migration (i.e., for the initial data load). In the following sections, we describe the standard data import processes, key fields, and settings, and share with you the considerations to make when importing data. The structure of the CSV import templates depends on the configuration of the target SAP SuccessFactors system and is offered to you as a template that corresponds to your system configuration. The following data imports are used to import Employee Central data:

- Foundation object data and general (generic) object data
- Personal or employee data

All imports work according to the same patterns. Let's go into further detail next.

### 4.5.1 Procedure for Importing Employee Central Data

**Employee Central data import**

The individual imports are partly based on data that must already exist in SAP SuccessFactors at the time the import is performed. Otherwise, the import will not work. As a rule, you want to proceed in the following order:

1. Selection lists must be available in SAP SuccessFactors prior to data imports. This requirement is especially true for selection lists referenced in CSV import templates.
2. Once the selection lists are available in the system, you want to start importing the data of the foundation objects (e.g., legal entity, location, etc.).
3. Since the employee data uses foundation object data, the employee data can only be meaningfully imported in a subsequent step.

[«]

**Sequence and Dependencies of Data Imports**

Since the order also depends on your system configuration, an important step is to test the import sequence. SAP SuccessFactors immediately alerts you when loading or validating data if a different date needs to be loaded first.

### 4.5.2 Provisioning Import Templates

How you generate import templates depends on the corresponding type of template. The relevant types include the following:

- Templates for foundation objects
- Templates for generic objects
- Templates for users and employees

In the following sections, we'll describe these individual types in more detail.

#### Templates for Foundation Objects

**Templates for foundation objects**

You can access the relevant CSV import templates by typing "Import Foundation Data" in the action search.

CSV import templates can be downloaded by clicking **Download a blank CSV template** and selecting a template from the list, as shown in Figure 4.15.

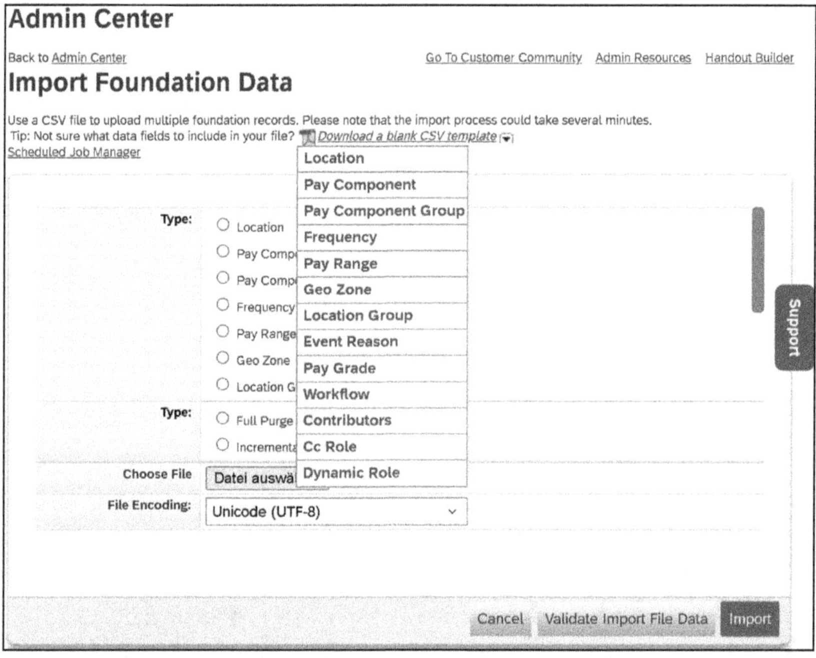


Figure 4.15 Importing Basic Data

### Templates for Generic Objects

**Templates for generic objects**

Download CSV import templates by typing "Import and Export Data" in the action search. In the window that appears, select the **Download Template** option in the **Select the action to perform** dropdown list, as shown in Figure 4.16.

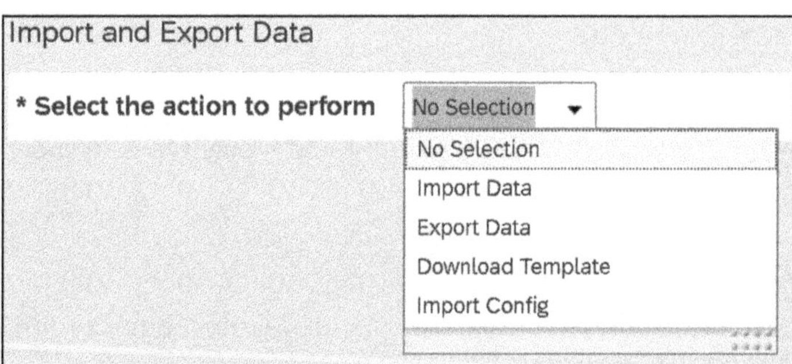


Figure 4.16 Downloading a Template

Several options are available for downloading templates, as shown in Figure 4.17, such as the following:

- **Select Generic Object**: This setting allows you to download the generic object template.

- **Include dependencies**: This setting specifies whether to include CSV import templates for dependent objects, such as linked objects, pick lists, and business rules.

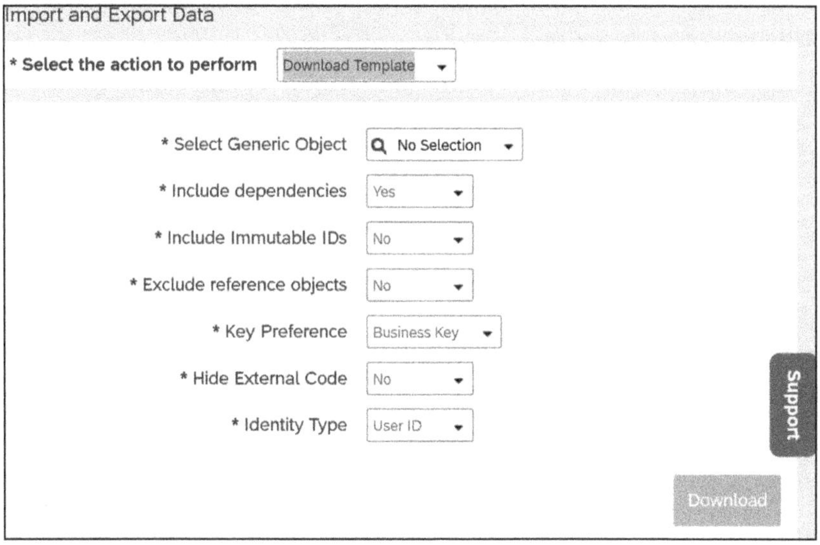

**Figure 4.17** Importing and Exporting Data

After maintaining these settings, click **Download** to download the CSV import template. Note that, while there is a CSV import template for a generic object, there is still a separate CSV import template for each connection between the object and the connected generic object. This CSV import template defines only the connections; a separate CSV import template exists for each associated object. For example, the **Department** object has a CSV import template for department object data, a CSV import template for department-business unit mappings, and a CSV import template for business units. However, these associations are basically dependent on how they have been configured in your system.

**Rules for Filling Import Templates**

Some tips for working with generic objects when filling in import templates include the following:

- Remember to use the `external code` of the target object when filling in linked objects or references.
- For the selection lists, use the `external code` of the value.
- Make sure that the data of the linked objects exists in the relevant time period.

### Templates for Users and Employees

Templates for users and employees

You can access the templates for employee data via the action search by typing "Import employee data." You can get the corresponding CSV templates on this page by selecting the **Download Template** action from the **Select the action you want to perform** dropdown list, as shown in Figure 4.18.

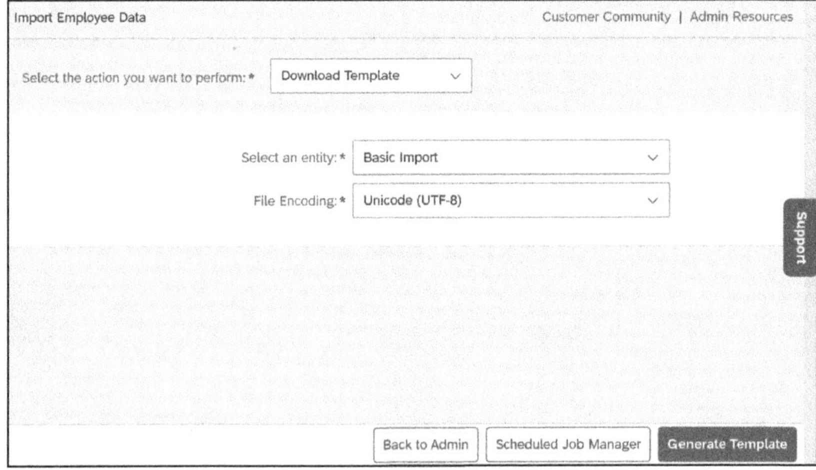


**Figure 4.18** Downloading a Template: Basic Import Option

You'll then see different entities to choose from. For some import templates, you can only select the file format, for example, the template for the basic import of employee data (**Basic Import**, as shown in Figure 4.18).

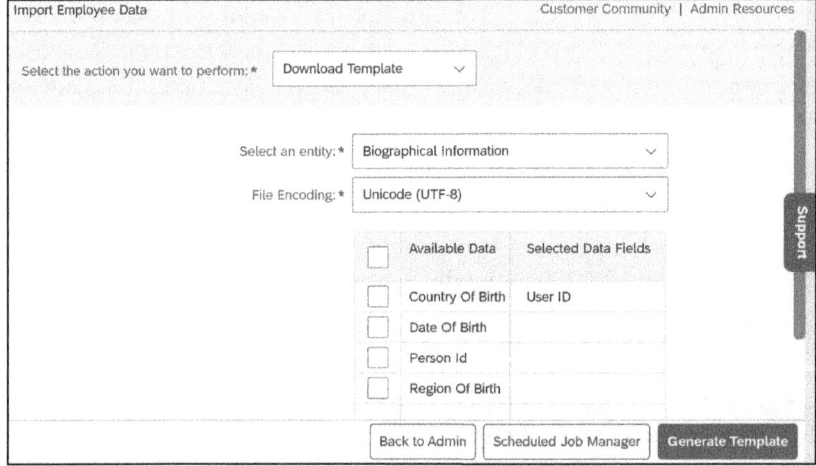


**Figure 4.19** Downloading a Template: Biographical Information

For other import templates, you can decide which fields you want to include in the import template. Figure 4.19 shows an example of the available data fields. However, you cannot remove mandatory fields, such as the user ID. Which fields are mandatory depends on your configuration, which you can view and modify in **Manage Business Configuration** in the action search.

[«]

**Customizing Import Templates According to the Configuration Changes**

If the data model changes in the system for which you want to import data, you must download new CSV import templates or adapt existing CSV import templates to match the new configuration.

### 4.5.3 Different Import Procedures

Provisioning import templates

When provisioning import templates, the import procedures are similar for all three object or data types (foundation objects, generic objects, and employee data):

- Data imports for foundation objects
- Data imports for generic objects
- Data imports for users and employees

In the following sections, we'll describe these import approaches in more detail.

#### Importing Foundation Object Data

To import the completed CSV import templates, you must navigate to **Import Foundation Data** and, as shown in Figure 4.15, perform the following steps:

Importing foundation data

1. Select the foundation object you want to import.
2. Select the type (**Full Purge** or **Incremental Load**).
3. Select the file you want to import by clicking the **Select File** button.
4. If necessary, change the file encoding in the **File Encoding** dropdown list. (We recommend the **UTF-8** option/)
5. Click **Validate Import File Data** to validate the file or click **Import** to import the data.

Validate the import file prior to the actual import to make sure it is free of errors. If errors exist, process the feedback and validate again. Repeat the correction until you see no more indications of inconsistencies or errors. Now, load the data.

Once the data has been loaded, check that it has been loaded correctly. For this task, navigate to **Manage Organization, Pay and Job Structures** in the **Admin Center**. You'll see all the data records, and you can change individual records directly in the **Manage Organization, Pay and Job Structures** screen.

[»]

**Importing the Data for Foundation Objects**

Note the following instructions for importing the data for foundation objects:

- **Objects with parent–child relationships**
  The objects containing a parent–child relationship must be loaded at least twice—the first time without the relationship, so that all records can be created. This first load is then followed by a second load with the relationships that can be created in this import run. An example is the hierarchical relationship of positions or departments.
- **Reference to employee data**
  If employee data is referenced in foundation objects, for example, in the case of cost centers to the cost center manager (`costCenterManager` field) or in the department to the head of department (`headOfUnit` field), this data must also be loaded at least twice—first the pure object data and after the employee imports the same data again with the information of the relevant co-working person or the relevant manager.

### Importing Generic Objects

Importing generic objects

For generic objects, you can use the **Import and Export Data** function in the action search to perform a range of different imports. CSV import templates are imported into the system by selecting **Import Data** from the **Select the action to perform** dropdown menu on the **Import and Export Data** screen, as shown in Figure 4.20.

You can choose which type of import you want to use from the following three options:

- **CSV file**
  Import a single CSV import template with data.
- **ZIP file**
  Import a ZIP file that contains multiple files.
- **SuccessStore**
  Import the data with the content provided by SAP SuccessFactors. This option is not relevant for data migration.

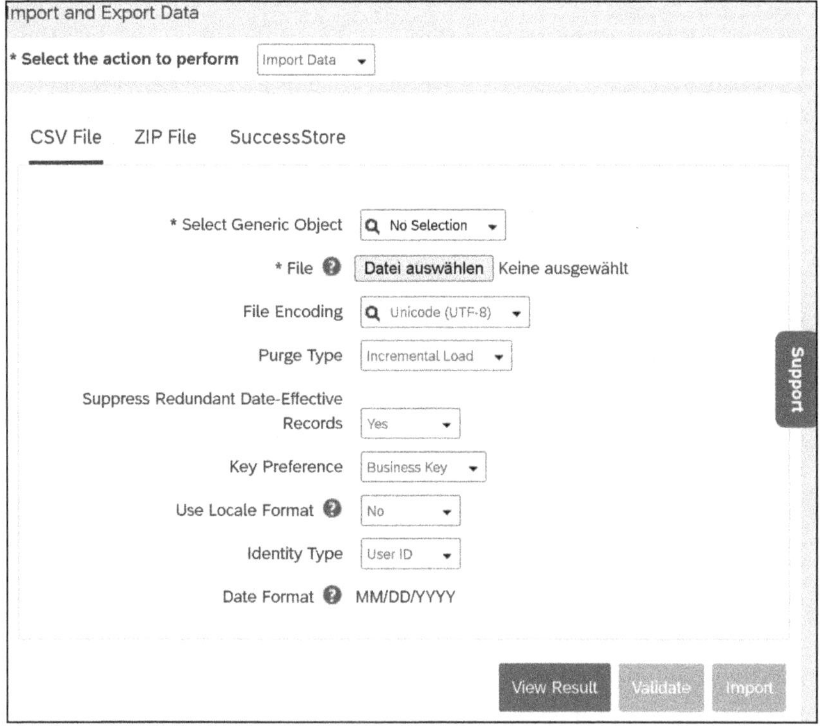


**Figure 4.20** Importing Data: Generic Objects

The **CSV file** tab provides the functions necessary for data import, such as the following:

- **Select Generic Object**
  This field contains the generic object for which you want to upload data.
- **File**
  This field contains the filename of the file to be uploaded.
- **File Encoding**
  The file encoding is entered in this field. (We recommend **UTF-8**.)
- **Purge type**
  In this field you have the choice between the options **Full Purge** and **Incremental Load**.
- **Suppress Redundant Date-Effective Records**
  The setting in this field is **Yes** by default to prevent duplicate insertions and unchanged updates.
- **Key Preference**
  This field contains the method for setting the business key; select either the **Business Key** setting (the default method) or the **External Code** setting.

- **Use Local Format**
  Select **Yes** to enable this feature (the **Locale** field will be displayed to select the locale). Select **No** to use the locale of the user performing the import.
- **Identity Type**
  The options for the identity type are **User ID** or **Assignment ID**. If you're using more than one employment information, you should use the correct employment info ID for each.

Data imports are processed asynchronously. You can download the results of imports via **Scheduled Job Manager** in the action search.

> **Secured Generic Objects**
>
> If you have activated the security functions for the generic objects in the object definition, the import/export authorization for each generic object must be granted to the user who is to perform the action. Often this circumstance is mapped via a technical user.

***Importing CSV Templates with Data***

When using the **ZIP** file option, a ZIP file gets uploaded, as the name suggests. The import is controlled by the *import.properties* file. This file contains the information you configure for the **CSV file** type in the options. The *import_sequence.csv* file is used to define the import sequence of the data delivered via the ZIP file. Figure 4.21 shows a ZIP import file for importing business unit information.

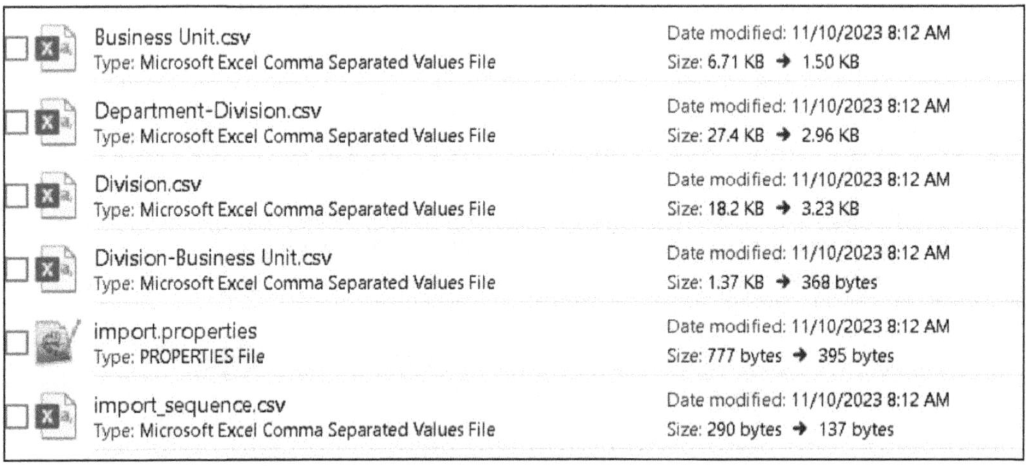

| | |
|---|---|
| Business Unit.csv<br>Type: Microsoft Excel Comma Separated Values File | Date modified: 11/10/2023 8:12 AM<br>Size: 6.71 KB → 1.50 KB |
| Department-Division.csv<br>Type: Microsoft Excel Comma Separated Values File | Date modified: 11/10/2023 8:12 AM<br>Size: 27.4 KB → 2.96 KB |
| Division.csv<br>Type: Microsoft Excel Comma Separated Values File | Date modified: 11/10/2023 8:12 AM<br>Size: 18.2 KB → 3.23 KB |
| Division-Business Unit.csv<br>Type: Microsoft Excel Comma Separated Values File | Date modified: 11/10/2023 8:12 AM<br>Size: 1.37 KB → 368 bytes |
| import.properties<br>Type: PROPERTIES File | Date modified: 11/10/2023 8:12 AM<br>Size: 777 bytes → 395 bytes |
| import_sequence.csv<br>Type: Microsoft Excel Comma Separated Values File | Date modified: 11/10/2023 8:12 AM<br>Size: 290 bytes → 137 bytes |

**Figure 4.21** Import Definition for a ZIP File

Use the **Validate** button to validate the data, as shown in Figure 4.20. The results of the validation and import are available under **Scheduled Job Manager**, as shown in Figure 4.22.

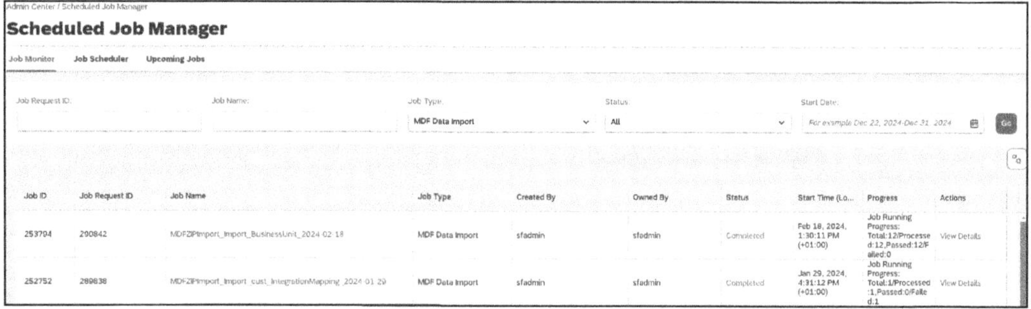

**Figure 4.22** Scheduled Job Manager

[!]

> **Recommendation for the Data Migration Import**
>
> We recommend that you use the **CSV file** option when migrating foundation object data. For other imports, the choice depends on the data record. Complex objects and contexts require multiple files and require a ZIP file.

### *Importing Attachments*

**Attachment data type**

It may be useful to include attachments when migrating data. You can include attachments in a data import for generic objects if the objects have one or more fields with the **Attachment** data type. Attachments can only be uploaded as ZIP files. All attachments must be placed in a folder named *attachments* inside the ZIP file. The file names must be properly referenced in the appropriate columns in the attachment.

The contents of the ZIP file are similar to what was shown earlier in Figure 4.21. The biggest difference is that the ZIP file has a subfolder where the files to be imported are located. As shown in Figure 4.23, notice the *attachments* folder. The individual elements have the following functions:

- *attachments*: Folder containing the files, in this case, pay slips.
- *import.properties*: Definition of the import behavior.
- *import_sequence.csv*: Definition of the import structure.
- *Payslip-PaySlips.csv*: Definition as to which file should be uploaded for which employee. The file name used in this case is specific to the content being imported.

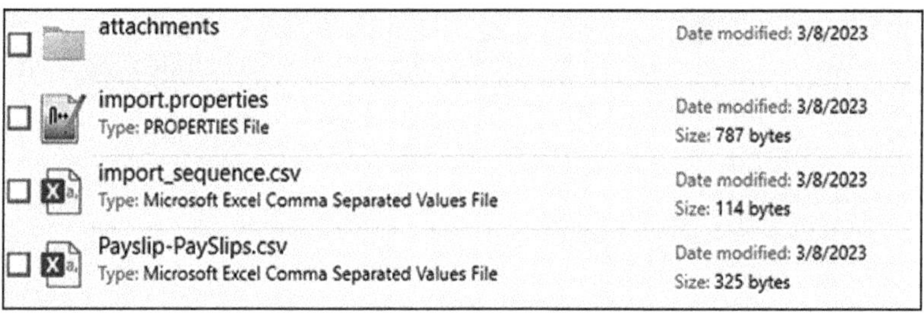


**Figure 4.23** CSV Import File for Attachments

### Importing User and Employee Data

**Importing employee data**

If you have already imported foundation data, you'll quickly find your way around the **Import Employee Data** section, as shown in Figure 4.24. Notice a few small differences, however: Additional options are available for basic import and employment details as well as some additional options for importing consolidated records.

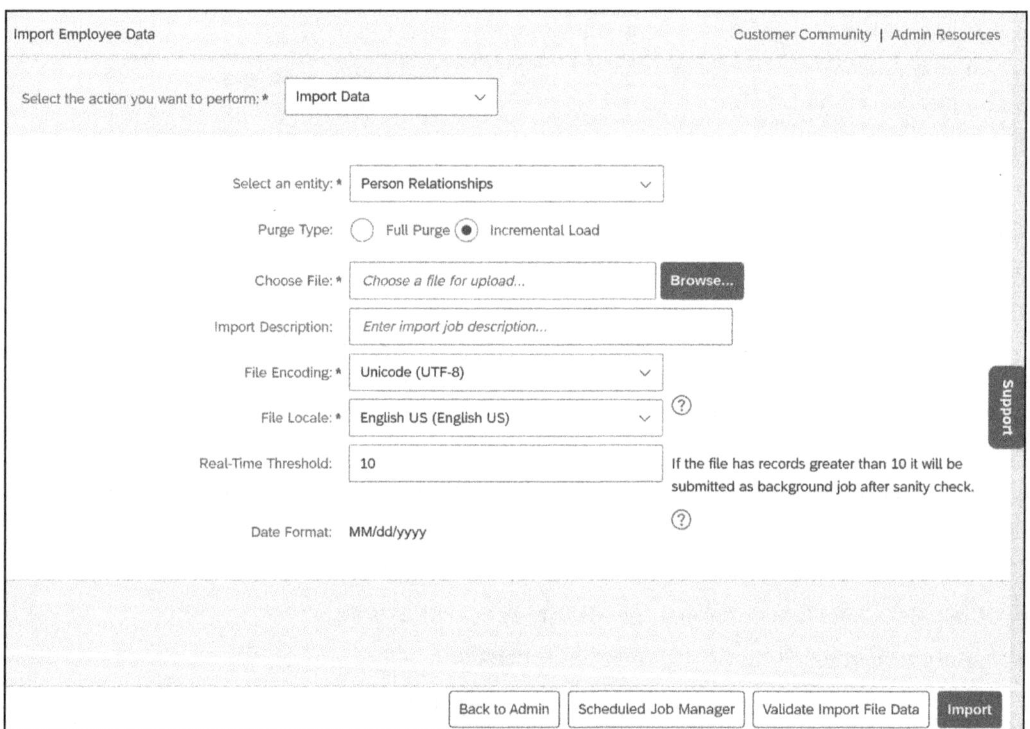


**Figure 4.24** Importing Employee Data

[«]



**Minimum Import of Employees**

If you want to load employees via import, you must successfully perform at least the following imports:

- Basic import
- Biographical information
- Personal information
- Employment information
- Job history

This data is a prerequisite for managing employees in Employee Central. Further imports can also be executed in parallel.

The following list presents the imports and options available to you under **Import Employee Data** that are relevant to Employee Central where they are displayed in the employee profile:

- **Basic Import**
  You create user accounts by importing the *user data file (UDF)*. This file is the basis for the employee record in SAP SuccessFactors. Note that you need this import even without Employee Central for the talent modules and the employee profile. Several indicators are available for this import, as shown in Figure 4.25, of which **Process Inactive Employees** is of particular interest. Select this option if you want to import inactive employees.
- **Biographical Information**
  Use this function to import the data displayed in the employee profile under **Biographical Information** in the **Personal Information** section. With this type of import, a complete cleansing is always carried out.
- **Personal Relationships**
  Use this function to import the relationship data displayed in the **Relatives** portlet in the **Personal Information** section. The **Personal Information** and **Addresses** imports are required to import personal information about the relatives. Alternatively, you can use the **Consolidated Relatives** import option to import all the data you have on employees' relatives.
- **Employment Information**
  This function imports the employment information.
- **Personal Information**
  This function imports the data contained in the **Personal Information** section, such as salutation or nationality.

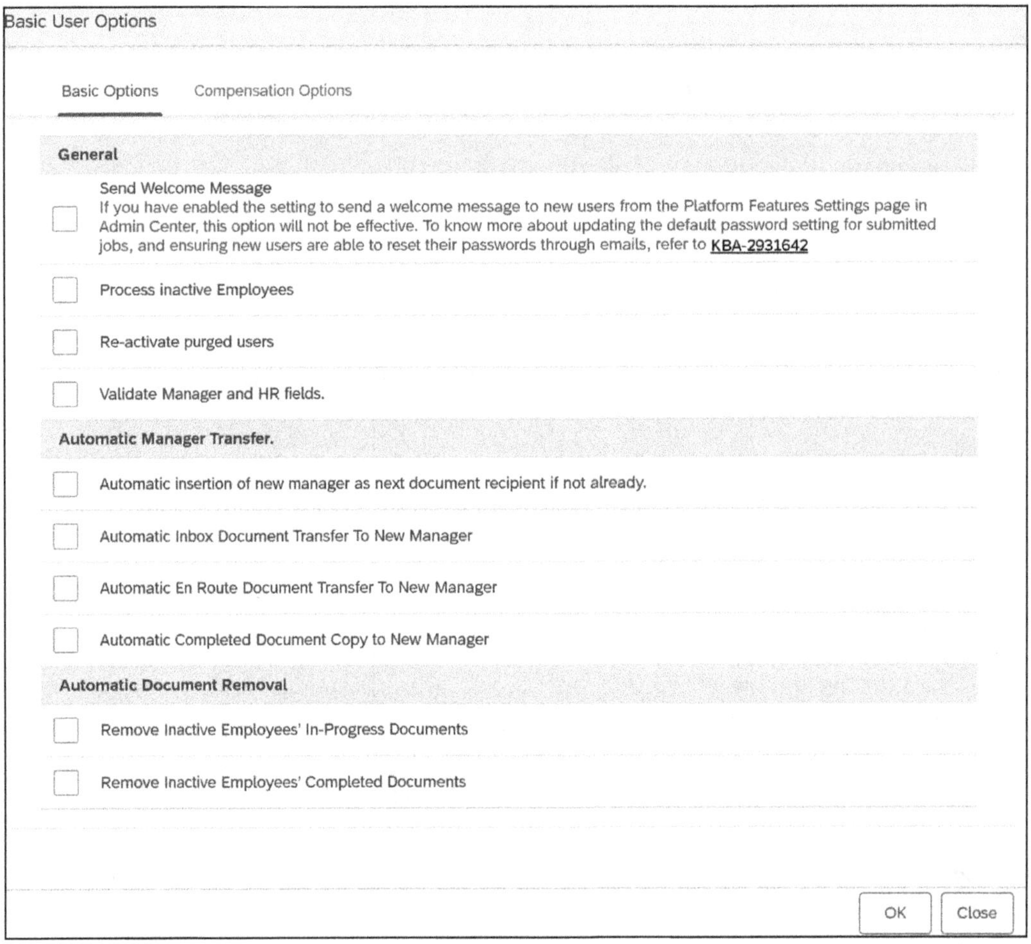

Basic User Options

Basic Options Compensation Options

General

Send Welcome Message
If you have enabled the setting to send a welcome message to new users from the Platform Features Settings page in Admin Center, this option will not be effective. To know more about updating the default password setting for submitted jobs, and ensuring new users are able to reset their passwords through emails, refer to KBA-2931642

Process inactive Employees

Re-activate purged users

Validate Manager and HR fields.

Automatic Manager Transfer.

Automatic insertion of new manager as next document recipient if not already.

Automatic Inbox Document Transfer To New Manager

Automatic En Route Document Transfer To New Manager

Automatic Completed Document Copy to New Manager

Automatic Document Removal

Remove Inactive Employees' In-Progress Documents

Remove Inactive Employees' Completed Documents

OK Close

**Figure 4.25** Selecting Basic Import Options

- **Global Information**
  This function imports employee restriction information and associated data.
- **Termination Information**
  This function imports the termination details of an employee's employment. This information is displayed in the **Employment Details** portlet in the **Employment Information** screen. With this type of import, a complete cleansing is always carried out.
- **Job History**
  This function imports data displayed under **Job information**. Note that, while both the **Full Purge** option and the **Incremental Load** option are available for this import, the incremental load option cannot be used for records with an event reason that uses the **Hire** or **Re-Hire** events.

- **Compensation Information**
  This function imports information related to **compensation**.
- **Phone Number Information**
  This function is used to import phone numbers and related information.
- **Email Information**
  This function imports emails and related information.
- **National ID Information**
  This function imports data from the environment of different ID documents. In the United States, for example, this includes the social security card.
- **Addresses**
  This function imports different information from the address data area.
- **Emergency Contact**
  This function imports information about the contact person in case of emergency.
- **Personal Documents Information**
  This function imports work authorization data and, if desired, document attachments, which are displayed in the **Work Authorization** area under **Personal Information**.
- **Pay Component Recurring**
  This function imports information on the data in the recurring salary components area.
- **Pay Component Non Recurring**
  This function imports spot bonus data, which is displayed in the **Spot Bonus** section.
- **Job Relationships**
  This function imports information on the relevant job relationships.
- **Composite (ZIP) Data Upload**
  This option allows you to upload a ZIP file containing several CSV import files. The system processes the files in the order you specify. Each file within the ZIP file can be monitored in the **Admin Center** via **Monitor Job**, and an email notification is sent for each file. For new employees, files must contain the CSV import templates **Basic Import**, **Biographical Information**, and **Employment Details**. For existing users, there is no requirement as to which import files must be included.
- **Consolidated Dependents**
  This function imports the relationship data of the relatives and the personal data of the relatives of an employee. Note that this type of import always performs a full cleanup.

[!]

**Rules for Filling Import Templates**

The following general notes will assist you in completing the import templates:

- The maximum number of data records you should enter into a CSV import template is 50,000.
- The **User ID** is the user ID defined in the basic import or during the direct creation in SAP SuccessFactors.
- During the initial import, the `Person-id-external` field usually contains the same value as the **User ID**.
- For an employee's first or only (hire) record, the date in the **Start Date** field in all CSV import templates should match the date in the **Start Date** field in the CSV import template **Employment Information**.
- The end date column can be left empty for the last record.
- For multiple records with the same effective date, you can use the `seq-number` field to specify the order within the date.

To import the completed CSV import templates, enter "Import Employee Data" in the action search and follow these steps, as shown in Figure 4.24:

1. Choose what kind of data you want to import.
2. Select the purge type (**Full Purge** or **Incremental Load**) if it is available in the context.
3. Select the file you want to import by clicking the **Browse** button.
4. Optionally, you can describe the import for better identification in the **Monitor Job** section.
5. If necessary, change the file encoding in the **File Encoding** dropdown list. (We recommend **UTF-8**.)
6. If necessary, change the **File Locale** in the dropdown list. (We recommend **en_US**.)
7. Change the threshold value in the **Real-time threshold** field if you want to trigger an asynchronous import.
8. Click **Validate Import File Data** to validate the file or click **Import** to import the data.

We recommend that you validate the import file before importing it to ensure that it does not contain any errors. If errors occur after validation or after trying to import the file, fix them and try again.

## 4.6 Tips and Tricks

Tips and tricks

If you only focus on the provisioning of data when talking about a data migration to SAP SuccessFactors, you're falling short. We hope this section gives you some food for thought so that you can optimize your data migration projects.

### 4.6.1 Relevant Environments in the Project Process

Environments in the project process

The preparations and the execution of the data migration take place in different environments over the project period.

Which environments are relevant may vary from implementation to implementation. Figure 4.26 shows how a three-tier environment can enables extensive testing of the data migration and other go-live–related steps. For example, one of these steps could be testing the relevant integrations of Employee Central to local systems, such as its integration with a local SAP ERP HCM system.

[!]

**SAP ERP HCM System**

For the integration of Employee Central with an SAP ERP HCM system, a three-tier environment is strongly advised and even essential for minimizing risk at go-live.

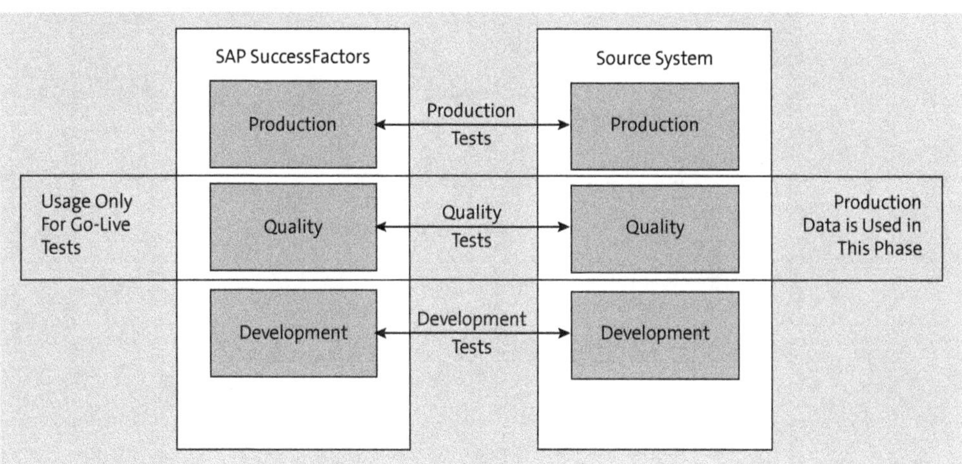


**Figure 4.26** Employee Central: Project Environments

The purposes of activities and also the data used are different for individual environments:

- **Development environment**
  This is where the definition of the Employee Central data model takes place. This environment is used for initial functional testing. The employee data used here is purely test data.

[»]

**Special Procedure for the Basic/Organizational Data**

A special approach for the basic/organizational data could be to provide the relevant production data in Employee Central already in the development environment. The approach is then to prepare the productive basic/organization data for import and then import it into the development environment for a final review.

For the other environments, the data prepared in this way can be exported and imported from the Employee Central development environment. This special procedure, shown in Figure 4.27, does not have to be used for all basic/organizational data. You might also find it useful to apply this approach only to more static data (e.g., legal entity, locations, etc.) and use the same approach for the other data as for employee data.

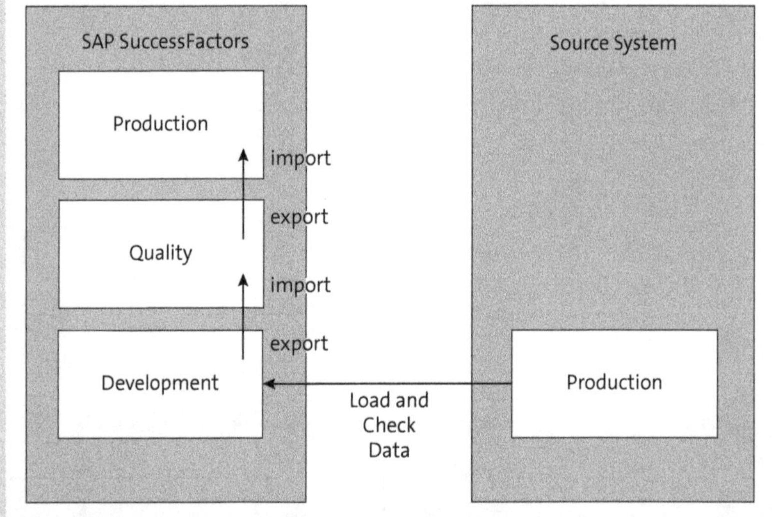


**Figure 4.27** Special Procedure for Basic/Organizational Data

- **Quality environment**
  This environment is used for the verification of previous preparations. You could also regard it as an environment in which the go-live is rehearsed. The data used for the tests is production data. Due to this fact, the same authorization framework conditions must apply to this environment as to the production environment. If this matching is not pos-

sible or desired, personal employee data must be anonymized, but this can lead to increased complexity of the data volume.

[«]

**Go-Live Rehearsal**

When we talk about a "rehearsal for go-live," we mean, on one hand, the preparatory activities such as the transfer of the Employee Central configurations, the provision of the basic/organizational data, and the execution of the complete employee data loads are rehearsed.

Aspects like the execution times for each data load are included. The results of the execution duration can then be incorporated into the go-live schedule and thus optimize or make more concrete the duration of a data freeze time, for example.

Once the data has been successfully loaded, the quality of the imported data is checked. Results of this audit can provide evidence for improvements and corrections in the provisioning process.

**Production environment/ quality check**

In the different environments relevant to the implementation of an Employee Central project, different requirements will arise for the way in which tests or data quality checks are performed.

As far as requirements are concerned, your tests should be the same in the quality environment as in the production environment. So you can perform reproducible test results, some tools should be implemented. A useful step is to set up defined quality checkpoints for defined blocks, which enable you to assess the quality of the completed step. Not until a quality checkpoint has been successfully completed should the next step begin.

**Quality checkpoints**

Figure 4.28 shows the interaction of individual quality checkpoints for the provision of data in an Employee Central system, where a subsequent integration with an SAP ERP HCM system is relevant.

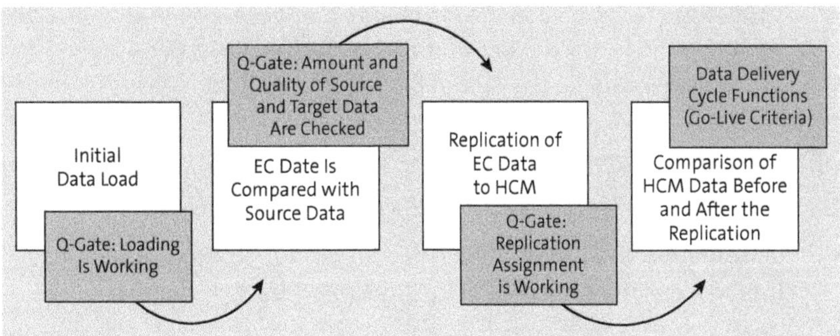


**Figure 4.28** Quality Checkpoints for Employee Central Data Load: Replication to SAP ERP HCM

How an approach affects the risk at go-live is shown in Figure 4.29.

Go-live risk

The highest quality level or the lowest go-live risk is certainly achievable with a tool-supported web service data migration with a three-tier landscape. This approach is also the most extensive in terms of the effort to be invested. Using Infoporter, such an approach is feasible for SAP. A good cost–benefit ratio is offered by the approach of creating CSV files automatically and then testing and optimizing them in a dedicated conversion and integration test phase in a quality environment.

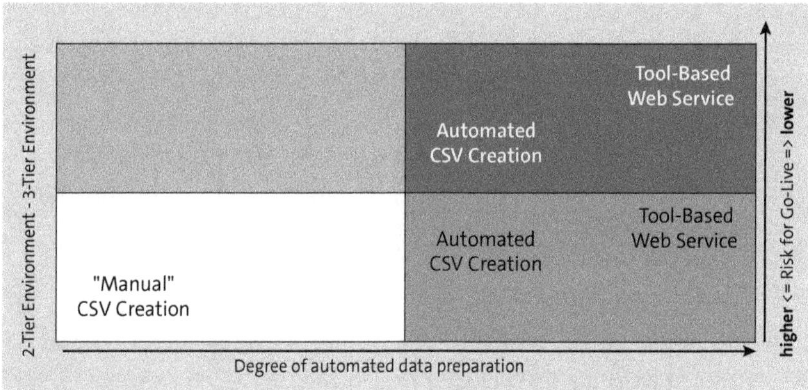


**Figure 4.29** Risk Assessment for Go-Live in Different Environments

The question as to which approach makes sense in your specific case also depends on the number of records or employees to be migrated. In some cases, a (partly) manual creation of CSV files using Microsoft Excel and its popular `Vlookup` function might be useful.

### 4.6.2 Identifiers in Data Migration

Identifiers in data migration

Employee Central has different identifiers/IDs. To prepare and execute a data migration, knowing these identifiers and having a clear understanding of their functions is useful. Identifiers exist for basic/organizational data and for employee data. In the case of basic/organizational data, an important ID exists for data migration called the *external code*. Different identifiers are used for employee data.

The most important personal identifiers include the following:

- **PersonIdExternal**
  A unique identifier for a person in Employee Central.
- **UserId**
  A unique identifier of a user unit. One or more `UserIds` can be assigned to a person, depending on whether the person has or has had one or more jobs.

- **AssignmentId**
  The `AssignmentId` correlates with the `UserId`. The big difference is that the `AssignmentId` can be changed after the initial creation, whereas the `UserId` cannot. A person can have one or more assignment IDs, again depending on the number of jobs.
- **PayrollId**
  `PayrollId` is a field in Employee Central to store another identifier. In the standard system, no automatic value is generated for the field, for example, when setting, unlike the fields mentioned above. This ID is an example of the ability to store additional IDs.

**Unique company ID**

An important step in any Employee Central project involves defining the format of the IDs, such as specifying that the `UserId` should only follow an 8-digit numeric format. It may happen that IDs need to be transferred from source systems during data migration. One reason may be that a unique company ID already exists for the employee in the company, and this should be continued, or that continued use makes sense due to corresponding applications.

**IDs of the organizational data**

An application of this type could be one or more SAP ERP HCM systems, in which the IDs of the organizational data (e.g., organizational unit, position, job) and the IDs of the employees (e.g., personnel number PRNR) are transferred, in order to maintain them in Employee Central after the go-live. During the transfer, you must check in advance that no overlaps arises among your existing number ranges, a situation that must be rectified.

It makes sense to list the individual identifiers in a Microsoft Excel file, for example, and to run through their processing during a data migration, when an object is generated for the first time, or when a new object is created. Figure 4.30 shows some example organizational data, and Figure 4.31 shows some example employee data in Microsoft Excel tables.

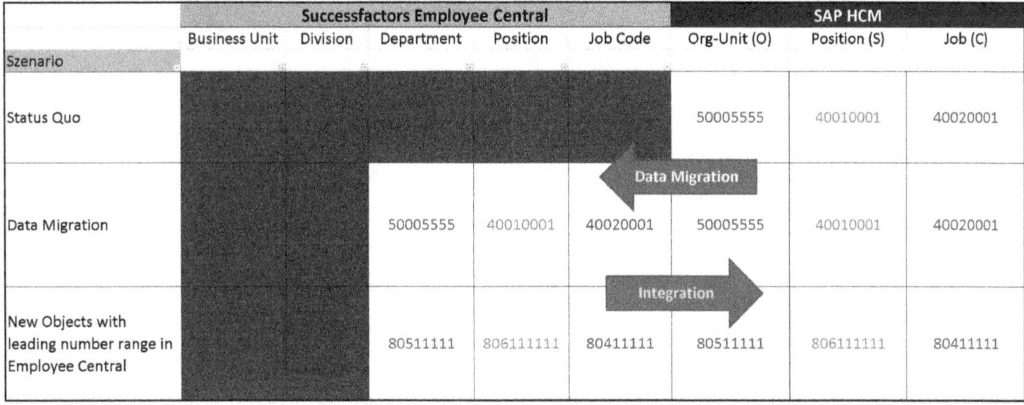

| | Successfactors Employee Central | | | | | SAP HCM | | |
|---|---|---|---|---|---|---|---|---|
| Szenario | Business Unit | Division | Department | Position | Job Code | Org-Unit (O) | Position (S) | Job (C) |
| Status Quo | | | | | | 50005555 | 40010001 | 40020001 |
| Data Migration | | | 50005555 | 40010001 | 40020001 | 50005555 | 40010001 | 40020001 |
| New Objects with leading number range in Employee Central | | | 80511111 | 806111111 | 80411111 | 80511111 | 806111111 | 80411111 |



**Figure 4.30** Analysis and Transfer of Identifiers Using SAP ERP HCM Organizational Data

| Szenario | Successfactors Employee Central | | | | other Payroll System | SAP HCM |
|---|---|---|---|---|---|---|
| | PersonIDExternal | UserID | AssignmentID | PayrollId | ID | PRNR |
| Status Quo before Migration | | | | | 91212 | 50014528 |
| Data migration (the data was initially loaded into the EC system) | 70200001 | 50014528 | 50014528 | 91212 | 91212 | 50014528 |
| New Hire - with leading number range in Employee Central | 70500001 | 70000001 | 70000001 | 95001 | 95001 | 70000001 |

Data Migration
Integration

**Figure 4.31** Analysis and Transfer of Employee Identifiers

### 4.6.3 Key Date for the Cutover Date

**Cutover date**

An important decision is the determination of the cut-over date, that is, the relevant cutoff date or cutover date, for the transfer of all relevant data. This date must be reflected in the data import files to be created, as shown in Figure 4.32.

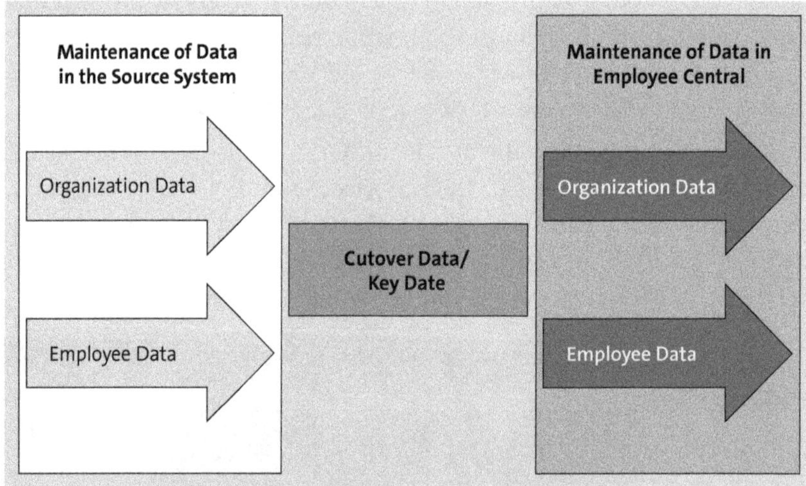


**Figure 4.32** Cutover Date for the Organizational/Employee Data

In all relevant time slice–based areas in Employee Central, such as employment information, a time slice is generated and loaded in the import files on the key date. For possible integrations and data maintenance, the entry under **Employment Information** is the hub for all other leading data maintenance workflows in Employee Central. Different cutover dates are defined in the course of the project and in the different environments. For example, one cutover date exists for the quality environment, and another cutover date, for the production environment.

### 4.6.4 Cutover Method



*Cutover* is a general IT project management term that refers to the activities that must be performed when moving from an old system to a new one.

In the context of Employee Central implementation projects, the a cutover typically comprises an extensive list of activities that begin with setting up the provisioning settings of a production instance, transferring data models, loading configurations into the production instance, connecting interfaces, loading certificates, etc. The actual master data (employee and organizational data) is loaded into the Employee Central system at a later point in the cutover process.

During a certain time slot during the cutover to Employee Central, HR data must be frozen in the source systems for which subsequent integration is envisaged, so some HR processes are temporarily unavailable. For this reason, you must keep this time slot as short as possible. An example of a rough cutover plan is shown in Figure 4.33.

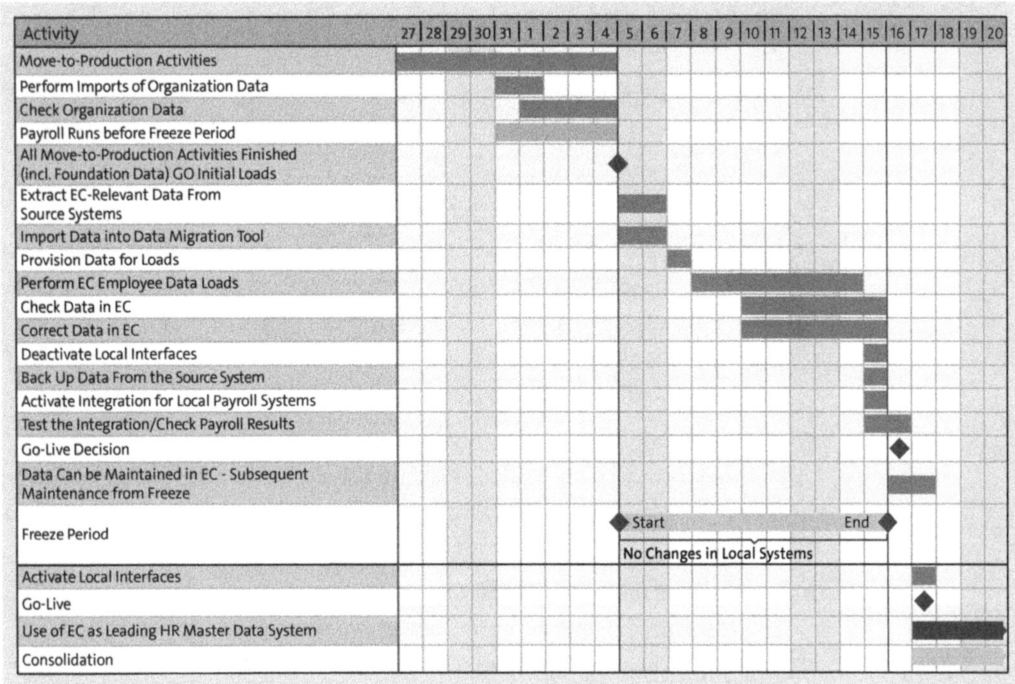


**Figure 4.33** Example of a High-Level Cutover Plan

Cutover plan

The individual activities include the following:

- **Move-to-production activities**
  Activities related to the transfer of configurations (Employee Central, integrations with both SAP and non-SAP systems, other corporate systems) are completed and functionally verified.
- **Carrying out the imports of the organizational data**
  All relevant organizational data is loaded to Employee Central in the correct order.
- **Review of the organizational data**
  After successfully loading the organizational data, the data is checked for quantity and quality.
- **Payroll runs before the freeze period**
  In the affected companies, the last settlement is performed before the start of the activities for the initial data loading.
- **Extraction of Employee Central-relevant data from the source systems**
  In this activity, you must ensure that all information required for the provision of the Employee Central load files is up to date and provided as defined.
- **Reading the data into the data migration tool**
  If up-to-date data has been provisioned in the previous step, it can be read into the tool and prepared for the next step.
- **Provisioning the data for the loads**
  From the tool, the individual Employee Central import files are provisioned for performing the loads.
- **Performing Employee Central employee data loads**
  The data provided by the tool is imported to Employee Central according to a predefined sequence. The files cannot always be loaded 1-to-1 in this case. A necessary step might be dividing the files into blocks or performing the loads twice because there are dependencies that do not yet exist in the first load (such as job information–manager relationships).
- **Verification of data in Employee Central**
  In this step, the data loaded into the Employee Central system is compared (in terms of quantity and quality) with the data from the source system.
- **Corrections of data in Employee Central**
  The results from checking the data in Employee Central are corrected in this step. You might undertake single corrections as well as mass correction imports.
- **Backup of data in the source system**
  Before transferring data from Employee Central to a local system (first

integration run), the local system should be backed up so that a fallback is possible.

- **Deactivating local interfaces**
  Before the first integration run of Employee Central into the local system, the outgoing interfaces should be temporarily disabled. They are reactivated after a successful check of the replication events.
- **Activating the integration for the local payroll system**
  Depending on the integration interface, this activity is different. Basically, you are performing an initial data replication of the involved systems, for example, Employee Central to the local system.
- **Testing the integration and verification of the payroll results**
  Using the data from the **Backup of data in the source system** step, a comparison takes place. No changes should be detected to employee data that is relevant to payroll and is in replication.
- **Go decision**
  The preceding tests were successfully completed, and possible errors were eliminated or will be finalized in the next step.
- **Data can be maintained in Employee Central: Post-maintenance from freeze**
  The Employee Central system is ready for use; last-minute changes made in the source system during the freeze phase may still need to be maintained in Employee Central.
- **Freeze period**
  No changes may be made in the target system in the replication-relevant areas during this period. These changes may be lost after the initial replication because the state of the data in Employee Central at the time of replication is relevant. The setting/termination should be performed before the initial data loads with a future date.
- **Activating the interfaces**
  After the successful verification of the data in the local system, the previously deactivated interfaces can be reactivated.
- **Go-live**
  Employee Central is used as the leading employee master data management system.

## 4.7 Summary

This chapter provided an overview of the challenges and specific considerations in an Employee Central data migration. We shared insights into the basics of Employee Central for a better understanding of the subject. Then, we explored some of the tools available for data migrations—from SAP ERP HCM or FI/CO to SAP ERP to Employee Central. A deeper explanation of how to use the standard Employee Central import capabilities will allow you to master the necessary steps. We hope the tips and tricks we've provided inspire ideas about what topics are relevant in addition to pure data migration topics.

Chapter 5
# Data Migration to SAP Ariba

*In this chapter, we describe the different options for data migration using SAP Ariba. You'll learn about the tools to perform the migration, and we'll give you some tips on where you can find further documentation on this.*



This chapter is about data migration and integration in SAP Ariba as part of SAP's procurement solutions. Since data migration requirements can vary greatly from customer to customer, we'll limit ourselves to a general overview at this point; knowledge of SAP Ariba is a prerequisite, however.

We deliberately speak of "data migration and integration" because the data in SAP Ariba and in SAP Procurement Solutions is integrated or migrated from different systems. The available interfaces are used for both operations. In the following sections, we'll speak in simplified terms of data migrations in accordance with the main topic of this book.

Depending on the source systems used, different options are available for data migration. In this chapter, we focus mainly on SAP S/4HANA systems, but integration from SAP ERP 6.0 is often possible in the same way, based on the procedures described. In each section, we also briefly review the available options for migrating data from a non-SAP backend system.

This chapter provides an overview of the different options and refers to further documentation on data migration and integration in SAP Ariba in SAP Help Portal or the SAP Ariba Connect portal.

We focus on four key areas in this chapter:

- Migration/integration of master data (Section 5.2)
- Migration/integration of suppliers (Section 5.3)
- Migration/integration of transaction data (Section 5.4)
- Migration/integration of user data (Section 5.5)

## 5.1 Challenges and Areas of Complexity

Our first challenge is to determine what data must be imported into SAP Ariba in the first place. The requirements may be different in each of your upstream and downstream processes.

[»]

**Upstream versus Downstream**

Before we go further, let's break down these terms:

- **Upstream**
  Upstream processes describe sourcing (tenders), the conclusions of contracts, spend analysis, and supplier management.
- **Downstream**
  Downstream refers to the actual process of purchasing (i.e., how and from which supplier orders are placed and in which system ordering and goods receipt take place). These processes also include contract fulfillment as well as invoicing and payment.

Think about what you want your integrated purchasing process in SAP Ariba to look like in the future. For example, the purchasing process can run completely and separately in SAP Ariba Buying without synchronizing orders in SAP S/4HANA or SAP ERP. In this case, the goods receipt is also posted in SAP Ariba and not in the SAP ERP system. For strategic (upstream) processes, for example, you should think about where the contracts with the suppliers are managed and in which system the management of the suppliers takes place. This may also depend on which and how many SAP ERP systems are involved in the processes.

**SAP Master Data Governance**

In such a multi-ERP landscape, supplier management can take place, for example, in a central SAP ERP system or in SAP Master Data Governance system as the master data management system. If suppliers are not harmonized between the systems involved, such harmonization may be required if you want to implement a solution such as SAP Ariba Supplier Information and Performance Management.

If you want to integrate an SAP ERP system with SAP Ariba, you need to use a new integration method that must first be set up technically. The integration method differs depending on whether you want to use SAP S/4HANA Cloud, SAP S/4HANA, or SAP ERP as the source system.

The add-on for SAP Integration Suite, managed gateway for spend management and SAP Business Network must be installed and set up by your Basis team in SAP S/4HANA or in SAP ERP. Secure communication with certificates and an associated cloud connector must also be installed and set up. To install and set up SAP Integration Suite, managed gateway for spend management and SAP Business Network, you can use the documentation from SAP Help Portal, available at *https://help.sap.com/docs/sisgw*.

**Cloud connector**

Details about the cloud connector can be found at *https://help.sap.com/docs/CP_CONNECTIVITY*.

If you want to integrate SAP Ariba with a non-SAP backend, a necessary step might be to bring the data from the source system into the format required by SAP Ariba through a middleware system by means of mapping. The development and testing of these mappings can be costly; however, numerous SAP partner companies offer prebuilt mapping content for various systems. File interfaces, generally for comma-separated values (CSV) files, and web service interfaces provided by SAP Ariba can also be used to facilitate this integration.



When migrating transaction data, you're first faced with the decision of whether you want to migrate the data at all or whether you would prefer to complete the relevant transactions in the legacy system. This way you can avoid the sometimes-complicated migration of transaction data. Interfaces are available for some data to be transferred to SAP Ariba. In this case, too, mapping is usually necessary during transfer. Furthermore, this data is then sometimes in a special state in SAP Ariba and rather for archiving purposes in the new system; transactions are thus no longer possible or only possible to a limited extent.

In Section 5.4, we cover the various options available, and the challenges, when migrating user data. In increasingly complex system landscapes, administration complexity also increases. You must ensure that changes in authorizations, substitute regulations, and organizational structures can be implemented equally in all systems.

## 5.2 Migrating Master Data

You must first migrate master data before you can integrate and process other data, such as supplier data or transaction data. Depending on the source system available, the procedures for migrating master data differ.

[«]

> **Leading System for Master Data**
>
> Note that SAP Ariba cannot be the leading system for master data because the leading system is always the connected backend or SAP Master Data Governance system. Only in the case of supplier data with SAP Ariba Supplier Lifecycle and Performance is there a way to update and integrate the data in both directions.

**Upstream processes**

The following master data is required for upstream processes:

- Organizational data
- Plant

- Plant purchasing organizations
- Company codes
- Purchasing organizations
- Purchasing group
- Cross-application configuration data
- Payment terms
- Trade clauses
- Material master
- Item category
- Enterprise resource planning (ERP) commodity code
- Purchasing info record

**Downstream processes**

The following master data is required for downstream processes:

- Account assignment categories
- Account category field status
- Asset and asset class
- Cost centers
- Company codes
- Currency conversion rates
- ERP material group
- Budget management data (such as budget, funds centers, financial management (FM) areas, functional areas, commitment items, earmarked funds, budget periods, and grants)
- General ledger accounts
- Internal orders
- Language-dependent names for data imports
- Payment terms
- Plants
- Plant purchasing organizations
- Purchasing groups
- Purchasing organizations
- Supplier-related information
- Tax codes
- Users and user groups
- Work breakdown structure (WBS) elements

For the integration of master data from the SAP S/4HANA cloud system, SAP Ariba runs via the *Master Data Integration* service in an SAP BTP subaccount.

SAP Business Technology Platform (SAP BTP)

For integration from SAP S/4HANA or SAP ERP 6.0, we recommend you use SAP Integration Suite, managed gateway for spend management and SAP Business Network (formerly SAP Ariba Cloud Integration Gateway).



[«]

**SAP Integration Suite, Managed Gateway for Spend Management and SAP Business Network - SAP Ariba Cloud Integration Gateway**

While we were writing this book, SAP changed the name of *SAP Ariba Cloud Integration Gateway* to *SAP Integration Suite, managed gateway for spend management and SAP Business Network*. Many of the documents we link still use the old name in their URLs or titles. This will have been updated in the latest release.

Alternatives are the CSV and web service interfaces for integration via a middleware system. This approach can be useful if only a small amount of master data needs to be integrated or middleware already exists. In this way, the effort can be kept lower. However, SAP recommends and supports the integration via SAP Integration Suite, managed gateway for spend management and SAP Business Network.

SAP S/4HANA for central procurement

You need a different approach to a master data concept if you want to implement SAP S/4HANA for central procurement in a multi-ERP landscape. In this scenario, the majority of your master data remains in your source system and is temporarily loaded into SAP S/4HANA for central procurement via OData application programming interfaces (APIs) as needed.

If you want to migrate master data from a non-SAP system, you must rely on the CSV and web service interfaces mentioned earlier. A manual upload of master data as a CSV file via the SAP Ariba user interface (UI) is also possible.

How often which master data is updated depends on how often certain data changes in your backend systems and therefore needs to be available updated in the purchasing processes.

### 5.2.1 Master Data Migration from SAP S/4HANA Cloud

If you use SAP S/4HANA Cloud, various business scenarios for purchasing are supported. An overview of the available business scenarios is provided by SAP Signavio Process Navigator at *https://me.sap.com/processnavigator/SolutionScenario*.

[»]

**Planned Replacement of SAP Best Practices Explorer**

At the time of this writing, the SAP Best Practices Explorer website announced its replacement by SAP Signavio Process Navigator starting June 30, 2023. For this reason, you should use the new search options for the SAP Best Practices package for SAP S/4HANA Cloud from mid-2023.

You can also find an overview of integration scenarios for SAP Ariba in the Overview Guide for SAP Ariba Cloud Integration Gateway. In the **Use** section you'll find an overview document, in the appendix of which the possible scenarios are listed: *https://help.sap.com/docs/sisgw*.

**SAP Master Data Integration**

Master data migration from SAP S/4HANA Cloud is performed via the SAP BTP service called SAP Master Data Integration. The use requires an existing SAP BTP global account and subaccount. Furthermore, a Cloud Foundry space must be created in this subaccount, and the necessary users must be added.

SAP Master Data Integration is available in the following regions:

- Europa (Frankfurt) - EU10
- USA East (VA) - US10
- Australia (Sydney) - AP10
- Singapore - AP11

**SAP One Domain Model**

SAP Master Data Integration is based on the SAP One Domain Model, which can be used as a common cross-system data model for the integration of master data. An extension of the data model with custom interpretations and attributes is possible. SAP Master Data Integration provides direct out-of-the-box integration between many SAP systems via available REST APIs. Using SAP Integration Suite, you can even integrate non-SAP applications.

To migrate master data from SAP S/4HANA Cloud to SAP Ariba, this service must be set up and configured. The following master data can be migrated from SAP S/4HANA Cloud to SAP Ariba via SAP Master Data Integration:

- Cost centers
- WBS elements
- Company codes
- Purchasing organization
- Suppliers (only from SAP S/4HANA Cloud to SAP Ariba)

**Master Data Native Integration Service (MDNI)**

Updating data in SAP S/4HANA Cloud triggers an automatic update of records in SAP Master Data Integration. Master data is transferred to SAP Ariba via the SAP Ariba service *Master Data Native Integration (MDNI)*, which is run as a job at regular intervals.

To learn how to configure SAP Master Data Integration, see the document "Master data replication using SAP Master Data Integration for SAP Ariba applications." This document can be found in the **Integration** section with the titles "SAP Ariba Strategic Sourcing solutions" and "SAP Ariba Procurement solutions." You can find the entry point on the SAP Help Portal at *https://help.sap.com/docs/ariba*.

### 5.2.2 Migrating Data from SAP S/4HANA and SAP ERP 6.0

The integration of master data from an on-premise SAP S/4HANA system and an SAP ERP 6.0 system can be performed by following these steps:

- Manual data upload in SAP Ariba UI
- Automatic upload using the SAP Ariba Integration Toolkit
- Real-time upload via web services with SAP Integration Suite, managed gateway for spend management and SAP Business Network

#### Manual Data Upload in the SAP Ariba User Interface

The function for a manual upload of master data in SAP can be accessed in the UI via the following path: **Manage • Administration/Core Administration - Site Manager**.

Click the **Data Import/Export** link, which enables you to import and export the data in CSV format. Figure 5.1 shows the list of different import files that can be created through the SAP Ariba UI.

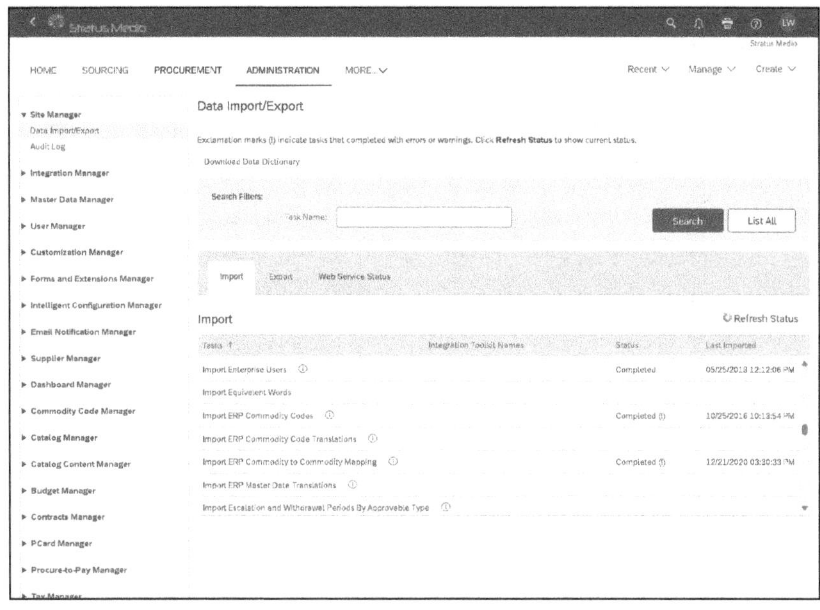


**Figure 5.1** Site Manager in the SAP Ariba Administration Area

Use the **Import** and **Export** sections to switch between importing and exporting data. You can also search for a specific master data file using the **Search** button.

**Data Dictionary**

The Data Dictionary serves as an overview file for the structure of the data in the SAP Ariba application. In the Site Manager, you can download the available Data Dictionary as a Microsoft Excel document. The Data Dictionary contains the structure of the available import and export tasks, with their descriptions. If you have questions when filling out your CSV files, you should consult the Data Dictionary spreadsheet.

**Site variant data model of your SAP Ariba site**

The Data Dictionary spreadsheet is always structured according to the variant of your SAP Ariba page (SAP data model or generic data model). It includes the unlocked features for your specific SAP Ariba site.

**The Data Dictionary Depends on the Group Membership of Your User!**

Keep in mind that the Data Dictionary is based on the group membership of your user. If you do not see certain functions in the UI with your user, they will not be visible as tasks in your downloaded Data Dictionary.

#### Automatic Upload Using the SAP Ariba Integration Toolkit

Using the SAP Ariba Integration Toolkit is the second way to migrate master data. Two tools are available for this purpose:

- The *SAP Ariba Data Transfer Tool* to transfer the CSV files in batch mode from your backend system to SAP Ariba
- The *Database Connector*, which allows for connecting to Java Database Connectivity (JDBC)-based ERP systems

**SAP Ariba Data Transfer Tool**

SAP Ariba Data Transfer Tool is an HTTP client that is operated and controlled from the command prompt. The SAP Ariba Data Transfer Tool can load the data itself to SAP Ariba in batch mode. The tool is set up on a machine within your network and can be run on a Windows or UNIX-based operating system.

The SAP Ariba Data Transfer Tool can also be replaced with a custom tool. Alternatively, you can use a middleware system of your choice to transfer the data to SAP Ariba via HTTP after previously loading the data from the source system.

**Database connector**

The Database Connector allows you to extract master data from your JDBC-based backend system database and save it to CSV files. In addition, you can extract data from backend systems or load it into backend systems that expose interface schemas. The database connector enters data directly into the interface tables and retrieves data by querying either interface or production tables.

For a description of the options and procedures, see the "SAP Ariba integration toolkit guide," available at SAP Help Portal under **SAP Ariba cloud integration**: *http://s-prs.co/v581780*.



#### SAP Integration Suite, Managed Gateway for Spend Management, and SAP Business Network (Formerly SAP Ariba Cloud Integration Gateway)

For the migration and integration of master data from SAP S/4HANA or SAP ERP 6.0 to SAP Ariba, SAP recommends the solution, SAP Integration Suite, managed gateway for spend management and SAP Business Network. This relatively new integration option is designed as integration as a service. SAP Integration Suite, managed gateway for spend management and SAP Business Network is based on SAP BTP running in a separate instance for SAP Ariba systems. It contains all necessary integration objects and mappings in a ready-made form and only needs to be set up by you. Thus, SAP takes care of the development and maintenance of the integration mappings that you as a customer can use.

[«]

**No License Costs for SAP Integration Suite, Managed Gateway for Spend Management, and SAP Business Network**

SAP Integration Suite, managed gateway for spend management and SAP Business Network is provided directly with the SAP Ariba license for your chosen SAP Ariba product. So, there is no need for additional licensing, and you don't need a license for SAP BTP.

For an overview of all functions and options, see the SAP Integration Suite, Managed Gateway for Spend Management and SAP Business Network Overview Guide, under the **Use** tab: *https://help.sap.com/docs/sisgw*.

SAP Integration Suite, managed gateway for spend management and SAP Business Network supports the following SAP Ariba solutions:

- SAP Business Network
- SAP Ariba Buying
- SAP Ariba Buying, Multi-ERP Edition
- SAP Ariba Catalog
- SAP Ariba Sourcing
- SAP Ariba Collaborative Sourcing
- SAP Ariba Commerce Automation

- SAP Ariba Discount Management
- SAP Ariba Supplier Lifecycle and Performance
- SAP Ariba Invoice Management
- SAP Ariba Buying and Invoicing
- SAP Ariba Contract Invoicing
- SAP Ariba Strategic Sourcing Suite
- SAP Ariba Contracts
- SAP Business Network for Supply Chain
- SAP Ariba Supplier Risk

**SAP One Support Launchpad**

For SAP Integration Suite, managed gateway for spend management and SAP Business Network, an add-on must be installed in SAP S/4HANA or in the SAP ERP 6.0 system. There is one version for SAP S/4HANA systems and one version for SAP ERP 6.0 systems. This add-on can be downloaded from the SAP One Support Launchpad in the software downloads.

For more information about setting up SAP Integration Suite, managed gateway for spend management and SAP Business Network, see the SAP Integration Suite, Managed Gateway for Spend Management and SAP Business Network Installation Guide, which you can find on at SAP Help Portal via this link: *https://help.sap.com/docs/sisgw*.

**Full load and incremental upload**

SAP Integration Suite, managed gateway for spend management and SAP Business Network supports the full load and incremental upload of master data from your backend system. You need to configure the master data integration using the add-on, which you can find after installation via SAP GUI in Transaction SPRO or via the IMG menu path **Integration with Other SAP Components • Managed Gateway for Spend&Network for Buyer**. Figure 5.2 shows an overview of the tasks for setting up the **SAP Ariba Cloud Integration Gateway** add-on in the backend for master data integration.

First, you need to maintain the general settings for the connection to SAP Integration Suite, managed gateway for spend management and SAP Business Network and the respective parameters. Next, in the **Maintain System ID for Master Data** section, set the system ID of the connected backend system. Via **Connection Parameters for Master Data,** you can then maintain the realm name of your SAP Ariba page. You can see this name in the top right-hand corner of the SAP Ariba UI and in the URL of SAP Ariba page. After that, a job can be set up in the system using Transaction SM36, which uploads the relevant master data to the selected solution via the ABAP program `ARBCIG_MASTER_DATA_EXPORT`.

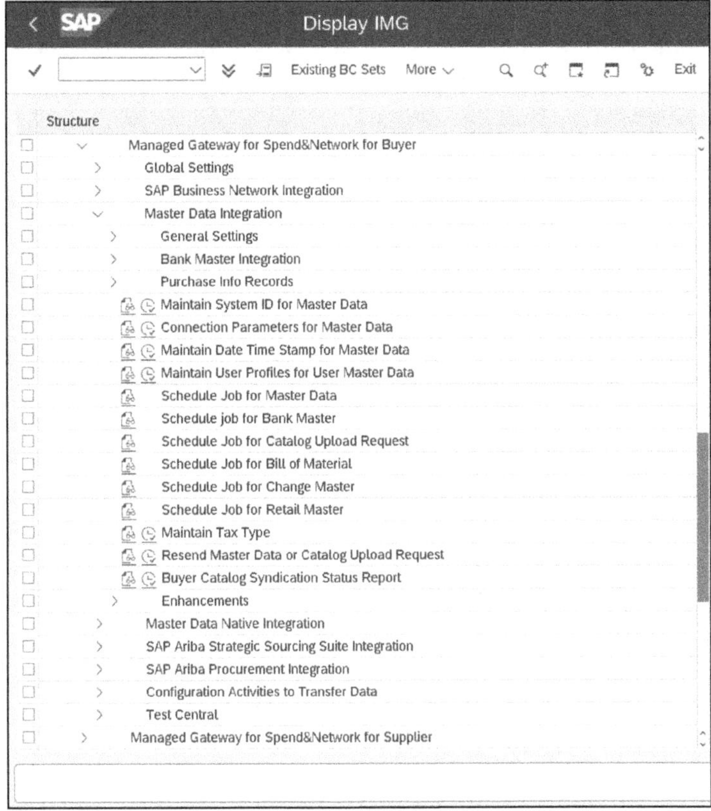


**Figure 5.2** Customizing Options for Master Data Integration with the SAP Ariba Cloud Integration Gateway Add-On

**SAP Ariba MDNI**

In addition, SAP Integration Suite, managed gateway for spend management and SAP Business Network supports the native master data integration to SAP Ariba MDNI. It provides the following functions:

- Transfer of master data from backend to SAP Ariba MDNI via SAP Integration Suite (SAP Ariba Cloud Integration Gateway) in XML format
- Multiple ERP systems are supported simultaneously.
- Each ERP system receives its own system ID.
- The standard web services in the SAP ERP system are used.
- There is no mapping in SAP Integration Suite, managed gateway for spend management and SAP Business Network.
- A full load as well as an incremental upload are supported.
- For a full load, all objects are loaded. If an object already exists, it is first deactivated before it can be loaded again.
- In an incremental upload, only incremental changes are uploaded. If an object does not exist, it will be created.

However, the integration with SAP Ariba MDNI also has some limitations that you should consider. For details, see the SAP Integration Suite, Managed Gateway for Spend Management and SAP Business Network Overview Guide at *https://help.sap.com/docs/sisgw*.

### 5.2.3 Master Data Migration From Non-SAP Systems

SAP Integration Suite, managed gateway for spend management and SAP Business Network are not yet available for migrating master data from non-SAP systems. Instead, you can upload this data via the UI functions presented in SAP Ariba and via the SAP Ariba Integration Toolkit (Section 5.2.2).

Moreover, you can set up your own integration of master data with existing middleware. See an overview of available interfaces at SAP Help Portal via this link: *https://help.sap.com/docs/ARIBA_APIS*.

## 5.3 Migrating Supplier Data

Depending on your scenario, the migration of supplier data differs for SAP Ariba (Section 5.1). If only SAP Ariba Buying and SAP Ariba Guided Buying are used without upstream processes, suppliers are integrated in the format of a supplier organization.

**SAP Ariba Supplier Management**

If your solution also uses SAP Ariba Supplier Management, the supplier data must be loaded via the existing native interfaces in business partner format.

**Business Partner**

The *business partner* designation is a central master data object for all natural persons and legal entities with whom the company has a business relationship. These business partners can be suppliers or customers as well as contact persons.

Managing supplier data as a business partner provides numerous other benefits in addition to native integration:

- Harmonized data architecture between applications
- Strategic data model in SAP S/4HANA
- Relationships between the individual business partners (e.g., "Is service provider of") are possible.
- Different addresses, bank data
- Different roles like *customer* or *supplier* are possible.

- Generic data (name, address, contact details) is shared between business partner roles.

### 5.3.1 Migrating Supplier Data From SAP S/4HANA Cloud via the SAP Master Data Integration Service

Using SAP S/HANA Cloud, supplier data can be integrated via the *SAP Master Data Integration* service from SAP BTP. The integration uses SAP Integration Suite, managed gateway for spend management and SAP Business Network; the architecture is shown in Figure 5.3.

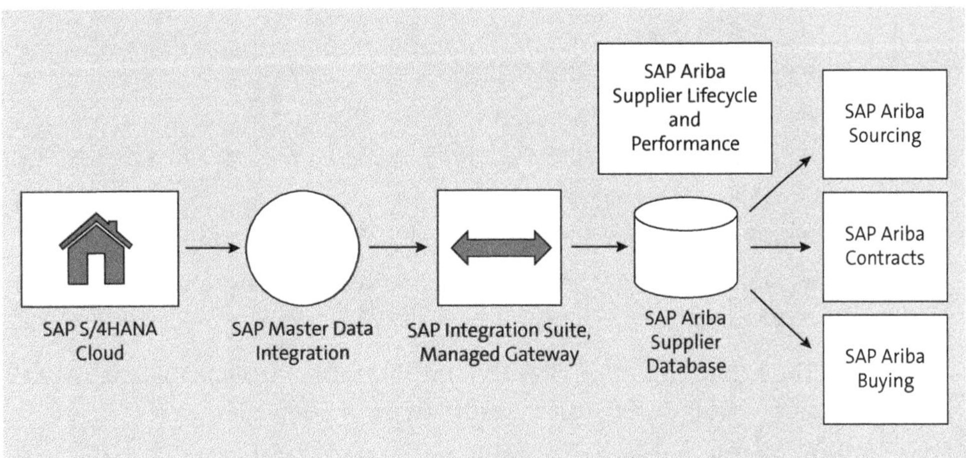


**Figure 5.3** Migration of Supplier Data via SAP Master Data Integration Service

**Pass-through scenario**

From the SAP S/4HANA cloud system, the first step is to define which suppliers should be loaded into SAP Ariba. The data is transferred to SAP Integration Suite, managed gateway for spend management and SAP Business Network (SAP Ariba Cloud Integration Gateway) via native interface in the SAP Master Data Integration service. No mapping takes place there (pass-through scenario); however, an integration scenario must be set up. The data is then stored in the SAP Ariba supplier database.

The migration of supplier data using SAP Master Data Integration also supports the upload of suppliers to SAP Ariba Supplier Lifecycle and Performance and SAP Ariba Supplier Information and Performance Management (New Architecture).

For a description of the configuration steps required for setup, see the document "Master Data Replication using SAP Master Data Integration for SAP Ariba Applications" on the SAP Help Portal pages for SAP Ariba solutions.

### 5.3.2 Migrating Supplier Data from SAP S/4HANA and SAP Master Data Governance, Supplier

SAP Ariba Supplier Management

If you want to manage and integrate your suppliers in SAP Ariba Supplier Management, you need to use the business partner data model to load your suppliers into SAP Ariba.

SAP Master Data Governance, Supplier

A supplier data can be loaded from an SAP ERP 6.0, SAP S/4HANA, or SAP Master Data Governance, Supplier system. The minimum requirements for this integration are the following systems:

- SAP Ariba Supplier Lifecycle and Performance, SAP Ariba Supplier Information and Performance Management (New Architecture), or SAP Ariba Supplier Risk
- SAP ERP 6.0 EHP8 with SP12
- SAP S/4HANA 1809
- SAP Master Data Governance 9.2 as add-on for SAP ERP 6.0 EHP8 with SP12
- SAP Master Data Governance as add-on for SAP S/4HANA 1809
- Newer versions of SAP ERP 6.0, SAP S/4HANA, or SAP Master Data Governance

The following web services are available for the integration with the SAP Ariba system, which can be used natively or via middleware:

- `BusinessPartnerSUITEBulkReplicateRequest_In`
- `BusinessPartnerSUITEBulkReplicateRequest_Out`
- `BusinessPartnerSUITEBulkReplicateConfirmation_In`
- `BusinessPartnerSUITEBulkReplicateConfirmation_Out`

Due to the bidirectional synchronization of SAP Ariba and the connected SAP backend or SAP Master Data Governance, Supplier system, different scenarios are possible.

#### Supplier Creation in SAP S/4HANA and Replication to SAP Ariba

The supplier manager creates a new supplier and wants to replicate it in SAP Ariba:

- You can create a new supplier in SAP S/4HANA and assign the responsible purchasing organization and company code to it.
- SAP S/4HANA replicates the supplier via the data replication framework (Transaction DRFOUT).

#### Supplier Update in SAP Ariba and Replication to SAP S/4HANA

For example, the supplier manager has updated the supplier's contact and wants to update these changes in SAP S/4HANA as well:

- Your employee updates a supplier registration questionnaire, which is then answered by the supplier.
- The update of the questions triggers an automatic replication of the data from this supplier to SAP S/4HANA.
- The SAP S/4HANA system updates the supplier data.

#### Supplier Creation in SAP Ariba and Replication to SAP S/4HANA

The supplier manager registers a new supplier and wants to replicate it to the SAP S/4HANA system after the entire process:

- *Creating a supplier in SAP Ariba*: After a request to create a new supplier is created in SAP Ariba, it is approved, and a new supplier is created with the default settings.
- *Registering the supplier*: A responsible supplier manager invites the supplier via SAP Ariba and sends them one or more registration questionnaires via the SAP Business Network.
  - The supplier confirms the registration and answers the questionnaires.
  - The defined persons approve the supplier's responses, and the supplier receives the **Registered** status.
  - Based on the set integration settings, the **Registered** status can trigger the initial replication of the supplier from SAP Ariba to SAP S/4HANA.
  - In SAP S/4HANA, the purchasing organization, company code, and partner function are assigned according to the defined best practices processes.

For information about the configuration via SAP Integration Suite, managed gateway for spend management and SAP Business Network of this integration, see the SAP Help Portal pages for SAP Integration Suite, managed gateway for spend management and SAP Business Network on the **Integrate** tab in the two configuration documents:

- Integrating SAP MDG-S with SAP Ariba Supplier Lifecycle and Performance
- SAP Integration Suite, managed gateway for spend management and SAP Business Network Configuration Guide

### 5.3.3 Supplier Management and Upload to SAP Business Network

SAP Business Network

To interact with suppliers via SAP Business Network (formerly SAP Ariba Network), suppliers must be enabled in the SAP Business Network by your supplier administrator or supplier manager and connected to your SAP Business Network buyer account.

Suppliers can be created in SAP Business Network (buyer account) by uploading a CSV file. This CSV file can also be used to update the data of existing suppliers in SAP Business Network. A template file can be downloaded from an SAP Business Network buyer account, under the **Supplier Activation** tab.

The conditions for using this CSV file are the following:

- You can have up to 100,000 entries in the file.
- The maximum file size is 10MB.
- You can also use a ZIP archive that contains only the CSV file.
- The file must comply with the syntax rules.

If you work in the administration of the SAP backend, you can also export the CSV files directly from the backend system in the correct CSV format. The syntax rules for this CSV file are as follows:

- The file must have the *.csv* extension.
- The coding must be stated in the first row.
- The values can contain strings, but these values must be enclosed in quotes.

The following fields are required in this CSV file:

- Supplier name
- Supplier ID
- Tax number (only required if no supplier ID has been provided)
- City or zip code
- Supplier country code
- State (only for suppliers in the US)
- System ID (only in multi-ERP configurations)
- Buyer ANID (only for external service providers with a provider account)

Data Universal Numbering System (DUNS)

For more detailed information and other possible fields such as Data Universal Numbering System (DUNS) numbers, see "Enabling Suppliers on SAP Business Network" from SAP Help Portal for SAP Business Network.

## 5.4 Migrating Transaction Data

The migration of transaction data to SAP Ariba is possible to some extent, but not always recommended. Interfaces are also not available for all objects. For example, no interface can import open or closed orders. SAP Ariba Procurement can import requirements from which purchase orders can be generated. However, it is not suitable for archiving data from legacy systems. We therefore recommend that you first close transaction data in the legacy system and/or create the new objects in SAP Ariba in parallel.

One conceivable way out is to create purchase orders from a legacy system via API in SAP S/4HANA. For this purpose, an OData interface in SAP S/4HANA can create, update, and delete purchase orders. You can access SAP API Business Hub at *http://s-prs.co/v581712.*

However, you can use some interfaces to load existing objects into SAP Ariba. One example is the Simple Object Access Protocol (SOAP) interfaces for importing tender projects and tender events to SAP Ariba Sourcing. This can then be combined with the SOAP interface for document uploads to existing tender projects. In this way, you can see and manage existing tenders in SAP Ariba Sourcing.

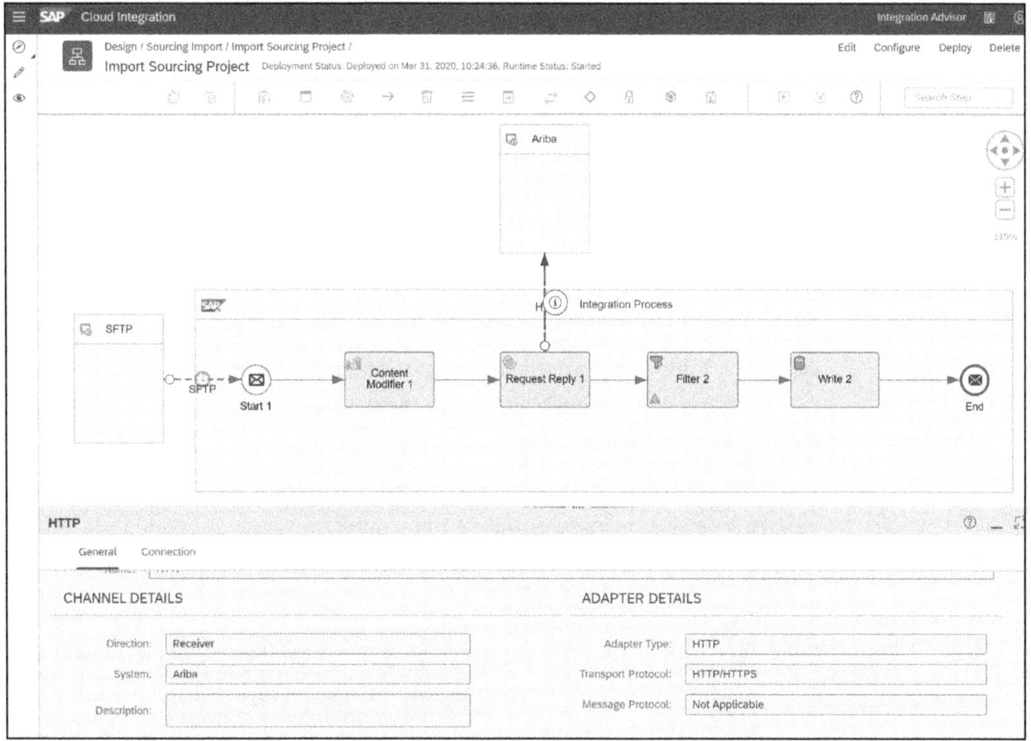


**Figure 5.4** Cloud Integration Service iFlow for Importing a Tender Project

This integration can be achieved relatively easily by using the Cloud Integration service as middleware. Cloud Integration brings all the necessary adapters and functions to create such matching iFlows.

**SAP Integration Suite: mapping iFlows**

Figure 5.4 shows an example of an iFlow that reads the files from a Secure File Transfer Protocol (SFTP) server and can create the objects in SAP Ariba via a request-reply HTTP POST connection. To connect to your SAP Ariba site, create an endpoint for this interface in SAP Ariba via **Administration • Integration Manager • Endpoint Configuration**.

So, in general, some work needs to be put into the middleware if you want to load existing objects into your SAP Ariba systems. SAP also offers its own migration service as an additional consulting service for a fee, which can take over such tasks with ready-made tools.

## 5.5 Migrating User Data

**User manager**

SAP Ariba includes its own user management (user manager) in the administration area. The user data comprises data such as the user ID, name, organizational affiliation, and address as well as contact details, but also the assigned user groups. You can load user data from different systems into SAP Ariba. You can load user data into SAP Ariba in three different ways:

- Upload via the SAP Ariba UI with the relevant CSV files
- Upload using SAP Integration Suite, managed gateway for spend management and SAP Business Network
- Upload and user management using the SAP BTP Service, SAP Cloud Identity Services - Identity Provisioning

### 5.5.1 Upload CSV File via the SAP Ariba User Interface

When you upload user data via the SAP Ariba UI, SAP Ariba provides several CSV files that need to be filled in with the data to be transferred. You can also load a consolidated file containing the other data in a simplified variant.

For a detailed description of the different CSV files and the consolidated variant, see the "Common Data Import and Administration for SAP Ariba Procurement Solutions" guide on SAP Help Portal.

### 5.5.2 Upload Using SAP Integration Suite, Managed Gateway for Spend Management, and SAP Business Network

SAP Integration Suite, managed gateway for spend management and SAP Business Network enables you to load user data and roles directly from SAP

S/4HANA or SAP ERP. For this purpose, an import event named **Import User Group Mapping** is available in the associated add-on.

The data exported here via the add-on for SAP Integration Suite, managed gateway for spend management and SAP Business Network contains the information you can obtain from Transaction SU01. The HR information on cost center, company code, HR nationality and language is also included.



However, with this method, all users you want to use in SAP Ariba must be maintained in SAP S/4HANA or SAP ERP.

### 5.5.3 Upload and User Management with SAP BTP and SAP Cloud Identity Services for Identity Provisioning

**SAP Cloud Identity Services**

The most convenient way to migrate user data is to use SAP Cloud Identity Services. With the Identity Authentication and Identity Provisioning services included, two cloud services are available to provide the authentication and distribution of user master data from different connected systems, such as Microsoft Active Directory, Microsoft Azure, or another identity provider (IdP) system. To integrate with the SAP Ariba system, Identity Provisioning uses SAP-internal SOAP interfaces to create, update, or delete the data.

The advantage of this approach is that you can use a central tool for user management and distribution. However, additional licenses must be purchased on the SAP BTP for SAP Cloud Identity Services, as they are not included in the SAP Ariba license.

For details on the different source and target systems, see SAP Help Portal's SAP Cloud Identity Services - Identity Provisioning page at *https://help.sap.com/docs/IDENTITY_PROVISIONING.*

## 5.6 Summary

In this chapter, you learned about the different options for importing data to SAP Ariba. With SAP S/4HANA Cloud and an SAP S/4HANA system, you have advantages over a manual or CSV file-based data migration due to native interfaces and a standard integration via SAP Integration Suite, managed gateway for spend management and SAP Business Network. Ultimately, you can also often choose a suitable integration variant that best suits your processes and system infrastructure.

Chapter 6

# Data Migration in SAP Business ByDesign

*The migration tool in SAP Business ByDesign offers step-by-step user guidance and a comprehensive simulation and preview functionality so you can easily assess the possible result of a migration in advance.*

6

This chapter presents the migration tool for migrating data into the SAP cloud solution SAP Business ByDesign. As a user, you'll be guided through the process step by step, mainly via dynamic task lists.

Many business objects are available for migration. The migration process is designed so that complex objects usually don't require programming. In most cases, the use of Microsoft Excel formulas is sufficient. However, be aware that the ease of use of these technical processes can lead to an underestimation of the complexity and consequences of details in the migrated data. This chapter should enable you to apply the features of the migration tool in a controlled and effective way and narrow down root causes during error analysis or even avoid errors up front.

In each section of this chapter, we'll discuss the following:

- Section 6.1: Object-specific migration template files
- Section 6.2: The features and functions of the migration tool
- Section 6.3: The central steps **Validation**, **Value Conversion**, **Simulation**, and **Import**, with a focus on identifying and fixing data errors
- Section 6.4: Migrating transactional accounting data
- Section 6.5: Enriching or modifying migration data
- Section 6.6: Migration as a subproject, with an emphasis on the test strategies for each phase
- Section 6.7: Tips and tricks from real-life projects, with a focus on parallelization and reconciliation after migration

## 6.1 Migration Templates

For the collection of legacy data, SAP Business ByDesign provides *migration template* files. These XML files open in Excel and can largely be maintained like any other spreadsheet file. Let's first look more closely at the migration templates and then discuss how to populate them with data.

### 6.1.1 Structure and Elements of Migration Templates

Technically, migration templates are files in the XML spreadsheet format. For every migration object (e.g., customers, materials, general ledger account balances), there is a separate migration template. Apart from the English language version, there are separate versions in many languages supported by SAP Business ByDesign, including Chinese, Dutch, French, German, Italian, Japanese, Portuguese, and Spanish.

**Structure of migration templates**

Every migration template consists of several sheets, between which you can navigate as usual via the tab strips at the bottom of the screen:

- **Introduction**
  Technical information on how to work with templates.
- **Data sheets**
  One or several data sheets for the actual data (including hidden data sheets).
- **Documentation**
  Object-specific information.
- **Field list**
  List of all data sheets and fields (including hidden ones).

**Data sheets**

Let's take a closer look at the data sheets, for example, the migration template for **Customer** master data. The **General** sheet contains the header part of the customer record, that is, exactly one line for every customer, as shown in Figure 6.1.

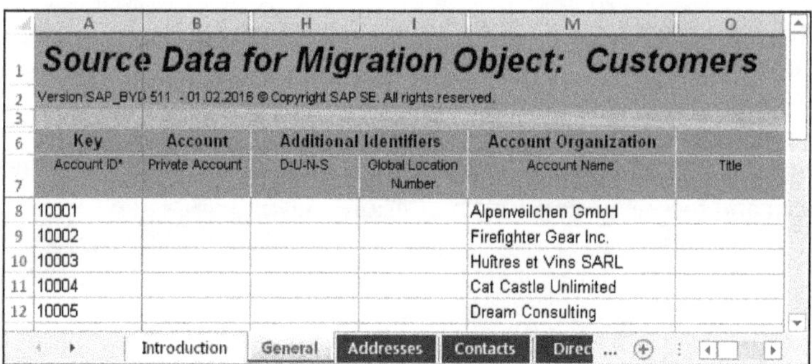


**Figure 6.1** Migration Template: Header Sheet General

A customer can have one contact person, several contact persons, or none at all. Correspondingly, on the next data sheet (**Contacts**), you enter one line, several lines, or no lines at all for each customer. With the customer ID in the first column, the lines are uniquely assigned to header records, as shown in Figure 6.2.

**Source Data for Migration Object: Customers**

Version SAP_BYD 511 - 01.02.2016 © Copyright SAP SE. All rights reserved.

| Key | | | | | Contact Name | | |
|---|---|---|---|---|---|---|---|
| Account ID* | Contact ID | First Name | Last Name* | Main | Title | Academic Title | Phone |
| 10001 | | Heidi | Maier | | Frau | Dr. | 09256-45485 |
| 10002 | | Susan | Wolve | | Frau | | |
| 10002 | | Homer | Smith | X | Mr | | |
| 10003 | | Jean | Luc | | M | | +33 1 4617- |
| 10005 | | Stefan | Moeller | | Mr | | +61 466 123 |
| 10005 | | Judy | Lopez | | Ms | | |

Introduction | General | Addresses | Contacts | Direct Resp ...

**Figure 6.2** Migration Template: Data SheetContacts

**Field help text in line 7**

Every field has input help in the form of a short field help text. You'll find this text in template line 7, underneath the field name. Enlarge the line height of line 7 to see the entire cell content, as shown in Figure 6.3.

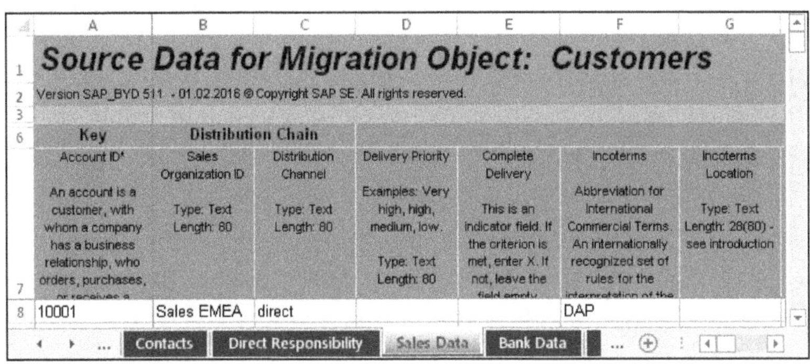


**Figure 6.3** Field Help Text in Line 7

**Mandatory data sheets**

Data sheets, for which the tab strip is highlighted in orange, must be filled in for every record. This requirement applies to the header sheet in this example; another example is the **Items** sheet on the **Sales Order** template.

You might have encountered some field names that are marked with an asterisk (*) in the column header and in the field list. These fields are mandatory, which means that if the current sheet is used by a record, then these fields must be populated.

Some exceptions exist, however, in which whether a certain field or sheet is mandatory for a given source record depends on certain conditions. These conditions can include the value of a neighboring field, the value of a fine-tuning entry in configuration, or the planned usage of this record in certain business processes. In such conditionally mandatory cases, you'll find further information in the field help text or in the object documentation (**Documentation** sheet).

Hidden sheets

Some sheets are used only rarely but are nevertheless necessary for some use cases. In favor of better usability with migration templates, these sheets are hidden by default. On the **Field List** sheet, these sheets are marked as **Optional, Hidden**. You can unhide these sheets easily by following these steps:

1. In Microsoft Excel, move your cursor over any visible tab.
2. Right-click and select **Unhide** in the context menu. The list of all currently hidden sheets appears.
3. Select which sheet to unhide and confirm by clicking **OK**.

Originally hidden sheets appear in as gray tabs. For example, the **Contact Hours** sheet in the **Customer** template is hidden so preferred visiting times for customers can be maintained.

Hidden columns

The same applies to columns within a sheet. Template columns that are rarely used are hidden by default. In the field list, these fields are marked as **Optional, Hidden)**, so you can spot them easily, as well as their next visible neighbors on either side. If you want to unhide such a field, open the corresponding sheet, and mark the column range between these visible neighbors. Move your cursor over the top end of the marked columns, right-click to open the context menu, and click **Unhide**.

Alternatively, you can mark a column range and follow the menu path **Home • Cells • Format • Hide & Unhide • Unhide Columns**. If you want to unhide all the columns of a sheet, mark all its columns.

If you want to copy and paste blocks of data values from a source file into the migration template, unhide all columns first to better control which columns are actually being populated.

For further processing, you can also unhide any columns and sheets that aren't needed in your case for the convenience of your teams.

### 6.1.2 User Tips

Avoid structure changes

Using migration templates downloaded from the same system is beneficial because it avoids potential compatibility issues both between system versions and when customer-specific fields are being used because they have different technical keys in different systems.

Don't change the structure of the migration template; in particular, don't add any columns or sheets manually. If you miss a field, check in the **Field List** sheet to see whether it's among the hidden fields. Avoid entering formulas, print area definitions, or similar Microsoft Excel functions in the file that you're using for upload; instead, make yourself a copy of the template file for these purposes.

File types

Always save the migration template in the current XML file format (Excel spreadsheet file). Don't save the file as an *.xls* or *.xlsx* file, even temporarily, if you still want to use this file for upload to the system. Again, use a copy of the template for your purposes.

[«]

**Warnings When Saving Templates**

When saving a template in Microsoft Excel, you usually receive a warning popup window, as shown in Figure 6.4. Particularly in Microsoft Excel 2010 and earlier, the wording of this warning is somewhat confusing to many users.

The warning refers to the fact that, while having the file open you could have theoretically made some changes that aren't compatible with the XML file format (e.g., defining a macro) and thus won't get persisted when saving. Because you haven't made such changes, answer the question of whether you want to stay with the XML spreadsheet format with the default option **Yes**.

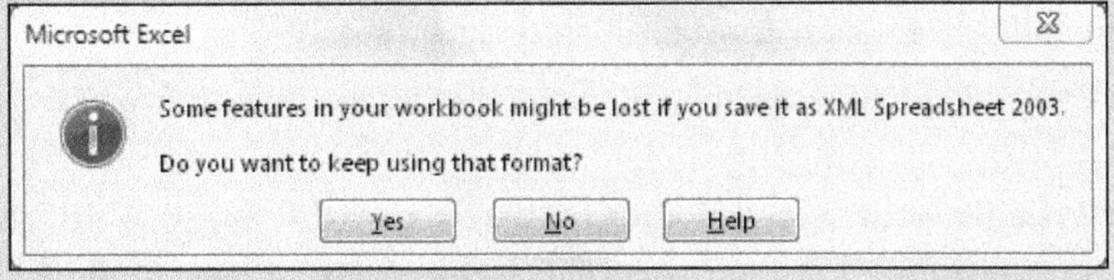


**Figure 6.4** Warning When Saving Templates

Customer-specific extension fields

In the application work center views (e.g., **Account Management • Accounts**), you can define customer-specific extension fields for the corresponding business objects. This task is performed in adaptation mode (Section 6.7.2 for step-by-step instructions). After you've "published" your field there (via the **Publish** button), the field is available in the newly downloaded migration template. In template line 7, the user will see the field label, while the technical key in the (hidden) line 5 is relevant for the import.

[!]

**Errors When Populating Migration Templates**

Issues during file upload can be caused by mistakes that during the population of the migration template with data, such as the following:

- Additional entries in columns outside the expected columns range (often manually added columns or some free text notes outside the data range)
- Formulas instead of values
- Lines with subtotals and totals that a user has added
- Definition of print areas or similar
- Migration template from an earlier system version
- Inadvertent data copied into some hidden columns

Invalid field formats are regularly a root cause for issues. The next section discusses the expected field formats.

### 6.1.3 Field Formats

Every field expects the data in a certain format. You'll find the information on the format in two places:

- Column header in line 7 as part of the field help text
- Field list on the last sheet

**Text fields** Most fields in the migration templates are text fields.

**Field length** For most text fields, an explicit length is indicated, for example, `City; Type: Text; Length: 40`. Longer entries will be truncated during import, and a corresponding warning is raised in all processing steps.

The field length of notes and long text fields is usually indicated as `unrestricted`.

**Length 10(80)** An exception are length indications of the form `Length: 10(80)`, which you find mainly in code list fields (e.g., **Language**, **Country**) and in some ID fields (e.g., **Customer ID**, **Contact ID**, **Material ID**).

What does `10(80)` mean, for example, for the **Contact ID** field?

- At first, it says that the permitted target length for migration import is 10.
- The `(80)` now says: If the length of your source contact ID is more than 10 characters, you can still use them in the migration template as long as they don't have more than 80 characters. However, the **Convert Values** migration step will then enforce a reduction to the right target length of 10 (depending on your parameter choice either by automatically assigning a new number or by manual conversion).

**Key text fields** Restrictions exist for key fields of this type (e.g., IDs, customer numbers, material numbers, document numbers, etc.). For key fields, you can only use the following characters:

- **Capital letters A–Z**
  Lowercase letters in key fields are automatically shifted to uppercase during processing. So if you're filling key fields with lowercase letters, take care that they remain unique for the file also when all entries are shifted to uppercase.
- **Numeric digits 0–9**
  Be aware that number range definitions are usually defaulted in a way that IDs such as `1ABC` or `1_2000` (character string starting with a digit) aren't expected.
- **Special characters**
  - Slash ( / )
  - Backslash ( \ )
  - Colon ( : )
  - Hyphen ( – )
  - Underscore ( _ )
  - Period ( . )

For some key fields, even stronger restrictions apply, as described in the field help text. Most prominent examples are accounting documents. **Document IDs** in accounting can have only uppercase letters and digits (e.g., 140000, A10023). Leading zeros aren't permitted.

Unicode

Fields for names and descriptions allow you to draw from abundant resources. SAP Business ByDesign is a Unicode system, which means that their normal character set includes characters from many languages (including Chinese, Japanese, and Cyrillic), even some symbols, with only some technical control characters excluded.

Number format

Quantities and amounts are expected in number format, for example:

- Full payment due dates: `Length 10` (without decimal places)
- Discount percentage: `Length: 5; Decimals: 3`

`Length: 5; Decimals: 3` means that in total five digits are available; of these, up to two are before the decimal point, and up to three are after the decimal point. For example, `12.345` is valid for this field.

Whether you should use a comma or a colon as a decimal marker depends on the settings on your PC. The basic setting is defaulted by the country version of your operating system, which is a period (`123.50`) for the United States, the United Kingdom, or Switzerland, and a comma (`123,50`) in the country versions for most other European countries.

However, you can override these defaults in the options in your local Microsoft Excel installation by following these steps:

1. Open the Microsoft Excel options menu (depending on your version, via the **File** menu or the **Office** button).
2. Choose **Advanced • Editing options** and deselect the **Use system separators** checkbox.
3. Maintain the values for the **Decimal separator** and **Thousands separator** fields.

**Amounts**

Amount fields always refer to a currency field that is either adjacent or inherited from a superordinate sheet. In the template, a generic number definition is given, usually with five decimal places. Note that, even if the template fills every number up with trailing zeros, it depends on the currency as to how many of these decimal places you can really use. Amounts in the euro or US dollar allow only two decimal places. For example, 123.12 (displayed as 123.12000) is a valid amount for these currencies; neither 123.12900 nor 100.230001 are valid.

**Date format**

The date format is likewise defined in the settings in your Excel options, with the default value derived from your operating system. For example, MM/DD/YYYY, (12/31/2024) is used for the US version, and DD.MM.YYYY (31.12.2024) is used for many European versions. Internationally, there are more formats in use. You can override these default settings on your PC. In Microsoft Windows 7 and newer, simply choose **Control Panel • Region and Language** and select the value for the **Short date** entry.

Independent from the local settings, there is one format that always can be entered into date fields in the migration templates: YYYY-MM-DD (e.g., 2024-12-31). Microsoft Excel recognizes this format in all cases, independently from your settings (unless the cell format has inadvertently been overwritten).

[!]

**Suspect Left-Aligned Dates and Amounts**

If a date or amount field appears left-aligned in a migration template, then this is evidence that the field content might not have been recognized by Microsoft Excel as a correct date or amount. Check these fields for potential errors, like the following:

- Invalid values; for example, 30/02/2024 and 02/30/2024 aren't valid dates.
- Format doesn't correspond to the settings (e.g., wrong decimal separator, confusion of a thousands separator in the source data).
- The Microsoft Excel cell format in the migration template has been overwritten, for example, by an incorrect copy–paste operation (paste instead of paste values, the replacing the original format with text format).

Time format

Provide the time in the country-independent format HH:MM:SS, for example, `14:52:40` (24-hour format) if possible. If you have a different setting on your PC, then the corresponding time values should also be recognized by Microsoft Excel when populating the template (e.g., `02:52 PM`).

Date/time format

The format `Date/Time` combines date and time into one field. Format both parts as just described, separated by a space; for example, the US country version format `12/31/2024 02:52:40 PM` can be used. You can also stay on the safe side with the country-independent format YYYY-MM-DD HH:MM:SS (e.g., `2024-12-31 14:52:40`).



### 6.1.4 Export Data from the Source System

After you've become familiar with the information that is collected in a template, the next task is data retrieval from the source system. Some recommendations for this step include the following:

Use source values

- Use the source values from the legacy system as they are. In the **Convert Values** step, the migration tool offers you a value mapping from your source values to valid target values. Doing the value mapping here and not outside of the system offers several advantages:
  - The value mapping is persisted so that later runs with other source files for this object will benefit from these value conversions automatically.
  - The tool ensures that other objects will also use the same value mapping, which supports effectiveness and data consistency. This principle applies also for certain key fields such as **Customer IDs**.
  - You can check the field help texts in line 7 for documented object-specific exceptions.

Use descriptions

- If possible, you can export not only a code (e.g., `SI` or `0003`) but also a meaningful description (e.g., `Slowenia`, `Trading Goods`). The use of semantic values gives more control during the work with migration templates and prevents misunderstandings. Your company wouldn't be the first to migrate its Canadian suppliers (code `CDN` in some registers, but International Organization for Standardization [ISO] country code `CA`) not only to a different system but also to a different continent, if the value mapping decisions are made based on codes, not on descriptions. (`CD` is the ISO country code for the Democratic Republic of the Congo.)

  If the descriptions from the source system aren't unique, then a combination of both will do, for example, `SI_Slowenia` or `0003 - Trading Goods`.

Use UTF-8

- If your source system allows you to choose the character encoding, then use UTF-8.

### 6.1.5 Filling Data into Migration Templates

Frequently, the reporting and download functionality in the source system is used to export data into a format that can be opened with Microsoft Excel and to transfer data block-wise or column-wise into the template sheets.

If you're looking rather for a direct upload of comma-separated values (CSV) files into the target system using field mapping, then read Section 6.1.6 and Section 6.2.6.

Even then, you might be using the copy-paste method for some test data. If so, watch out for the following:

Paste Special: Values

- When transferring data via copy and paste, don't use the simple **Paste** option; instead, use the Microsoft Excel function to paste only values. For this task, right-click and select **Paste special • Values.**
- Pasting values only prevents the original template formats from being overwritten and thus avoids potential issues (e.g., a wrong interpretation of thousands separators).

Column sequence

- If the source system allows you to define the column sequence during export, it's helpful to choose a sequence that matches the column sequence in the template.

Hidden columns

- Some columns are hidden by default in the migration template. Unhide all columns before pasting data values into the template sheet. If possible, a beneficial step might be to create empty columns in the source export that allow you to transfer larger blocks of columns in one step.

**Copy Data into the Migration Template**

A few recommendations for the export and transfer of data into migration templates include the following:

- Avoid value mapping outside the migration tool.
- Use descriptions and self-explanatory field contents where possible.
- Unhide columns in the template sheet.
- Define a smart column sequence for export and insert empty columns where applicable.
- Copy the data block-wise.
- Always use **Paste Special: Values**, not **Paste**.

### 6.1.6 Alternative Options to Populate Migration Templates

Uploading CSV files

The most important alternative to using the local template files is the direct upload of CSV files. This functionality allows you to upload flat list files into the migration tool. During this step, a flexible field mapping is available, which allows you to store the mapping information for later reuse.

Both field mapping (Section 6.2.6) and value mapping (Section 6.3.2) are parts of the migration functionality.

**Manual entry**

Of course, you can enter a data record manually into a migration template, but this approach takes about the same effort as entering the record directly into the system. Still, several good reasons to use a migration template include the following:



- The option to reuse the template file pays off on several occasions—and thus the time investment pays off. This reuse can be in another system (e.g., an additional test system or your production system) or even within the same system, if you decide to go back to a restore point.
- The file allows you to quickly re-run some test variations of the same record.
- Reusing the file ensures the exact reproduction of the same data, without the risk of typing errors. So, the manually entered data can even be tested in a test system without additional manual effort.
- In some objects, there are capabilities that are open to migration only, for example, allowing historic dates that the operational work center won't allow.
- Imported migration files remain available as documentation for auditors.
- For accounting data, the migration tool gives you a posting and balance preview across objects, proving even before loading that the collection of accounting files is consistent. To make use of this check, all files need to participate, including files that by nature don't have many records, such as the list of balances of your home bank accounts.

**Programming approaches and ETL tools**

Additional options to provide data depend on the export functionality of your source system. Migration templates are XML files of a certain format, thus it's technically possible to generate a text file that matches the XML source code of a populated migration template exactly, with all data values in the right XML tags. However, we advise against such an approach because it attracts your resources toward the generation logic and away from the central point, which is the business data content that is to be migrated.

A similar concern applies to the usage of a separate extract, transform, and load (ETL) tool. While in some groupings such tools can add values to the data collection (e.g., by joining source data from different systems), we're hesitant to put effort into an external field and value mapping. The migration in SAP Business ByDesign is designed to push forward and confront the business data with the target system's reaction as early as possible, so that the business topics (data as well as configuration) quickly become the center of attention.

### 6.1.7 Data Cleansing

In Chapter 1, we discussed some fundamental considerations for data cleansing. Users sometimes experience field checks in their SAP solutions to be stricter than in their legacy system.

SAP Business ByDesign, for example, doesn't allow you to store *www.sap.com* as an email address. The SAP system's data quality paradigm aims at smooth operational usage later—at the expense of higher effort up front during migration. In view of the ongoing issues in future marketing campaigns, a weaker data quality check during migration of this field isn't sustainable.

This raises the need to cleanse certain data before or during migration. The following aspects should be respected during data cleansing:

**Data cleansing examples**

- Unnecessary records (e.g., former test data, obsolete records, etc.) should, if possible, already be filtered during data export from the legacy system. Of course, you must ensure that these records are really unnecessary, that is, that no other dependent data still have a reference to this record.
- Unnecessary spaces or blank lines in text fields can lead to a visually undesirable result.
- Numeric fields must not contain letters.
- For negative numbers, check the field help texts in line 7 of the migration template to understand how the values need to be provided. This advice applies particularly to transactional accounting objects, where some fields expect only positive values without signs and collect the correct sign from the debit/credit information in a neighbor field in the same template sheet.
- Postal codes are checked against the valid format definitions in their respective country. This check is very exact.
- Email addresses must be formally correct.
- Phone numbers cannot have characters.
- Watch out for fields in your legacy system that had been used for other purposes than their original intention. Ensure that the content of these fields is moved to the appropriate target field. Create a customer-specific extension field if necessary.

Our approach to data cleansing is deliberately somewhat conservative. If an entire network of dependent data is being migrated (transactional data, accounting data, higher master data such as production models), a double cleansing might quickly create inconsistencies. If only fundamental master data is being migrated, you have more room to maneuver.

**Risk and benefit**

A strong argument for data cleansing is that the data can be changed fairly easily outside of the systems. On the other hand, a data cleansing subproject binds resources that might be overcharged already by other implementation activities and parallel operational duties. We recommend balancing risks and benefits for your individual situation.



## 6.2 Migration Tool

In this section, we show you step by step how to migrate data to SAP Business ByDesign by following a concrete, practical example. We'll start by showing how a migration is integrated into the task list used by the system to guide you through the entire implementation project. After addressing early questions like where to get the migration template files that we presented in Chapter 5 and where to get more detailed information, the central part of this chapter is a presentation of features the migration tool offers to process your migration template files. As an alternative to the usage of migration templates, we'll also cover uploading data from CSV files, including the field mapping functionality. We'll finish this chapter by familiarizing you with the data editor, which helps you efficiently work with the uploaded data as well as helps you analyze and resolve potential data issues.

[«]

**Note**

If you haven't had the opportunity to work in these systems so far, we're going to focus on the terms and options that you need in migrations. A more general introduction to SAP Business ByDesign isn't the intention of this book.

### 6.2.1 Navigation in SAP Business ByDesign

As soon as you've logged on to the system, the SAP Business ByDesign homepage will welcome you.

**Navigation between work centers**

Using the entries on the left side of the screen (e.g., **Home**, **Business Configuration,** etc.), you can navigate to the *work centers* of the various business applications. For every work center, you can choose between several *views*, focusing on certain business data objects or offering direct navigation to the most common tasks for the current context.

As soon as your mouse cursor hovers over a work center name, a list is displayed of all the views assigned to your user account, as shown in Figure 6.5.

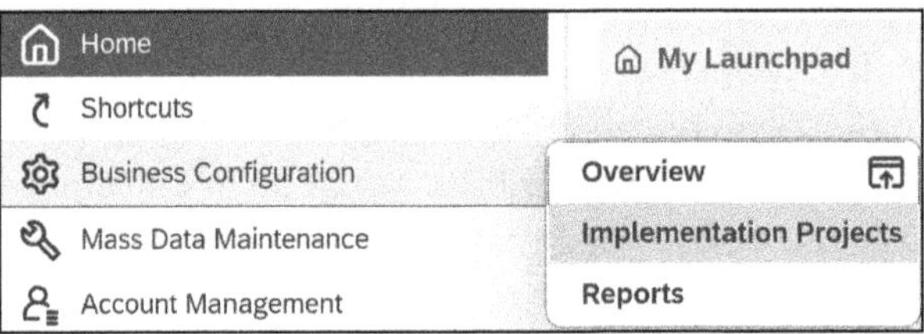


**Figure 6.5** Navigating to Work Center Views

For views that you need often, click on the little **Add to Shortcuts** icon at the end of the view name. This feature allows you to jump directly to this view next time, as shown in Figure 6.6.

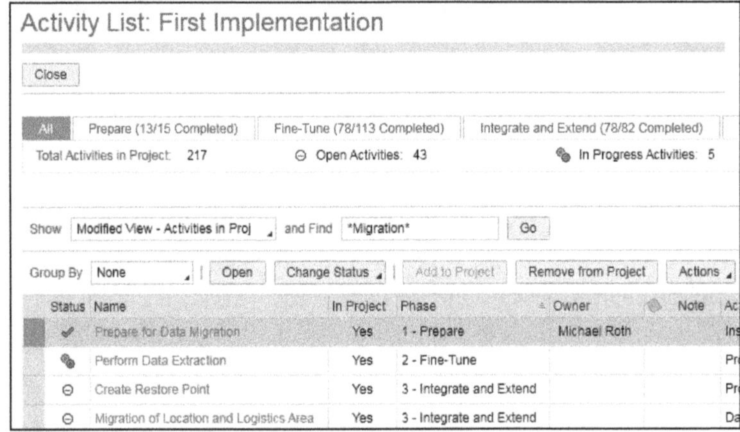


**Figure 6.6** Navigation via Shortcuts

Alternatively, you can use the click the icon (**Start**) in the bottom-left corner to navigate to a specific work center view.

You'll find the data migration in the **Business Configuration** work center in the **Implementation Projects** view. Open this view now.

[+]

**When the Business Configuration Work Center Isn't Visible**

If the **Business Configuration** work center isn't offered in the list on your screen, check the assignment of **Access Rights** in the **Business User** view of the **Application and User Management** work center. Make sure the **Business Configuration** work center with the **BC_DATAMIGRATION** view has been assigned to your user account.

### 6.2.2 Migration Tasks in the Implementation Project

Initially, your list of implementation projects initially contains only a single entry: the **First Implementation** project.

**Activity list**

Your entire implementation is driven by a task list, which has been generated according to the chosen *project scope*. The data migration is represented by several separate task activities, which are embedded into the sequence of other implementation tasks.

Navigate to the activity task list by clicking the **Open Activity List** button. The tasks are structured into five project phases; the default view shows the tasks of the first phase: **1 - Prepare**.

For a list of all migration-related tasks, select the **All** tab and filter by description by choosing **All Activities** in the **Show** list and entering "*migration*" in the **Find** field. Then, click **Go**, as shown in Figure 6.7. The **In Project (Yes/No)** column informs you which tasks are currently assigned to your implementation project. Tasks can be changed using the **Add to Project/ Remove from Project** buttons.

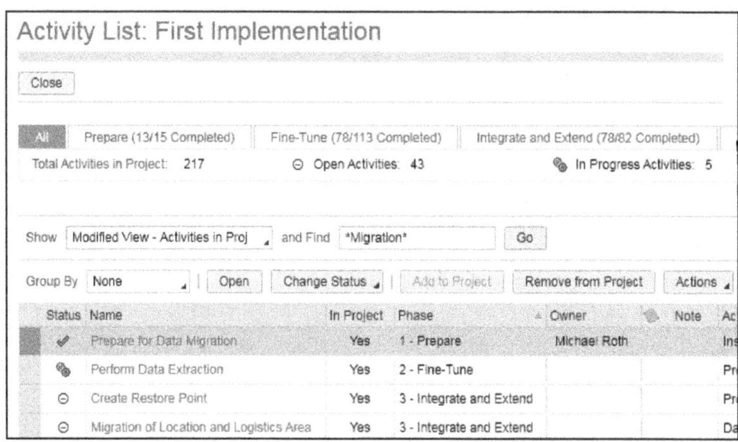


**Figure 6.7** Migration Tasks

In this section, we'll focus on tasks necessary for the actual data migration work:

- **Prepare for Data Migration** (Phase **1 - Prepare)**
- **Migration of <Name of Migration Object>** (Phase **3 - Integrate and Extend),** with a separate task for every relevant data object.

Some additional migration-specific tasks will be discussed in Section 6.6, including embedding the migration into a larger context as a central subproject of the entire implementation project.

Let's get an overview of your migration tasks using the filter **Show Activities in Project** and entering "*migration*" in the **Find** field. By selecting **Actions • Change Owner**, you can pick this task or assign it to a colleague.

### 6.2.3 Prepare for Data Migration

Open the **Prepare for Data Migration** activity to see its tasks, as shown in Figure 6.8.

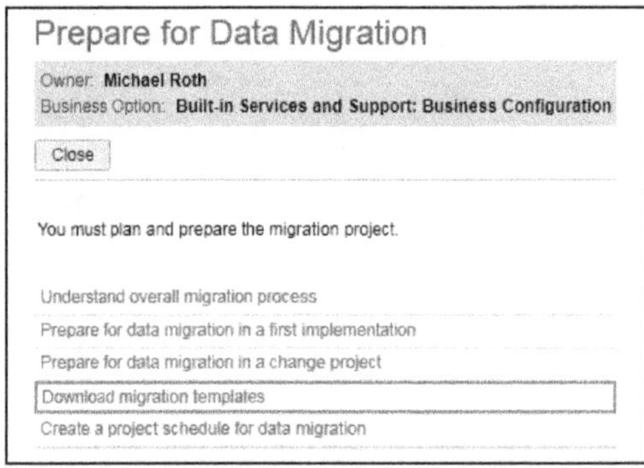


**Figure 6.8** Prepare for Data Migration

The first three tasks branch to detailed descriptions and instructions:

- **Understand overall migration process**
  A summary on migration functionality with links to further migration-related articles in the help center.
- **Prepare for data migration in a first implementation**
  A short help text.
- **Prepare for data migration in a change project**
  A detailed article on the aspects of change migrations, including links to object-specific information. We'll come back to this topic in Section 6.5.

**Download migration templates**

Open the **Download migration templates** task. These migration template files are offered for download covering all objects corresponding to your project scope, as shown in Figure 6.9. The objects are structured into four groups:

- Master data
- Transaction data
- Transaction data for Financial Accounting (FI)
- Basic data

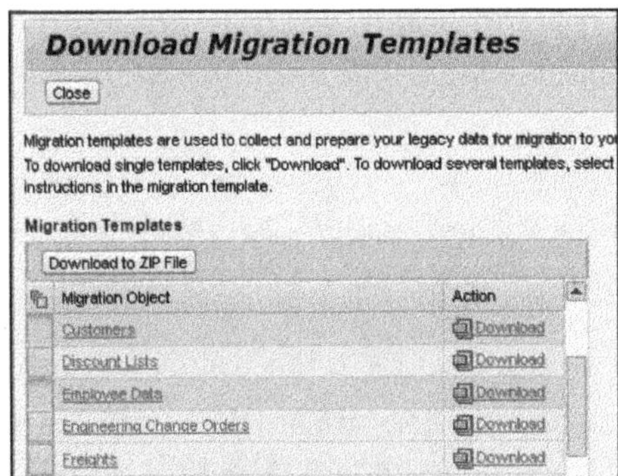


**Figure 6.9** Downloading Migration Templates

Use the **Download** button to store single template files locally on your PC. Alternatively, select an entire group or several objects by holding down the [Shift] or [Ctrl] key and clicking the **Download to ZIP File** button.

Clicking on the object name in this list brings you to an object-specific migration documentation. (The same text is also contained in the template file on the second to last sheet but without the hyperlinks that you have in the online version.)

In our example, we're storing the migration template for customer data using the local file path *C:/Templates/Customer.xml*.

**Customer-Specific Extension Fields**

Most migration objects support the extensibility functionality. A downloaded migration template has all extension fields that have been defined and published in the system so far, so these steps need to be timed to each other.

The template contains not only the label of your extension fields but also (in a hidden header line) their technical keys, which are specific to your system. Therefore, in migrations with extension fields, it's important to use a template file only in the actual system from which you've downloaded it.

In the implementation activity list in the later phase **2 – Fine-tune**, you'll eventually find the **Perform Data Extraction** activity, which allows you to navigate to the template download screen as well.

[+]

**Object Missing in Template Download List**

If you're missing an object in the **Migration Templates** download list, check the decisions that have been made during the project scope definition: Go to the **Business Configuration** work center, select the **Implementation Projects** view, click the **Edit Project Scope** button, and mainly verify steps **3 – Scoping** and **4 – Questions**.

> If, for example, in step **3 – Scoping,** the element **Human Resources • Personnel Administration** hasn't been selected, then the corresponding migration object won't be available.
>
> In step **4 – Questions** (located in the **Dependency** tab accessed by choosing **Built-In Services and Support • Business Configuration • Data Migration • Detail**), you can see which scoping decision triggered the selection or deselection of a migration object, including last-changed-by information.

**Download project schedule template for migration**

The **Create a project schedule for data migration** task provides you with a spreadsheet to help you plan and track the migration tasks on a detailed level. In Section 6.6.1, we'll take a closer look at this topic.

### 6.2.4 Migration Objects

The implementation phase **3 - Integrate and Extend** mainly contains a separate task activity for every relevant migration object, for example, **Migration of Customers**. Similar to the **Migration Templates** download list in the previous section, only object migrations are offered that match the selected scope. In addition, the activity list allows you to exclude activities manually—for example, because a certain object is in scope, but for some reason there is no data to migrate for it. Open the **Migration of <Name of Migration Object>** activity (e.g., **Migration of Customers**) to get to a fact sheet with a more detailed view, as shown in Figure 6.10.

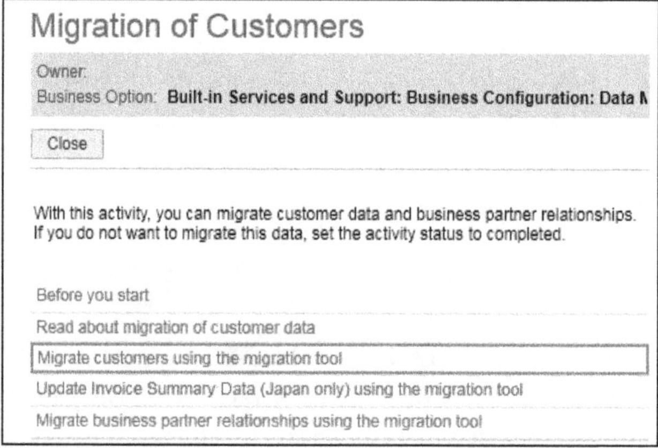


**Figure 6.10** Migration of Customer Data

This fact sheet has basically same structure for every object. In any case, there will be at least two entries:

- **Read about migration of <migration object>**
  In our example, this option is **Read about migration of customer data**. With this option, you'll find documentation on this migration object and

on potential dependencies to other objects. The same text is also provided offline in the downloaded migration template file (second to last sheet); however, the online version has the benefit of many hyperlinks to other help center articles.

A few objects have an additional **Before you start** article (e.g., **Customer**, **Supplier**, and **Employees** on available business partner types).

- **Migrate <migration object> using the migration tool**
  This function is of course the main link on this page, which opens the actual migration tool for this object.

For certain objects, further types of functions are offered on the fact sheet.

**Manually enter <migration object>**

For most business objects, you can always enter data records using the regular work center, which will also be in use during the operation of the system. The migration object fact sheet doesn't contain a separate link in such cases. If you want to enter a customer record manually, then you would simply open the **Business Partner Data** work center and use the **Create** function there.

However, in certain cases, the functions in the operational work center don't cover the requirements of a legacy data migration. Let's consider the example of *open purchase orders*. If you use the operational **Purchase Requests and Orders** work center to manually enter a released purchase order from your legacy system, then the entire purchasing process will be triggered with all side effects; that is, an order confirmation is sent to the supplier, which certainly isn't intended because that had been done in the legacy system already.

In such cases, during the implementation phase, specific UI screens for the manual transfer of certain data records are offered, which react in a different way compared to their operational counterparts. In these exact cases, a separate task for manual data transfer is offered in the migration activity. For example, in **Migration of Purchase Orders,** you'll find an additional **Manually enter purchase orders** entry that will bring you to such a specific entry screen.

This principle applies to the following migration objects: employees, responsibilities, bank account balances, petty cash, projects, inventory, purchase orders, and purchasing contracts.

**Migrate <Supplementary object> using the migration tool**

Finally, some objects (e.g., customer, supplier, business partner, service agents, and employees) have a related object that is technically a separate migration object but is so closely linked to the main object that their migration task is offered in the fact sheet of the main object. Some prominent cases of this scenario include the following:

- **Account determination group assignments**
  You can assign account determination groups to your suppliers using the supplementary tool migration object that is offered in the **Migration of Suppliers** activity. (Note: The corresponding assignment for customers is already part of the customer object itself.)
- **Business partner relationships**
  The migration task for business partner relationships is in the fact sheets of the various business partner migration objects (e.g., customers, suppliers, service partners, and general business partners).
- **Employee data**
  The employee data object and related objects are often country specific. You'll find several country-specific migration activities, each bundling the objects for this country. Depending on the country, this area covers the migration objects for employee basic data, personnel administration, time and labor management, compensation, and payroll data.

[+]

**If You Miss Objects**

If you miss an object in the activity list, consider the following:

- The activity list opens with the default filter **Activities in Project**, thus hiding activities that haven't been assigned to the current implementation project. Switch to **All Activities** and refresh the view by clicking the **GO** button. The **Add to Project** button allows you to include currently hidden migration objects to the active worklist.
- Use the **Find** function to search by object name. This approach will guide you also to migration objects that are part of an activity with a different title. (For instance, the migration of *lease contracts* is found in the activity **Transition of Lease Contracts**.)
- Check whether the object is contained in the **Download Migration Templates** task in the **Prepare for Data Migration** activity (refer to Section 6.2.3). If the object was missing there already, then the root cause is presumably in the scope definition (**Business Configuration • Implementation Projects • Edit Project Scope**).

### 6.2.5 Migration Cockpit

The starting point for most migration work is the *migration cockpit*, the entry screen of the migration tool. From the migration cockpit, you can upload migration files into the system, edit and administer these uploaded files, and start the migration processing. Let's proceed with our example **Migration of Customers** activity.

Choose **Migrate customers using the migration tool** to open a separate window with the migration cockpit, as shown in Figure 6.11.

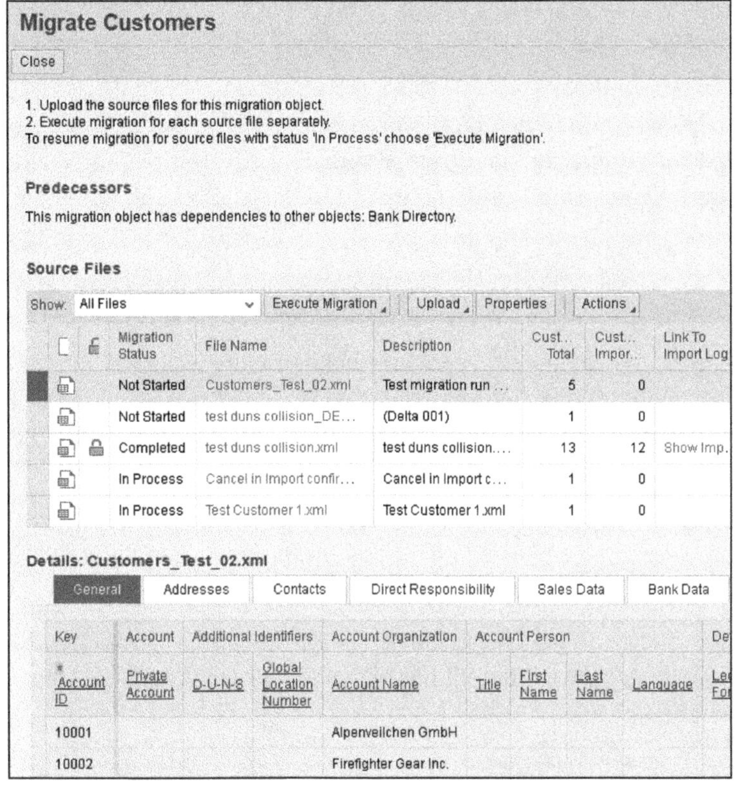


**Figure 6.11** Migration Cockpit for the Customers Object

In the upper area of the screen, you'll see some quick instructions on the main migration process as well as information on predecessor objects. In our example, the **Bank Directory** object is listed as a prerequisite. Any bank information in the customer data file is checked against the bank directory; therefore, the **Bank Directory** object needs to go first.

The **Source Files** screen area has the list of migration files that have been uploaded so far, showing the following details:

- **Migration Status:** Possible status values are the following:
  - **Not Started:** Freshly uploaded file, no processing yet.
  - **Completed:** Processing has been finished after final import step, and the file is now locked against changes 🔒.
  - **In Process:** Anything in between the preceding two statuses.
- **File Name**
- **Description**
- **Customers Total/Customers Imported** (number of total records/number of imported records)
- **Link to Import Log** (for completely processed files)

The bottom part of the screen displays a preview of the content of the file that you marked in the file list. You'll immediately recognize the structure of the migration template with the same tabs and column headers.

Upload files

Choosing **Upload • Data from Migration Template** opens a dialog box where you can *upload* a migration template file with data from your PC to the file list. **Browse** for the local file and maintain the **Description** and **Comment** fields.

Following a good naming convention for file names and file descriptions helps you distinguish different versions and different migration waves.

Although the **Description** and **Comment** fields are technically not mandatory, we recommend investing the time to fill in these fields. Doing so will make reconciliation easier and help potential auditors who might be reviewing imported files years later.

If no file has been uploaded yet for an object, the file upload dialog box appears automatically when the migration cockpit is opened for the first time.

The migration tool can load migration template files also in a compressed ZIP file format, as long as the ZIP file contains no more than one single migration template file.

**Use ZIP Files!**

While the usage of a zipped file is mandatory for populated migration templates larger than 100MB, the use of ZIP files is almost always beneficial. Migration templates usually have very high compression rates, thus speeding up the upload process significantly.

As an alternative to migration templates, you can upload data from text files (CSV files). This technical approach requires more user interaction and will be explained in the Section 6.2.6.

Column check

During file upload, the system verifies the file version and the formal consistency of the migration template file. This verification process determines whether the actual columns in the migration template file match the expected columns. While deviations don't inhibit the file upload, warning messages will inform you of the system reaction if column deviations have been detected:

- If the uploaded migration file contains columns that are unknown to the system, these columns will be ignored during upload.
- If the uploaded migration file lacks expected columns, these columns remain empty during the upload. If feasible, you can populate these

empty columns later using the integrated data editor that will be presented in Section 6.2.8.

This check applies to customer-specific extension fields as well. In migration template files, extension fields are identified by a system-specific technical key that is contained in the template (in a hidden header line). Therefore, the attempt to upload a template with extension fields in a different system, other than the one from where it had been created, results predictably in both messages, even if the other system has extension fields with the same field labels.

[«]

**Errors During File Upload**

Errors during file upload are mostly raised by formal inconsistencies. Potential root causes include the following:

- Data has been copied into the template sheets using **Paste** instead of **Paste Values**, overwriting field formats.
- Data from HTML type sources has been pasted into the template sheets in a way that HTML code is ending up in the field content.
- The error message **Incompatible migration template** appears if you try to upload a migration template that is actually for a different migration object.

Consider the following self-help strategies:

- Download a fresh migration template. Transfer the data of one record (or few records) into this template by strictly using **Paste Values**.
- Avoid potential file confusion with a good naming convention for your files so you can recognize the migration object, creation time, and data type that is in the file, for example, *Customer_Test2_20240730*.

If you're interacting with product support regarding an error during file upload, don't forget to attach a zipped copy of the file because that is the key information for any root cause analysis.

**Data editor**

In the migration cockpit, clicking on a file name opens the file in the integrated *data editor*. The data editor enables you to enter data into an uploaded migration file or modify it in multiple ways. The functionality of the data editor will be covered in detail in Section 6.2.8.

Let's first round off the list of functionalities that the migration cockpit offers.

**Execute Migration option**

You can start the migration process by choosing **Execute Migration • Execute Migration Step by Step**. On the first execution of this process, the

**Validate File** step starts automatically. On re-entries, the processing starts at the processing step where the file was left in the previous session.

Properties

Certain properties of the file can be maintained by clicking the **Properties** button. In addition to the already familiar **Description** and **Comment** fields, you can maintain the **Migration Mode** field. The default mode **Add new records** is the choice for the usual initial data migration. The other modes are needed to enrich or modify already existing data records. Section 6.5 discusses these possibilities in detail.

Finally, the properties view contains information about when the file has been uploaded or changed and by which user.

Actions: Create

Choosing **Actions • Create** creates a new empty migration file directly in the migration cockpit. As soon as you've maintained the requested **File Name** and **Description** fields, the file opens in the data editor, ready for data to be entered. (See Section 6.2.8 for details.)

[»]

**Migration Files for First Test Migrations**

The **Create** function is an efficient way to run your first tests with manually entered data records. You can quickly enter test records with certain variations in the field values and then observe the resulting system reaction.

Actions: Copy

Select a line in the file list and then choose **Actions • Copy** to create a copy of the selected file. As during creation of a file, maintain the **File Name** and **Description** fields. In combination with the data editor's capabilities, you can quickly test some data variations in a temporary copy without spoiling your main file.

[»]

**Narrow Down Errors**

Sometimes, it isn't clear at first sight which field or which combination of field values is causing an incorrect data constellation. Use a temporary copy to narrow down data errors. Iterate the same migration process step (**Validate File**, **Simulate File**) using a copy, which you're simplifying step by step, until the root cause factor becomes clear.

Actions: Delete

Choosing **Actions • Delete** deletes the marked file irrevocably. Files in the **Completed** status have already been imported and are secured against changes or deletions for auditing reasons.

Actions: Edit

Choosing **Actions • Edit** opens the marked file in change mode in the data editor (Section 6.2.8 for details). Alternatively, you can click on a name in the file list to open the data editor in display mode.

When using the data editor, you might want to save a local copy of the edited migration files occasionally. Choose **Actions • Download** for this purpose.

Actions: Download

One of the most important interactions is the value conversion that you enter or select from target picklists during the **Convert Values** step (Section 6.3.2 for details). For reconciliation or auditing purposes, you might want to export the value conversions that had been used for a certain file. For files in **Completed** status, choosing **Actions • Download** allows you to download this information to a local text file.

Actions: Download Value Conversions



### 6.2.6 Upload CSV Files

Exported data from the source system of a migration is often provided as a *CSV file*, that is, a flat text file in which each line contains one data record and the different field values are separated by a certain character (e.g., a comma or a tab).

As an alternative to collecting data into a downloaded migration template first, you can also upload such CSV text files directly in the migration tool.

The best approach is to begin with a new migration file (**Actions • Create**), which automatically opens in the data editor view. Now, you can populate this file tab by tab using the upload of text files approach—one file for every tab.

Choose **Upload • Data from Text File** to open a dialog box where you can indicate details regarding how exactly your CSV file is structured, as shown in Figure 6.12:

- **Source File**
  Select the local source file, which must be of file type *.txt* or *.csv*.
- **File Type**
  The file type is defaulted accordingly to *.txt* or *.csv*.
- **Import Data to Tab**
  Select into which target template tab you want to upload this text file. Each template tab needs a separate source file. The selection is defaulted to the first tab. Forgetting to change this default when adding data for other tabs is the most common mistake users make here.
- **Character Encoding**
  The default is the **Unicode UTF-8**. If you can, save your local text file as a UTF-8 file. Otherwise, change the setting here accordingly to match the character encoding of your local text file.
- **Field Delimiter**
  This character separates the different field values in your text file. If your

file is using a comma, semicolon, or tab character, then you don't need to change the default setting; the system analyzes the first lines automatically to recognize which of these three is actually used.

- **Text Qualifier**
  Any text values in your source file should be kept together by a text qualifier at the beginning and the end of the text entry. This avoids characters within the text being wrongfully interpreted as field delimiters. The default setting is a double quote (").
- **Decimal Separator**
  Determines how numbers with digits are represented in your text file (e.g., 3.14 or 3,14?).
- **Field Names in First Row**
  Determines whether your file starts with a header line holding the field names or starts immediately with the first data record.

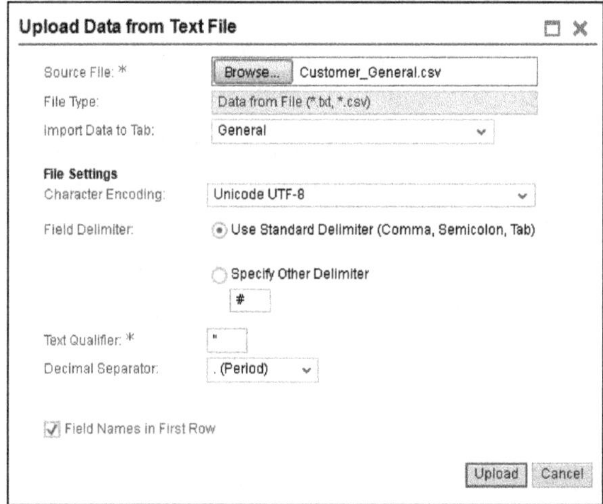


**Figure 6.12** Upload Data from Text Files Screen

How do the fields of your file correspond to the data in the selected tab? Click the **Upload** button to advance, after a short information message, to the next step: the *field mapping*.

The screen shown in Figure 6.13 enables you to assign the fields of your source text file to the corresponding target fields in the selected tab.

The upper screen area displays the first lines of your text file. In the lower part of the screen, you see the field names of the target tab. Move your mouse cursor over the field name, and a short field help text is displayed that you've seen already in the migration templates.

Directly underneath every target field name, a dropdown menu offers the available source fields as shown in Figure 6.13. Select one of these source fields to establish the field mapping. Immediately, as a result of this field mapping, a preview of the first data lines is displayed.

In both the upper and lower screen areas, you can use filter functions to narrow down the columns that are displayed. We recommend using **Unmapped Columns** for a systematic field assignment.



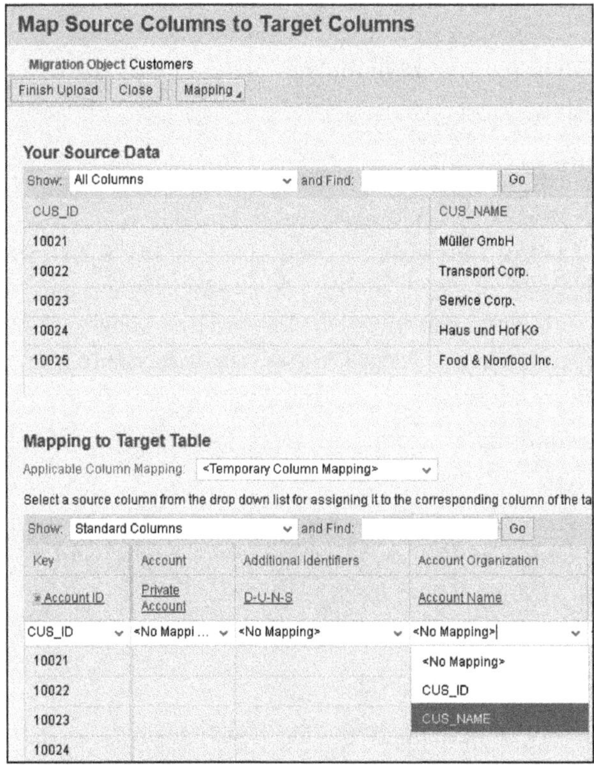


**Figure 6.13** Field Mapping

**Automatic field mapping**

If your text file already contains column header names using the correct target field names, then a mapping is proposed automatically for these fields. For this proposal, the system evaluates all available languages as well as the technical field names that are in the migration templates in the hidden header line 5. You can download a list of all technical field names also in the migration cockpit by choosing **You can also • Download CSV-Migration-Template.**

You can save a field mapping (choose **Mapping • Save**) and reuse for a later CSV file upload (select **Applicable Column Mapping** from the dropdown menu in the lower part of the screen).

You can also export a saved field mapping to your local PC (**Mapping • Export to Local File**) and reimport (**Mapping • Import from Local File**), which allows you to use your field mapping in another system.

Choosing **Mapping • Reset** clears all field mappings and choosing **Mapping • Reset and Rebuild Automated Mapping** clears all field mappings *and* reinitiates a new automated field mapping based on the field names.

The **Finish Upload** button finally uploads the text file data into the **Migration** tab, which will then be shown in the data editor. If you started the upload from the data editor, and your target tab already had some data, you're asked to decide how the system should proceed:

- Append the new lines to the existing lines?
- Replace the existing data entirely?
- Merge the new lines with the existing lines?

Merging means that, if a record already exists with the same key, then only those fields will be changed for which the new record actually has a data value; all other field values remain untouched. If the key isn't known so far, the system simply appends the new record. The system asks you to select the key field to be used.

A message in the data editor finally confirms the result of the upload. Save the file before continuing with the next data tab.

Actually, you can also start the text file upload directly from the migration cockpit (**Upload • Data from Text File**), which uploads the data to a new temporary migration file that you can save after the upload.

### 6.2.7 Migration Cockpit: More Features

The **You Can Also** menu function in the migration cockpit gives you access to the migration object documentation and to some more advanced functions, as shown in Figure 6.14.

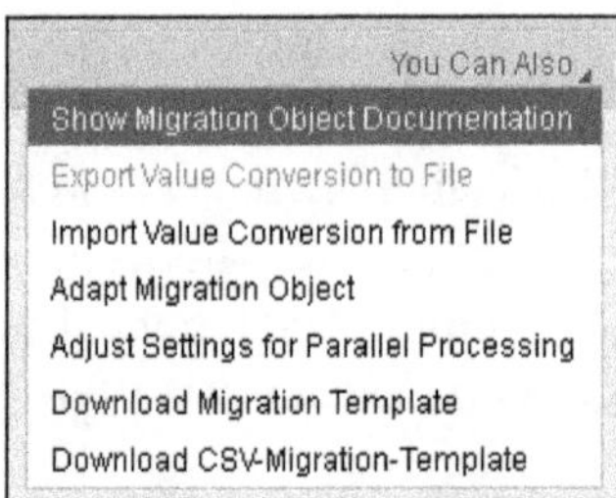


**Figure 6.14** You Can Also Menu

The following options appear in the **You Can Also** menu:

- **Show Migration Object Documentation**
  This option is another way to display the migration object documentation that you've met already in other places. These display methods include the following (with our customer data sample):
  - In the **Migration of Customers** activity
  - In the **Prepare for Data Migration** activity
  - As offline versions in the downloaded migration template **Documentation** tab
- **Export/Import Value Conversions**
  One of the major interactions during data migration is the value conversion that you define in the **Convert Values** migration step (Section 6.3.2 for details). A typical example is the value "UK" in a **Country** field in your migration template. The first time such an entry value is processed, it needs to be mapped to the country code **GB** because SAP Business ByDesign uses the international ISO code for countries.

  For potential reuse, you can export value conversions into a local file of type TRV and import this information into another system accordingly. However, some important aspects need to be respected if you want to avoid unexpected results. Use this transfer only after you're familiar with the details described in Section 6.3.2.
- **Adapt Migration Object**
  We recommend setting up required customer-specific extension fields early in the project. At the least, all migration relevant extension fields should have been defined in the system when you start working with migration. The parallel usage of migration templates with different versions of extension field lists is a potential source of errors.

  If you define a customer-specific extension field (or delete an existing one), after you've uploaded migration files into the migration tool, then the structure of these files in the system becomes outdated. The **Adapt Migration Object** function triggers an update of the field structure. Corresponding columns will then be added (empty) to the already uploaded files (or deleted, respectively).

  Extension fields are identified by a technical key that is generated by the system. If you delete an extension field from the system, and then create a new extension field with the same field label, this field will be a different field, making the use of a fresh migration template mandatory.
- **Adjust Settings for Parallel Processing**
  During migration, the data is finally imported using service calls to the system. Many objects allow a parallelization of these service calls. This

advanced function allows you to specify the parallelization factor that is being applied. Section 6.7.1 addresses this topic in more detail.

- **Download Migration Template/Download CSV-Migration-Template**
This option is another place where the download of empty migration templates is offered to you. In addition, you can download a text file containing the technical field names of all tabs. This option can be helpful for the creation and uploading of CSV files as discussed in Section 6.2.6.

[»]

**Migration of Transactional Accounting Data**

For accounting objects, the migration cockpit and the entire process of migration looks somewhat different. For the migration of accounting data, the files of the involved objects (general ledger account balances, open items, fixed assets, etc.) are being grouped into a *migration unit*.

While the initial processing steps are similar to other data objects, some additional steps are performed (e.g., a check summarizing the postings of all files in the unit), and other steps are different (e.g., import can be triggered only on the level of the entire unit, not for a single file).

Section 6.4 will guide you through these specialties in more detail.

### 6.2.8 Integrated Data Editor

The migration tool in SAP Business ByDesign has an integrated data editor. You'll likely use it during the **Validate File** and **Import Simulation** steps to analyze and fix encountered data errors. Becoming familiar with the capabilities of the data editor will make your migration work more efficient.

In the data preview in the bottom of the migration cockpit (shown earlier in Figure 6.11), you can already see that the structure of the uploaded migration files corresponds to the downloaded migration templates.

Let's view the data first in a read only mode, before we are going to switch to the edit mode that will allow us to change the data before processing it, for example to resolve data issues.

#### Display Mode

Click on the file name in the migration cockpit to open the data editor full screen in display mode, as shown in Figure 6.15. The same tabs and columns as in the downloaded template are displayed. A quick spot check will show more similarities.

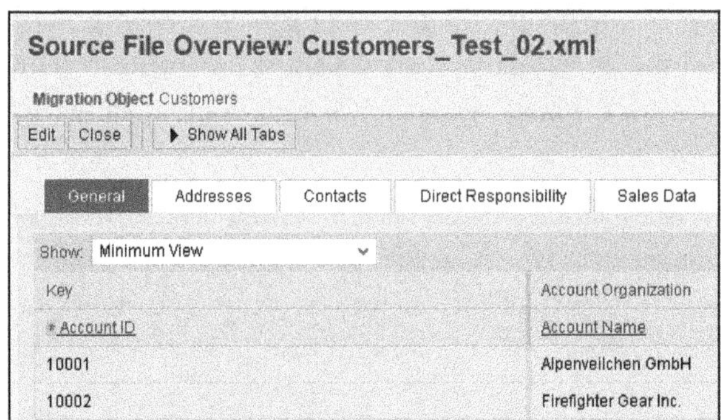


Figure 6.15 Data Editor in Display Mode

**Navigate between Tabs**

If more tabs are unhidden than the screen can accommodate, use the icons next to the tabs to navigate to tabs outside the screen.

Initially, not all data tabs are displayed. You can choose between two views: To start, the **Standard View** is shown. With this view, all data tabs are visible that are also visible by default in the migration template. In addition, those tabs are displayed that actually hold data. Using the **Show All Tabs/Show Standard Tabs** buttons enable you to toggle between the standard view and a maximum view with all tabs visible.

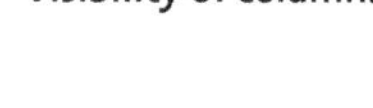


As in the migration template and also online, only frequently used columns are visible by default. Using the **Show** dropdown, you can choose between three column views, as shown in Figure 6.16.

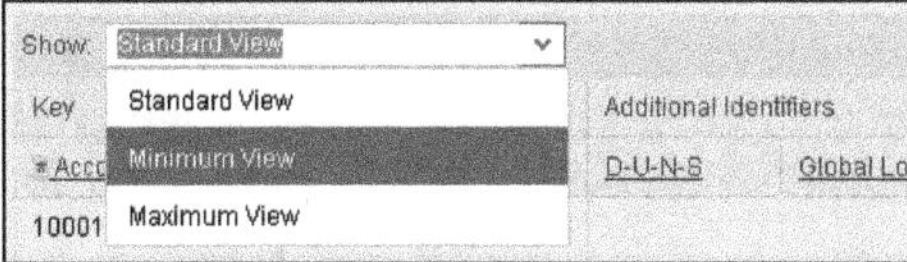


Figure 6.16 Column Views

The column views are distinct in the following ways:

- **Standard View**
  The same columns as in the original migration template are visible, plus columns that actually hold data.
- **Minimum View**
  Only columns marked as **Mandatory** are visible, plus columns that actually hold data.
- **Maximum View**
  All columns are visible.

Line filter

A filter function enables you to select the records displayed. At the right side of the screen, you'll find the icon (**Set Filter**), which brings up an additional filter line (the nongray line) beneath the header line, as shown in Figure 6.17.

If you enter the value "1" in this filter line under the **Account ID** column and then press Enter, only records whose account ID starts with 1 are displayed. You can use an asterisk (*) as a wildcard character. For example, with the selection ***111**, only those records remain visible whose account ID contains 111.

**Records with Errors**

During the processing of error messages, the view is automatically restricted to the erroneous data records.

Clicking the **Undo Filter** icon at the right edge of the screen finally brings all records back to the view.

| Key | | | |
|---|---|---|---|
| * Account ID | Contact ID | First Name | * Last Name |
| | | J* | |
| 10003 | | Jean | Luc |
| 10005 | | Judy | Lopez |

**Figure 6.17** Data Editor: Filter with Wildcard

### Edit Mode

Choosing **Actions • Edit** in the migration cockpit (shown earlier in Figure 6.11) opens the data editor as expected in edit mode. Not surprisingly, edit mode allows you to manually change or add data and save the changes using **Save** or **Save and Close**.

You've already seen the tab views, the column views, and the filter function earlier.

In addition, functions are available that affect rows (e.g., **Add Row**, **Copy**, **Delete**) or columns (e.g., **Mass Change** functions).

Line functions

**Add** is used to add lines above the current lines for as many as you've currently selected.

You can select lines using the button area at the left end of the line. As in many other applications, you can select an entire block of lines by holding down the Shift key during line selection. Similarly, you can collect nonadjacent lines by holding down the Ctrl key.

In addition, to the left of the line headers, a small icon [icon] is visible that allows you to mark all lines (**Select All**) or remove the selection from all lines (**Deselect All**).

The line functions **Copy** and **Delete** are effective for all selected lines. **Copy** doubles the selected lines by creating a neighboring line of identical content for each line.

The **Delete** function has a special feature that enables you to delete while preserving the structural logic of the record. In addition to the selected line on the visible tab, all dependent lines on hierarchically subordinated tabs are automatically deleted as well.

Let's consider a specific example from customer migration. When you choose to **Delete** the header line of a certain customer master, let's say with Account ID 10099, on the **General** tab, then automatically all subrecords for the account 10099 are deleted on all tabs, including contact addresses and other subrecords.

This logically consistent **Delete** function across tabs allows you to effectively create smaller test file copies from large uploaded files. This feature allows you to identify potential issues and mass errors using small files with short runtimes.

Column functions: mass change

The purpose of the **Mass Change** function is to change the content of an entire column. Follow these steps:

1. If applicable, use the **Filter** functionality to reduce the displayed lines, narrowing down the subset that needs to be changed.
2. Use the **Select All** icon to mark the resulting lines. The selection applies only to the visible lines. Lines that are hidden due to filtering won't be affected by the mass change.
3. Click the **Mass Change** button and choose the field that needs to be changed, as shown in Figure 6.18.

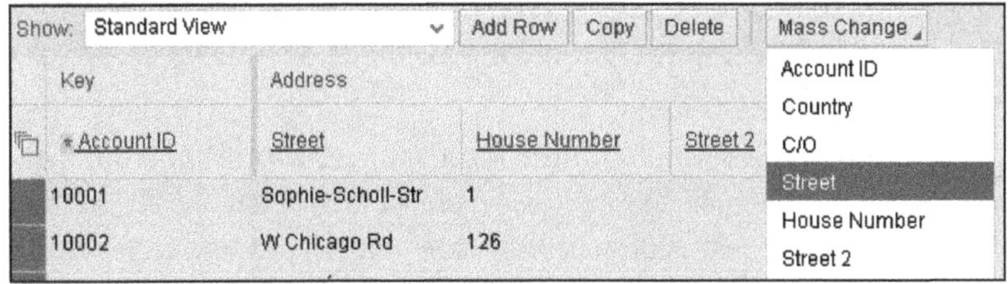


**Figure 6.18** Mass Change: Choosing the Field to Be Changed

4. Choose the change function that you want to apply and fill in the required parameter values, as shown in Figure 6.19.

[»]

**Mass Change**

The most common root cause for an unexpected mass change result is a forgotten or incorrect line selection. If you're uncertain about the status, you can always close the file without saving to return to the last saved version.

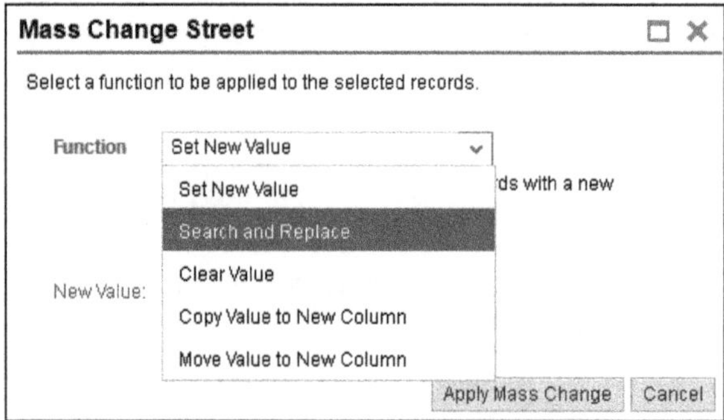


**Figure 6.19** Mass Change: Available Functions

Change functions in mass change

**Mass Change** offers several functions to pick from. In all cases, the change is made to one column and becomes effective only for selected lines:

- **Set New Value**
  The entire column is populated with the new value, overwriting any current value.
- **Search and Replace**
  In all entries in this column, all occurrences of the search string are replaced by the new string. This rule is also true if the search string occurs multiple times in the same field value.
- **Clear Value**
  The values in the entire column are deleted (in all lines of the selection).
- **Copy Value to New Column**
  The content of the selected column is copied into another column.
- **Move Value to New Column**
  The content of the selected column is moved into another column. This function helps to correct situations where a block of data has been copied into a spreadsheet without taking care of hidden columns, resulting in the wrong columns being populated.

Now that you've populated your migration templates according to the information you collected from Section 6.1 and have become familiar with the tool in Section 6.2, you're well prepared for the core part of a data migration in SAP Business ByDesign.

## 6.3 Migration Process Steps

In this section, you'll process the main steps of a data migration, as shown in Figure 6.20:

- **Validate File**
- **Convert Values**
- **Simulate Import**
- **Execute Import**
- **Confirmation**

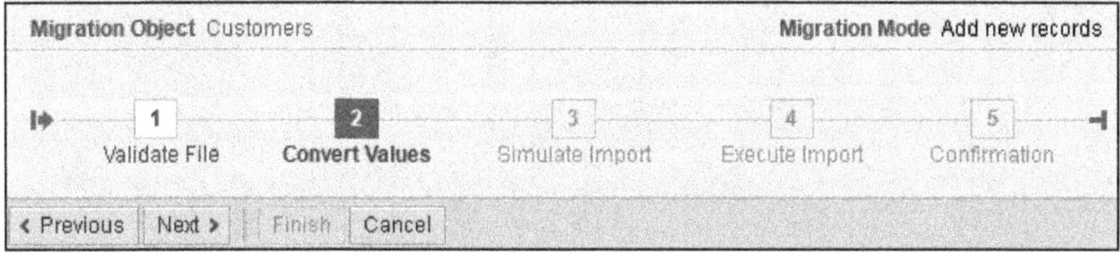


**Figure 6.20** Main Steps of Migration

Several elements are alike in all steps, so we'll explain them once up front:

- Use buttons to navigate: **Previous** gets you to the previous step, and **Next** takes you to the next step. **Cancel** stops the processing of the file and returns to the migration cockpit. Clicking on the object name opens the object documentation.
- The different steps contain additional buttons with specific functionality, for example, for the repetition of the same step. These functions are described in their corresponding sections.
- The functionality is presented following the specific data example *customers.xml* as shown in Section 6.1.4. We placed some typical mistakes into the data example (at places not visible in any screenshots so far) to help you see typical system reactions and effective strategies for error handling.

### 6.3.1 Validate File

In the migration cockpit for customers, select the template file *Customers.xml* and trigger processing by choosing **Execute Migration • Execute Migration Step by Step**. This action starts the first processing step: **Validate File**.

**Validation**

In the **Validate File** step, the *formal* correctness of records and their field contents are being checked, verifying the following in the data:

- Are all fields populated that are marked as mandatory?
- Have field length limits been respected?
- Are the data values valid?

- Are the date values formally correct?
- Are the amount values in the correct format?

In addition, certain checks are applied to target fundamental data errors. For example, as shown in Figure 6.21, the repeated migration of an already migrated record is caught.

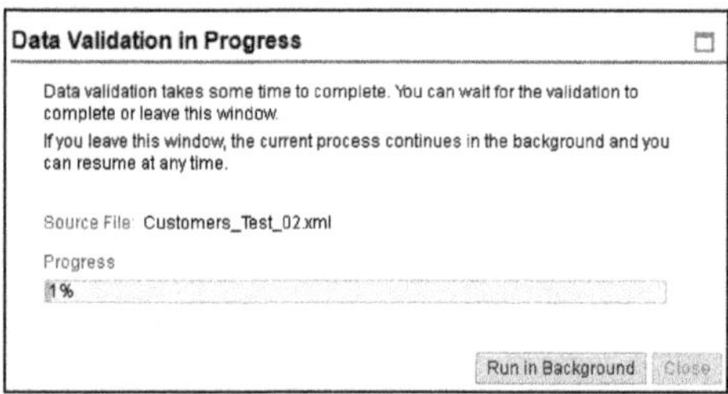


**Figure 6.21** Validate Data: Progress Indicator

The progress bar allows you to monitor the progress.

Clicking the **Run in Background** button allows you to return to the migration cockpit, while the actual file processing continues in the backend.

After you reenter the same file from the migration cockpit via **Execute Migration**, you return automatically to the progress indicator.

The **Close** button becomes active as soon as all records have been validated, and clicking it brings you to the result of the validation.

**Validation log** A validation log captures all identified errors in a structured list, as shown in Figure 6.22.

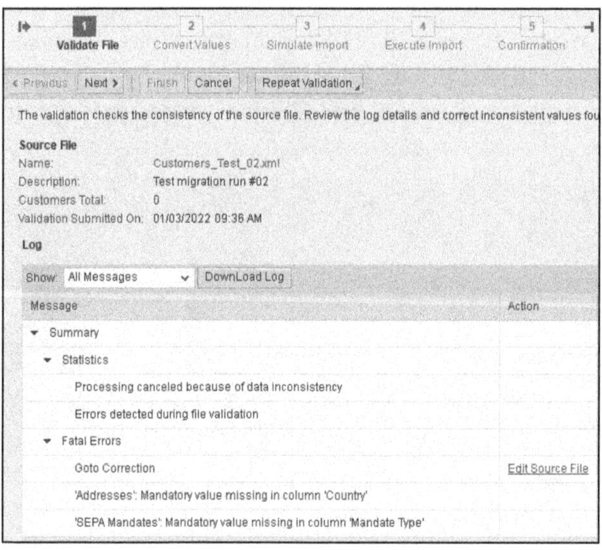


**Figure 6.22** Validate File: Prescriptive Error Log

Download log

The **Download Log** button saves the error log file locally to your PC. The log is truncated after 200 errors or warnings. If there are many warnings, you might want to apply the message filter first, that is, selecting **Show: Error messages**, which gives you a better overview of any actual errors.

If there are no errors, the log rewards you with the message **No errors detected during file validation**. However, you'll usually find a work list of some length. The resolution of this error log is the actual task in this migration step.



Iterative processing

Follow these steps to resolve the error log:

1. Fix one or several reported data errors, if possible, using the data editor.
2. Check the error status by re-running validation (**Repeat Validation • Validate All Records**), as shown in Figure 6.23.

Repeat this cycle until all data errors have been identified and fixed locally. Keep records of errors and fixes systematically so that you can apply them to a newly exported source file.

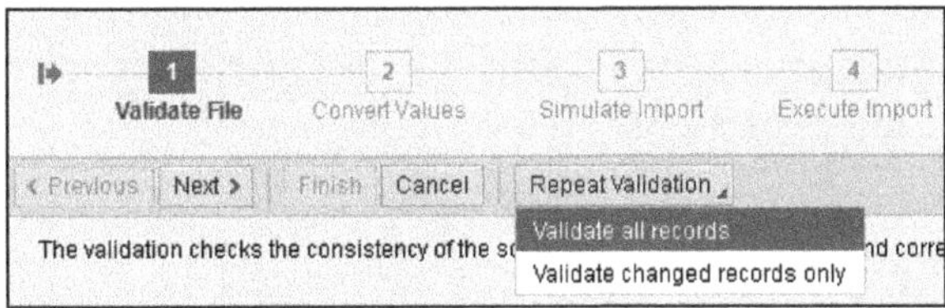


**Figure 6.23** Repeat Validation

For the re-run, you can alternatively choose **Repeat Validation • Validate changed records only**. It's faster because it's processing fewer records. However, validation is usually very fast anyway, so that the benefit is limited during this step. In any case, you should run a final repetition of validation using **Validate all records.**

Inconsistent values

Certain data errors are classified as *inconsistent values*, for example, missing keys or the wrong key relation between data tabs. If you encounter such an error already during the validation run, the following message is issued in the popup window: **Inconsistent values found in your source file. Correct all errors first.** Close the popup window to proceed to the next window, which contains details on the error. Fix these inconsistency errors and re-run the validation.

[«]

**Repeat Validation**

The purpose of the revalidation is to verify that the data change has solved the reported error and to complete the validation run if the previous error prevented completion up front; thus, additional issues might be recognized.

There are two versions of revalidation available through the **Repeat Validation** dropdown list:

- **Validate changed records only**
  Only records that have been changed since the last run are validated. The benefit of this version is shorter runtimes (although validation runtimes are very short usually anyway).
- **Validate all records**
  The validation run has another important purpose beyond error recognition: The run collects all values that need a value conversion in the next step. Changes in the file might make some of these value conversions obsolete, which can't be recognized by a validation of the changed record only. Therefore, after changes are made in the file, a final validation run across *all* records is recommended in any case to consolidate the value task list.

We recommend using the data editor for data corrections if possible. You can navigate from the error message log directly to a purpose-made view in the data editor using the **Edit Source File** link. The editor supports data error handling with two convenient functions:

- The header area repeats the error messages from the log, where applicable, together with a hyperlink that allows navigating directly to the affected subrecord.
- The erroneous fields can be spotted quickly because they are marked in red color where possible.

The editor offers two views to structure the work, which you'll see when processing the practical examples: a *collective view* for errors classified as *fatal* errors and a *single record view* for all other errors. These errors can be described in the following ways:

Fatal errors

- **Fatal errors**
  Fatal errors include the already mentioned inconsistent values but also some other types. Error messages of this type are printed in a separate section at the beginning of the error log. In this log, every error message is only listed once, independent of how many records are affected. Follow the **Edit Source File** link to navigate to a filtered view in the data editor showing only the affected records, as shown in Figure 6.24.
- **Nonfatal errors**
  The list of nonfatal errors in the validation log is grouped by records. For every record, you can directly branch to a single record view that lists all error messages for this record.

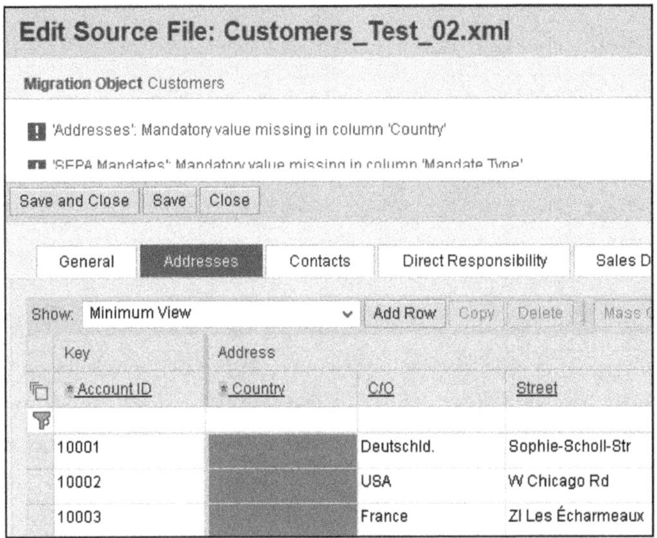


**Figure 6.24** Error Handling in the Data Editor: Fatal Errors

**How to Avoid Data Errors in Validation**

Errors in validation often are caused be one of the following data errors:

- **Fatal errors with inconsistent values:**
  - Key values are missing.
  - Mandatory fields aren't populated.
  - The file contains two or more records with the same key.
  - Mistakenly, some fields below the actual line range are filled.
- **Fatal errors:**
  - A record with the same customer number has already been migrated (and the current run isn't a change migration).
- **Other errors:**
  - Text exceeds the defined field length.
  - Invalid date or time format is used, or the entry hasn't been recognized as a date or time value.
  - Amounts have invalid formatting.

Let's look at the error analysis steps in a practical example. We've mounted some data errors into our file that typically occur in real-world migrations.

The first validation run responds with the log shown in Figure 6.25, reporting inconsistent values.

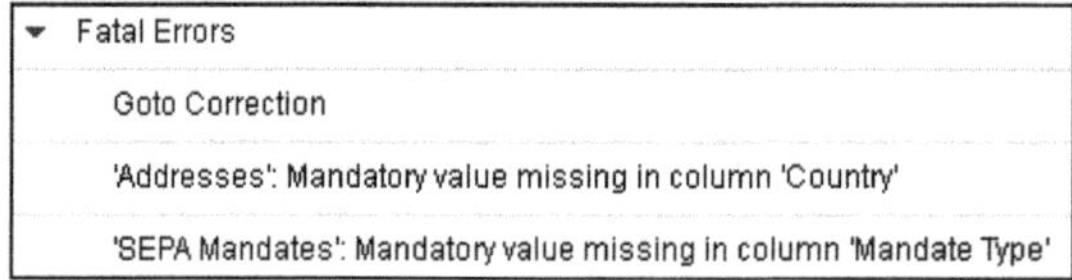


**Figure 6.25** Example: Inconsistent Values

Navigate to the data error using **Edit Source File** for an overview of the error and the affected data records.

As shown in Figure 6.26, obviously some columns have been mixed: The data for the **Country** field is in the neighboring **C/O** field. The values can quickly be moved to the right column using the **Mass Change** function in the data editor. Click the **Select All** icon to define the line range for the change and open the **Mass Change** dialog box, choosing the column that needs to be moved (**C/O**), as shown in Figure 6.26.

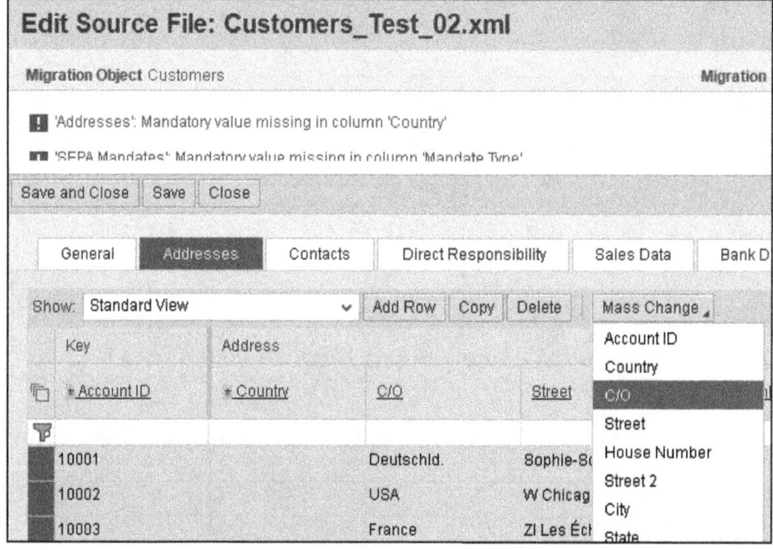


**Figure 6.26** Data Editor: Move Columns

Proceed with the change by choosing the **Move Value to New Column** function and indicating the target column (**Country**). Click the **Apply Mass Change** button to execute the change, and the values are now in the right column.

In our example case, the header area of the data editor lists another error message. Click on this message to jump directly to the affected place in the data, which, in this case, is the **Mandate Type** field on the **SEPA Mandates** tab, as shown in Figure 6.27.

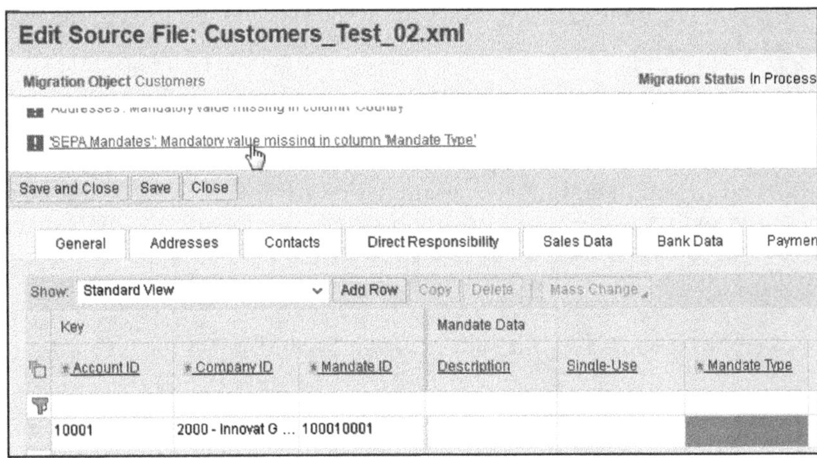


**Figure 6.27** Navigation via Error Message

Fix the data flaw and return to the error log by clicking **Save and Close**. Trigger a revalidation by choosing **Repeat Validation • VALIDATE changed Records Only** and address any additional data errors if applicable.

Dependencies between objects

The validation step doesn't depend on any other data objects. You can, for example, validate a template with sales orders independently from the status of the migration progress in any master data such as customers and materials used in these orders. This flexibility allows at least a parallelization of initial work steps across objects.

As soon as all data errors have been resolved, we recommend running a final revalidation by choosing **Repeat Validation • Validate all records**. Then, advance to the next step by clicking **Next**.

### 6.3.2 Convert Values (Value Mapping)

Value mapping

The migration in the SAP Business ByDesign system is designed in such a way that you can populate the migration files with the code and the ID values exported from your source system. During the **Convert Values** step, a value mapping functionality allows you to map the data values used in your template onto these target values, automatically ensuring consistency across objects.

Typical examples include the following use cases:

- Code list fields such as country codes, in which the SAP system is working internally with 2-digit ISO-3166 codes (i.e., AU for Australia, CA for Canada, etc.)

- IDs that you defined during configuration or in your basic data, such as sales organization IDs
- IDs of precedent business objects such as supplier IDs and customer IDs

**Converting values**

A worklist guides you through this step, offering all open value mapping decisions in a systematic way. Automatically generated target value suggestions and value help picklists support your decision process. Of course, every value needs to be mapped only once, no matter how often it occurs in the template.

Let's take a closer look at this worklist, as shown in Figure 6.28.

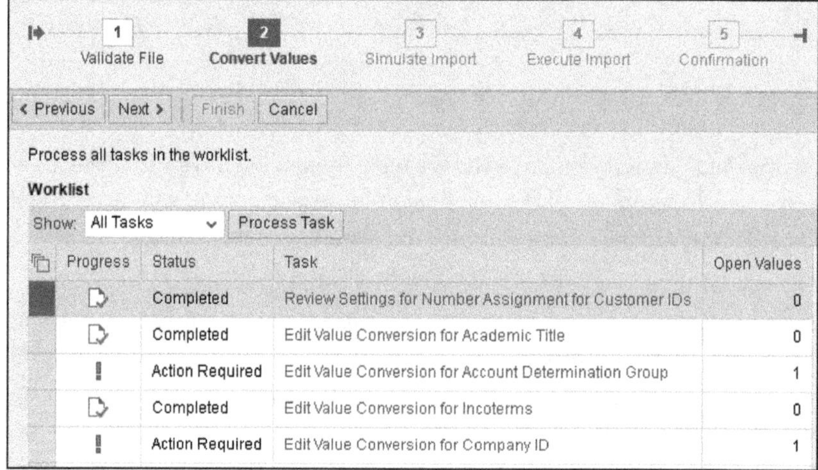


**Figure 6.28** Convert Values: Task List

Usually, every task refers to the value conversion for a specific field, for example, the **Country** field. However, in some objects, the list also contains additional tasks such as the setting for certain control parameters. We'll discuss this exception at the end of the section.

The task status is either **Action Required** or **Completed**. The **Open Values** column shows how many value mapping decisions are still open within this task.

Using the **Show** dropdown list, you can filter the listed tasks with the following options: **All**, **Open Tasks,** or **Completed Tasks**. At the bottom of the screen, the last-changed-by information is logged.

To process a task, mark the line and click **Process Task** or click on the task name. In our example migration, open the **Edit Value Conversion for Country** task to see the dialog box shown in Figure 6.29.

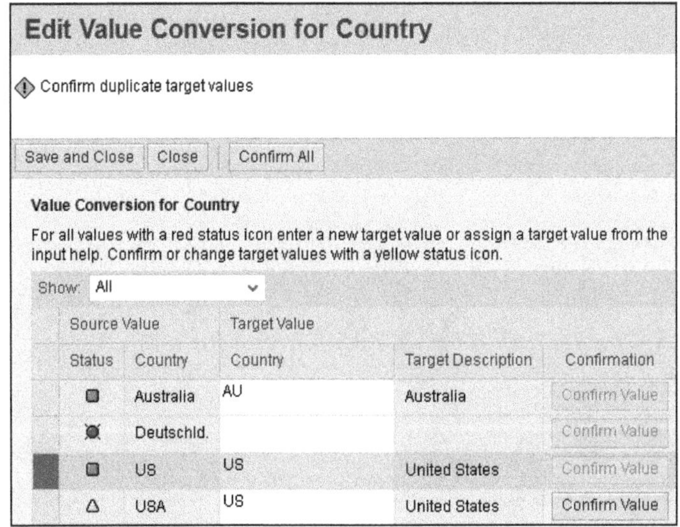


**Figure 6.29** Edit Value Conversion

**Status of value conversion**

A traffic light icon represents the conversion status of a value:

- **Green square** [■]**: OK**
  This value has already been mapped to a target value, either manually or by the automated value mapping recognition. For target fields with code lists, the system compares the field value from the template with code and descriptions in the target picklist (in all active languages). If a unique hit in a description is found, the value mapping is established automatically (e.g., `Deutschland` à `DE`). In addition, you benefit from any confirmed value mapping that had been created previously while working with any migration object in your system that contains the same field.
- **Yellow triangle** [△]**: Confirmation Required**
  If the automated value mapping suggestion has been made solely based on a code (not on a description), then the proposed mapping needs to be confirmed. Confirm the suggested value mapping with the **Confirm Value** button or select the right target value from the picklist. In both cases, the status will switch to green **OK**.
- **Red circle** [●]**: Conversion Required**
  This value hasn't yet been mapped. Open the target value picklist, for instance, pressing F4, and choose a valid target value, as shown in Figure 6.30. This step will make the status change to green **OK**.

This list can also be filtered. Use the **Show** dropdown list to restrict the view to the yellow or red lines only.

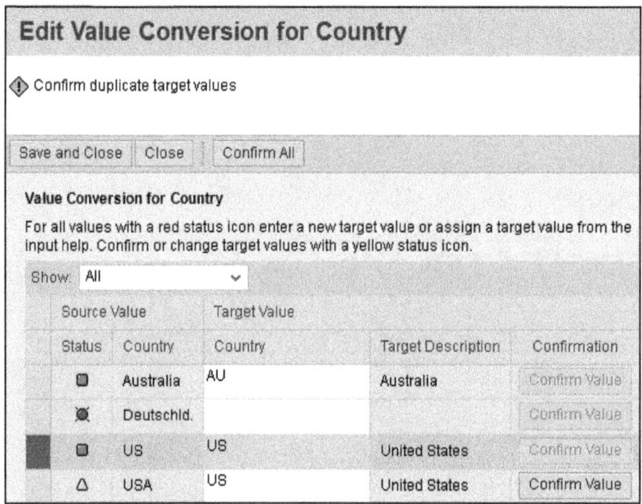


**Figure 6.30** Edit Value Conversion: Target Value Help

After you've completed the value conversion for the current field, return to the list of tasks by clicking **Save and Close** (or discard your changes by clicking **Close**).

The value conversion has some detailed functionality:

**Duplicate target values**

- Sometimes, you might want to assign the same target value to two different source values, maybe because in the source data two different spellings have been used for the same value. To ensure that the duplicate assignment isn't by mistake, the status initially changes to yellow, and you're requested to explicitly confirm the assignment using the **Confirm Value** button.

  While some identifier fields such as **Customer ID** and **Material ID** don't support target value help lists, protection against inadvertent duplicate target values is also in place with those fields.

**Multilevel value conversions**

- Regions are a good example for a multilevel value conversion. Region codes are unique only within a country. In the ISO region code list, the region code WA represents both the US state Washington and the state of West Australia; only the context of a country allows you to define value assignments in a reasonable way.

  Therefore, the relevant country values must be value converted first, before the **Region** field is addressed. A popup window will guide you through the process of region conversion without having finished the relevant country conversions.

  This approach also applies to the automated target value recognition. After having assigned the relevant country value conversions, re-run the validation if you want to use the automated target recognition for your region values.

There are even some three-level value conversions, for example, the US tax jurisdiction codes on the county level, for which the value conversion of the states must have been completed already.

- During migration of transactional data such as purchase orders, the value conversion of the master data (e.g., material IDs and supplier IDs) is already known to the system because this information has been established automatically during master data migration—if this master data was migrated as recommended. If migration hasn't been used for this master data, then the assignments are missing, and the purchase order migration fills up the target values with a 1:1 assignment that is marked with status **Confirmation Required**. Watch out especially for entries without target descriptions (e.g., supplier name, material description) before doing a quick **Confirm All**.



Missing target values

Are you missing entries in the target value help? During the **Value Conversion** step, the business configuration that you've defined so far is thoroughly challenged by the reality of the data in your templates.

While, ideally, the business configuration should have been widely completed at the time of migration, often the **Convert Values** step brings up some missing elements. This occurrence applies particularly to business configuration settings made in the **Fine-Tune** phase of the activity list and to basic data such as organizational structures.

Because the migration tool is an integrated part of the system, the new values will be available in migration directly after you've added them.

### Handling Missing Target Values

Some typical examples of how to handle missing target values in a migration value mapping include the following:

- Several migration objects have fields for payment terms (e.g., suppliers, purchase orders, or purchasing contracts). To avoid misunderstandings, the template comes ideally with semantically transparent entries such as "14 days 2%, 60 days due net." During migration, these values can easily be mapped to the payment terms defined in the SAP system, which comes with a predefined set and allows adding custom terms.
- If during value mapping, a term is missing in the target value help, then just open a different tab or window for the **Fine-Tune** activity (in our case, **Suppliers • Define Payment Terms for Procurement**) and add the required entry. Go back to the still open migration window, and

open the target value help again to find your new entry now ready for mapping.

- The source data sometimes contains references to a cost center that was missed during the initial setup of the organizational structure. This deficiency will become transparent during the **Convert Values** migration step because there will be no target value to map to. In a separate tab or window, open the **Perform Organizational Structure Setup** business configuration activity and add the missing organizational unit. As soon as you've activated the new unit, go back to the still open **Migration** tab and open the target value help for cost centers again to find your entry now ready for mapping.

If you want to use automated target value recognition after making changes in business configuration or basic data, then navigate back to the **Validate File** step and choose **Repeat validation • Validate all records**.

**Unused Source Values**

After you've changed the content of a partly processed file in the integrated data editor, any navigation between migration steps will automatically trigger a validation of changed records (unless already done) and will add new values to the mapping lists.

In this situation, you might still see a value mapping task from some "phantom" source values that are, due to your edits, no longer occurring anywhere in the file. In such a situation, run a *full* revalidation by navigating back to the **Validate File** step and choosing **Repeat Validation • Validate all records**.

Why is this? Only a full validation can recognize that a mapping value is really not used any longer in any record in the file.

**Control parameters for migration**

In some objects, the worklist in the **Convert Values** step contains a task of type **Review Settings for**, which allows you to set control parameters controlling the behavior of the migration tool. Let's look at a couple of examples:

- **Migration of customers; migration of suppliers**
  In the **Review Settings for Number Assignments** task, you can choose whether you prefer the migrated records to keep their legacy IDs or to get new IDs (including the creation of corresponding value conversion information).
- **Migration of opportunities**
  The **Review Settings for Create CRM document flow** parameter controls

whether, during migration of leads, opportunities, activities, and sales orders, the dependencies between these records should be collected and made visible, in addition to the actual documents, in the migrated records on the **Document Flow** tab. Activating this parameter enables a convenient navigation along the document flow for migrated records at the expense of additional consistency checks during migration and a more rigid migration object sequence (such as 1. leads and sales orders, 2. opportunities, 3. activities).



[«]

**Using Value Mapping Effectively**

When collecting the source data for migration, we recommend the following guiding principles:

- Use source IDs and source values wherever possible. Don't try to do an external value mapping toward target IDs or target codes in SAP Business ByDesign if that can be avoided.
- If possible, export not only pure code values (05) from the source system but also meaningful entries such as descriptions or combinations of codes and descriptions (e.g., `05 - Raw Materials`).

The main benefits of value mapping are robustness, efficiency, and consistency, along with the following:

- The risk of misunderstandings is reduced.
- No synchronization effort is required on configuration changes; new values are available in the migration tool as target values automatically.
- The same mappings are used across objects.

**Export and import of value mappings (TRV file)**

Depending on the quantity of value mappings that you've maintained in your test system, you might not want to enter everything again into your production system. The migration tool in SAP Business ByDesign allows you to export value conversions into a local file on your PC and import it from there. You find this feature in the migration cockpit, in the upper-right corner, under **You Can Also • Export Value Conversion to File** and **Import Value Conversion from File**. The local file is of file type TRV. As stated in Section 6.2.7, some important details of this feature include the following:

- During export, only value conversions of already imported files are taken into account (migration status **Completed** in the file list) and also only those value mappings with conversion status **OK** [■].
- If you want to reimport the exported TRV file, take care that the value mappings have been created in appropriate quality. You need to be sure that

no "dummy" mapping has been created during some quick migration. Another challenge comes from missing configuration settings. Although, technically, they could be fixed immediately as shown earlier in this section, you see "temporary" makeshift assignments to just any available target value in some test migrations. Certainly, you don't want to reuse such mappings in your production system.

- The import of a TRV file is overwriting single mappings. Assume that, in the system already, the value mapping `In-house Production à 001` has been defined. If the TRV file contains the mapping `In-house Production à 002` for the source value, then the import of this TRV file will overwrite the previous mapping and replace it with `In-house Production à 002`. The status **OK** remains unchanged.
- Value mappings for source values that aren't contained in the TRV file remain unchanged.
- The value mapping import also affects other exports because the value mapping is applied in a consistent way to all objects that contain the corresponding field.

[!]

**Prudence When Using Value Conversion Import**

Export and import of value mappings can be an effective tool when used in a correct way. Used in the wrong way, for example, with an improper value mapping file, this utility can cause damage, including a data import that is formally correct but with wrong data due to flawed value mappings. Therefore, this method is a tool for experienced users and requires prudence.

As a technically skilled user, at this point, you might have opened a TRV file already using some text editor (e.g., Notepad). You certainly have recognized the transparent structure of this text file. If you preserved the tab-separated structure of the lines in this file, then you could use the TRV file import to upload locally defined new value mappings to your system.

### 6.3.3 Simulate Import

In this section, we'll walk you through simulating an import and then provide you with some best practices for a smooth and efficient simulation. Finally, we'll help you determine the root cause and resolution for some common errors that may pop up during your simulation.

### Running the Simulation

After processing the value mapping worklist, click **Next** to access perhaps the most helpful feature of data migration in SAP Business ByDesign: the ability to perform a full simulation of the import without actually writing records to the database.

What happens during simulation?

In the **Simulate Import** step, the same routines are being called as later during the **Execute Import** step—just without saving. Most importantly, during import simulation, the same set of business checks is being processed as during the actual import.

For migrations into SAP Business ByDesign, the simulation run has a particular relevance because, for many object types, the import process can't be undone.

While the earlier **Validate File** step was checking mostly for formal data issues, the import simulation applies the full variety of business checks to the data. As a consequence, the spectrum of possible error messages is considerably higher than during the previous steps.

The higher complexity is also reflected in a longer runtime compared with validation.

Object dependency: prerequisite data must be loaded

Most objects contain references to other objects and thus depend on these objects (e.g., sales orders have customers and materials as prerequisites). While the more formal checks during the **Validate File** step were standalone, this doesn't apply to simulation. Here, any data that the current record is referring to needs to be actually in the system. For example, if you want to run the simulation for a sales order, the customer master records and the materials used in this sales order need to be in the system already. In the migration object documentation, the prerequisite objects are listed (also in the migration template on the second to last sheet).

When starting the simulation, a short reminder comes up (which you can hide permanently if you want) listing the prerequisite objects. If you've missed a prerequisite, expect to get some additional error messages during simulation.

During simulation, just like in validation, you can decide to leave the window and have the process continue in the background (**Run in Background** button) and do something else in the meantime, as shown in Figure 6.31.

As soon as the entire file has been processed, click **Close** to display the results of the simulation run, as shown in Figure 6.32. The log provides the following pieces of information:

Simulation log

- A short summary informing you about the total number of records processed and how many of these raised errors

- A list of error messages and warnings
- Instructions for actions

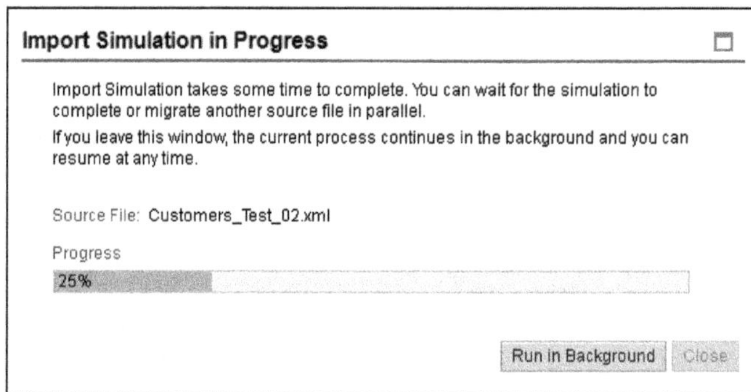


**Figure 6.31** Import Simulation: Run in Background

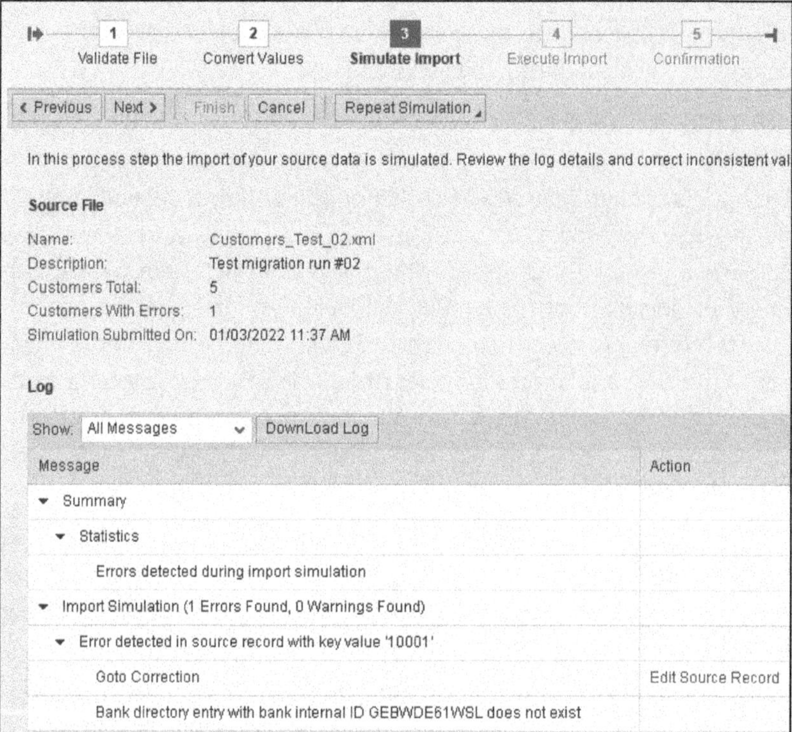


**Figure 6.32** Results of Simulation Run

For every erroneous record, you can directly navigate to the data editor (**Edit Source Record**). The view is filtered so that only the current record is displayed. In addition, in the header area, the error messages affecting this record are repeated.

**Display Help: Use Long Texts Where Offered**

Often enough, the error messages in the **Simulate Import** step are pretty short. Luckily, some error messages come with a **Display Help** link. Make use of this help feature whenever it's offered. It not only shows a more detailed description but also includes some context parameters that allow you to better identify the erroneous part within the record. This advice applies particularly to the error message **Additional message generated; see help for more information**, where the actual error description is fully in the **Display Help** text.

Click the **Download Log** button to save the error log to a local file on your PC.

**Error Message or Warning?**

The log contains not only error messages but also warnings. If there are several messages listed for one record, then it isn't transparent which of these are errors and which are warnings.

If the entry for this record starts with **Warning detected...**, then this record comes with warnings only. But if it says **Error detected...**, you could actually face a mix of errors and warnings.

To prioritize your work, use the **Show** filter and toggle between **Warning Messages** and **Error Messages** to understand the type of the listed messages.

**Repeat simulation**

The **Repeat Simulation** button allows you to re-run the simulation. Like in the validation phase, you can choose between a re-simulation of all records or only of records that have been changed since the previous run.

The iterative identification of error messages is the actual purpose of the **Simulate Import** migration step. Let's discuss how simulation as a tool can be used most effectively.

### Simulation Best Practices

**"Detector" files**

We recommend you start working with small "detector" files. Use these small files to detect and remove as many obstacles as possible before you increase the data variety and bring in a larger file.

It pays off in terms of efficiency to run the simulation with small files containing a pilot set of a few different records. This approach allows you to quickly identify and remove obstacles that would have affected many records long before you start working with larger files (with correspondingly longer runtimes). Because the files are small, you can fix potential data issues manually and immediately drill deeper with the next re-simulation that might provide you with additional error messages.

Iteration

Follow an iterative approach:

- Fix one or few issues that have been reported in the log.
- Re-run the simulation to reevaluate the result of these fixes.

Don't be surprised if during the first cycles, the number of errors found increases. Some more complex checks of business consistency are only processed after the basic issues have been resolved.

After the small file has detected all possible errors, advance to a larger file. As long as you're still in the "error detecting mode," size doesn't actually matter, but a wide variety of business data constellations does.

### Error Analysis

Depending on the root cause for the resolution of simulation issues, usually up to three places might need to be adjusted:

- The data that you're going to migrate
- The business configuration settings
- The value conversion

Let's take a closer look into these cases and the corresponding fixes:

Error in the data

- For the correction of data that is to be migrated, use the integrated data editor whenever you can. This approach allows you to directly proceed with the processing of this file and gives you a chance to identify any additional issues quickly. Some tips:
  - Choose **Edit Source Record** to open the file in single record mode, which allows you to fix data errors conveniently without scrolling.
  - If the same error affects many or all records, click on the file name, which opens the entire file in the data editor. Use filtering and the **Mass Change** functions shown in Section 6.2.8.
  - In this context, it really pays off if you started with smaller files
- Maintain an action log where you document the changes that you've directly applied to the file. This approach allows you to systematically address the corresponding changes in future files if applicable (e.g., to a later production cutover file version or to the full file if you worked with a subset of your data as recommended).
- To check the success of your changes immediately, use the re-simulation with the faster option: **Repeat Simulation • Simulate Import for Changed Records Only**.

[Ex]

**Typical Data Errors Detected during Simulation**

Some typical data errors often encountered during a simulation include the following:

- Some field entries are mandatory only in certain constellations, and therefore, they aren't marked as mandatory in general. The necessity of such a *conditionally mandatory* field can depend on a value in a neighboring field or a business configuration setting (e.g., whether segment reporting is active or not).
- Incorrect postal codes are used. (Accurate country-specific format checks are available.)
- Data constellations are inconsistent from a business point of view. For example, a sales order can refer to a project task but only to project tasks of a certain type.



**Errors in business configuration settings**

Configuration settings can be another potential root cause for migration simulation errors if they don't support the full variety of the migrated data.

As part of the implementation activity list in the **Business Configuration** work center, migration relies on the completeness of various settings made in previous implementation activities.

Some typical examples of configuration issues that can lead to problems during migration step simulation include the following:

- The definition of organizational structures is incomplete. Have all required cost centers, and so on, been defined and flagged correctly? Is the validity interval of the organizational units large enough to support the migrated data?
- Business configuration settings made in activities in the **Fine-Tune** phase aren't complete; for example, asset classes aren't sufficiently defined, or a default definition for depreciation methods is missing.
- Basic data isn't complete; for example, banks are missing in the bank directory, or logistics are missing.

After fixing any configuration issues, trigger a new check using a full simulation (**Repeat Simulation • Simulate Import for All Records**).

**Value mapping issues**

Some migration simulation errors have their origins in incorrect value mappings, as described next:

- In practice, occasionally you encounter value conversions that haven't been set with due care. A typical real-life situation might be that, during early tests, due to a lack in business configuration or basic data, a right target value wasn't available. Then, instead of fixing the root cause, an incorrect makeshift value was assigned. If this value isn't questioned later on, then you'll be lucky if it only leads to a simulation error, and not a silent import using a wrong value.

- In the **Convert Values** migration step, several objects offer certain control parameters to be set. In some projects, alternative settings are tried out, but then the right value is never restored.

If you returned from the **Simulate Import** migration step to the **Convert Values** step to apply some fix, then the simulation is started automatically, after you leave the **Convert Values** step, when you click **Next**.

**Unclear root cause**

We don't want to finish this section without sharing with you some general recommendations regarding how to address simulation errors for which the root cause is difficult to identify.

**Seven Methods to Address Unclear Root Causes**

From the text of the simulation error message, the root cause isn't always obvious; sometimes, it's even unclear which field or which subrecord triggered the error message. If you decide to track down the root cause without waiting for support incidents or answers to community requests, some methods worth trying include the following:

- **Comparison method**
  Open the corresponding work center in your test system and try to enter the affected record with all its field values manually. If the issue is in the record (or the consistency of the record with the system), then the manual creation will trigger an error reaction as well—often with a different wording that might give you some additional information to spot the root cause.
- **Community method**
  Use the SAP Community network. Maybe the same question has even already been discussed there; otherwise, give it a try and raise the question yourself. The keyword search in the Help Center in SAP Business ByDesign searches community resources automatically.
- **Reread method**
  Even if a message text is somewhat cryptic, it usually contains a valid core. Read the error message again in detail. Always check the **Display Help** link if the error message comes with this link.
- **Reduction method**
  In the migration cockpit, create a copy of your file. Now, open this copy in the data editor. Remove all but a handful of records for which the error message has been sent. Simulate this file, which should still raise the error message. From the simulation screen, now repetitively open the data editor, simplify the data, for example, by clearing a subordi-

nated sheet, and re-run simulation. Repeat this until the error message doesn't occur any longer. This procedure gives you the part of the record that has triggered the error message, that is, the part that you cleared or simplified directly before the error vanished.

- **Bisection method**
  In some rather rare situations, it might be unclear which record actually triggered an error message. While this certainly justifies a support incident, depending on the situation, you might prefer to try helping yourself with the following procedure: In the migration cockpit, copy your file twice, and use the data editor function to delete half of the records in each copy (remember, the **Delete** function on the main sheet will consistently delete the record across all sheets). Run the simulation for the new files and repeat the step for that half file that is still triggering the error message. Always following only the half with the error, this bisection should hunt down one erroneous record in a finite amount of time.
- **Correlation method**
  If there are several records with the same error message, download the template to your PC, and highlight some records that are correct and some that do throw the error message, for example, by using color. Compare the two groups and try to recognize patterns where the two groups differ.
- **Bottom-up method**
  In the migration cockpit, create a copy of your file. Now, open this copy in the data editor. Reduce the number of records in the copy so that at least one correct record and one erroneous record are left. Start changing the correct record in a way that, step by step, it becomes identical to the error record. Re-run the simulation and repeat the cycle until the built-up record is throwing the same error message. Now, you've tracked down the field that triggered the error.

While running test simulations in your test system, at any stage during the iteration, you can of course opt to push forward and actually import the current test file.

The review of the resulted data in the work center and the usage in business user tests might reveal additional data issues where data is, although consistent, not as expected. An example is nonmandatory fields that have been forgotten during the export from the legacy system.

At some point, you should run a simulation (and later import) of the full data volume that is to be imported during production cutover. It's a well-deserved milestone to be welcomed by the message shown in Figure 6.33:

**No errors detected during import simulation.** However, this isn't even so much targeting data quality; after the previous iterations, all but some exceptional error constellations should have been caught anyway.

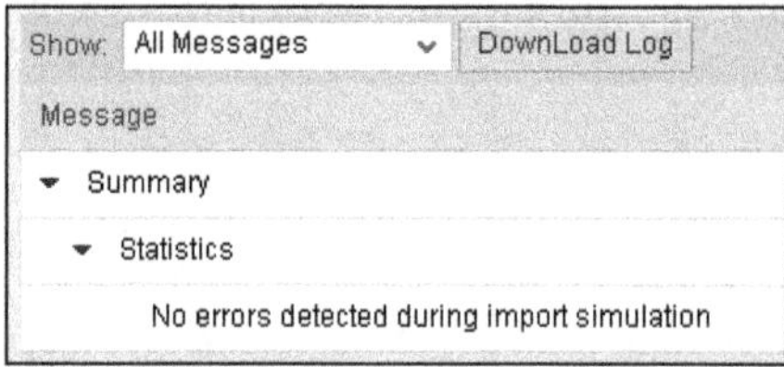


**Figure 6.33** Import Simulation without Errors

The main purpose of the full volume test is rather to check the velocity and volume readiness. Systematically take notes of the runtimes of your file during the different migration steps and use it to set up a cutover plan that respects the object dependencies. If your plan collides with given time constraints for your cutover interval, then the collected runtime information is the best source to look for objects where potential improvements will be most effective for the entire schedule.

Be aware that you don't need to have a simulation without errors to advance to the actual import. As you'll learn in the subsequent section, any records with simulation errors will be skipped for import, but this doesn't affect the correct records in your file. The skipped records will be copied into a new file for later processing.

Click **Next** to advance to the **Execute Import** step.

### 6.3.4 Import Data

On entering the **Execute Import** migration step, the warning message shown in Figure 6.34 is raised, requesting confirmation that you want to run the actual import of this file. Take this warning seriously because for most objects, the import of records can't be undone.

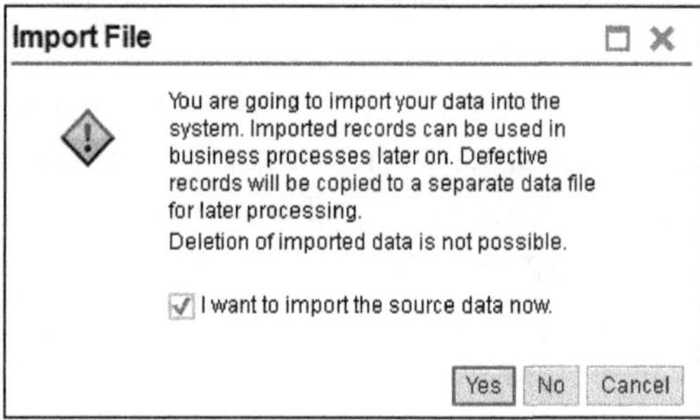


**Figure 6.34** Import: Warning and Confirmation

If you navigated here by mistake, click **No** or **Cancel**. Then, no data will be imported, and you get to a screen from where you can return to the simulation window using the **Previous** button.

Immediately after you've approved the import request by selecting the **I want to import the source data now.** checkbox, the actual import run starts. Like the validation and simulation runs, this process can also be run in the background without you needing to watch the progress bar. After the import run has finished, the import log is waiting for you, as shown in Figure 6.35.



The most important information is listed at the top:

- Total number of records processed
- Number of records imported successfully
- Number of records not imported due to errors

Like the simulation log before, the import log informs you of the detailed error messages, if data records have triggered errors during import. You already know most of the features of this log from the simulation log. You can navigate to a detailed help text if available, you can open and change the single record in the data editor, and you can export a local copy of the log using the **Download Log** button.

Using the **Display Imported Records** and **Display Not Imported Records** links, you can display a list of the records of these two sets. After you've advanced to the final migration step, **Confirmation**, an icon to download this file to a local template file will become available as well.

The **Cancel** button brings you back to the migration cockpit, but clicking this button doesn't undo the import, of course.

Choosing **Import File • Import Not Imported Records** starts the import run for records that haven't been imported yet due to errors. Of course, these records failed for a reason, so unless you're able to fix the error at that stage, the **Import Not Imported Records** function will just repeat the failure of these records.

Compared to simulations, the ability to fix issues is restricted. Only the failed records can be edited in single-record view. For auditing reasons, the system can't allow you to edit any other, already imported, records. And most of the errors that are listed now you should have encountered anyway during the previous simulation run. So, you should only encounter very limited use cases when you'll want to work on data fixes here. The much more convenient way to address not yet imported records is to process these using the delta file technique that you'll learn about in a minute.

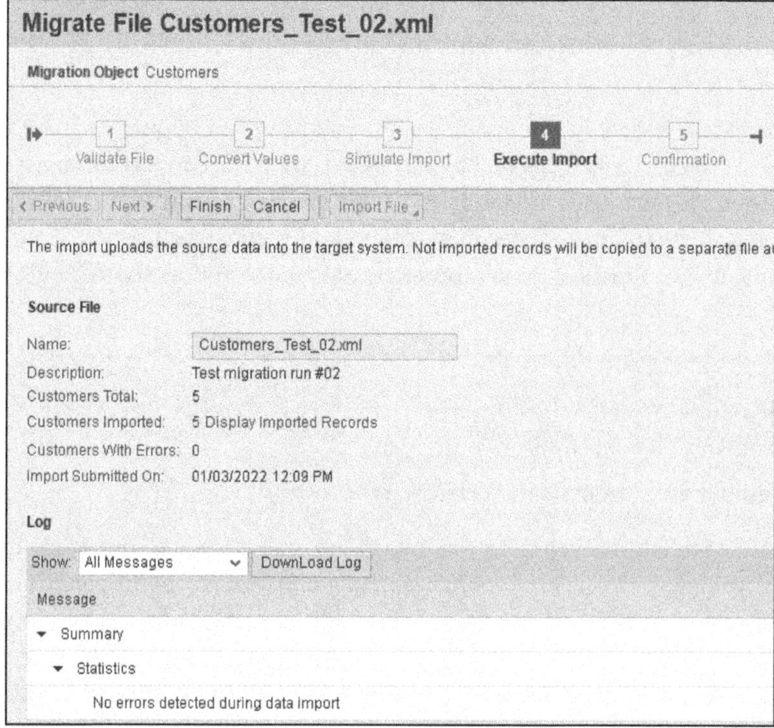


**Figure 6.35** Results of the Execute Import Step

Click the **Finish** button to finalize the migration of this file, bringing you to the **Confirmation** step. The summary of migration results repeats the record numbers that you know already from the **Import** step.

However, while finishing the file, two important processes are triggered automatically:

- The *delta file* is being generated.
- A separate log with additional export options is provided.

**Not imported records**

How about records that haven't been imported? The erroneous records fall into one of two groups:

- In most cases, these records had already failed during validation or simulation. The delta file technique allows you to advance with the file to the migration import anyway by skipping these erroneous records.
- There are only a few reasons why a record can fail during import after it had successfully passed the validation and simulation steps. These exceptions are usually a kind of collision between two records. A typical example from the customers object is a duplicate assignment of the Dun & Bradstreet (D&B) D-U-N-S® ID (a unique identification number from an international company directory).

**Processing a delta file**

At the end of the processing chain, all records that couldn't be imported due to errors are transferred into a new delta file. The delta file is generated and added to the file list in the migration cockpit. You can spot this file easily with the file name. It has the same name as the original file, with the suffix *_DELTAXL_0000000001* added to the file name.

With this delta file, you're usually running through all processing steps separately, starting with **Validate File** and ending with **Execute Import** and **Confirmation**.

Because the delta file has only erroneous records, you can address any issues in a condensed way, which is often more effective. During the import of the delta file, if there are still some erroneous records left, then the delta file mechanism applies again: The remaining records that couldn't be imported are transferred into a new delta file, this time with the suffix *_DELTAXL_00000000012* instead of *_DELTAXL_0000000001*.

**Download imported records**

In the file list in the migration cockpit, you'll find the imported file doesn't have the **Completed** status. You might need to toggle your file filter settings to see this file in the list (choose **Show** and then **All Files**).

The **Show Import Log** link appears at the end of the list. You know this log already from the **Execute Import** step. The only difference is that there are now two new icons: the **Download** option for importing records, as shown in Figure 6.36, and the other the download for records that were not imported.

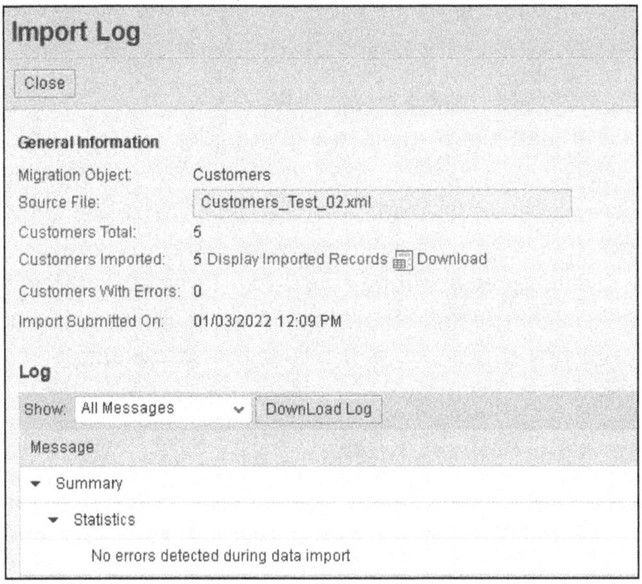

**Figure 6.36** Import Log: Download Imported Records

The import of the migration templates is finished, so you've reached the target of this important section. You've learned how to process the central steps of **Validate File**, **Convert Values**, **Simulate Import**, and **Execute Import**, as well as the corresponding error handling.

## 6.4 Migration of Transactional Accounting Data

This section describes the migration of transactional accounting data into an SAP Business ByDesign system. The goal of this migration is to transfer the complete and consistent accounting data of one company by a certain key date. The most important features supporting this paradigm include the following:

- **Migration unit**
  This element consists of several migration objects that together describe the complete transactional accounting data of your company. For migration, these objects are bundled into a *migration unit*.
- **Migration posting date**
  The files in your migration unit represent the accounting data of your company based on one certain *key date*—namely, the *migration posting date*.
- **Posting preview with balance cross check**
  To ensure consistency across all files in your migration unit, there are additional reconciliation steps placed before the import run.

Let's start with the migration posting date, which you can define as part of the implementation activities. After a condensed overview of preliminary steps in your source system, we are going to introduce the migration unit, which collection of all accounting migration files of your company, ensuring consistency from an accounting point of view across all these files. After processing the single files of the unit mainly in the same way that you have seen in the precedent chapter, you'll be introduced to some additional steps that allows you to control the completeness and consistency of your accounting data using predefined migration reconciliation accounts.

### 6.4.1 Preparation in SAP Business ByDesign

A fundamental decision when planning the go-live date is the choice of the key date at which the transactional accounting data will be transferred. You'll also come across the short-term migration date, which in this context also refers to the migration posting date.

Usually, this date is the last day of a month, for example, 5/31/2024. Of course, the actual migration cutover is some days later due to the time for the preliminary closing and the data export from the source system.

There are mainly three approaches to choose the migration posting date:

Migration posting date

1. **Migration for a month-end that isn't the end of the fiscal year (e.g., May 31st, depending on your fiscal year)**
   This is by far the most common strategy. Staying away from the fiscal year end has substantial benefits; for example, often the data can be provided faster.

   However, the main benefit is that both a system implementation and a fiscal year closure are usually coming down on your teams as a double burden on top of regular business operations. If you avoid making this a triple burden, you'll be compensated with a stronger focus and a higher quality in tests and reviews.
2. **Migration for fiscal year end**
   An argument for this approach is that all postings for the fiscal year of migration can be accessed directly in the same system. A risk is that this approach doesn't allow you to push out the go-live date by, for example, one month, without also questioning the chosen approach.
3. **Migration during the month**
   Although this approach is possible, its main disadvantage is that tax reporting from two source systems must be merged.

In the **Business Configuration** work center, set the migration posting date in the activity list of your implementation project in the **Integrate and Extend** project phase. Open the **Set Migration Date and Process Control** activity, select your company, and choose **Actions • Set Posting Date for Migration**, as shown in Figure 6.37.

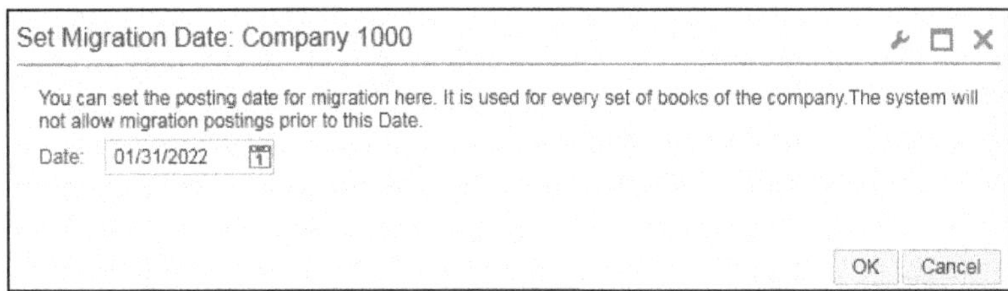


**Figure 6.37** Setting the Migration Posting Date

For test migrations in your test system, choose the migration posting date by following a strategy. You want to provide a test migration environment where the teams can become familiar with the migration reconciliation features. In addition, you need to be aware that the migration posting date for a company can be chosen freely only as long as no postings have been made for this company so far. In particular, you can't set the migration posting date to a date before the posting date of an unplanned manual test posting; you can only set it to a later date, which might easily disturb your migration test plan.

For this reason, it's important that you keep control of which periods are open at which time for postings. You can use a two-level concept to control these periods.

**Open and close posting periods**

To begin with, every single accounting period is protected separately. You can open and close periods in the **General Ledger** work center by choosing **Common Tasks • Open and Close Periods.**

**Set the process control**

In addition, the process control allows you to block or allow posting transactions for a period depending on the type of business process. Two alternative ways exist to access the process control settings in your system: either use the **Set Migration Date and Process Control** implementation activity, which we discussed earlier in this section, or use the **Set Process Control** button on the **General Ledger** work center, and choose **Master Data • Companies.**

The time interval up to and including the migration posting date should be reserved for the migration only. Most migration postings will actually post to the migration posting dates, although there can be certain exceptions in which earlier periods of the current fiscal year will see migration postings (e.g., certain postings from fixed asset migration). The time interval after the migration posting date is reserved for postings that are being entered manually or triggered by operational business transactions.

As soon as you've set the migration posting date, the system automatically adds a corresponding line for migration postings in the process control. Be sure to review the defaulted time interval.

In any case, you want to keep your migration interval free from uncoordinated manual postings. This wouldn't be the first journal that has been inadvertently initiated by some user test. Even if you can reverse the posting, both will leave unaesthetic marks. At least in your production system, you want to manage user access in a restrictive way before the result of the accounting migration has been officially accepted.

Add this access restriction by adding another entry in the process control, now for the business process group `01 - All` with status `3 - Blocked` and a sufficiently large time interval (e.g., `1/1/1900 - 12/31/9999`).

Regardless of this logic, the user authorizations granted in the user management offers a separate control that you should use anyway. The menu path in the work center is **Application and User Management • User and Access Management • Business Users**. Then, choose **Edit • Access Rights**.

### 6.4.2 Preparation in the Source System

The data export from the source system is a key element in the migration—one that you should plan for with several waves of exports.



To start with, the test focus is mainly on isolated test data for the various objects. These files are run in an isolated way through validation, value mapping, and simulation, including the posting preview feature that we'll explore in the following sections. Depending on the findings, you might repeat an export or start with subsets first. The focus at that stage is to identify and overcome any object-specific obstacles and misunderstandings; however, you might not actually import these first data files because consistency across objects might not have been a point in the object-focused work.

As soon as the object level is under control, you proceed to the next level, targeting a test migration import of an entire migration unit. For this test, a balanced and consistent set of accounting data is required, including general ledger account balances and corresponding subledger data. It's important to note that all data needs to come from a defined point in time, for example, the precedent month-end closure. Plan for the necessity to capture complete test datasets for all relevant objects from your legacy systems during your implementation phase, especially from month-end closings.

Take the opportunity to rehearse the key date process already in place during test migration by using the month-ends from which the test data has been exported. The following tasks apply to test and production cutovers correspondingly, each with its migration posting date, which we assume here to be a month-end date.

**Preparation in the source system**

Before you start exporting transactional accounting data, you should ensure the following prerequisites have been met:

- All available posting documents up to the key date have been posted in the source system.
- For fixed assets, all depreciation runs up to the key date have been processed.
- Open taxes from your legacy system have been reported and paid for.
- Your books are balanced, and you have run a preliminary month-end closing.

- Because migration cutover is usually performed during the first week of the new months, ensure that you have collected the posting data separately that is meant for the new period so that you may enter these documents directly after the go-live in SAP Business ByDesign.

**Production cutover: freeze the source system**

You need to ensure the consistency of the exported data. The exported account balances and subledger data must correspond exactly to the same status of the source system.

To perform reconciliation, capture the data and all relevant evaluations in a way that they don't change while migration is still on the way to avoid a moving target situation.

During production cutover, you can block the source system entirely against changes to ensure the ability to compare and reconcile the migration result.

**Test migration: frozen copy of source system**

During test migration, the operational usage of the source system continues; therefore, additional steps are required to ensure their comparability. The most convenient approach is, of course, a copy of the source system, which is frozen against changes until the migration results have been verified.

This approach helps you in two ways:

- This approach allows you to compare the migration result by running the same business reports and evaluations in parallel on the frozen source system copy and on the SAP Business ByDesign system. Because both systems are available, you can dynamically expand your analysis depending on your findings and drill down to more detailed levels as you go.
- The data for the migration templates should be exported from the system copy. On one hand, this tactic ensures that the ongoing operational usage in the original source doesn't cause any data changes while the export is still running. On the other hand, this tactic allows you to repeat the export from the same data source if an export error has been detected (e.g., if a field is missing).

If the approach of a frozen copy of the source system isn't feasible, then you need to compensate for the disadvantages as much as possible using static information, in the following ways:

- Decide up front which reports and business evaluations you want to use for the assessment of the migration result, including the granularity level that you want to apply. Typical candidates are the following:
  - Trial balance per ledger account
  - List of open payables and open receivables

- Asset history sheet
- Detail lists to asset movements (e.g., acquisitions, retirements, transfers)
- Inventory lists, including price information at that point in time

All these reports must be run and exported now to be available for comparison later.

- It's important to block the source system against changes while the export of data and reports is still running. Otherwise, the consistency of the data isn't ensured, and the evaluation might make you chase phantom issues if the data and reports don't reflect the exact same system status.



### 6.4.3 Migration Unit

The transactional accounting data of your company is described by several migration objects and includes general ledger account balances as well as their corresponding subledger data. A migration unit is a bundle of migration files, with one file from each migration object.

In the migration process, all files of a migration unit are only imported together.

**Transactional accounting objects in migration**

The following objects are available:

- General ledger account balances
- Fixed assets (two versions)
- Inventory
- Open items receivable
- Open items payable
- Bank account balances
- Open goods receipts
- Open outgoing checks

Depending on the scope that you've selected for your solution, the less frequently used objects, such as bill of exchange payables and bill of exchange receivables, may be added to the list.

[!]

> **Collaboration with the Business Department**
>
> For some migration objects, an incorrect value mapping might appear plausible, although it's actually not correct. The wrong accounting migration can have severe implications.

> Because the guided step-by-step procedure in the SAP Business ByDesign is easy to use, a certain risk exists of underestimating the business complexity that is hidden behind some data.
>
> Especially when the migration tasks are centralized in your project, we recommend establishing a close collaboration through all steps, with sign-offs also for intermediate steps such as the decision on which template fields to use (e.g., Is the low-value indicator in fixed assets necessary?) and the value mapping decisions.
>
> Major test migrations also should be used to rehearse the sign-off procedure in the same way that it will be done during the production cutover.

**Procedure**

During migration of transactional accounting data, the migration tool also guides you step by step through the process.

For every relevant object in the migration unit, you upload a migration template. In the first part, you're then processing these files separately in the validation, value mapping, and import simulation steps. After that, you'll encounter a new step—a posting preview that makes transparent which postings on which accounts will be triggered by the single file. If this is as expected, you can release the single file and do the same for the other objects of this unit.

The remaining steps are no longer on the level of the single files but only on the level of the entire migration unit. A posting preview collecting all release files of the entire unit provides a balanced cross-check that can detect inconsistencies between main ledger and subledger data. If reconciliation of the results with the legacy system is successful, you can start the import of the migration unit, which will then run the import for all release files of this unit.

**Completeness and consistency**

To audit the requirements during migration of accounting data, a special focus needs to be directed to the completeness and the consistency of the involved data. The migration unit offers a bracket that supports you in this task. For this task, as we've mentioned, two important features have been added to the known import simulation: a posting preview (for single files) and a balance cross-check (for the entire unit). The guided step-by-step procedure in the migration tool has been enhanced accordingly:

- For every file, a special posting preview has been added to the step-by-step migration process.
- In addition, a balance cross-check for the entire unit is introduced as a mandatory step before the import of the unit can be triggered.

The migration of accounting objects can be entered via the **Migration of Accounting Transaction Data** implementation activity. Entering via **Migration of Accounting Transaction Data** using the migration tool opens the (initially empty) list of migration units.

As shown in Figure 6.38, the list contains statuses (**Not Started/In Process/Finished**), information about which company the unit belongs to, and the migration posting date for this unit.



For the marked migration unit, the bottom part of the screen shows the processing status of the enclosed migration templates.

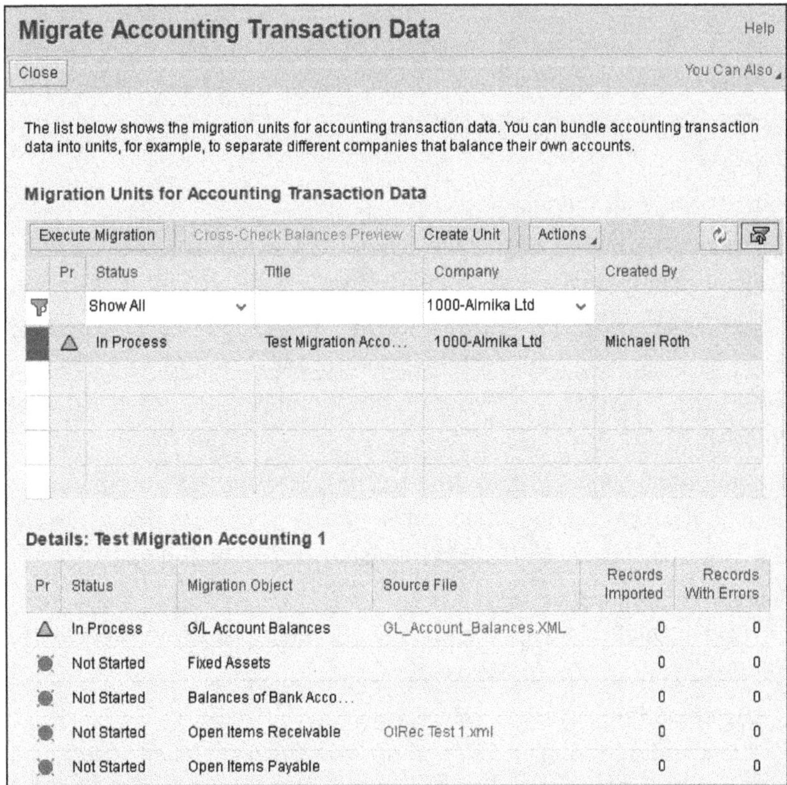


**Figure 6.38** Migration of Accounting Data: List of Migration Units

On this screen, nearly all buttons refer to migration units, such as the following:

- Click **Execute Migration** to open the selected migration unit in the migration cockpit, where you then can upload files and process the migration steps.
- You'll see in a minute which information to enter when using the **Create Unit** button. As long as the actual import of the unit hasn't been per-

formed, you can change this information by choosing **Actions • Edit Unit** or discard the entire unit with all its files by choosing **Actions • Delete**.

- As soon as you've processed the first files in a unit far enough to release these files for import, the **Cross-Check Balances Preview** button becomes active and opens a preview of the balance list based on the simulated postings of all released files. In the same place, this preview will be documented permanently for already imported units.
- The remaining user interactions (e.g., open files in the integrated data editor or functions in the **You Can Also** menu in the upper-right corner) are identical with the corresponding functions presented in Section 6.2.5.

**Create Unit button**

During creation of a migration unit, three pieces of information are requested, as shown in Figure 6.39:

- **Title**
  Enter the title for the migration unit.

**How to migrate several companies**

- **Company**
  Select the company for which you're migrating the accounting data.

  If you've set up several companies in your system, every company requires a migration unit of its own. Except for shared value mappings, there is no dependency between the accounting migrations of different companies.

  The migration posting date for your unit is displayed automatically, derived from the settings that you've made for your company. Review the date, and, if necessary, adjust it as described in Section 6.4.1.

**Fixed assets: two alternative migration templates**

- **Fixed Assets Migration**
  Decide which of the two available migration templates you want to use for the migration of your fixed asset data:
  - **Use default migration template...**
    The standard template offers more convenience. Usually only one sheet needs to be populated. To supply the additional valuation views used by this company, some processing logic is performed automatically, regulated by a parameter that you can set during the **Convert Values** step. If this logic meets your requirements and if you have only routine movements (i.e., acquisition, depreciation, and retirement), then this template option is for you. Also, if your legacy system has only one valuation view, and you're introducing additional views only with this implementation, then you might want to use this option.

- **Use extended migration template...**
  While the extended template looks at first glance somewhat scary with its four sheets, it isn't that complicated. Instead of relying on some logic to populate the secondary valuation views, you're expected to explicitly list the postings in the various valuation views. The higher effort buys you the maximum direct control over these postings. If you can export the postings for all secondary valuation views efficiently from your legacy system, then you might want to consider using the extended template. The more explicit control over every entry is also beneficial if your assets have a history of exceptional transfers and revaluations that must be retained. Note that the extended template offers more opportunity to enter data errors.

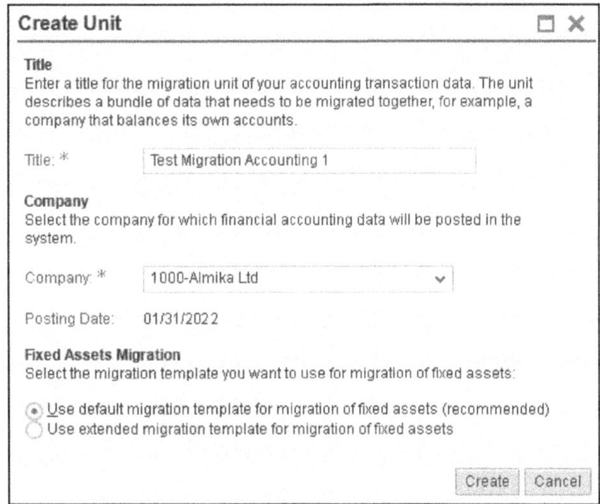


**Figure 6.39** Creating a Migration Unit

Click the **Execute Migration** button to start the processing of a migration unit, as described in Section 6.4.5. Before we turn to this process, however, we want to briefly discuss a few specialties of accounting migration templates and their corresponding data requirements.

### 6.4.4 Migration Templates in Accounting

Accounting documents usually refer to all prominent master data objects, particularly the "big five," which must be migrated before starting an accounting migration:

**Master data prerequisites**

- **Customer**
- **Supplier**
- **Employees**

- **Materials** (including **Inventory Valuation UoM** field and the **Financial Organizations** sheet)
- **Services** (including the **Valuation UoM** field and the **Valuation** sheet)

In addition, the data that controls the account determination must be in place in the following ways:

- **Customer**
  The account determination group assignment is part of the **Customer** template (**Payment Data** sheet).
- **Supplier, Employees**
  The assignment to account determination groups is in the separate **Account Determination Groups Payables** template.
- **Materials**
  The assignment to account determination groups and further accounting details are in the separate **Material Valuation Data** migration template.
- **Services**
  The assignment to account determination groups and further accounting details are in the separate **Service Valuation Data** migration template.

For migration templates in accounting, refer to the recommendations we gave in Section 6.1.4, which also apply in this context.

If the organizational structure in your system contains several companies, then provide a separate migration unit with a separate set of migration templates for each of these companies.

After the simulation run has been processed successfully for every file of a migration unit, use the preview functionality to check the consistency of your main ledger and subledger files on a detailed level. We recommend doing a first quick plausibility check for consistency based on your template files.

[»]

**Consistency of Accounting Data**

For a first quick consistency check of your templates, focus on two fundamental principles of bookkeeping:

- The general ledger is always balanced; that is, the total of the balances of all general ledger accounts must be zero.
- The total of a subledger must match with the balances of the corresponding general ledger accounts.

**Visual inspection of migration templates**

With a few fairly simple checks, you can perform a first assessment for consistency in your accounting migration files. You can apply these checks using some basic Microsoft Excel functions. A best practice is to use a temporary copy of your migration template files for this assessment so you can "unprotect" the data sheets and enter formulas without interference with the actual migration files.



Follow these steps:

- Start with a quick visual inspection of the amount fields. Take a closer look into entries that appear left-aligned, which may be evidence of an incorrect cell format.
- When using the general ledger account balances migration template, use the Microsoft Excel function `Sum()` to calculate the total sum of all entries in the **Amount in Company Currency** column. If the result isn't zero, either some accounts are missing from your list, or some balances are wrong. A potential issue in this template is an incorrect sign in the amount fields.
- For the fixed assets migration template, the entries must have the right signs, usually positive signs for acquisition costs and negative signs for depreciations. Depending on the complexity of your data, you might want to check subtotals of selected asset classes (e.g., buildings, cars, machines) and compare with corresponding general ledger accounts if a 1:1 relationship exists.
- For the open items receivables and open items payables migration templates, watch out for values with signs. In this context, all amounts should be entered without signs; instead, the sign is derived automatically from the **Document Type** column. You can check the subtotals of the records for a certain document type (e.g., invoices or credit memos) using sorting or the Microsoft Excel function `sumif()`in your temporary file copy, of course. If the granularity of your general ledger account allows, compare these subtotals with the corresponding accounts in your general ledger account balances template.
- For the open outgoing checks migration template, the total of the **Payment Amount in Company Currency** column must match the balance of the corresponding general ledger account.

### 6.4.5 Accounting Migration: Main Steps

Let's go through the migration of accounting data in detail, focusing on the steps that are specific to these objects and building upon the general migration tool functionality presented in Section 6.2 and Section 6.3.

You can enter the migration of a transactional accounting object as usual from the **Business Configuration** work center. To access a list of migration units created so far, follow the navigation path **Open Activity List • Integrate and Extend • Migration of Accounting Transaction Data • Migrate Accounting Transaction Data Using the Migration Tool.**

Create a new migration unit as described in Section 6.4.3 and choose **Execute Migration** to process the selected migration unit in the migration cockpit.

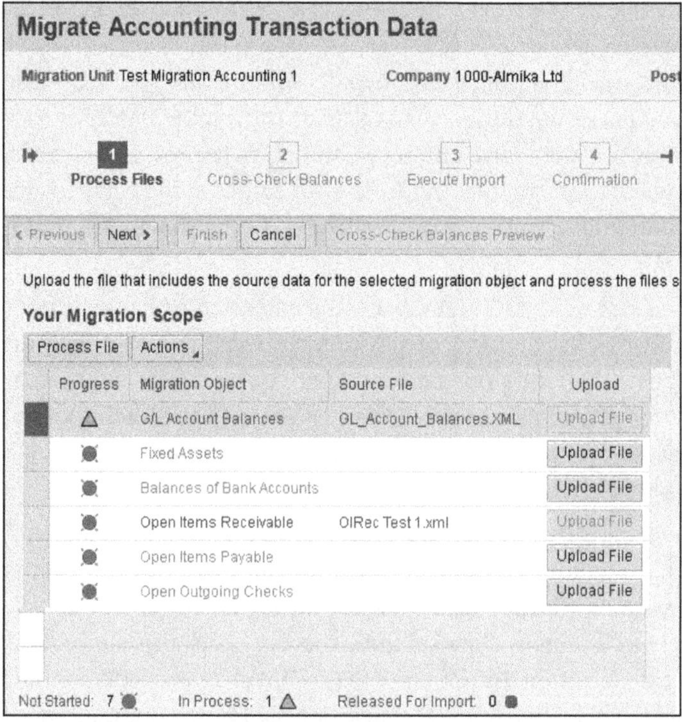


**Figure 6.40** Migration Cockpit for Accounting Data

As shown in Figure 6.40, the basic migration cockpit we described in Section 6.3 has been adapted to process migration units in accounting. After you enter the migration of a unit by clicking **Execute Migration**, you'll immediately encounter the main differences:

**Main steps during migration of accounting units**

- The validation, value conversion, and simulation steps have been grouped into a collective step called **Process Files** because these steps are still done on the level of a single file, not the entire unit.
- In the lower part of the screen, the files of the selected unit are listed. In this section, you can **upload** templates. Using the **Actions** button, you can edit, download, and delete the files as usual. If you want to perform a quick isolated test in a temporary unit, choose **Actions • Create** to create

some test data. A status flag shows to what extent every file has been processed.

- One by one, you can prepare the individual files of the unit. To enter this preprocessing, click on the object name or use the **Process File** button. There, you'll find the **Validate File, Convert Values,** and **Simulate Import** steps, followed by two new elements: In the **Verify Posting Preview** step, you can see the postings this file will trigger, and finally, you can finish this preparation work by releasing the file to import.
- After you've released all files of the unit, click **Next** to push the entire unit forward to the new **Cross-Check Balances** step.
- Until then, you can use the **Cross-Check Balances Preview** button to get a preview of this evaluation for those files that have been released up to that moment.
- During the **Execute Import** step, all released files of this unit will be imported.

Now, let's see this procedure in practice. Click the **Upload File** button to upload a migration template with some data.

### 6.4.6 Simulation of Single Files

Click the **Process File** button to start working on the selected file. As shown in Figure 6.41, the guided step-by-step procedure should now seem familiar. As substeps **1a** through **1c**, you'll see the steps **Validate File, Convert Values,** and **Simulate Import**, which we covered in Section 6.3.

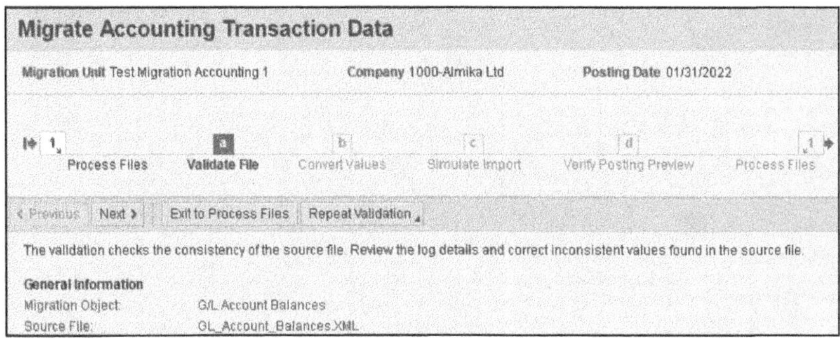


**Figure 6.41** Preprocessing of the Single Files

**No Delta Files in Accounting Objects**

During validation and simulation of transactional accounting data, you can't proceed to the next step unless all error messages of the current step have been resolved successfully.

As a consequence, the delta file technique presented in Section 6.3.4 isn't supported for transactional accounting objects because doing so would undermine the paradigm of a complete and consistent migration.

The new substep **1d**, **Verify Posting Preview**, sounds like a promising prolongation of the simulation—for good reason.

### 6.4.7 Posting Preview

**Posting preview**

In substep **1d**, **Verify Posting Preview**, all single posting entries are listed that the currently processed file will trigger during import.

Figure 6.42 shows an example for the migration object **GL Account Balances**. For every set of books used in your company, the corresponding data is listed, initially aggregated with subtotals. Expand the subtotal lines to drill down to the actual postings. In the preview, you can see the general ledger accounts that the system has derived for each posting with the account determination logic using the actual data, the value conversion, and the business configuration settings that have been made in your system.

**Review of account determination settings**

Compare the listed postings with evaluations from your source system. You can also export this preview to a local Microsoft Excel file. In the hands of accounting specialists, this preview serves as a valuable tool not only to reconcile the data but also to review the effect of your account determination settings.

**General Information**

Migration Object: G/L Account Balances
Source File: GL_AcctBalances_Testmig1.XML
Reporting Date/Time: 2022-01-03 22:21

**Posting Preview**

Show: Standard View Export

| Company | Set of Books | G/L Account | G/L Acct Old | G/L Acct Mapped | LC | LC Amount | G/L Acct Type Desc. |
|---|---|---|---|---|---|---|---|
| 1000 | 7000 | | | | USD | 0,00 | |
| 1000 | 7000 | 117000 - Other Current Assets | | | USD | 371.529,37 | |
| 1000 | 7000 | 117000 - Other Current Assets | 117000 | 117000 | USD | 371.529,37 | Other assets |
| 1000 | 7000 | 151500 - Organization Costs | | | USD | 1,00 | |
| 1000 | 7000 | 500100 - Marketing Expenses | | | USD | -100.023,00 | |
| 1000 | 7000 | 500100 - Marketing Expenses | 500100 | 500100 | USD | -100.023,00 | Costs or expenses |
| 1000 | 7000 | 500300 - Travel Expense | | | USD | -24.020,00 | |
| 1000 | 7000 | 500300 - Travel Expense | 500300 | 500300 | USD | -24.020,00 | Costs or expenses |
| 1000 | 7000 | 517900 - Other Expense | | | USD | -1,00 | |
| 1000 | 7000 | 599999 - Salary, Wages and Health Clearing | | | USD | -1.835.458,42 | |
| 1000 | 7000 | 990000 - Initial Data Carryforward G/L Account | | | USD | 0,00 | |
| 1000 | 7000 | 990010 - Initial Data Carryforward A/P Account | | | USD | -800.589,22 | |

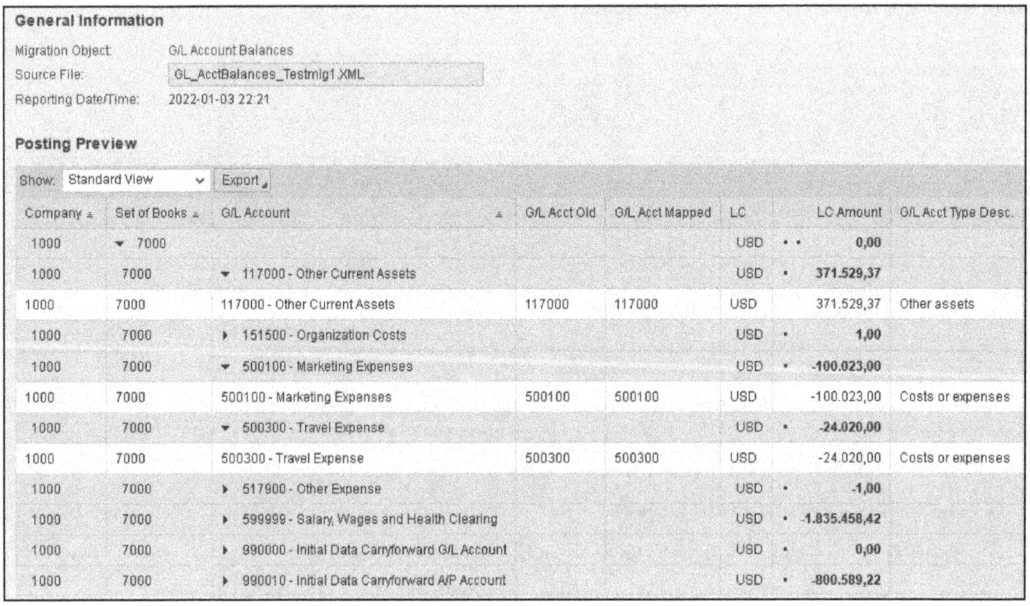

**Figure 6.42** Posting Preview

In particular, the posting logic of migration becomes visible, which uses some migration reconciliation accounts to support the reconciliation between the different objects in a unit.

You can always navigate back to the unit level using the **Exit to Process File** button.

Release file for import

After the correctness of the posting preview has been verified, you can release the file for further processing at part of the entire unit by choosing **Release File for Import.** This step will bring you up to the unit level view, where the progress status of your file is now switched to green (**Released for Import**).



If you change your mind, click the object name to re-enter the file processing, which brings you back to substep **1d**, **Verify Posting Preview,** and navigate to a previous substep (by clicking on the step or by using the **Previous** button). A popup window asks you to confirm that you want to reset the file status.

### 6.4.8 Cross-Check of Balances

Cross-Check Balances Preview button

After you've released the first migration files, you can click the **Cross-Check Balances Preview** button to evaluate the balances on the general ledger accounts that will result from the import of the files released so far.

This preview is the core element of accounting migrations. Now, the responsible project members from the accounting department are required to check the result in detail and reconcile the balances with the source system.

For the review, a helpful tactic is to export the balances preview into a local Microsoft Excel file.

**Export and Check for Messages**

The balances preview uses the **Message** column to display error messages and warnings. Particularly when scrolling through a preview, such messages are often missed. For long previews, we recommend getting into the habit of exporting these messages to an Microsoft Excel file to check the **Message** column for entries.

In the balances preview, as shown in Figure 6.43, all general ledger accounts are listed that receive postings from any of the released files. You can expand these lines to see which objects are actually contributing to this account. To further drill down into the actual postings that came from this account, click the hyperlink at the end of the line.

Unexpected balances are often evidence of an issue in your account determination settings. In your implementation activity list, you should have several activities that influence the account determination—most prominently, the fine-tuning activity **Charts of Accounts, Financial Reporting Structures**, and **Account Determination**, which has an entire section for **Account Determination** tasks. Other activities, as well as wrong account determination group assignments in your master data, can also be a root cause for unexpected target accounts.

If you're using several sets of books, then the evaluation lists all of them.

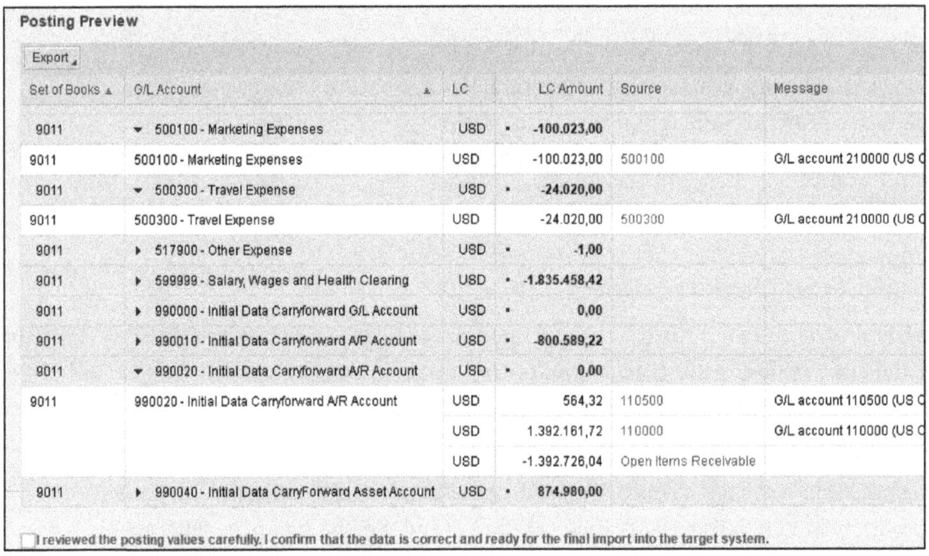

**Posting Preview**

Export

| Set of Books | G/L Account | LC | LC Amount | Source | Message |
|---|---|---|---|---|---|
| 9011 | ▾ 500100 - Marketing Expenses | USD | **-100.023,00** | | |
| 9011 | 500100 - Marketing Expenses | USD | -100.023,00 | 500100 | G/L account 210000 (US C |
| 9011 | ▾ 500300 - Travel Expense | USD | **-24.020,00** | | |
| 9011 | 500300 - Travel Expense | USD | -24.020,00 | 500300 | G/L account 210000 (US C |
| 9011 | ▸ 517900 - Other Expense | USD | **-1,00** | | |
| 9011 | ▸ 599999 - Salary, Wages and Health Clearing | USD | **-1.835.458,42** | | |
| 9011 | ▸ 990000 - Initial Data Carryforward G/L Account | USD | **0,00** | | |
| 9011 | ▸ 990010 - Initial Data Carryforward A/P Account | USD | **-800.589,22** | | |
| 9011 | ▾ 990020 - Initial Data Carryforward A/R Account | USD | **0,00** | | |
| 9011 | 990020 - Initial Data Carryforward A/R Account | USD | 564,32 | 110500 | G/L account 110500 (US C |
| | | USD | 1.392.161,72 | 110000 | G/L account 110000 (US C |
| | | USD | -1.392.726,04 | Open Items Receivable | |
| 9011 | ▸ 990040 - Initial Data CarryForward Asset Account | USD | **874.980,00** | | |

☐ I reviewed the posting values carefully. I confirm that the data is correct and ready for the final import into the target system.

**Figure 6.43** Balances Preview

A yellow warning icon [⚠] marks all migration reconciliation accounts that aren't balanced. Migrations are performed through regular double-entry booking, so there is no posting without an offset posting. Auxiliary migration reconciliation accounts are used to monitor the consistency across migration files. During processing of a migration template, offset postings are made to a migration reconciliation account.

You can maintain the migration reconciliation account for the various subledger objects as part of the account determination settings in the mentioned fine-tuning activity **Charts of Accounts, Financial Reporting Structures, Account Determination**. In the **Account Determination** section, use the **Data Migration** entry.

The postings for the general ledger account balances object also have offset postings. The target account for these objects is maintained in a different, somewhat less obvious place. During the migration of general ledger account balances, in the file processing step **1b – Convert Values**, you'll find a task for the value conversion of **GL Account IDs**. After you've run the simulation for a general ledger account balances template in your system for the first time, this task will contain an entry for the technical source value SAKOMIG, which is already defaulted to one of the migration reconciliation accounts (e.g., 990040).

[+]



**SAKOMIG**

Depending on the set of books that you're using, the default target value of the technical general ledger account ID SAKOMIG can sometimes improve to facilitate the analysis of postings. Proceed as follows:

- In the migration of general ledger account balances, in the value conversion for the **GL Account ID** field, check the target value for SAKOMIG. (There is one per chart of accounts.)
- In the *Charts of Accounts, Financial Reporting Structures, Account Determination* implementation activity in the **Account Determination** section, open the settings for **Data Migration**. For every chart of accounts review, the reconciliation accounts are assigned to the various subledger objects (usually 990010, 990020, etc.).
- The target value for SAKOMIG shouldn't be among these subledger reconciliation accounts. If you have a collision in this case, we recommend changing the target value for the SAKOMIG value conversion. Use the target value help to pick a free 99... account (best to use 990000 "Initial Data Carryforward GL Account", if available) or create your own. Take care that the new target account has the same *general ledger account type* as the current account (usually OASSET – Other assets).

This change will ensure that the offset postings of the general ledger account balances object will use a separate account. The benefit is that the postings of the subledger migration that was affected by the collision become easier.

**Migration reconciliation account with balance 0**

If the main ledger data exactly matches all subledger data, then every migration reconciliation account will be balanced. Thus, any migration reconciliation account with a balance other than 0 in the balances preview is evidence of an inconsistency in the provided data. (The only exception is the migration reconciliation account for open goods receipts, which by nature of this business transaction will only become balanced by the outstanding supplier invoices that are expected to arrive in the weeks after migration.)

The balance preview in the cross-check is based on data collected during the latest simulation of the contained files. In practice, after the first cross-check

(maybe even as a result of the cross-check analysis), some changes will be relevant to the account migration, for example, in the fine-tuning or in account determination data. In these cases, explicitly repeat the simulation for all records so that the preview will be regenerated accordingly.

**Typical Errors during Migration of Accounting Data**

The following typical errors are usually found during cross-check analysis of the balances preview and the posting previews:

- Incomplete files
- Source files that have been exported from the source system at different points in time
- Not all required sets of books or valuation views addressed in the template data (e.g., general ledger account balances, fixed assets)
- Incorrect value conversions (particularly for fields that affect the account determination, such as document types and fixed asset movement types)
- Incorrect signs or the wrong digits in amount fields
- Incorrect account determination assignments in the master data (e.g., customer classification foreign versus domestic)
- Error in the account determination setup in fine-tuning

**Accounting department and tax counselor error analysis**

Usually, the error analysis is performed in collaboration with representatives from the accounting department. Also, further inquiries with tax counselors regularly can be beneficial to clarify uncertainties in real-world projects.

After the reiterations, if the preview doesn't show any differences anymore, click the **Next** button to go to the **Cross-Check Balances** migration step, so that the cross-check for the balances is executed for the last time. The correctness of this evaluation must be confirmed explicitly in a checkbox at the bottom of the screen: **I reviewed the posting values carefully. I confirm that the data is correct and ready for the final import into the target system.**

We recommend that you define early in the project the accountability for this decision and the way it's documented, for example, with signatures using the printout of the downloaded preview files.

When advancing to the **Execute Import** step, you're approaching the finishing line. The system will import all released files of the unit and trigger the corresponding postings.

### 6.4.9 Acceptance of Imported Data

After the import run has come to an end, you've reached an important milestone in your project. Nevertheless, the celebration should wait until the business departments have signed off on the imported data. Until this signoff has happened, both the source system and target system must remain blocked for postings so that you still have all available options.

One important activity must be performed before business departments can start with the reconciliation work. In some error situations, the import has triggered a posting or other follow-up process that can't be completely processed. In certain constellations, this situation can't fall back upon the original record; instead, this partly processed document is automatically pushed to the worklist of a business specialist. These worklists must be checked and cleaned up if not empty; otherwise, data will be missing in the reconciliation work. Check the following two places for failed documents that had been triggered by migration:

- **Application and User Management** work center, in the **Business Data Consistency • Process Communication Errors** view
- **General Ledger** work center, in the **Closing Cockpit • Closing-Relevant Tasks** view

Key user authorization is required to see these lists.

**Reconciliation and sign-off**

Now, your accountant must reconcile, in detail, the migration result against the source system. During the production cutover, the comparison is performed against the production source system, which must have been frozen against changes since the data had been exported. For test migration, ideally, a frozen copy of the source system had been created, preserving the system status at the time of data extraction. Usually, different tools are used to support this comparison, including (but of course not limited to) the trial balance and the inventory lists.

### 6.4.10 Postprocessing

After the business department has signed off on the data, you can start the activities required to make the system ready for live usage.

The first important tasks include performing a month-end cutover and then securing the system against any further postings that could jeopardize the migration result. Follow these steps:

1. Run the month-end closing (if the migration posting data is the last day of an accounting period, usually a month, as recommended).

2. Close all accounting periods up to the period containing the migration posting date (**General Ledger • Common Tasks • Open and Close Periods**).
3. Adjust the process control for the migrated companies accordingly (choose **General Ledger • Master Data • Companies** and then click the **Set Process Control** button). Block the periods against (inadvertent) additional migration postings by setting the entry for the business process group **Migration** to status **3 – Blocked** for a sufficiently large time interval.

In Section 6.4.1, we proposed establishing a block against postings from other usages. This block can now be reduced appropriately in the following ways:

- In the blocking for the business process group `01 - All`, reduce the upper end of the time interval in a way that only the interval up to and including the migration posting date remains to be blocked.
- After the migration result has been secured, the focus advances to tasks in user administration to open the system for operational usage.

## 6.5 Change Migration

In earlier sections, we focused on the initial data migration during productive go-live of your SAP Business ByDesign system.

**Use cases for change migrations**

In real life, in several constellations, you might want to change your business data later, such as in the following typical instances:

- **Latecomer**
  Add data records that have been postponed or forgotten.
- **Rollout project**
  The usage of the system is being extended to include additional organizational units (e.g., companies, departments) in a separate implementation project. Often, existing master data needs to be enhanced with business information, for example, payment terms in customer master data that applies specifically to an additional sales organization.
- **Data correction**
  Incorrect data or missing data is encountered in the migrated data after go-live. For example, perhaps a particular field wasn't populated in the migration templates. Now, you want to fix this issue.
- **Mass change**
  For some data objects, the migration tool offers comprehensive change functionality.

For most of these changes, you'll apply the different options for the migration mode as discussed in Section 6.2.5. Depending on the data object, two different options are available, which we'll discuss in the following sections:

Migration modes

- **Add new records**
- **Enhance already existing records**



Additionally, we'll look at the **Mass Data Maintenance** work center, which is based on the migration functionality and complements the change functionality with an interesting export function.

### 6.5.1 Add New Records

In the simplest use case, you want to add some completely new data records, such as additional customer master records. For nearly all objects, you just need to migrate a new migration template using the default **Add New Records** migration mode.

This migration mode usually doesn't support to changing or overwriting existing data. That said, in some exceptional cases, you can effect certain extended changes with this migration mode.

Let's look into the main use cases, namely, *material valuation data*, *service product valuation data*, *sales price lists*, *organizational structures* and *customer hierarchy* data, and see what changes can be made here.

#### Material Valuation Data

Material valuation data is based on validity intervals, which allows you to migrate an additional validity record with a new validity interval using the **Add New Records** migration mode. If validity data is already available for this material, then the validity interval for the existing data is shortened accordingly to make room for the new record. The same applies to the service valuation data migration object.

This capability makes migration an equivalent alternative to the mass change that you find in the **Inventory Valuation • Master Data • Material Unit Costs** and **Product Data • Services** work centers.

Use cases

Typical use cases for this kind of migration include mass changes of amounts or validity intervals and the introduction of new sites.

#### Price Lists

To add new price list items, you basically have two options:

New price list

- **Option 1**
  Limit the validity interval of the current price list (via **Product and Service Portfolio • Pricing • Price Lists**). Now, migrate a new price list containing the current items (with new prices if applicable) as well any new items. Let the validity interval of the new price list seamlessly connect to the end of the old price list.

  The benefit of this approach is that you can migrate a new price list with a future validity start. Although you release this price list now, it doesn't become effective prematurely before the defined validity start date. Until then, the old price list is used automatically.

Add new items

- **Option 2**
  Keep the existing price list and just add a migration template with new product items. It's a specialty of the price lists migration object that it even allows price changes of existing items.

For this task, follow these steps:

1. For the **General** migration template sheet, populate the **List ID**, **Valid From**, **Valid To**, and **Currency** fields with the current values (usually the values from the initial migration).
2. For the **Items** migration template sheet, it's sufficient to add only lines for the new items and for items that need to be changed. Be aware that only a change of the **Amount** is allowed here. All other fields in the changed records (specifically, **Price List** and **Unit of Measure**) must be populated exactly as in the original migration.

   The same applies to the **Scales** data sheet.
3. Migrate the migration template using the **Add New Records** migration mode.

Be aware that these changes become effective immediately after the migration import, and the changed items records will only have the new price.

You can also mass change prices in the **Product and Service Portfolio** work center (in the **Pricing • Price List** view). Click the **Download** button to start this change, which is based on downloading and uploading Microsoft Excel spreadsheets. After some improvements in recent versions, some differences remain between the two approaches, especially in terms of performance and the way scales are handled, but are rather minor.

### Organizational Structures

If you want to add an entire company to your existing system, you can migrate it as a completely self-contained organizational unit together with its substructures (e.g., divisions, cost centers, etc.) by using the implementation activity **Perform Organizational Structure Setup • Import org structures using the migration tool**. After import, the new units will appear in the **Organizational Management** work center in the **Planning Area**. Follow the menu path **Common Tasks • Edit Org Structures** and use the **Select Org Structures** filter if necessary.

**New substructure**

A reorganization or a functional rollout to departments not yet reflected in the system may require you to add new substructures into an already existing company (e.g., a new sales organization). You can achieve this goal in two steps: First, migrate the new substructure like a fully independent organizational structure. Then, use the graphical UI in the **Planning Area** to move the new structure into the right place under an existing company (choose **Edit Org Unit • Cut and Paste**).

#### Customer Hierarchies

In this task, you can proceed in a similar way. Additional customer hierarchies that are fully self-contained can be migrated with a new file as usual.

**Migrate a new subtree**

However, to extend an existing customer hierarchy, first migrate the added subtree as if it will be a self-contained customer hierarchy. Then, use the graphical UI in the **Account Management • Account Hierarchies** work center to move the migrated subtree into the right place within an existing hierarchy tree (choose **Edit • Advanced Mode**).

More detailed object-specific information is listed in the **Prepare for Data Migration in a Change Project** article in the **Prepare for Data Migration** implementation activity.

### 6.5.2 Enhance Existing Records

The **Enhance Already Existing Records** migration mode enriches records in a specific way. In view of the migration template file, one or several lines on a subordinate data sheet are added to an already existing record.

**Examples**

Typical examples in the customer master include the following data sheets:

- **Contacts**
  Add new contacts to an existing customer record.
- **Sales Data**
  Add information for a new sales organization (e.g., delivery terms, payment terms, default currency).
- **Payment Data**
  Add information for a new company that has been defined in the organizational structures.

Thus, the migration mode is the ideal approach for rollout projects. For master data records that are used across all organizational units, by nature, the task arises to add information that is specific for a newly added organizational unit.

See the following boxes for the migration objects that support this migration mode.

**Enhance Existing Records: SAP Business ByDesign**

The **Enhance Already Existing Records** migration mode can be used in SAP Business ByDesign in the following objects:

- Customers
- Suppliers
- Service products
- Materials

**Supported sheets and fields**

Depending on the object, not all sheets and fields are supported by this functionality. The most effective way to see the available sheets and fields is to create a migration template file in the file list and select the **Enhance Already Existing Records** migration mode. The view is automatically reduced to the supported sheets and fields.

**Templates for Enhanced Mode Migration**

When working with the **Enhance Already Existing Records** migration mode, don't use the standard migration templates from the **Prepare for Data Migration** activity. It's better to start in the migration cockpit, create a file with this mode (**Actions • Create**), and then download it (**Actions • Download**).

The local file stores the mode under which it had been created and will propose that mode as the default when uploading the populated file later.

**Hidden fields**

You'll immediately see the changes in the visibility of sheets and columns. Now, every sheet or field is hidden that can't use this migration mode; also, the **Field List** sheet is adjusted accordingly. Don't use sheets or fields hidden in this version. When uploaded using this migration mode, data from such sheets and fields will be ignored. The same warning applies when you start with a file that had a different mode and then change the mode in **Properties**. When confirming the warning shown in Figure 6.44, the content of the affected fields will be cleared.

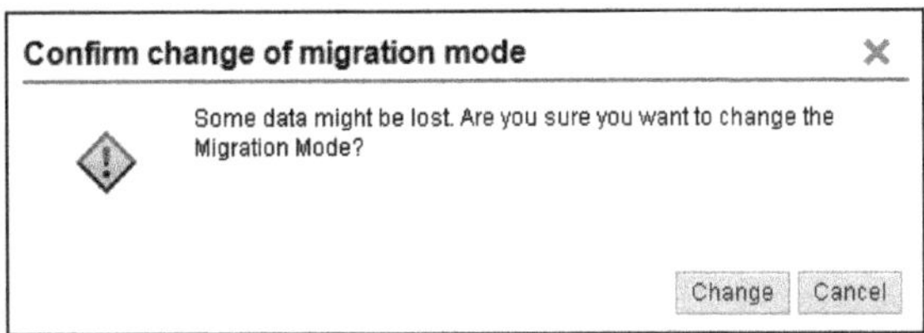


**Figure 6.44** Changing the Migration Mode



This discussion also revolves around a frequent misunderstanding: The enhanced mode template expects only the data that is to be added—not the header data or any other data that is already in the system.

**Use the same keys in enhanced mode**

The following consideration is important for any record that received a new key assignment during the initial migration: For a change mode migration, you must use exactly the same keys in the template that you used in the initial migration template. The reason for this requirement is that this migration mode is also reusing any value conversions that were established then and automatically assigns the right target key if the same source key is used again.

Let's apply the enhancement of records in typical a real-life example: the rollout of an SAP Business ByDesign system to another company. Let's assume that company A has been live for a while but has not used logistics so far. Now, the solution is introduced to the sister company B, including logistics processes. Both companies share the same material masters, which were migrated during the implementation of company A.

**Master data**

The generic parts of the material masters are already in place. While generic elements of the material master such as texts and base unit of measure (UoM) are in place already, we expect company B to require some specific additions.

Other entries might also be required for company B that haven't been required for company A and thus haven't been imported yet, such as specific UoMs and certain material properties or other site-specific data.

We're downloading a freshly created template from the migration cockpit as recommended earlier. Fill in the template for the enhanced mode migration in the following way:

- Enter the assignments and company-specific information for company B on the relevant template sheets, most prominently **Financial Organizations**, **Logistics**, and **Sales**, as shown in Figure 6.45.
- For materials, even the **General** sheet is visible, although with very few fields. Because company A hasn't yet used inventory functionality, the **Inventory UoM** field must now be populated to accommodate company B.

- On the **General** sheet, you'll also find the extension fields.
- If both this **Inventory Valuation** UoM and the extension fields have been maintained already during initial migration, then the **General** sheet simply remains empty.

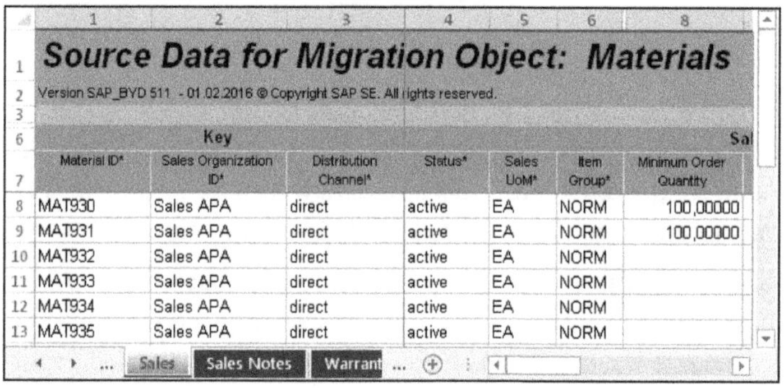

**Source Data for Migration Object: Materials**

Version SAP_BYD 511 - 01.02.2016 © Copyright SAP SE. All rights reserved.

| Material ID* | Sales Organization ID* | Distribution Channel* | Status* | Sales UoM* | Item Group* | Minimum Order Quantity |
|---|---|---|---|---|---|---|
| MAT930 | Sales APA | direct | active | EA | NORM | 100,00000 |
| MAT931 | Sales APA | direct | active | EA | NORM | 100,00000 |
| MAT932 | Sales APA | direct | active | EA | NORM | |
| MAT933 | Sales APA | direct | active | EA | NORM | |
| MAT934 | Sales APA | direct | active | EA | NORM | |
| MAT935 | Sales APA | direct | active | EA | NORM | |

Sales | Sales Notes | Warrant...

**Figure 6.45** Migration Template: Subordinate Data Sheet

After you've finished the enhanced mode migration for material master, migrate the material valuation data. As discussed in Section 6.5.1, you can use the **Add New Records** migration mode for this task.

**Accounting data**

The migration of accounting data is done company by company using company-specific migration units (Section 6.4). Migrate the accounting data for company B using the **Add New Records** mode.

With the **Enhance Already Existing Records** mode, we've presented a migration mode that can add new subrecords but can't modify any existing field values. The next migration mode is just the other way round in that it's specialized to modify existing field values but can't add new subrecords.

### 6.5.3 Repeat Migration for Objects with Mass Deletion

For the objects *opportunities* and *activities*, migrated records can be mass deleted, as long as no follow-up data has been created. This enables you to repeat the migration of these records if necessary.

To use this approach, open the **New Business • Opportunity List** or **Account Management • Activities** work center, respectively. Mark all relevant lines using the `Shift` key (and filtering if required) and then click **Delete**.

In the case of completed activities, change the status first by choosing **Change Status • In Process**. You can change the status of all the records you've marked at once.

Now, migrate a new migration file in the regular way, using **Add New Records**.

We've now presented several types of change migration. A huge variety of potential use cases for mass changes exist. Accordingly, no hard boundary divides a data migration with change capabilities on one hand and a mass data change in operational business on the other hand.

Several application work centers in SAP Business ByDesign offer individual capabilities to trigger certain mass changes. This functionality is complemented by the **Mass Data Maintenance** work center, which we'll discuss in the next section.



### 6.5.4 Mass Data Maintenance

The **Mass Data Maintenance** work center is a central UI that allows you to manage mass changes in different applications and objects. The most important features, at a glance, are the following:

**Mass Data Maintenance work center**

- Mass enabled
- Supported objects
  - Customer
  - Supplier
  - Materials
- "Roundtrip" procedure
  - Export (download sheet by sheet into CSV files)
  - Modify CSV files locally
  - Process the new CSV files: file upload, validation, data import (no import simulation)
- Change capabilities similar to a combination of migration modes enrich and modify:
  - Change field value (but no deletion of field values)
  - Enrich a record with new subrecords (e.g., add new addresses and contacts to the customer master)
- Doesn't use the ID conversion from migration, instead uses the actual IDs used in the system
- Object structure mainly a subset when compared to migration templates:
  - With few exceptions (e.g., payment methods), same sheet structure
  - Some sheets and fields omitted (e.g., **SEPA mandates** sheet omitted; on **Addresses** sheet, usage fields and main indicator omitted)

An important feature is the integrated data export functionality. The exported file uses the keys actually used in the system and, if required, contains a unique technical key (UUID, e.g., for address lines).

Now, let's explore this functionality using a real-world case. Imagine our task is to update the email addresses of customer contacts and the contents of a customer-specific extension field.

The **Mass Data Maintenance** work center screen is divided into several parts, as shown in Figure 6.46. The upper area holds the links to start the export (left side) and the reimport (right side) for the supported objects.

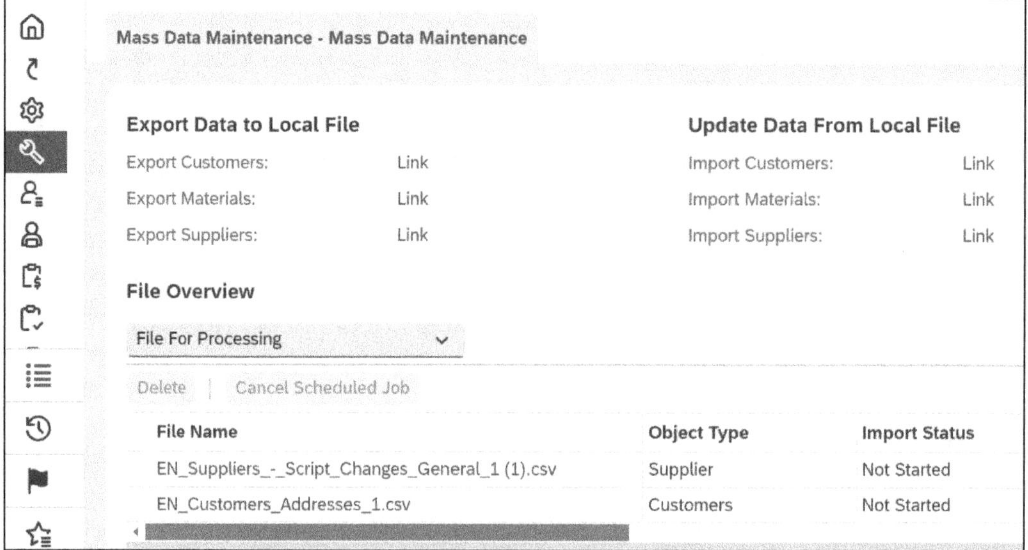


**Figure 6.46** Mass Data Maintenance Work Center

The table in the lower part of the screen lists details on the import files across all objects:

- Object and file name: Click on the file name to resume processing of a file.
- Import status: For instance, **Not Started**, **Finished with Errors**, **Finished**).
- Record counter: For instance, **Imported**, **Failed**, **Total**).
- Link to import log.

A double filter functionality is available: the usual **Show** function on the left and an icon on the far right to open a filter line directly under the header line.

Use the link next to **Export Customers** to open a selection screen for the export. Enter optional selection criteria and click the **Go** button to actually start the data collection process, as shown in Figure 6.47. The result will be displayed in the bottom part of the screen, with multiple data sheets populated with export data.

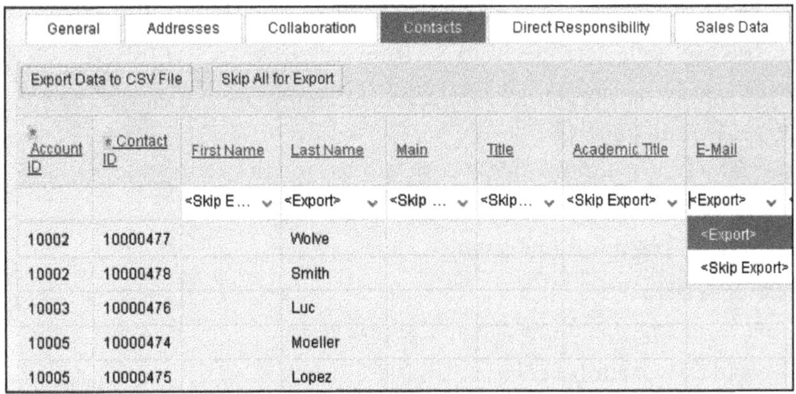


**Figure 6.47** Selecting the Fields to Be Exported

For our example, the email addresses need to be changed, so switch to the **Contacts** data sheet.

Before exporting the data, you can use the selection line under the header to narrow down the data volume. For every column, you can choose whether its content should actually be exported (**<Export>** or **<Skip Export>**). For this example, you need only a few fields. Click the **Skip All for Export** button to deselect all columns and then manually select only those actually needed (**E-Mail**, our extension field, the key information **Account ID**, and **Contact ID**). The **Export Data to CSV File** button starts that actual export. Save the exported CSV file to your PC.

Modify data in a local file

Assume that the new values for the two example fields have been provided in a local source file. Whichever way is easier, you now bring these values into the exported CSV file (e.g., via Excel using a `Vlookup` formula), or you modify the content of the source file in a way that its structure matches the exported CSV file so you can use the modified file instead.

You might need to covert the contact IDs in the local files because the source IDs in the source file might deviate from the contact IDs in the system, which are also in the exported CSV file. Now it pays off to follow our recommendation from Section 6.7.2 and always migrate the source contact ID into a separate extension field during initial migrations.

Reimport

As soon as the CSV file with the new values is available, reenter the **Mass Data Maintenance** work center. Start the import (**Import Customers: Link**) and choose the local CSV file. The dialog is the same as in Section 6.2.6. Don't forget to pick the right target sheet for the upload (**Import Data to tab: Contacts**). Start the **Upload** of your file, which will also trigger the **Data Validation** and open the main screen of the mass data import, as shown in Figure 6.48.

A data editor displays your data records, each with an **Update Status** (in this case, **Not Started**). The buttons in the header area give you control over the process:

- **Execute Data Import**
  Optionally, start immediately or schedule a job to be run later.
- **Close**
  Close the screen without saving.
- **Validate Data**
  Save all changes you've made in the data editor.
- **Upload Data from Text File**
  Populate additional sheets from CSV files.

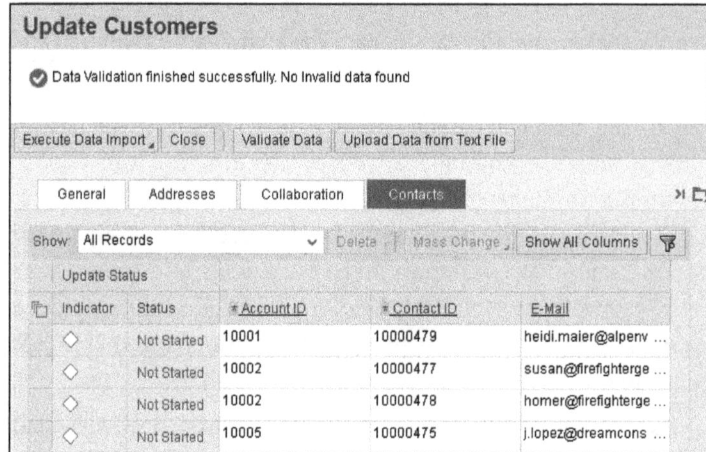


**Figure 6.48** Update Customer Data

If, instead of the validation screen, you see the field mapping screen we described earlier in Section 6.2.6, then your CSV file contains some field names that are unknown in the target sheet.

At this point, you can alternatively directly upload a source CSV file even if it has a slightly different column structure and do the field mapping now.

In our example, where we stayed close to the exported CSV file, all field names should be known, and the mapping screen shows clearly that we forgot to pick the right target sheet during upload.

**Data Editor: How to Save**

The data editor in **Mass Data Maintenance** doesn't have a **Save** button; instead, the save function is an implicit part of the validation. So, after any changes, be sure to run **Validate Data** before you **Close** the data editor window.

Data validation

The data validation applies a formal check, including field format and—in fields with picklists (**Codelist** field)—permitted target values. On errors, a particular error handling view opens, in which only the erroneous records are displayed with invalid field values highlighted in red. Click in this field value to open a little dialog box that allows you to replace the current value and, optionally, apply the same change right away to all other data records. If available, you can use value help lists.



The **OK** button closes the error handling view. Repeat **Validate Data** to save your changes and evaluate the remaining errors.

Data import

The **Mass Data Maintenance** process doesn't offer a simulation mode. Choosing **Execute Data Import • Immediately** directly starts the import run. For every record, the result of the import run can be seen in the **Status** column.

If there are errors, click on the **Failed** record status to see detailed information on the error. After changing the data records accordingly, you can trigger the import of the file again. Now, the entire file will be reimported, including the records that had been successful in the first place.

Be aware that imported files aren't protected against editing, so clear documentation for reconciliation or auditing purposes needs to be established in a different way.

To see the import log, navigate to the file overview by clicking the **Close** button and follow the link at the end of the file entry.

Alternative approaches

As an alternative to the normal "roundtrip" strategy in our example, the functionality also makes more sophisticated approaches possible. The import feature offers a file interface with integrated field mapping, so that, depending on the use case, CSV files also can be used that didn't start as export files. Likewise, the export functionality can be used to establish files for other purposes, such as a change migration (e.g., to benefit from the larger field coverage). But within the roundtrip paradigm, more advanced approaches are also possible (e.g., to use the exported address UUID information to provide contact persons with address assignments).

## 6.6 Migration as a Subproject

Let's now place the migration in the wider context of the entire implementation project. We are raising the awareness for dependencies between migration and other items of the implementation project and browsing through the main administrative tasks that you meet during a migration project. Finally, in perhaps the most important part of this section, we discuss how the test strategy evolves as the implementation project proceeds.

### 6.6.1 Project Management

**Numerous dependencies between tasks**

In projects with a medium number of objects, migration faces numerous dependencies between tasks both inside and outside the data migration subproject. Let's consider a few examples:

- For objects with multiple prerequisites, it's important that the main master data objects are loaded as early as possible. This rule applies, for example, to transactional data and higher master data such as production models.
- At the same time, master data migration is often waiting for other areas, for example, field extensions. Especially for migration-relevant fields, stability in extension field definition is key.
- Configuration changes in fine-tuning often set back the migration, causing repetition. In turn, often, a data migration subproject that keeps revealing shortcomings in configuration and data quality.
- Due to the various dependencies, you achieve transparency on the data quality often with delay, as issues are often shadowed until the prerequisites are in place, which can lead to unpleasant surprises.

**Project plan template**

Becoming more and more important, effective project management ensures that the different activities are synchronized like clockwork. We recommend adding the project plan template to your migration project tools, which you can download from the **Prepare for Data Migration** implementation activity, as shown in Figure 6.49. In this project plan, the migration objects are grouped by sequence to reflect their object dependencies.

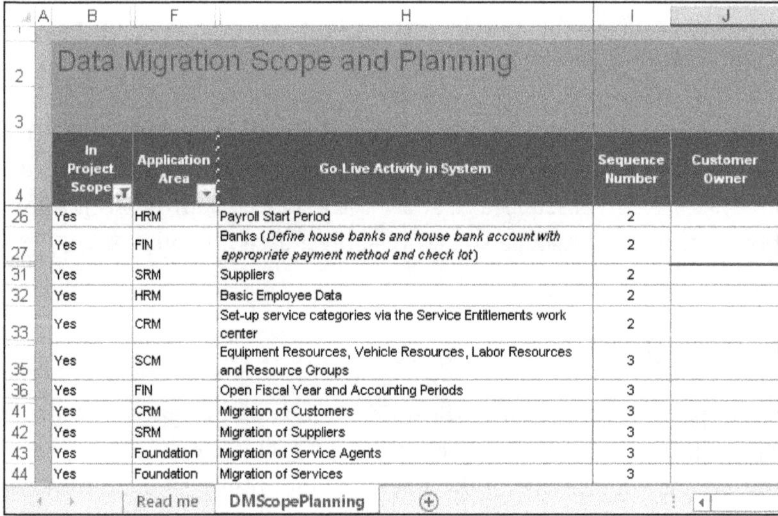

Data Migration Scope and Planning

| | In Project Scope | Application Area | Go-Live Activity in System | Sequence Number | Customer Owner |
|---|---|---|---|---|---|
| 26 | Yes | HRM | Payroll Start Period | 2 | |
| 27 | Yes | FIN | Banks (*Define house banks and house bank account with appropriate payment method and check lot*) | 2 | |
| 31 | Yes | SRM | Suppliers | 2 | |
| 32 | Yes | HRM | Basic Employee Data | 2 | |
| 33 | Yes | CRM | Set-up service categories via the Service Entitlements work center | 2 | |
| 35 | Yes | SCM | Equipment Resources, Vehicle Resources, Labor Resources and Resource Groups | 3 | |
| 36 | Yes | FIN | Open Fiscal Year and Accounting Periods | 3 | |
| 41 | Yes | CRM | Migration of Customers | 3 | |
| 42 | Yes | SRM | Migration of Suppliers | 3 | |
| 43 | Yes | Foundation | Migration of Service Agents | 3 | |
| 44 | Yes | Foundation | Migration of Services | 3 | |

Read me | DMScopePlanning

**Figure 6.49** Migration Project Plan Template

In addition to the actual migration import tasks, this project plan also lists important tasks from neighbor areas, for example, the fine-tuning activities for number range definitions.

Adjust the migration project plan to your situation and integrate it into your master project plan. Start early in the project to work on migration because late surprises on migration issues can potentially jeopardize the success of the entire project.



Include representatives from the business department from the beginning in the migration work stream, assigning responsibilities for data quality in a transparent way. No head of accounting would allow control of the data for the annual balance to go out of his hands. The migration of these data needs to be rated in the same way. If necessary, demand the responsible participation of the business departments. Migration is not just a technical transport of bits and bytes; instead, it's a powerful business process.

Many decisions that must be made during migration require a deep understanding of the business operations that will be affected.

### 6.6.2 Migration Tasks in Implementation Projects

Relevant aspects

Apart from the actual migration import tasks such as **Migration of Customers**, the following other tasks are relevant for migration:

- **Migration project plan**
  Include representatives of your business departments. We recommend requesting statements on the data quality from the different business departments. Make agreements to safeguard these dependencies. Early in the project, agree upon a list of concrete reconciliation checks for every business area, which will serve as fixed targets for the subproject. Document all agreements in a transparent way.
- **User administration**
  Provide the migration team with user accounts that have sufficient authorization in implementation and migration tasks, including at least the following work centers:
  - **Business Configuration** with the **Data Migration** view
  - All relevant application work centers for the application area this user is migrating in, so that the migration result and prerequisite data can be checked
  - **Organizational Structures**

[+]

**Temporary User Accounts for the Migration Team Members**

Members of the migration team often also participate in application tests in their operational business home area. In this case, every member should have two user accounts:

- A permanent application business user with authorization for the operational business tasks in the home area. This user is best derived from the employee master record of this team member. Ensure that this user is logged on for application tests that reflect the member's regular business tasks.
- A migration user with comprehensive migration-relevant authorizations, which is provided only temporarily during the implementation project. This user is best derived from a temporary service agent master created for the project member and that will be deactivated after the project is completely finished. This user that should be used for migration work.

- **Fine-tuning**
  During the **Fine-Tuning** phase of the implementation project, various configuration decisions are made with significant consequences for migration. Prominent examples include the definition of organizational structures, account determination settings, and number range settings.
- **Request migration test system**
  Always use a migration test system. These systems are provided with the same configuration (solution profile) as your production system and therefore serve best for comprehensive tests. Although you can fix configuration issues locally just to proceed with the test, ensure that every identified issue is collected on a highly visible central issue list, from which the configuration master system is fixed in an organized way.
- **Plan migration in production system**
  This task is the task everything else is running toward. Uncompress the migration work as much as possible. Together with your business representatives, identify data objects that can be migrated in advance and determine how this can be achieved. For instance, can we freeze certain master records in the source system against changes for some days? How, technically or organizationally?

  If you have objects with high volumes and long runtimes in your migration portfolio, run volume tests to collect experiences on the runtime, possible parallelizations, and the optimal file sizes if the data needs to be split into several files. Make tests using the entire volume, as extrapolations can miss important facts (e.g., a nonlinear behavior).

The more objects and the higher the volume, the more detailed your cutover plan must be. For critical volumes, establish a detailed cut choreography down to every single hour and every single file. Plan for restore points to safeguard important milestones, including the time required to take it.

### 6.6.3 Test Strategy



Test migrations not only serve as tests for the migration, but they also create a data environment from which all application tests can benefit. Inevitably, they also rehearse the data export from the legacy system, which is required to ensure constant data quality.

As the implementation project proceeds, the focus of the test strategy changes:

- **First steps: early migration tests**
  Get familiar with every migration template you plan to use. Start with a small example file in which you enter a few records manually. Process the records at least up to a successful simulation and gradually also import these test records (at least for master data objects). Use IDs that won't collide with the number ranges that you'll need for larger tests, so that these early tests don't interfere with later tests. Review the imported records in the application work center to encounter major misunderstandings (e.g., not all address fields show up in all screens, which might influence the way you populate the address fields).
- **Expand your footprint: data variety**
  Early tests usually start with simplified parameter values, already due to the fact that the business configuration is still being worked on. Parallel to the fine-tuning progress, migration also can use more fields with a broader variety of values. Create test files containing actual source records from the legacy system. Modify the IDs (e.g., with different prefixes) to enable a repetition of the tests.

  The more test records represent actual source data, the more the migration challenges the configuration settings and the basic data (e.g., organizational structures) to ensure that these are prepared for the actual data variety in the source system. The reality of the source data often reveals some exception cases (sometimes historic) that no one was aware of.

  Typically for this stage, you'll fine-tune and reiterate migration tests. At this stage, high variety is more important than high volume. During this test, you should also start to use the export functionality that you want to use to export all your data.

- **Early test environments for application tests**
  Advance to records that reflect the actual data complexity in your source system. Even if some details are still being worked on, your business departments will be grateful for a test environment where most of the available data is familiar. The migration delta file technique presented in Section 6.3.4 can help you in this regard. As soon as you've resolved the errors classified as inconsistency errors, you can push through and import all good records. This step provides your colleagues with test data, while all records with issues are automatically collected into the delta file. Thus, you can work on these issues without holding up application tests more than necessary.
- **User acceptance tests**
  We recommend performing a complete test migration here for several purposes. After all records have been migrated without error messages, you'll need user tests to identify remaining issues by challenging the data in business processes and staging a day in the life of your company. At the same time, this test verifies the steps from the source system to the migration template regarding correctness, completeness, and efficiency.

  For transactional accounting data, we recommend using a precedent month-end closing to mimic the production cutover situation best. This approach of course needs to be planned in time, so that the exported source data is consistent across all objects. It's best to have a frozen copy of the source system on one hand to compare the migration result with and on the other hand to have time to export all data according to the same timestamp without interfering with regular business operations. In this way, you can also repeat the export from the same source situation if required.

  Request your business departments officially approve the migration result using the same quality criteria that will be applied after production migration.
- **Volume test**
  Whether for the user acceptance test as recommended or separately, the higher your data volume is, the more important it is to capture the runtimes of every file in detail, that is, capturing for every file the number of records, the times required to process the different migration steps, the parallelization factor, and the error rate (which needs to be very low at that point in the project). If you decide to extrapolate the expected runtimes from an extrapolation of a subset (not recommended), then be aware that the runtime of a file doesn't necessarily depend in a linear way on the number of records.

- **Dress rehearsal**
  A best practice is to run a complete migration at least once under the same time constraints as in the go-live situation to verify your cutover choreography. Especially in large migrations, the damage could be significant if later during production cutover, any assumption in the cutover schedule turns out to be unrealistic.



## 6.7 Tips and Tricks

We want to finish this chapter on migration in SAP Business ByDesign by sharing some experiences that helped us successfully overcome some challenges in our real-world projects.

The first section highlights a feature for high volume migrations that is often missed in projects, while the second part helps you to reconcile the results of your migration.

### 6.7.1 Volume Performance by Parallelization

During migration, the data is finally imported using service calls to the system. As mentioned in Section 6.2.7, many objects support a parallelization of these service calls so that the migration tool automatically feeds the service interface over several streams, always having four or more service calls in process.

Parallelization factor

Objects supporting this feature run by default with four-fold parallelization. In addition, these objects allow you to specify this parallelization factor in the migration cockpit using the advanced function in **You Can Also • Adjust Settings for Parallel Processing.**

Due to competition for certain system services, the runtime of each file increases with higher parallelization factors, so that from some point on, a higher factor is counterproductive. The optimal factor depends on the object and the specific data constellation. If you have an object with very high volume, then a couple of systematic sample imports into your test system with similar test files of 500 records and different parallelization factors each can help you evaluate good settings for your situation. In several objects, we've had good experience with 8 to 12 factors.

Be aware that your change of the parallelization factor becomes effective only during the actual import, not during import simulation. The following box shows which objects support parallelization.

[+]

**Parallelization in SAP Business ByDesign**

The following migration objects support parallelization:

- Customer
- Supplier
- Account determination group (assignments) payables
- Customer hierarchies
- Service products
- Materials
- Warranties
- Registered products
- Production bill of materials (BOMs)
- Engineering change orders
- Production bill of operations
- Knowledge base
- Projects
- Opportunities
- All activities (e.g., appointment, task, email, phone, etc.)
- Service requests
- Service quotes
- Sales quote
- Quote for contracts
- Sales orders
- Service orders
- Customer contracts
- Bills of exchange
- Transport zones

But even if your object isn't on this list, some objects can handle a manual parallelization by starting two to four files shortly one after another. Test this feature in a test system to verify your constellation. After triggering a simulation or an import for one file, you can navigate back to the file list by clicking the **Run in Background** button and start a second file. In any case, ensure that your data is divided into the parallel files in a way that the records in different files don't write to the same data elements (e.g., all material valuation records for the same material should be in the same file).

### 6.7.2 Reconciliation of Migration Results

Receiving the **No errors detected during data import** message shown in Figure 6.50 at the end of a large migration import is without doubt a big relief during a production cutover. However, it's still a little early for the champagne.

The data migration isn't finished until your internal or external clients have accepted the result as successful and confirmed the handover of the data. The question is of course: What exactly is "successful"? A cumbersome experience lies ahead if you haven't clearly defined this criterion in advance. A best practice is to agree at the beginning of the project on a list of acceptance criteria that define in detail which quality and quantity checks the migrated data must fulfill. Use these criteria throughout the project in your progress status reports.

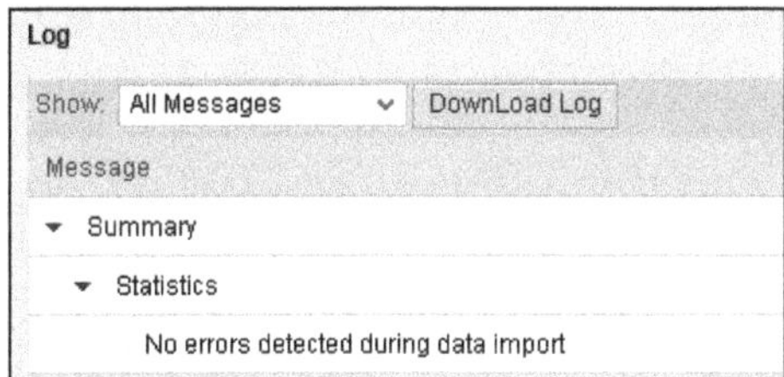


**Figure 6.50** Success!

**Success criteria**

At the project start, for every object, a list of target criteria necessary for successful acceptance should be agreed upon. The absolute minimum criteria include the following:

- The number of records on the header level are equal in the source and target system.
- Grouped by certain fields (e.g., countries), the number of records per group is equal in the source and target.
- A defined amount of spot check samples of source and target records are being compared field by field and proven to be equal.

This net is so wide meshed that most errors and gaps won't be caught; therefore, our recommendation is that you check for the following items:

- The number of subrecords on certain levels is equal in the source and target (e.g., customer addresses, customer contacts, etc.).
- The result of certain application reports is identical in the source and target. This rule applies particularly to transactional data and includes trial balances, open item lists, asset history sheets, and so on.

- For a certain defined set of fields with a high business impact, source and target values are being reconciled completely, across all records (e.g., using exported lists from source and target, which have been joined and compared, for instance, via Microsoft Excel's `Vlookup`).

**Test reconciliation** The reconciliation process itself also needs to be rehearsed so it will perform correctly when under time pressure. Use the test migrations to rehearse the reconciliation and the acceptance processes with your business departments. At cutover, time is rather limited, so any misunderstandings and evaluation gaps must be identified in the test phase.

**Key fields** A major challenge when comparing migration results between a source and its target are the key fields. For several reasons, such as the following, a record can have a different ID on the target side:

- During import, the migration automatically assigned new numbers (e.g., contacts and activities; in SAP Business ByDesign, also opportunities).
- You've opted for new numbers during the migration (e.g., new customer ID, supplier ID).
- The ID is being used in a different way in the target system (e.g., for fixed assets).
- During export, the key information had to be modified to avoid collisions with data collected from different source systems.

We've shown how value conversion information can be exported from the migration cockpit (**Actions • Download Value Conversion**). This information can be used for the reconciliation, that is, for joining exported lists from source and target via old and new IDs with the limitation that, for some objects, the new ID is hardly transparent (e.g., activities). Therefore, we propose an approach that can avoid this additional effort.

**Extension fields for legacy ID** One tip that we can provide is to create customer-specific extension fields in the relevant objects and populate them with information that helps you reconcile the migration result. In addition, this approach can be a great help when analyzing issues during test phase.

We recommend using this feature for the following pieces of information:

- **Source record ID**
  A helpful approach is to have a link to the legacy ID not only in a central value conversion list but also in every record.
- **Source system ID**
  If the source data is being collected from several systems, then we recommend an additional extension field to capture the source system this record is coming from.

- **File name (or at least a unique shortcut for it)**
  In high-volume migrations, often the number of records is distributed across several files to limit the file sizes. During error analysis work in the test phase, knowing which file brought a certain erroneous record into the system can be helpful.

As soon as the reconciliation has been confirmed, in just one click, you can hide the extension fields in the work center views (via the **Visible** checkbox).

6

[+]

**How to Create Extension Fields**

Extension fields are beyond the scope of this book, so we'll only provide some brief instructions now:

1. Navigate to the work center view of the relevant object (e.g., **Account Management • Accounts**).
2. In the menu bar, choose **Adapt • Enter Adaptation Mode**. Confirm the popup window. At the top of the page, a white and yellow bar appears.
3. In the menu bar, choose **Adapt • Edit screen**.
4. In the bottom right area of the screen, click **Extension Fields**.
5. Under **List of Fields**, choose **Add**.
6. Choose a **Business Context** (usually the header level, e.g., **Account – General Information**) and define a field name. Click **Save**.
7. In the field list, check the **Visible** column.
8. Click the **Publish** button and confirm the subsequent message (i.e., don't select the checkbox).
9. In the menu bar, choose **Adapt • Leave Adaptation Mode**.

Document results

We recommend documenting the migration result, together with a local copy of the relevant files (i.e., migration templates, import logs, downloaded value conversion information, and work files from the reconciliation), and using this documentation in the handover acceptance process.

## 6.8 Summary

In this chapter, we presented the data migration process in SAP Business ByDesign in detail. You've seen how the guided procedure with dynamic task lists leads you through all steps of the migration. The step-by-step user guidance opens up data migration work to representatives from the business departments who aren't based in IT (e.g., sales, purchasing, accounting, etc.). This highly beneficial feature enhances the quality and the acceptance of the migration result, as we've seen that complex business processes and options haven't simplified in the same way as the technical migration processes have.

PART III

# Alternative Methods

# Chapter 7
# Batch Input

*With the batch input technique, the data to be entered is transferred to the fields of the screen templates of an SAP transaction. We'll provide an overview of how you can use this technique for data migrations.*

7

In this chapter, we'll familiarize you with the batch input technique, a basic data entry technique of SAP systems. Today, the technology would be classified under the term *process automation*. The batch input technique is the oldest technique but is still often used to record a process and then play it back automatically.

[!]

**Why Batch Input?**

In general, we recommend using the batch input technique only if all other tools, such as the *SAP S/4HANA migration cockpit* (see Chapter 3) or *SAP Legacy System Migration Workbench* (see Chapter 9), do not provide suitable migration objects or solutions for your data migration.

To create batch input sessions, you can either use the standard batch input programs already delivered by SAP, or you can create your own recording. For the latter, you'll manually enter a sample data record in the SAP system as part of a transaction recording. All fields relevant for the data transfer must be supplied with values for this data record. Afterwards, you can, for example, have an ABAP program generated from the recording at the push of a button. This program can then be used for a data transfer after a few small changes.

## 7.1 What Is Batch Input?

**Most used method for data transfer**

In the early days of SAP systems, batch input processing was the only technique for transferring external data into an SAP system. In the meantime, other techniques such as OData, Simple Object Access Protocol (SOAP), Business Application Programming Interfaces (BAPIs), and IDocs have been added. Nevertheless, batch input processing is still a much-used technique today. On one hand, batch input has a natural affinity to dialog transactions, which are still frequently used in the on-premise world, and on the other hand, a user receives extensive support from the SAP system. We look more closely at both aspects in the following sections.

**Operating principles** The principle of batch input processing is to "feed" the input fields of the screen templates (screens) of SAP dialog transactions with provided data in the background. This concept ensures that all input and authorization checks are run through that would also be carried out if the data were entered manually. Data imported via batch input is therefore just as correct and consistent as data entered manually via this dialog transaction.

## 7.2 How Does Batch Input Work?

In this section, we'll explore the batch input technique in more detail. We'll explain the central concept of the *batch input session* and then demonstrate how such a batch input session can be created and processed.

### 7.2.1 What Is a Batch Input Session?

The core of batch input processing is what is called a *batch input session*. A batch input session consists of one or more calls of SAP transactions, along with the data to be processed by these transactions. For each transaction call, each screen template, and each input field, the value to be transferred to the corresponding field is specified. A batch input session is, in a sense, an ordered sequence of work instructions to the SAP system: "Provide field Z with value A in Transaction X and screen template Y." You can use this feature to run an entire business process, or just parts of one, automatically.

**ABAP Dictionary structure BDCDATA** A batch input session is stored in an SAP system according to the ABAP Dictionary structure BDCDATA (*BDC, Batch Data Communication*), which has the following fields:

1. PROGRAM: This field stores the name of the ABAP program to which the current screen belongs.
2. DYNPRO: The 4-digit number for the current image is stored in this field.
3. DYNBEGIN: This field marks whether a new screen template (value "X") or a new transaction (value "T") starts.
4. FNAM: This field contains the technical name of the input field.
5. FVAL: This field contains the input value.

A section of a batch input session is shown in Table 7.1, which calls SAP Transaction KS02 (Change Cost Center) and supplies it with data, to give you a first impression.

| No. | PROGRAM | DYNPRO | DYNBEGIN | FNAM | FVAL |
|---|---|---|---|---|---|
| 01 | | 0000 | T | KS02 | BS |
| 02 | SAPLKMA1 | 0200 | X | | |
| 03 | | 0000 | | BDC_CURSOR | CSKSZ-KOSTL |
| 04 | | 0000 | | BDC_OKCODE | /00 |
| 05 | | 0000 | | CSKSZ-KOKRS | A000 |
| 06 | | 0000 | | CSKSZ-KOSTL | 0010100000 |
| 07 | SAPLKMA1 | 0299 | X | | |
| 08 | | 0000 | | BDC_OKCODE | =BU |
| 09 | | 0000 | | BDC_SUBSCR | SAPLKMA1.... .............. .............. 0300SUBSCREEN_ SINGLE |
| 10 | | 0000 | | BDC_CURSOR | CSKSZ-KTEXT |
| 11 | | 0000 | | CSKSZ-KTEXT | Standard |
| 12 | | 0000 | | CSKSZ-LTEXT | Dummy cost center |
| 13 | | 0000 | | CSKSZ-VERAK | Finkbohner |
| 14 | | 0000 | | CSKSZ-KOSAR | W |
| 15 | | 0000 | | CSKSZ-KHINR | 0001 |
| 16 | | 0000 | | CSKSZ-BUKRS | 1010 |
| 17 | | 0000 | | CSKSZ-PRCTR | YB000 |

**Table 7.1** Excerpt from a Batch Input Session

Table 7.2 explains what lines 01 to 17 mean in detail.

| No. | Explanation |
|---|---|
| 01 | Transaction KS02 is called. |
| 02 | Screen 0200 from program SAPLKMA1 is called. |
| 03 | The cursor is positioned in field CSKSZ-KOSTL (cost center). |

**Table 7.2** Explanation of the Batch Input Session Specified in Table 7.1

| No. | Explanation |
|---|---|
| 04 | The 00 function code (OK code) is triggered. This code is equivalent to pressing [Enter].* |
| 05 | The value "A000" is entered in field CSKSZ-KOKRS (controlling area). |
| 06 | The value "0010100000" is entered in field CSKSZ-KOSTL (cost center). |
| 07 | Screen 0299 from program SAPLKMA1 is called. |
| 08 | The BU function code (OK code) is triggered. This code is equivalent to pressing [Save]. |
| 09 | The 0300 subscreen dynpro is called. The field value is the string of subscreen program name (40 characters), subscreen dynpro number (4 digits), and subscreen name in the calling screen (30 characters). Missing digits in the program name must be filled with blanks.<br>The dots (...) listed above represent the filled spaces. |
| 10 | The value cursor is positioned in field CSKSZ-PRCTR (profit center). |
| 11 | The value "Standard" is entered in field CSKSZ-KTEXT (designation). |
| 12 | The value "Dummy cost center" is entered in field CSKSZ-LTEXT (description). |
| 13 | The value "Finkbohner" is entered in field CSKSZ-VERAK (person responsible). |
| 14 | The value "W" is entered in field CSKSZ-KOSAR (type of cost center). |
| 15 | The value "0001" is entered in field CSKSZ-KHINR (hierarchy area). |
| 16 | The value "1010" is entered in field CSKSZ-BUKRS (company code). |
| 17 | The value "YB000" is entered in field CSKSZ-PRCTR (profit center). |
| * The function code (OK code) is, of course, not triggered until the input fields of the screen are supplied with values (i.e., after line 06). | |

**Table 7.2** Explanation of the Batch Input Session Specified in Table 7.1 (Cont.)

You now have an idea of what a batch input session does and how it is structured. Based on this basic information, two main questions may arise:

- How can you create a batch input session?
- How do you process a batch input session?

Let's start with the easier one, which is the second question.

### 7.2.2 How Do You Process a Batch Input Session?

Batch input processing modes

In connection with the processing of a batch input session, one often speaks of "playing back" the batch input session. The SAP system provides three different options (called *modes*) for processing or playing back a batch input session:

- **Process/foreground**
  In this mode, each screen is displayed with the set data. You can move from one screen to the next by pressing Enter, and you can correct incorrect transactions interactively. The OK code /n enables you to exit the current transaction and go to the next one. You can use the OK code /bend to cancel processing the batch input session and resume this session later if necessary. This mode is not suitable for a large number of transactions.
- **Display Errors Only**
  This mode is similar to the **Process/foreground** mode, except that transactions that have not yet been executed and do not contain errors are not visible (interactive), but run in the background. If an error occurs, the background processing is interrupted and the screen where the error occurs is displayed. Once the error has been corrected, the system switches from dialog processing back to background processing until the next error situation occurs or the batch input session has been processed. You can switch to the **Process/foreground** mode by using the OK code, /bda. /bde then allows you to return to the Display Errors Only mode.
- **Background**
  In this mode, a batch input session is scheduled for background processing.

Principles of batch input processing

Basically, the following principles apply to the processing of a batch input session, as shown in Figure 7.1:

- Transactions that could be processed successfully are removed from the batch input session.
- Incorrect transactions remain in the batch input session and must be corrected and processed manually.
- If all transactions of a batch input session could be processed successfully, the entire batch input session is deleted.
- You can specify that the session should be "kept" when you create a batch input session. In this case, the batch input session will not be deleted from the batch input session overview even if processing is error

free. This option can be used for documentation. In general, however, a batch input session can only be processed once.

- The processing of a batch input session is recorded in a detailed log.

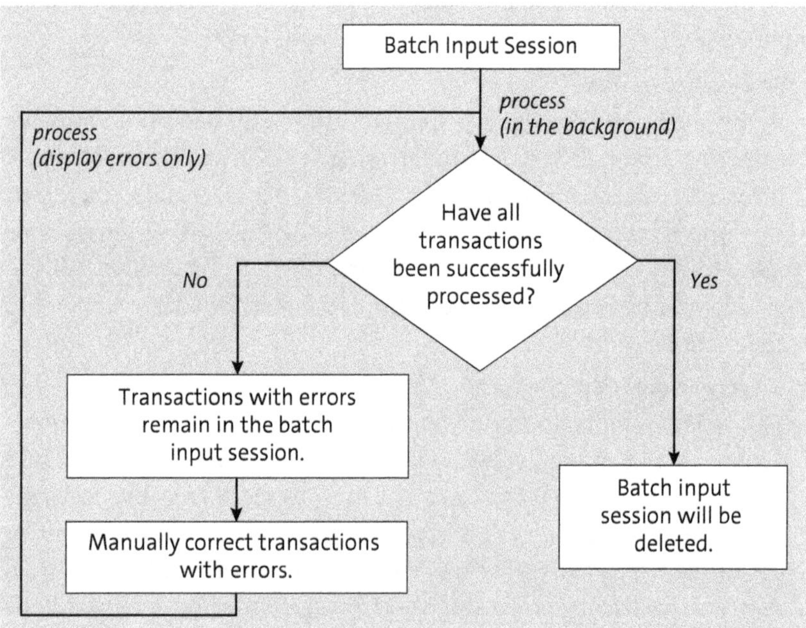


**Figure 7.1** Processing a Batch Input Session

To see, at a glance, the state of processing a batch input session is in, you must pay attention to its status, which can take the following values:

- **New**
  Batch input sessions that have the **New** status are ready for processing but have not yet been processed.
- **Incorrect**
  If a batch input session has the **Incorrect** status, either processing could not be completed due to incorrect transactions, or processing was terminated prematurely in the **Process/foreground** or **Display Errors Only** mode.
- **Processed**
  As mentioned earlier, successfully processed batch input sessions are usually deleted: However, if the **Keep session** option is set for a batch input session and the session has been successfully processed, it is assigned the **Processed** status. Processed batch input sessions remain in the SAP system until they are deleted by a periodic reorganization run. Incorrect batch input sessions are not covered by this reorganization run.

- **Being Created**
  Batch input sessions have the **Being Created** status only in the phase of their creation.
- **In Process**
  A batch input session is in the **In Process** status while it is being processed.
- **Locked**
  It is possible to lock a batch input session to protect it from unwanted processing.



**Other options**

Further control options are available for processing a batch input session, which we'll only briefly describe in the following list:

- **Automatically process batch input sessions**
  If, for example, batch input sessions are generated regularly (e.g., every night) by the relevant programs as part of a periodic data transfer to the SAP system, they should usually also be played back automatically. The automatic playback of one or more batch input sessions is achieved by a suitable call of the ABAP program RSBDCSUB. You can pass the following selection parameters to this program:
  - Name of the batch input session(s)
  - Date and time of creation
  - Status (**New** or **Incorrect**)
- **Delete batch input sessions**
  If you no longer need a batch input session in the list, you can delete it manually. However, do not delete batch input sessions that still contain unprocessed transactions (i.e., those with the **New** or **Incorrect** status). You must first correct and process such transactions or enter the transaction data into the SAP system some another way.
- **Lock and unlock batch input sessions**
  You can lock a batch input session to prevent the system from playing it before a date you specify.

**Transaction SM35**

These and other functions related to batch input sessions can be found in Transaction SM35 (Batch Input: Session Overview). You can access this transaction via by following the menu path **System • Services • Batch Input • Sessions**. Transaction SM35P (Batch Input: Log Overview), through which you can see the logs of processed sessions, can be called via **System • Services • Batch Input • Logs**.

Let's now turn to the more difficult of the two questions.

### 7.2.3 How Can You Create a Batch Input Session?

Batch input sessions can basically be created in three different ways:

- From a batch input recording
- By means of an ABAP program
- From within a standard batch input program

Figure 7.2 shows these three approaches to creating a batch input session.

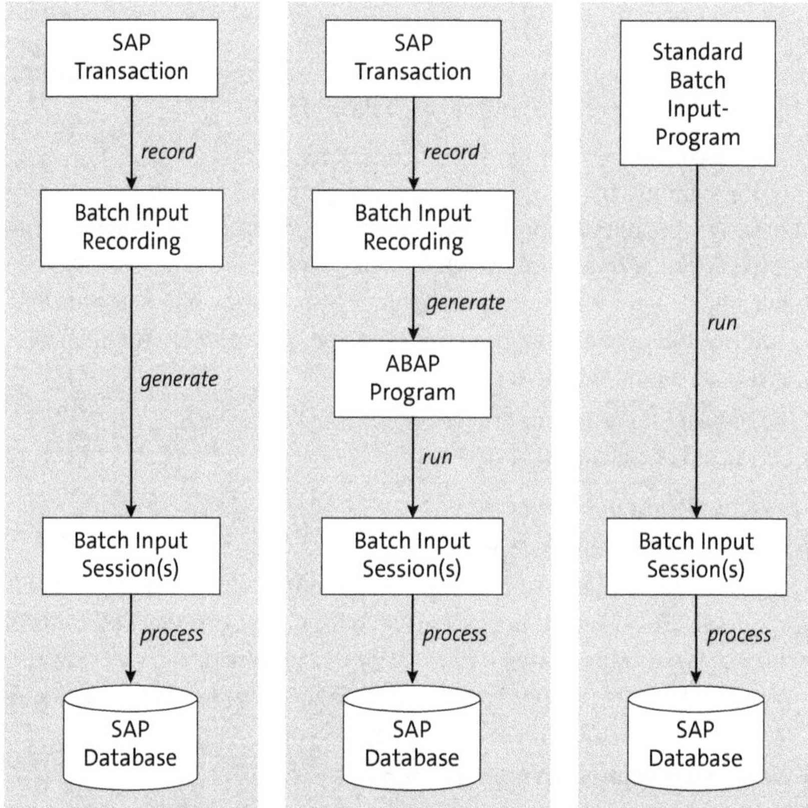


**Figure 7.2** Procedures for Creating Batch Input Sessions

**Batch input session from the recording**

You can record transaction processes using the SAP transaction recorder. This approach is referred to as *batch input recordings* or simply *recordings*. A batch input session can be generated from such a recording at the push of a button. The corresponding transaction in the SAP system is Transaction SHDB (Transaction Recorder Batch Input).

*SAP Legacy System Migration Workbench* (introduced in Chapter 9) contains its own transaction recorder.

Batch input sessions can be created using a suitable ABAP program. In this context, we can distinguish between two types of programs:

Batch input session from within the program

- The SAP system allows you to generate an ABAP program from a batch input recording at the push of a button. As a rule, a program generated in this way must be post-processed manually.
- Included in the SAP system, you also receive a number of what are called *standard batch input programs*.

[«]

7

**Standard Batch Input Programs in SAP S/4HANA**

In SAP S/4HANA, you should only use these programs if they have also been released for SAP S/4HANA. Some of these programs are based on SAP ERP transactions and cannot be used at all or only to a limited extent in SAP S/4HANA. You should therefore check if these reports or transactions are included in the simplification list. You should also look for recent SAP Notes or Knowledge Base Articles (KBA) before you run into problems. Just because a standard program is listed in SAP Legacy System Migration Workbench doesn't mean that it has also been released for SAP S/4HANA.

### 7.2.4 Batch Input Recording: General Procedure

The procedure can be divided into the following steps:

Steps in data migration

1. The transaction with which the legacy data is to be transferred to the SAP system is identified.
2. The identified transaction is executed as an example with a typical dataset to be transferred and recorded by Transaction SHDB or by SAP Legacy System Migration Workbench.
3. From this recording, the ABAP source code (or, within SAP Legacy System Migration Workbench, the corresponding target structure) is generated at the push of a button.
4. The ABAP program generated in this way forms the basis for the data transfer. However, it must be manually adapted to the specific circumstances of the migration.

## 7.3 Advantages and Disadvantages of the Batch Input Process

During the batch input process, extensive checks are performed when the data is imported into the SAP system (when the batch input session is played back). These checks correspond to the checks in dialog processing,

Safe but slow

which minimizes the risk of incorrectly entered data records. However, these additional checks have a negative effect on performance. For example, depending on the hardware environment and system configuration, the batch input interface transfers only between 10 and 40 data records per minute to the database server of the SAP system. In particular, loading a screen into memory is one of the main factors behind the poor performance of this process.

**Convenient error handling**

Incorrect transactions that cannot be processed by the SAP system when the batch input session is processed remain in the batch input session (*error session*). Successfully executed transactions, on the other hand, are removed from the batch input session. Thus, during error processing (i.e., when you reprocess the batch input session, which is now visible), you're only confronted with incorrect transactions that require manual reprocessing. Whether you process the incorrect batch input session in its entirety right away, process only a portion of the session and leave some portion for further processing later, or have it processed further by someone else is irrelevant.

**Standard batch input programs**

The SAP system comes with standard batch input programs for a number of applications, which you can use for your data migration—for example, in combination with SAP Legacy System Migration Workbench. On one hand, you can thus benefit from all the advantages of batch input technology mentioned earlier; on the other hand, you can also profit from the clear user guidance provided by SAP Legacy System Migration Workbench.

Standard batch input programs normally can generate the appropriate screen sequence in the batch input session for a set of input data so that it can be played back without errors.

**Only screens are recorded**

A major drawback of the batch input method, due to its age, is its limitation to recording screens. In addition, not all types of "newer" screen elements are supported, and thus, dialog transactions in the "new" design of the 2000s cannot be recorded at all. Newer process automation solutions from SAP are much more flexible in this regard.

**Call transaction versus batch input**

As described in Section 7.2.4, you have basically two options for processing the provided data when executing a batch input program (individual standard batch input programs also provide this option):

- **Processing using call transaction**
  Using the call transaction option, the data is processed faster than with batch input sessions. However, unlike batch input sessions, the call transaction method supports only limited interactive correction or logging functions.
  In addition, this technology enables the synchronous processing of the

data. When you run the program, you can use the selection screen to determine whether you want a synchronous or an asynchronous database update. An asynchronous update is useful for a steady system load.

- **Combined processing using batch input and call transaction**
  The general advantages of the batch input process, such as data security and error handling capabilities, have already been discussed. If you want to combine these advantages with improved performance during the data transfer, you should consider combining the batch input and the call transaction methods. For this task, you must decide to use the call transaction procedure when you run the data transfer program, and you must also designate an error session in which all incorrect transactions are placed that the call transaction method is unable to process. As soon as the program is finished, manual post-processing of the error session can be started. This approach allows you to combine two data transfer goals that at first glance might seem incompatible: speed and data security.



**Programming required**

However, as soon as you want to transfer data for which no standard batch input programs are available, you won't be able to avoid programming if you follow the "traditional" approach. Even if these programs are largely generated automatically by the SAP system on the basis of an underlying recording, an extension of the code (and thus at least rudimentary ABAP knowledge) is required. For example, the program must be equipped with additional functionality that allows more than one data record (i.e., an entire transfer file) to be loaded and processed in the SAP system (Section 7.2.4).

**Special features of a recording**

A program generated from a recording exactly maps the sequence of this recording. However, SAP transactions may behave differently for different input data. In other words, the sequence of screens is not always the same, but often depends on the selected input data.

But a generated batch input program only knows the screen sequence of the underlying recording. Accordingly, so the program can react, in a program-based way, to the possibility of changing screens, you must store alternative screen sequences in the program logic, which are to be processed depending on a certain condition. In turn, therefore, you must create various recordings of a transaction, taking into account all contingencies, in order to have the corresponding source code generated from them. This task can therefore become pretty complex.

**Problem case**

You cannot assume that you can always use recordings to prepare a data transfer environment for all SAP transactions that produce the desired batch input sessions. For this reason, recordings should always be discarded if the number of item records per record may vary. Such a situation exists, for example, when documents are to be transferred. In extreme cases, you may have documents with only a single item as well as documents that consist of several hundred items.

Even if you can prepare the program generated from an appropriate recording for this range of contingencies with some programming skill, we don't intend to go into more detail. In such situations, instead, we recommend that you use the SAP S/4HANA migration cockpit, described in detail in Chapter 3. Using this tool and its more than 200 predelivered migration objects, you can import virtually any business object in a more elegant way, without any programming work.

[+]

**Further Information on Batch Input BDC**

In the SAP Community wiki, you can find more interesting information about Batch Input BDC at *http://s-prs.co/v581713*.

[»]

**The Future of Recordings**

With *SAP Intelligent Robotic Process Automation (SAP Intelligent RPA)*, SAP delivers a solution that enables for automating recurring processes. SAP Intelligent RPA is used, for example, in SAP S/4HANA Cloud to test business processes after a release upgrade. However, like before, the *Extended Computer Aided Test Tool (eCATT)* can theoretically also be used for data transfer. More information is available via SAP Community at *http://s-prs.co/v581714* or via SAP Help Portal at *https://help.sap.com/docs/IRPA*.

## 7.4 Summary

As we've seen, data transfers based on batch input recordings often cannot do entirely without any programming work, even though most of the code is generated automatically and only needs some adjustment. As an SAP data transfer tool, SAP Legacy System Migration Workbench (Transaction LSMW) provides the option of recording transactions, similar to the Transaction Recorder, and then integrating them into SAP Legacy System Migration Workbench completely without any coding work. Details on this procedure can be found in Chapter 9, Section 9.3.

# Chapter 8
# IDoc and ALE Distribution

*In this chapter, we introduce IDocs (Intermediate Documents) as an interface technology and data migration method. We provide examples of how this technique can be used via Application Link Enabling (ALE) distribution and present a tool for editing IDoc files.*



In this chapter, we first provide some introductory information on the topic of IDocs. We explain the origin and use of IDocs and how to set up IDocs in your system. You'll learn the most important transactions and typical terms related to IDocs. Then, we'll share important tips for optimizing the performance of your system.

The second part introduces you to using an *ALE* distribution in a data transfer project. For this purpose, we use the common scenario of distributing classes and characteristics and provide additional information on product data distribution. Finally, we list the advantages and disadvantages of using IDocs and the ALE method.

## 8.1 Introduction to IDocs

In this section, we introduce the IDoc interface technology and its background. IDocs are important interfaces in the context of SAP data transfers and are used both in master data distribution via ALE (Section 8.2), when using SAP Legacy System Migration Workbench (Transaction LSMW) described in detail in Chapter 9, and in a data migration via SAP Data Services. We also want to provide you the knowledge to perform all the necessary steps for setting up IDocs and optimizing their performance in your SAP system yourself.

### 8.1.1 Why IDocs as an Interface?

**EDI and ALE**

Historically, the origins of today's IDocs predate the use of XML structures in *Electronic Data Interchange (EDI) format*. EDI involved the exchange of files in a well-defined structure that was a predecessor of today's IDoc structure. This IDoc structure was later used within SAP communication also for

ALE (Section 8.2), where in this case the IDocs are exchanged via *remote function calls (RFCs)* instead of files. What both areas of application have in common is that the metadata is not sent along with the actual data each time, as in the case of XML. For this reason, in IDoc communication, a rather important point is that both communication partners know the same IDoc definition for a specific IDoc type in order to understand each other. This results in one disadvantage compared to XML: You cannot do anything with the data without knowing this "template." One advantage is that the size of the message only corresponds to the message content and has no overhead. Today, however, IDocs can also be represented in XML and exchanged between SAP systems both via XML files and via web services as XML strings.

### 8.1.2 Using IDocs

**Segments**

In a standard SAP S/4HANA system, numerous IDocs have been developed by the individual application areas within SAP using IDoc technology and IDoc development tools. An IDoc is a hierarchically nested structure, and the individual data records of an IDoc are referred to as *segments*. In addition to the hierarchical structure, an IDoc also always consists of an *outbound function module* for sending data and an *inbound function module* for receiving and further processing IDocs. During a data migration via SAP Legacy System Migration Workbench, as described in Chapter 9, inbound IDocs are stored as files in the target SAP system, and inbound processing of the file is triggered locally. In the case of *SAP Data Services*, the IDocs are sent to the target SAP system via RFC. In both cases, only the inbound IDoc in the target SAP system is important for the data migration.

**IDoc status values**

IDocs always have defined statuses in the SAP system. For data transfer in inbound IDocs, the most important status values are the following:

- **Status 64 (waiting status)**
  The IDoc is ready to be passed to the application.
- **Status 53 (IDoc successfully posted)**
  The application document has been posted.
- **Status 51 (IDoc incorrect)**
  The application document was not posted.

**Message type and basic type**

You can differentiate between *IDoc message types*, which indicate the semantics of IDocs, and *basic IDoc types*, which define their syntax. For example, the IDoc message type `DEBMAS` is responsible for customer data, while the basic types `DEBMAS08` or `DEBMAS07` as different versions specify the exact syntax of the segments and all fields in them. The version concept specifies that fields and segments can always be added but never deleted. Thus, DEBMAS08 includes the entire functionality of DEBMAS07. This version control makes sure that both a newer system can handle an older IDoc (upward compatibility) and that a newer system with a lower version can send IDocs to an older system (downward compatibility). However, the relationship between message type and basic type is not always 1 to *n*, as it may seem at first. For example, the basic IDoc type ORDERS05 transmits the logical message types ORDERS (purchase order) and ORDRSP (purchase order confirmation), which changes the meaning of the IDoc message.

**IDoc type**

An enhancement concept is available for IDocs, which can result in an IDoc extension ZDEBMAS08, for example, which combines additional customer-specific fields or customer segments in what's called the *IDoc type*.

We won't delve further into the IDoc technology beyond this overview. The available tools, such as SAP Legacy System Migration Workbench and SAP Data Services, will do the work for us. In both cases, the tools ensure that the IDocs are built with the correct IDoc control record and syntax.

**Transaction BDBG and BAPI**

Over time, IDocs have proven to be solid and consistent interfaces equipped with a sophisticated version concept. IDocs are also rather reliable for posting, as a complete *rollback* is always carried out in the event of an abort, and the IDocs are always available for reprocessing. For these reasons, SAP provides Transaction BDBG, which is a tool that can generate an IDoc interface from an asynchronous Business Application Programming Interface (BAPI) at the touch of a button. Asynchronous BAPIs are precisely those that can load data into a system instead of executing an operation and providing a response. They merely return information about success or failure, similar to a classic IDoc. Transaction BDBG allows you to extend the IDoc world considerably.

Most of the BAPIs are already delivered with the so-called *BAPI/ALE interface*, which you can use to generate an IDoc structure and an IDoc type from the BAPI. The IDoc will serve as a wrapper around the BAPI, and data will be sent as IDoc instead of directly calling a BAPI in a remote system. The inbound IDoc is then unwrapped, and the BAPI is called locally in the target system. In general, a BAPI is an SAP function module with a defined interface and documentation. The advantage of using BAPIs is that, as with ALE, the send process is decoupled from processing. Otherwise, the connection between the systems must remain open during the entire BAPI processing time.

The example shown in Figure 8.1 is the result of an ALE interface for a BAPI of object type BUS7050 and the CREATEFROMDATA method. This BAPI is used to transfer claims and is not delivered with a generated IDoc interface.

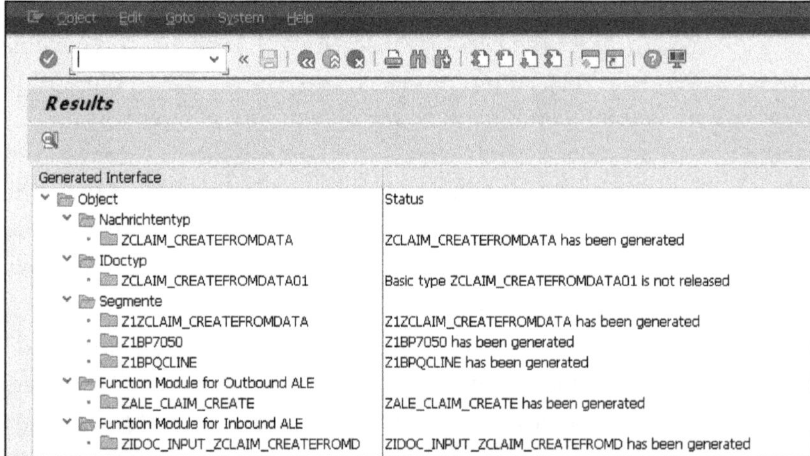


**Figure 8.1** Transaction BDBG: Result of a Generating an IDoc Interface

Most of SAP's business objects are covered by this technology, and it makes no technical difference whether an IDoc inbound function module or a BAPI is called at the end. A user wouldn't even notice this difference.

**Important IDoc Transactions in the Context of Data Migration**

The following IDoc transactions are available for data migration:

- SALE - IDoc Customizing
- WE19 - IDoc Test Tool
- WE20 - IDoc Partner Profiles
- WE21 - IDoc Port Maintenance
- WE30 - IDoc Type Overview
- WE31 - IDoc Segment Overview
- WE82 - Assign Message Types to IDoc Types
- WE57 - Assign Function Module to IDoc and Message Type
- WE60 - IDoc Documentation
- WE02 – IDoc Monitor
- WE09 - Advanced IDoc Search
- BD87 - IDoc Overview
- BDBG - BAPI/ALE Interface
- SLG1 - Error Logs in Application Log

[+]

**WEDI Area Menu for All IDoc Transactions**

The WEDI area menu (**IDoc and EDI Basis**) gives you a direct overview of all IDoc transactions. You can call the area menu like a transaction in the command field. The area menu will then appear in the menu tree below **SAP Menu**. If you exit SAP Easy Access via /n in the command field of the system toolbar and then enter "WEDI," the area menu does not appear as a tree but as a normal menu bar. You can navigate more quickly using the mouse than in the menu tree.



### 8.1.3 IDoc Customizing in the SAP System

To prepare your SAP system for receiving IDocs, you must some perform basic customizing there. All steps must be performed if no IDoc customizing has been done previously. All ALE-relevant customizing functionalities can be accessed in Transaction SALE. The customizing functionalities that are mandatory for data migrations include the following:

1. **Setting the IDoc administrator**
   Navigate to **Basic Settings • IDoc Administration**.
2. **Performing a workflow customizing**
   Navigate to **Basic Settings • Perform Automatic Workflow Customizing**.
3. **Creating a logical system**
   Navigate to **Basic Settings • Logical Systems • Define Logical System**.
4. **Creating a cross-system company code**
   Navigate to **Modelling and Implementing Business Processes • Global Organization Units • Cross-System Company Codes**.

Cross-system company code

The cross-system company code is an IDoc-/ALE-specific setting in the SAP system. Since ALE communication via IDocs may well take place between two remote SAP systems that hold data for the same company code, it was necessary to make use of a *cross-system company code* as a helper construct because the two self-sufficient systems can know the same company code under different company code numbers. Most often, this discrepancy has historical reasons. For this reason, a new, unique name must be used as an identifier for the transfer of company code-dependent data in an ALE scenario (e.g., via the DEBMAS IDoc).

Since we do not perform an ALE communication between SAP systems in the case of data migration, no further attention is needed on the topic of cross-system company codes. The only necessary action is to create the cross-system company code, since the IDoc always works with this internal value. Usually, it is sufficient for a migration to simply choose the name of the (local) company code for the cross-system company code, if the name has not already been assigned elsewhere or if different naming conventions must be met (such as using the "GL" prefix for global cross-system company codes). Figure 8.2 shows the relevant entries in Transaction SALE.

**IDoc partner profile**

Furthermore, you must make settings in the SAP system for the IDoc message types used to determine what should happen to the data once a specific IDoc is received. SAP Legacy System Migration Workbench (see Chapter 9) offers a generation option for this task; nevertheless, you can set up this so-called *partner profile* manually using the customer master as an example. For this task, open Transaction WE20 and define the following items, as shown in Figure 8.3:

1. IDocs from the specific partner **Logical System** (LS) with the partner number BOBJTFR may be received. For this task, you must create a corresponding partner entry under the **Partner type LS** node.
2. IDocs of message type DEBMAS are transferred to the customer generation. For this task, you must create a corresponding input parameter for the partner entry.

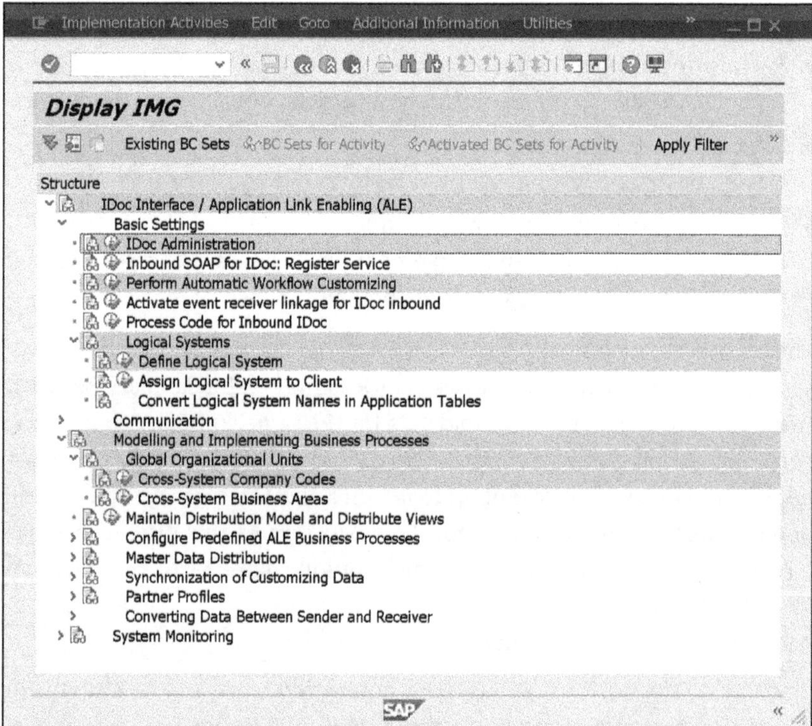


**Figure 8.2** Transaction SALE: Relevant Customizing

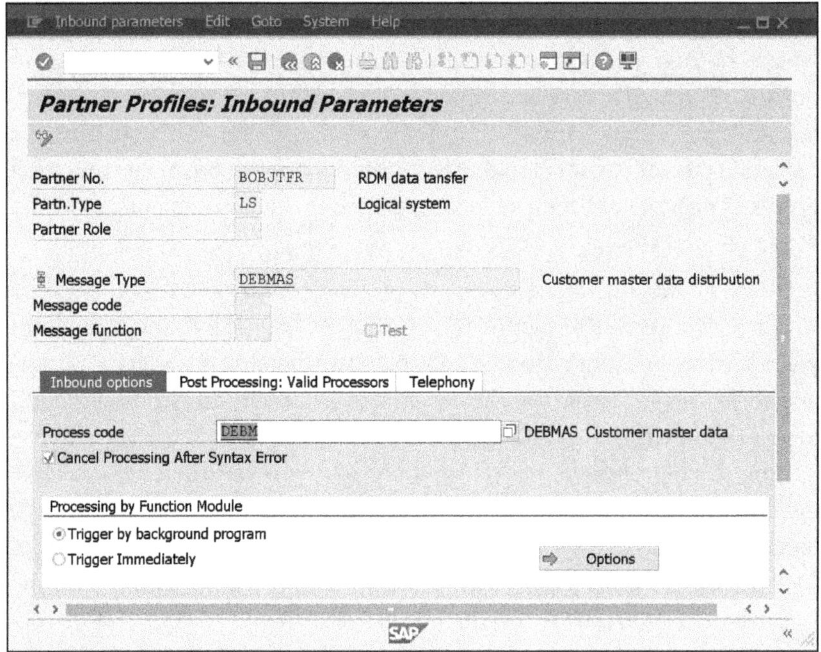


**Figure 8.3** IDoc Partner Profiles

In the **Process code** field, you define what the customer application is supposed to do with the data; in this case, the DEBM code is used to create the customer master data in the system. Generally, the application suggests a possible process code, but this suggestion may not always be suitable. For example, some applications may know different process codes for individual processing versus for bulk processing. When you create a partner profile manually, this setting can be selected via the input help. When generating them automatically (as described, for example, via the distribution model and in Section 8.2), manual adjustment may be required.

**Background Processing**

The default setting in the partner profiles is **Trigger Immediately**, which results in the synchronous processing of an IDoc after it has been received. However, because this processing constitutes a separate work process for each IDoc, this method is not always ideal, for example, for large data volumes where resources would quickly bottleneck. An alternative is background processing and triggering by the background program RBDAPP01 (Transaction BD20).

### 8.1.4 IDoc Performance Optimization

**Report RBDAPP01**

To achieve high performance when mass uploading IDocs for data migrations, background processing is mandatory. Only in this way can you transfer multiple IDocs as packages to a work process for synchronous processing. Ideally, you should use multiple SAP work processes that each process a package of IDocs, which will decouple the receiving process from the actual processing.

In direct processing, a single IDoc is received via an RFC and processed. However, if you collect inbound IDocs using the **Trigger by background program** setting in Transaction WE20, you can improve the update performance for IDocs. Generally, however, this triggering in the background means that the IDocs are "hanging" in status 64 for the time being; they are waiting to be processed. Background processing is not the only available option; you can also start the processing in a dialog box. The best way to process the IDocs is to schedule report RBDAPP01 (inbound processing of IDocs ready for transfer) either via Transaction SM36 as an ABAP job in the background or start the report directly via Transaction SE38 or via SAP Legacy System Migration Workbench in dialog mode.

**Package size**

On the selection screen of report RBDAPP01, the following items are critical:

- **IDoc selection tab**
  - **Package Size**: The package size controls the maximum number of IDocs to be processed in a logical unit of work (LUW) in a dialog work process. A large package reduces the number of required processes to a minimum but also requires a large roll area. Either the database commit is performed for the entire package, or the database is rolled back, and no package data is stored. In general, it can be said that a large package size lowers the degree of parallel processing.
- **Parallel Proc. tab**
  - **Parallel Processing Active**: Select this checkbox to activate parallel processing. If you select this checkbox, the application server uses a free dialog process for each IDoc package for inbound processing; the packages are thus processed in parallel. If you select too many packages, all dialog processes of the server will be occupied. You should therefore specify a server group that controls how work processes are assigned (e.g., parallel_generators) to avoid overloading the system. If the flag is not set, there is no parallel processing of the IDocs. In other words, each package is passed sequentially to the application. In total, only one work process will be occupied on the application server, as shown in Figure 8.4.

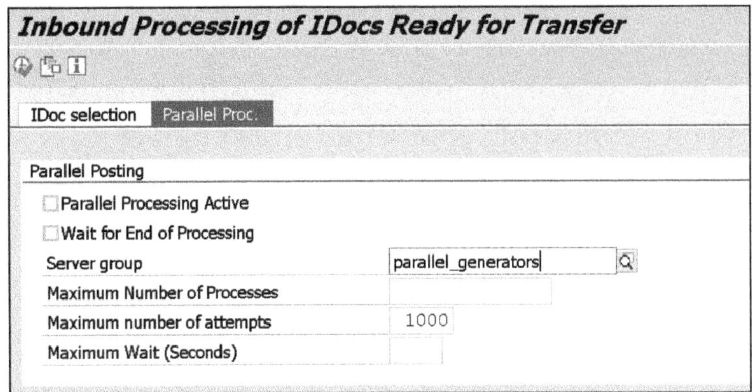




**Figure 8.4** Parallel Processing in Report RBDAPP01

- **Server Group**: The server group determines how resources are distributed across the existing work processes of the application server(s), that is, how many work processes are provided in each application server. You can adjust these settings in Transaction RZ12, as shown in Figure 8.5.

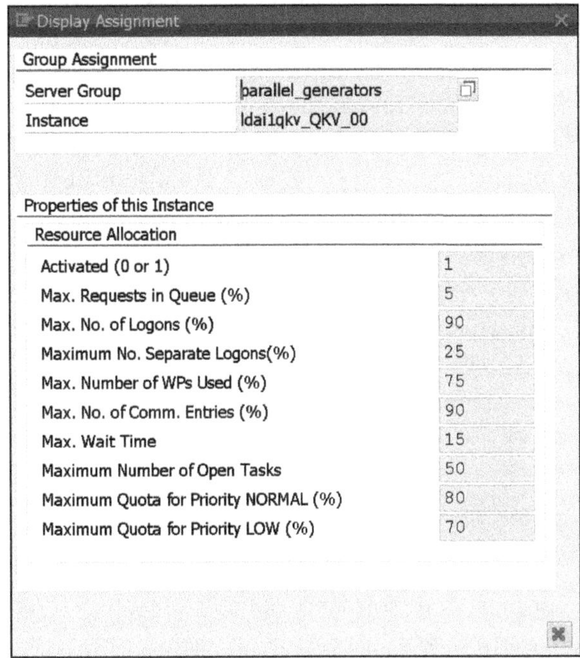


**Figure 8.5** Transaction RZ12: Server Group Maintenance

We can't make general recommendations about package sizes or the number of work processes in a server group for parallel processing. You should always use test data to determine the best values for you. This value depends on the IDoc types and IDoc sizes (number of segments) as well as database performance and server performance. For large IDoc segments, 50 IDocs per package and, for smaller segments, 100 IDocs per package are good starting values.

Tips Other important settings of program RBDAPP01 for additional performance optimization and to assist the migration project include the following:

- **IDoc selection tab**
  - The **Output list** flag controls whether a detailed list is output as a log at the end of IDoc processing. In general, this output is not required because all information can be called via the IDoc monitoring transactions and by confirming the IDoc status in SAP Data Services; the information is therefore available separately. Depending on the data volume selected, omitting this step can result in corresponding performance gains.
- **Parallel Proc. tab**
  - **Maximum number of attempts**: In this field, you can specify the time (in seconds) during which the system tries to occupy a parallel process with a package of IDocs to be processed. An attempt is made every second. If the attempt is unsuccessful, all IDocs of the package receive an error status after the specified time elapsed. If you use bulk parallelization and have not optimized the package size yet, you should select this value as high as possible (e.g., 99999) to avoid aborting the processing.
  - **Wait for End of Processing**: With this setting, you specify whether the run of program RBDAPP01 ends as soon as all IDocs have been processed (*synchronous*) or whether it should end right after all packages have been transferred to the application (*asynchronous*). Selecting this checkbox makes sense for the background mode to log the overall processing time. The log simplifies the optimization process because you can test with various settings and then compare processing times. For the live run, however, you should deselect this option to avoid the performance impact from this waiting process.

You can now adjust the settings for performance optimization on the SAP target side—also for bulk transfer, which is frequently required for some business objects.

## 8.2 Master Data Distribution

In this section, we cover ALE master data distribution in an SAP system for data migrations using the specific example of transferring classes and characteristics. This is usually a critical factor in data transfer projects, as most non-SAP source systems are not aware of these SAP-specific objects.

**ALE and distribution model**

The ALE technology also uses IDocs for data integration in a distributed system landscape. A good example is the distribution of master data via the IDoc message types `CLSMAS` (classes) and `CHRMAS` (characteristics), where one of the SAP systems assumes the role of master data management. This system for the central maintenance of classes and characteristics can then replicate all changes to remote systems (such as individual plants or subsidiaries) via ALE based on precise rules. For this automatic distribution based on change pointers and filters, SAP provides what's called a *distribution model* via Transaction BD64. However, you can use this model just as well for transaction data or for transferring data to external business partners.

### 8.2.1 Setting Up an ALE Connection in the Sender

**Transaction SALE**

You must always set up the ALE connection in the sending client of the source system. The first steps up to the "Define RFC connection" point are client independent, but from this point on, the entries become client dependent. The steps are as follows:

1. Call Transaction SALE.
2. If not already present, you need to set up the logical systems via **Basic settings • Logical systems • Define logical system**. Follow the convention `<system>CLNT<client>`, for example `ABCCLNT100` for client 100 in system ABC.
3. If not already done, you must assign the logical system to a client via **Basic Settings • Logical Systems • Assign Logical System to Client**.
4. Define the corresponding RFC target connections in the sending source system (connection type **3 ABAP connection**) using Transaction SM59. (Tip: You should test the connection after saving.)

### 8.2.2 Setting Up the Distribution Model in the Sender

In general, usually distribution models are set up for distributions between systems. As a result, separate models for the distribution exist within the development system landscape, the quality landscape, and the production landscape. In the case of data migration, however, we can deviate from this pattern since the data between the landscapes is also to be distributed via ALE.

**Setting up the distribution model using Transaction BD64**

1. Call Transaction BD64 to set up the distribution model.
2. Switch to change mode by clicking the icon and then filter the display by clicking Filter Model Display.
3. Under **Message type**, enter "CHRMAS" for characteristics and "CLSMAS" for classes. Click the icon to filter the display.
4. Then, create a model view by clicking Create Model View.
5. Assign a short text and a technical name for the model view.
6. Select your model view and add the message types by clicking Add Message Type.
7. Specify **Sender**, **Recipient**, and **Message Type**.
8. Repeat the entire process for all other message types and recipients.
9. After clicking the (**Save**) icon, have the partner profile generated automatically for the systems via **Environment • Generate Partner Profiles**. To achieve better performance results, you can again use the background processing option (Section 8.1.3).
10. Then, distribute the complete model view to the relevant systems via **Edit • Model View • Distribute**.

### 8.2.3 Setting Up an ALE Connection in the Recipient

**Generating a partner profile using Transaction BD64**

Similarly, you must generate the partner profile on the recipient's side. For this task, log on to the appropriate system and call Transaction BD64. The model distributed there should be available; for this reason, only the input parameters in the partner profiles need to be generated.

### 8.2.4 Distributing Characteristics

**Distributing characteristics using Transaction BD91**

To distribute characteristics using the distribution model you have just set up, use Transaction BD91 or the ABAP program, RBDSECHR. When you run the report in Transaction SE38, the selection screen is displayed, as shown in Figure 8.6. On this screen, you must specify what exactly should be distributed to which system. The program automatically selects the appropriate entries and sends them to the target system specified in the Transaction BD64 distribution model using CHRMAS IDocs.

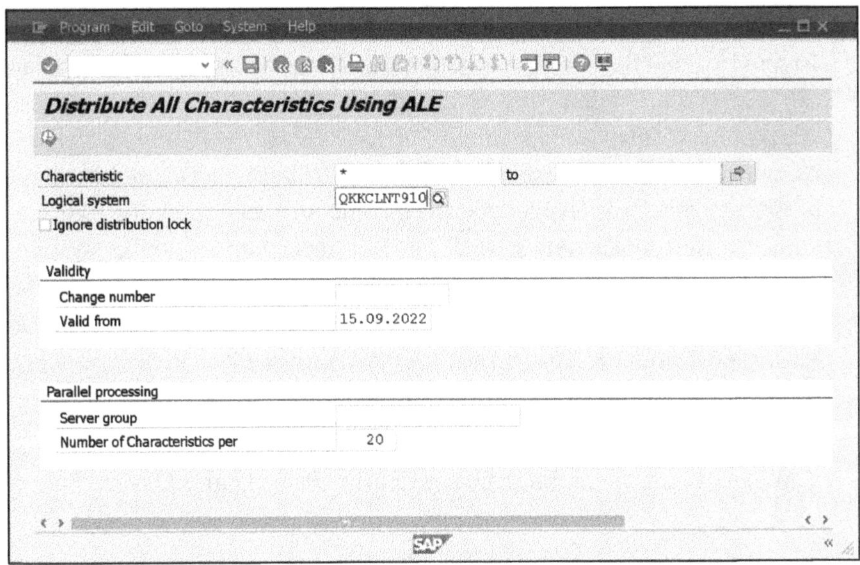


**Figure 8.6** ABAP Report RBDSECHR for Distributing Characteristics



### 8.2.5 Distributing Classes

**Distributing classes using Transaction BD92**

The distribution of classes works in the same way via Transaction BD92 or program RBDSECLS, as shown in Figure 8.7. Create your own variants for different class types such as batches and material classes. This report sends the IDocs of message type CLSMAS according to the settings in the distribution model.

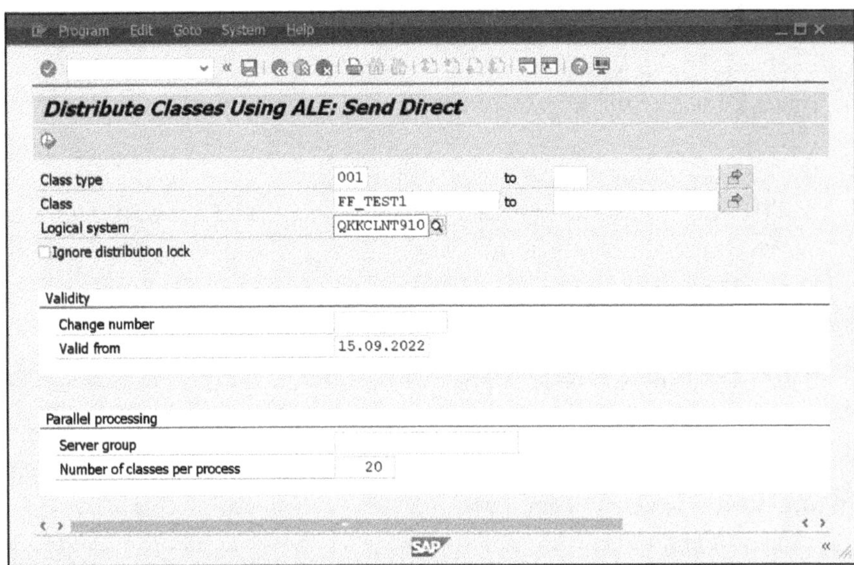


**Figure 8.7** ABAP Report RBDSECLS for Distributing Classes

### 8.2.6 Monitoring the Distribution

**Transaction BD87 (Status Monitor)**

To monitor the IDoc distribution once the IDocs have been generated, you can use certain monitoring transactions, namely, Transactions WE02, WE05 or BD87, introduced earlier in Section 8.1.2. However, we recommend using Transaction BD87 especially for ALE distributions. This status monitor for ALE messages enables you to conveniently check both the IDoc creation in the sender system and the IDoc receipt in the target system. Restrict the view to the `CHRMAS` and `CLSMAS` message types. Note that, in the inbound system, in the case of processing by the background program, as set in Transaction WE20 (partner profiles), as described in Section 8.1.3), program `RBDAPP01` must be started first. Otherwise, all IDocs remain in status 64 and await processing.

If all IDocs in the inbox show status 53, as shown in Figure 8.8, everything is OK.

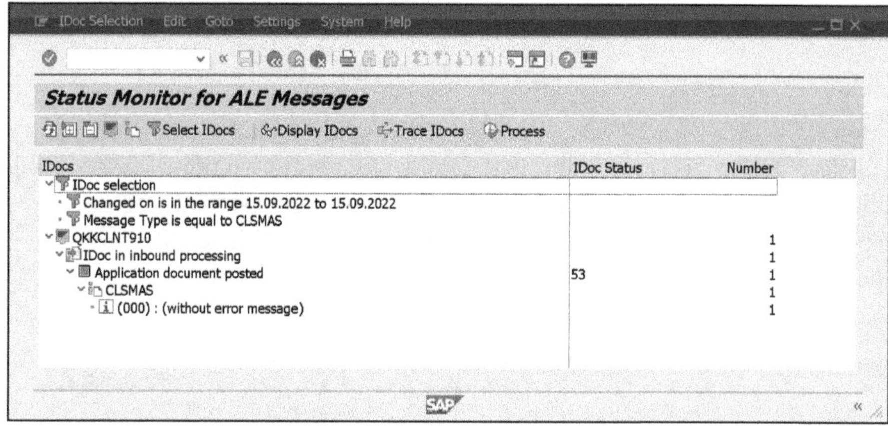


**Figure 8.8** IDoc Status Monitor BD87

**Analyzing the application log using Transaction SLG1**

If errors have occurred, you'll see a red traffic light icon in the IDoc monitor, and the IDoc will show status 51. The application could not post the IDoc, and you should investigate the cause in the *application log* (Transaction SLG1).

### 8.2.7 Replicating Product Data

Another example of master data distribution as an alternative to classic data migration is the *product data replication (PDR)*. It is preferred when you work with variant configuration. PDR is an application that allows you to distribute product configuration models between systems. It identifies all objects that belong to a product configuration model. Instead of then distributing objects using individual IDocs, these are combined in a *Uniform Packaging Service (UPS) package* and then distributed to the target systems. You can thus store your master data centrally and consistently in one system and avoid the redundant maintenance of your objects in different systems (see *http://s-prs.co/v581778*).

[«]

**Product Data Replication in SAP S/4HANA Cloud**

PDR can also be used in SAP S/4HANA Cloud. However, the cloud system can only be the recipient of the data. Either an SAP S/4HANA or an SAP ERP system serves as the sender. (See *http://s-prs.co/v581779* for more information.)



## 8.3 Advantages and Disadvantages of the Method

The advantages of using IDocs and ALE for data migration are manifold:

- **High performance through parallelization via dialog processes**
  One advantage is the use of dialog processes in parallel processing. Since usually many times more work processes of the **dialog** type exist than of the **background** type configured in a system, a higher degree of parallelization can be achieved. The SAP S/4HANA migration cockpit, for example, parallelizes via background processes and thus has a performance disadvantage with very high data volumes.
- **Asynchronous processing**
  An advantage, but also a disadvantage is the asynchronous processing. On the one hand, you can generate IDocs, but you don't have to process them right away, similar to a batch input session (see Chapter 7). On the other hand, the generation of IDocs is time consuming.
- **No or limited simulation**
  The simulation of data is given if the function module has a simulation parameter that can be addressed as an IDoc field. However, this option is of rather limited use in the case of IDocs. In this case, the migration cockpit has the advantage.
- **Compatibility**
  Before using the SAP ERP IDoc types you know for SAP S/4HANA, you should first read whether these IDoc types are still valid at all in SAP S/4HANA or SAP S/4HANA Cloud. The older IDoc type may not be useful for transferring data at all. Some IDoc types have been adapted for use in SAP S/4HANA, but others are not usable at all or only to a limited extent.

**Migrating Purchasing Info Records**

An example of the limited usability is the INFREC IDoc type for transferring purchasing info records. SAP recommends the use of function module ME_INFORECORD_MAINTAIN_MULTI. The latter is also used by the migration cockpit for migrating purchasing info records. SAP Note 327083 provides more detailed information.

- **Packaging IDocs**
  Packaging IDocs, as you have learned, is another factor in speeding up performance. Finding the perfect package size requires a lot of testing. If the package size is too large, an unfortunate error distribution can result in, for example, not being able to post a single package, and then all IDocs in the package, both faulty and error-free ones, fall into status 51. Here, a smaller package size such as 25 often makes much more sense than a package size of 1,000. This increases the probability of having packages that don't contain any faulty data records.

**Garbage Collector Job Reposts Error-Free IDocs**

A sensible strategy when posting IDocs in packages is to schedule a garbage collector job depending on the volume of data. Make sure you start this job a little time after the other jobs (e.g., 30 to 45 minutes later). Report RBDMANI2 is scheduled as a job for all IDocs in status 51 without package processing and then starts the corresponding IDocs sequentially again. As a result, all error-free IDocs are properly posted in status 51, and in the end, only the really faulty ones remain. Depending on data volumes, you may also need to schedule more than one garbage collector job if the first one has already finished, and the parallel jobs are still busy producing faulty IDocs.

## 8.4 Summary

In this chapter, you learned the basics of the IDoc technology as well as explored several application examples of its use. You now have the necessary tools to use this technology for data transfers. A good example—and already frequently mentioned—is the use of IDocs as an import technique in SAP Legacy System Migration Workbench, which we introduce in detail in the next chapter. IDocs are also popular for data transfer projects using SAP Data Services.

Chapter 9

# SAP Legacy System Migration Workbench

*In this chapter, you'll learn about a flexible and powerful tool that you can use to transfer data from non-SAP systems to an SAP S/4HANA system largely without programming.*

9

The use of *SAP Legacy System Migration Workbench* (often colloquially shortened to its transaction code *LSMW*) is always a good idea in SAP S/4HANA if there is no corresponding migration object for the SAP S/4HANA migration cockpit or if adapting an existing migration object or creating a new one using the SAP S/4HANA migration object modeler is too time consuming or not possible. This chapter presents the ways in which you can perform data migrations using SAP Legacy System Migration Workbench.

After an overview of the basic functionality, you'll learn how to use SAP Legacy System Migration Workbench in detail, step by step, based on an example. Then, you'll learn how to use recordings of transactions in this context. Finally, we'll explain how to transport migration projects between SAP systems and the use of the IDoc interface to illustrate the wide range of functions in SAP Legacy System Migration Workbench. From Section 9.8 onwards, you'll find further functions and usage examples for advanced users, and in Section 9.9, we'll cover the transfer of long texts.

[«]

**SAP S/4HANA Release 2022**

All the information and examples in this chapter have been verified and tested in SAP S/4HANA 2022 FPS00. They should also be usable in underlying releases.

## 9.1 Overview of SAP Legacy System Migration Workbench

**History of development**

Historically, SAP Legacy System Migration Workbench emerged from the R/2-R/3 Migration Workbench; essential concepts and experiences from the R/2-R/3 migration were included into the development of SAP Legacy System Migration Workbench. It is a powerful tool based on the SAP

NetWeaver technology that supports you in the individual and periodic transfers of data from non-SAP systems (*legacy systems*, here called *source systems*) to SAP systems that are based on the SAP NetWeaver technology. So, SAP Legacy System Migration Workbench is actually a data migration tool for the classic SAP Business Suite, but it is also included in the *ABAP platform* for SAP S/4HANA.

[»]

**Restricted Use at Your Own Risk**

You may have heard that SAP Legacy System Migration Workbench is on the *simplification list for SAP S/4HANA* and that it is therefore "supposedly" forbidden to use it. However, the simplification list and the corresponding SAP Note 2287723 only describe that SAP no longer supports the tool in SAP S/4HANA and that its use is therefore *restricted*. For this reason, the use of this tool is not recommended and can only be done at your own risk. Due to changes and simplifications in SAP S/4HANA, we can no longer ensure that the transfer methods offered for the SAP ERP data migration work properly, for instance, Business Application Programming Interfaces (BAPIs), IDocs, and standard programs. For this reason, you should make sure that the transfer methods used are also released for SAP S/4HANA and determine which functional scope of SAP S/4HANA they support. Use SAP Legacy System Migration Workbench only as a supporting tool alongside the SAP S/4HANA migration cockpit and the SAP S/4HANA migration object modeler.

**Functional scope**

SAP Legacy System Migration Workbench conveniently supports the import of data from files of non-SAP systems, the conversion of this data into SAP formats, and the import of the converted data into an SAP system using the standard SAP interfaces, batch input, direct input, BAPI, and IDoc.

In addition, SAP Legacy System Migration Workbench is equipped with the recording functionality of SAP transactions (batch input). This capability enables you to record a data entry or change transaction and use the resulting recording as the basis for a data migration project.

**Principles**

The development of SAP Legacy System Migration Workbench has been guided by the following principles:

- No individual tables or field contents are migrated, only related business data objects such as customer masters, material masters, financial documents, etc.
- The entire functionality is located in the SAP system. In the end, you'll avoid having a collection of scattered, disjointed, and difficult-to-maintain programs on different platforms.
- The quality and consistency of the data imported into the SAP system is given a higher priority than the speed of the data migration itself.

- No ready-made data conversion programs are delivered. Instead, the required programs are generated from predefined conversion rules. These conversion rules can be used multiple times (reusable) and thus ensure a high degree of data consistency.

These considerations resulted in the concept presented in Figure 9.1.

**Core functions**

The following are the core functions of SAP Legacy System Migration Workbench:

- **Importing data**
  This function involves importing data from the source system, which is stored in files on the frontend or on the application server.
- **Converting data**
  This function refers to the conversion of the data into SAP format.
- **Importing data**
  Importing data means posting the data to the database of an SAP system.

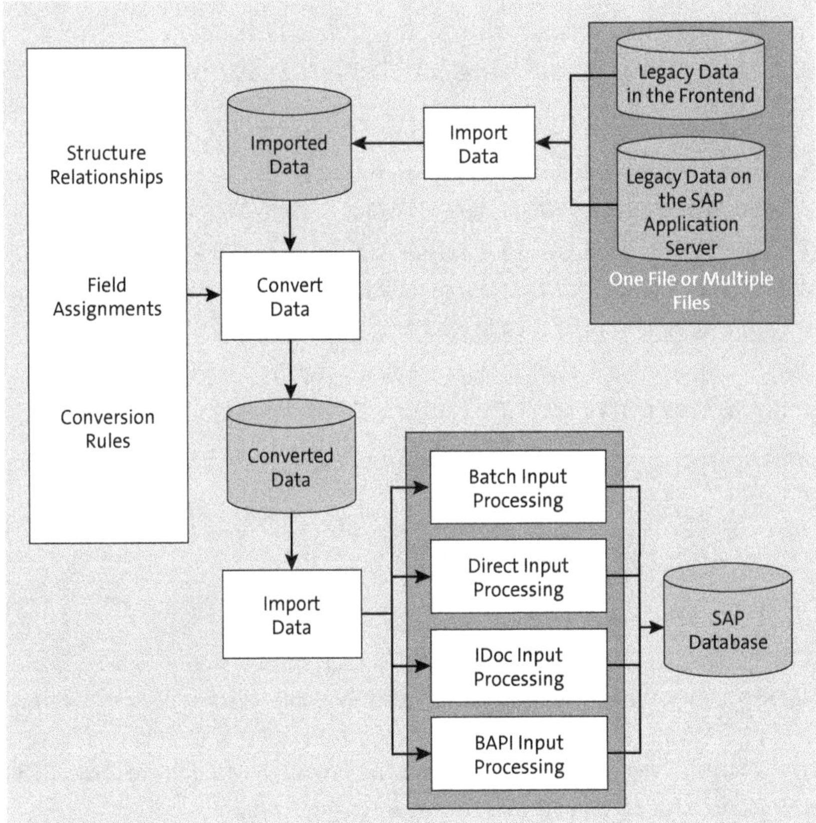


**Figure 9.1** Outline of a Data Migration Process Using SAP Legacy System Migration Workbench

The main advantages of SAP Legacy System Migration Workbench can thus be summarized in the following ways:

- SAP Legacy System Migration Workbench is part of the SAP system and is therefore platform independent.
- It offers extensive technical capabilities for data conversion.
- It ensures data consistency by using standard SAP import methods (standard interfaces).
- It generates ABAP programs from defined conversion rules.
- It provides clear user guidance.
- It supports importing data from both the frontend and the application server.
- It can record SAP entry or change transactions and use the resulting recording as the basis for data migration.
- It is available free of charge to SAP customers and SAP partners.

## 9.2 Migrating Data Using SAP Legacy System Migration Workbench

In this section, we'll describe step by step through a concrete example how you can migrate your data using SAP Legacy System Migration Workbench.

**Migrating customer master data**

Our task is to transfer customer master data from a source system to an SAP system. We assume that the source data has already been extracted and is available in two worksheets of a Microsoft Excel file named *Customers.xlsx*.

The Customer_Header worksheet contains a header record for each customer, as shown in Figure 9.2.

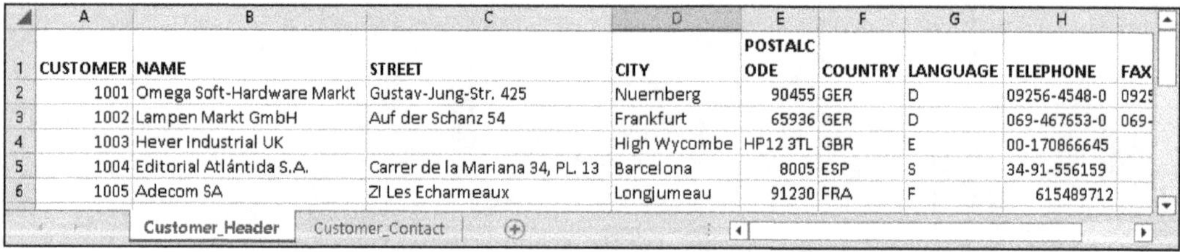

| | A | B | C | D | E | F | G | H | |
|---|---|---|---|---|---|---|---|---|---|
| 1 | CUSTOMER | NAME | STREET | CITY | POSTALC ODE | COUNTRY | LANGUAGE | TELEPHONE | FAX |
| 2 | 1001 | Omega Soft-Hardware Markt | Gustav-Jung-Str. 425 | Nuernberg | 90455 | GER | D | 09256-4548-0 | 0925 |
| 3 | 1002 | Lampen Markt GmbH | Auf der Schanz 54 | Frankfurt | 65936 | GER | D | 069-467653-0 | 069- |
| 4 | 1003 | Hever Industrial UK | | High Wycombe | HP12 3TL | GBR | E | 00-170866645 | |
| 5 | 1004 | Editorial Atlántida S.A. | Carrer de la Mariana 34, PL. 13 | Barcelona | 8005 | ESP | S | 34-91-556159 | |
| 6 | 1005 | Adecom SA | ZI Les Echarmeaux | Longjumeau | 91230 | FRA | F | 615489712 | |

Customer_Header | Customer_Contact

**Figure 9.2** Migrating Customer Master Data: Example Header Records (Excerpt)

The Customer_Contact worksheet contains one or more data records with contact data for each customer, as shown in Figure 9.3.

| | A | B | C | D | E | F | G |
|---|---|---|---|---|---|---|---|
| 1 | CUSTOMER | NAME | FIRSTNAME | TITLE | DEPARTMENT | POSITION | TELEPHONE |
| 2 | 1001 | Schmidt | Ralf | Herr | Einkauf | Leiter Einkauf | 09256-45485 |
| 3 | 1001 | Huber | Luise | Frau | Einkauf | Einkäuferin | 09256-45486 |
| 4 | 1002 | Luchs | Harald | Herr | Organisation | Marketingleiter | 069-467653-145 |
| 5 | 1002 | Wolf | Susanne | Frau | Einkauf | Leiterin Einkauf | 069-467653-146 |
| 6 | 1003 | Winchester | George | Mr | Vorstand | Vorstand Einkauf | 170877755 |
| 7 | 1003 | Smith | Judy | Ms | Finanzen | Leiterin Finanzbuchhaltung | 170877756 |
| 8 | 1004 | Lopez | Juan Carlos | Sr. | Geschäftsführung | Vorstand Einkauf | 34-91-556153 |

Customer_Header | Customer_Contact

**Figure 9.3** Migrating Customer Master Data: Example Contact Records (Excerpt)

9

## 9.2.1 Getting Started with SAP Legacy System Migration Workbench

**Start transaction**

Use Transaction LSMW to launch SAP Legacy System Migration Workbench. When you call this transaction for the first time, you'll see a welcome message in a dialog box. You're informed that SAP Legacy System Migration Workbench is a tool that supports you in transferring data from non-SAP systems.

After confirming this dialog box, you'll be taken to the initial screen, as shown in Figure 9.4.

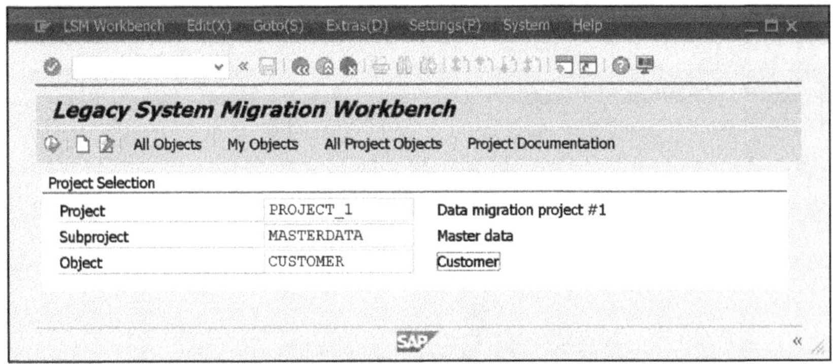


**Figure 9.4** SAP Legacy System Migration Workbench: Initial Screen

**Project, subproject**

On this initial screen, you can see that the *data migration objects* (i.e., the data units to be migrated) in SAP Legacy System Migration Workbench are organized by project and subproject. A project can contain any number of subprojects, and a subproject can contain any number of data migration objects. You can use this structure according to your individual requirements.

We cannot provide a general recommendation for the use of this structuring option. However, it does not make sense to generate many micro-projects and thus expose yourself to the risk of "uncontrolled growth."

Example

For our example, let's choose the following structure:

- **Project**
  - Identifier: PROJEKT_1
  - Name: Data Migration Project #1
- **Subproject 1**
  - Identifier: MASTER DATA
  - Name: Master data
- **Subproject 2**
  - Identifier: TRANSACTION DATA
  - Name: Transaction data

Object

A subproject can contain any number of objects. *Object* refers to a business data unit (e.g., customer master, material master, financial document, etc.) including all definitions required for data migration (e.g., source, target, field assignments, conversion rules, etc.).

Creating a project, subproject, object

So, for our example, you first need to create a project with the identifier PROJECT_1 and the name "Data Migration Project #1". For this purpose, you need to place the cursor in the **Project** field and click the button (**Create Entry**). Enter the data in the dialog box that follows, as shown in Figure 9.5.

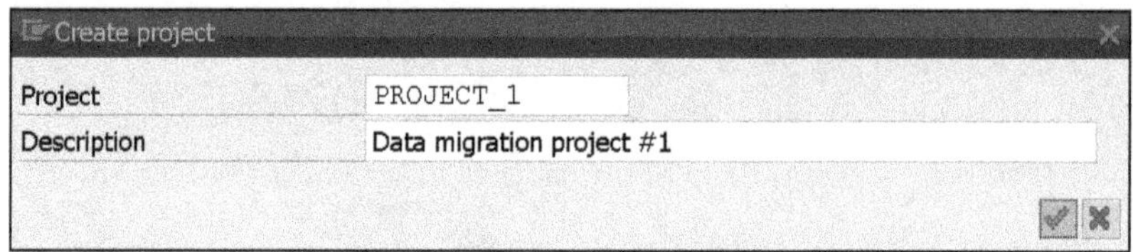


**Figure 9.5** SAP Legacy System Migration Workbench: Creating a Project

Then, proceed in the same way to create the subproject with the identifier MASTERDATA and the name "Master data" as well as create the object with the identifier CUSTOMER and the name "Customer." The identifiers for projects, subprojects, and objects can be chosen completely freely within the maximum permissible length of 15 characters.

Useful features in the initial screen

Before we continue, let's briefly review a number of useful features available to you on the initial screen of SAP Legacy System Migration Workbench (shown earlier in Figure 9.4):

- The **All Objects** button takes you to an overview of all projects, subprojects and objects already created.
- The **My Objects** button provides an overview of all objects created by you (under the current user ID).

- The **All Project Objects** button takes you to an overview of all subprojects and objects of the selected project. From this overview, you can navigate to an object, for example, if you double-click on the name of an object. Figure 9.6 shows this overview for our example project PROJECT_1 with all its subprojects and objects.
- The **Project Documentation** button—if available—shows all the documentation you have created for the individual dialog boxes and work steps. You can print, send, or save the project documentation in various file formats.



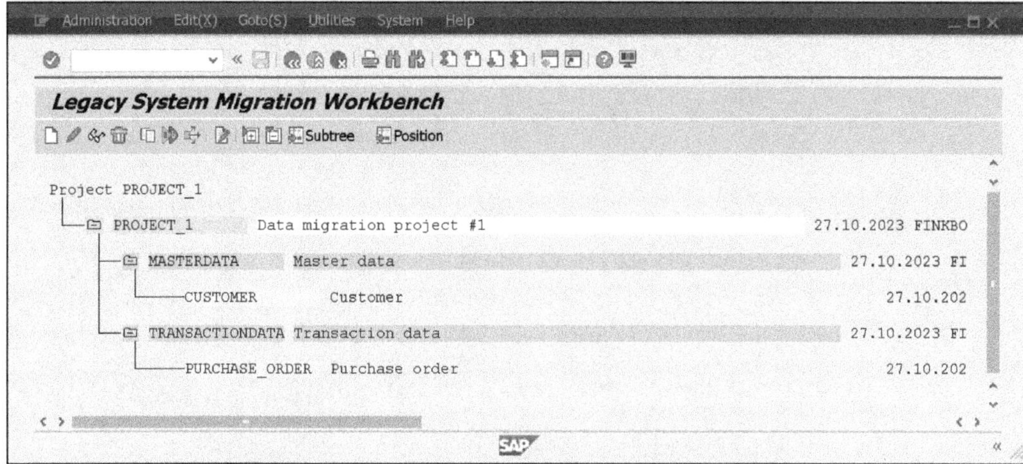


**Figure 9.6** SAP Legacy System Migration Workbench: Overview of a Project

- The icon (**Documentation**) allows you to create your own annotations. In the dialog box that is displayed, you can store your personal documentation.

Administration

The **Administration** function, which you can access from the start screen by following the menu path **Goto • Administration**, goes a step further than the **All Project Objects** function. With this function, you'll see an overview of all projects defined in SAP Legacy System Migration Workbench.

Furthermore, you can create, edit, display, delete, copy, or rename projects, subprojects, objects, and reusable rules (Section 9.2.8). Double-click on an entry to go to the display or edit mode of an entry.

As the name suggests, the administration function is used for the general administration of projects and their components.

When you position the cursor on an entry, you can create a note by clicking the icon (**Documentation**). For each edit, the name of the person who made the last change and the date of the last change are recorded.

**Authorization concept**

Since SAP Legacy System Migration Workbench ultimately manipulates the SAP system's dataset, the access to this functionality must be controlled; however, because SAP Legacy System Migration Workbench is integrated into the SAP authorization concept, this control can be easily managed. Two authorization objects are available for this purpose, which are explained in Table 9.1.

| Authorization Object | Authorization Fields | Function |
|---|---|---|
| B_LSMW<br>"LSMW: Transaction and Activity" | TCD | For TCD, always specify the value "LSMW". |
| | ACTVT | The ACTVT field accepts four values:<br>▪ "02" Modify; change and copy objects<br>▪ "03" Display; show projects and work steps<br>▪ "06" Execute; authorization to read, convert and import data<br>▪ "36" Extended maintenance; authorization for all SAP Legacy System Migration Workbench functions |
| B_LSMW_PRO<br>"LSMW: Project" | PROJECT | You can restrict access to the projects you name in the Project field. |

**Table 9.1** Authorization Objects of SAP Legacy System Migration Workbench

[!]

**No More Authorization Profiles in SAP S/4HANA**

The four authorization profiles delivered with SAP R/3 and SAP NetWeaver, namely, B_LSMW_SHOW, _LSMW_EXEC, B_LSMW_CHG, and B_LSMW_ALL, no longer exist in SAP S/4HANA.

[!]

**Additional Authorizations**

In addition to the general authorizations for SAP Legacy System Migration Workbench, you also need other authorizations:

- For field mapping, you need the development authorization (S_DEVELOP) for local development classes/packages ($TMP). You need this authorization, among other things, to change code in the field mapping and to generate the conversion programs.

- To read and write SAP Legacy System Migration Workbench files on the application server, you need the S_DATASET authorization object.

Especially for production systems, the S_DEVELOP authorization object should always be viewed critically. The person responsible for the authorization concept should limit the authorizations to only the most necessary ones.

### 9.2.2 User Guidance: Primary Steps in Data Migration

Once you have created or selected your project including its subprojects and objects, press Enter or click **Continue** to access SAP Legacy System Migration Workbench user guidance, as shown in Figure 9.7. Note that, even though the range of functions is identical, the appearance of your personal menu may differ slightly depending on how it is displayed.



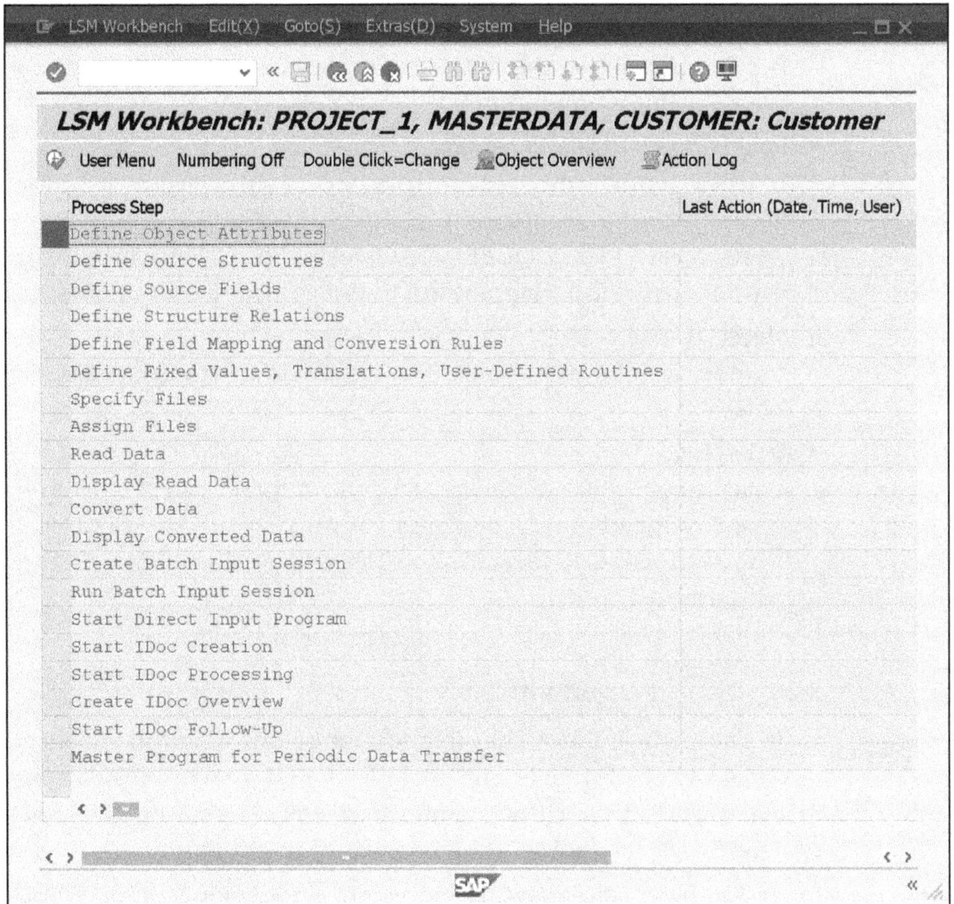


**Figure 9.7** SAP Legacy System Migration Workbench: User Guidance

Navigation

From this screen, SAP Legacy System Migration Workbench guides you through each step of the data migration: You can navigate to the individual process steps and then return here again. The selection button on the left side of the screen jumps to the next step to be performed. However, this button should only be regarded as a non-binding recommendation. You can return to any already executed step at any time, if necessary.

The following functions are available on this screen:

- (**Execute**)
  This function executes the step where the selection button is currently located. Alternatively, you can double-click on an entry.
- **User menu**
  You can make an individual selection of the displayed work steps. Of course, you can show hidden steps again at any time. The **Primary steps** button automatically activates all the steps that are mandatory for converting data. See also Section 9.6.1.
- **Numbering On/Off**
  This function allows you to number the selected work steps consecutively or to hide the numbering.
- **Double-click = Display/Change**
  You can determine whether the display mode or the change mode is to be selected in the further process by means of a double-click. When you create an object, it makes sense to select **Double-click = Change** here. In this way, you can avoid having to switch to change mode in every work step.
- **Object Overview**
  You'll see all information about the selected object at a glance (Section 9.2.18).
- **Action Log**
  A detailed overview (including date, time, and user name) of all work steps already performed is displayed. You can reset the action log via the menu path **Extras • Reset Action Log**. This action is noted with reference to the user and the date.

### 9.2.3 Defining Object Attributes

In the **Change Object Attributes** step, you can specify what data should be migrated and how it is to be imported into the SAP system. In SAP Legacy System Migration Workbench terminology, you are selecting the *object type* and the *import method*.

Object type and import method

In the **Attributes** screen area, all entries are already preset. In the **Object type and Import method** screen area, you want to choose the **Standard**

**batch/direct input** import method and use the [F4] input help to select the **Customer master** object (identifier 0050). After a selection has been made, you must select the method using the [F4] input help. If the object has only one method, this method is automatically selected. In our example, two methods exist for the **Customer master** object. We'll choose the default method 0000, which uses the standard batch input program RFBIDE00. This report is displayed after you make a selection in the **Program name** field with the program type B for batch input, as shown in Figure 9.8.

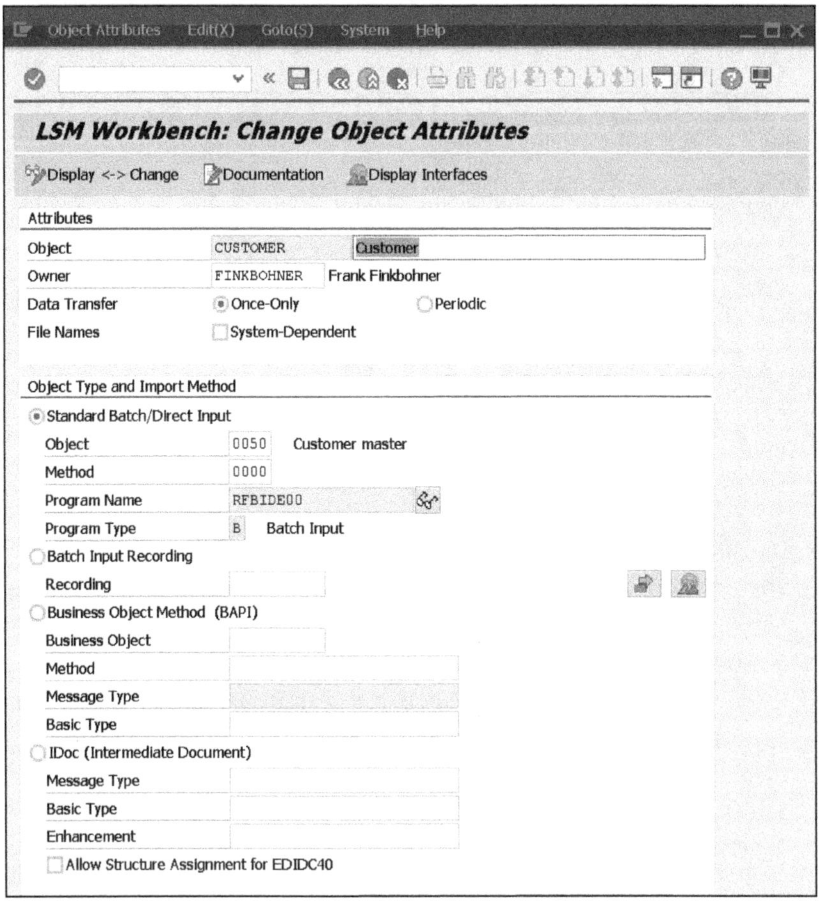

**Figure 9.8** SAP Legacy System Migration Workbench: Changing Object Attributes

Save your changes by clicking the **Save** button.

In the case of batch input and direct input, you can click the icon (**Display**) to navigate to the ABAP Editor, view the documentation of the program, and access detailed information about the functionality of this program.

**Other import methods**

SAP Legacy System Migration Workbench supports other import methods, such as batch input recordings, the business object method (BAPI) and IDoc (Intermediate Document). For this reason, using different import methods for different data objects may make sense.

**Criteria for selecting the import method**

The following aspects may play a role in the selection of an import method:

- **Availability**
  Not all data import methods are available for every data object. In the **Define Object Attributes** step, you can click the **Display Interfaces** button to display all the interfaces available for a data object and transfer the corresponding information, as shown in Figure 9.9.

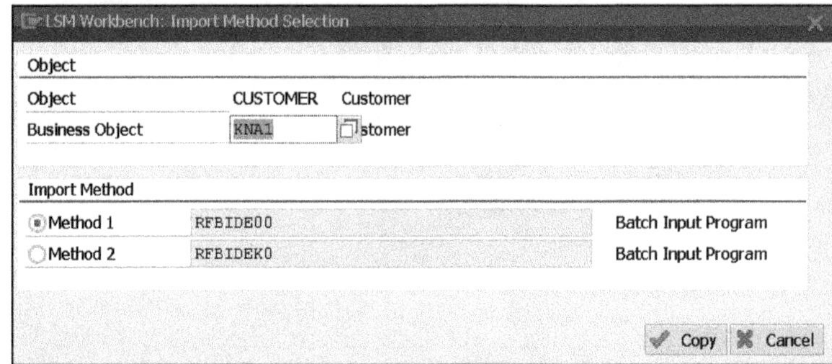


**Figure 9.9** SAP Legacy System Migration Workbench: Available Interfaces for a Business Object

- **User comfort**
  The batch input technique provides convenient options for reprocessing faulty data.
- **Runtimes**
  If the volume of data is very large, direct input is preferable to batch input because the batch input technique may result in excessively long runtimes. A rough guide for estimating the time required is 3,000 to 5,000 transactions per hour, although this value can vary greatly depending on the hardware.

**Recording only for simply structured source data**

- **Complexity**
  If the data from the source system is structured simply enough to be stored in a table, the recording technique may be preferable. With this technique, field mapping (Section 9.2.7) usually does not cause any difficulties, since normally the number of SAP fields to be populated is manageable.
- **Flexibility**
  However, the recording technique should only be used if the associated SAP transaction always provides the same screen sequences for different data records. If your source data is structured in such a way that a header

record includes a variable number of item records, the recording technique is definitely out of the question.

Note the following: If you choose the BAPI or IDoc import method, the system checks whether a *partner profile* already exists for the preset partner (Section 9.5) and the selected message type when you save. If this is not the case, the system tries to create it. A partner profile is a technical link between a sender or recipient of a message (partner) and a message type. Accordingly, the partner profiles define who is allowed to send or receive which types of messages.

**Other attributes of the object**

In addition to the essential information about the object type and import method, you can define a number of other useful attributes in this step (shown earlier in Figure 9.8). For example, you can enter a name for the object or change it.



An entry in the **Owner** field adds the project to the list of all projects you have created. You'll then find the project in the initial screen under **My objects.**

**Periodic data transfer**

Moreover, you can specify whether the data transfer should take place once or periodically. The normal case is a one-time data transfer. For periodic data transfers, no files can be read from the frontend. In this case, you'll be offered the additional **Master program for periodic data transfer** step in the list of work steps.

**System-dependent file names**

In addition, you can determine whether the file names should be defined as *system-dependent*. If so, you can later enter your own file names in each SAP system. This option is quite useful when you migrate data to multiple SAP systems.

Now that we have determined, in this step, that we want to migrate the customer master data using the batch input technique, we can turn to defining the source data.

### 9.2.4 Defining Source Structures

With the methods presented so far, as the user, you had to convert the data into the SAP format. The SAP system does not need any information about the format of the data exported from the source system. Now, SAP Legacy System Migration Workbench is supposed to take over the task of conversion and must therefore "know" how the data of the source system is structured.

**Two types of data records**

The source data for an application object usually consists of one or more *record types*. Typical record types are header records and item records. These record types are referred to in SAP Legacy System Migration Workbench as *source structures*. In our example, there are two record types: the header record per customer (shown earlier in Figure 9.2) and the contact record per contact (shown earlier in Figure 9.3), which formally is an item record, since there can be any number of contact persons per customer.

In the **Define Source Structures** step, the source structures of the object are defined with a name, description and hierarchical relations. From the navigation screen (shown earlier in Figure 9.7), you call the **Define Source Structures** step, click on the button (**Create a Structure**), and create a source structure with the HEAD identifier and the name "Customer - Header." Then, position the cursor on the source structure you just created and click the icon (**Create a Structure**) again. You'll be asked whether the second source structure should be on the same level or on a lower level. Choose **Lower Level** and enter the CONTACT identifier as the source structure and the name "Customer - Contact." The result is shown in Figure 9.10.

**Equal or subordinate?**

When should you choose the same level, and when should you choose a lower level? In general, the following applies: Record type 2 is subordinate to record type 1 if exactly one record of record type 1 exists for each record of record type 2. In our example, we have exactly one associated header record for each contact record.

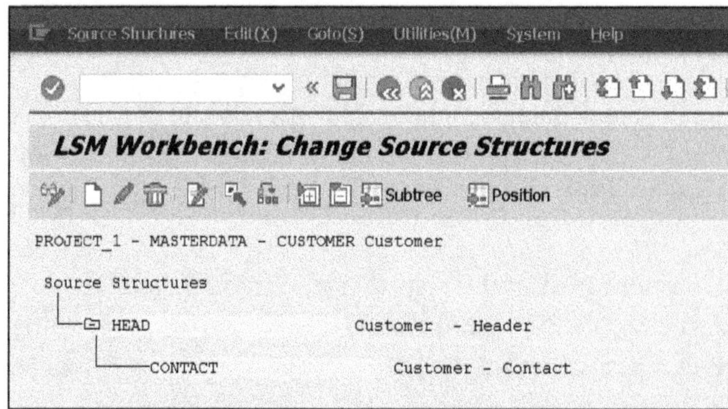


**Figure 9.10** SAP Legacy System Migration Workbench: Changing Source Structures

**Other functions**

Furthermore, in this step, you have the option to create new source structures and to change, reassign, or delete them. Buttons are available for all these functions.

### 9.2.5 Defining Source Fields

A source structure normally consists of several fields. These fields are referred to as *source fields* in SAP Legacy System Migration Workbench.

**Defining source fields and properties**

In the **Define Source Fields** step (shown earlier in Figure 9.7), the corresponding source fields are created for the source structures defined in the previous step and provided with the desired properties. In our example, the fields CUSTOMER NUMBER, NAME, STREET, CITY etc. are to be defined for the HEAD source structure. Thus, we map the structure of our source data (shown earlier in Figure 9.2 and Figure 9.3) into SAP Legacy System Migration Workbench.

**Attributes of a source field**

A source field is described in SAP Legacy System Migration Workbench by the *field name*, *field label*, *field length*, and *field type*. This information can be entered individually (i.e., field by field).



For this task, you must position the cursor on the HEAD source structure, click the [icon] icon (**Create field**) and enter the required attributes for the CUSTOMER field. Then, position the cursor on the CUSTOMER source field, click the **Create field** button again, and enter the required attributes for the FIELD NAME field, etc. to achieve the result shown in Figure 9.11.

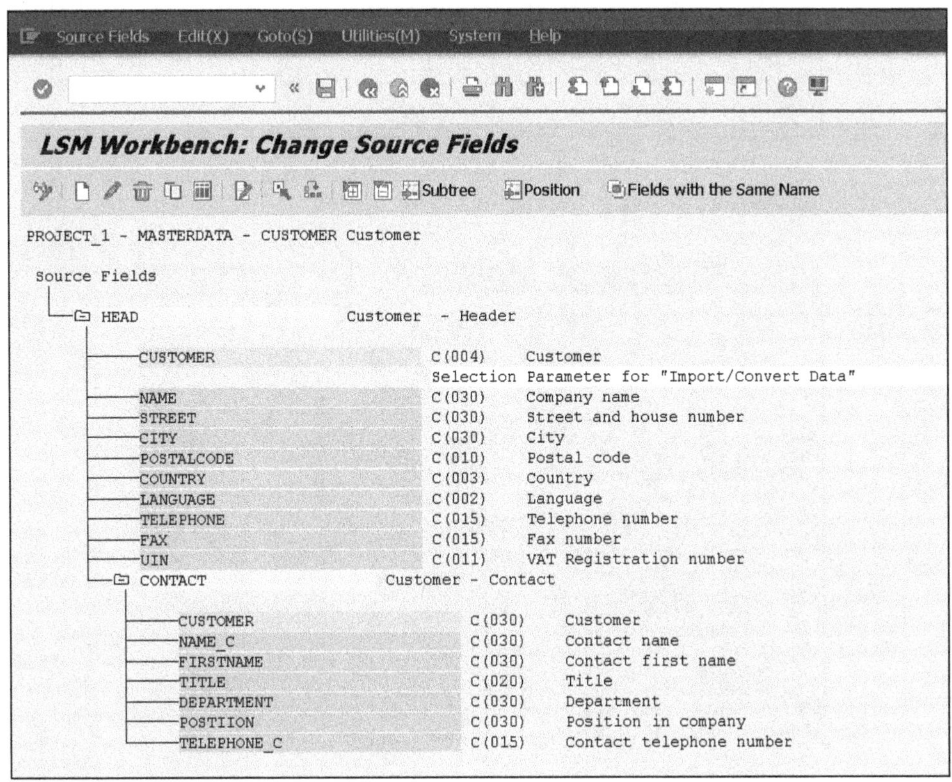


**Figure 9.11** SAP Legacy System Migration Workbench: Changing Source Fields

We'll describe the individual options for defining and maintaining source fields in the following sections.

### Creating Source Fields Individually

As described earlier, position your cursor on a source structure or on an already created source field and click the icon (**Create Field**). This step will open the dialog box shown in Figure 9.12. You can define the **Field Length** field as you like, and you can select the **Field Type** using the F4 input help, as shown in Figure 9.13.

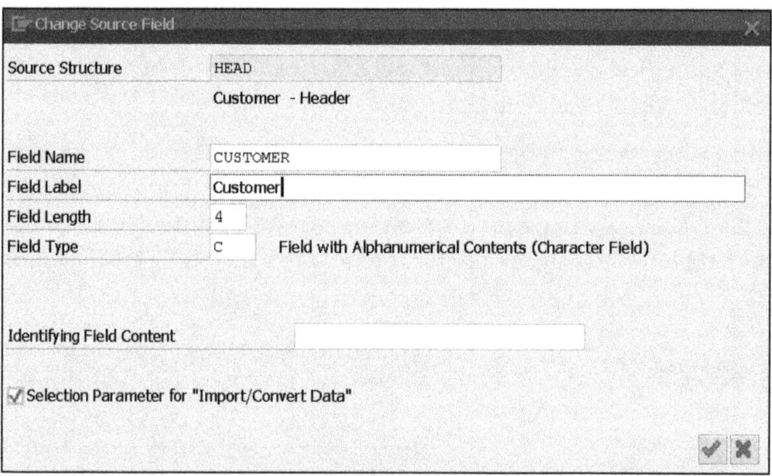


**Figure 9.12** SAP Legacy System Migration Workbench: Creating Source Fields Individually

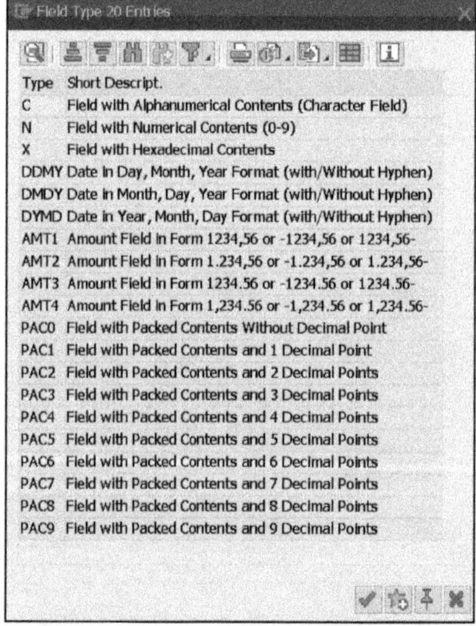
Field Type 20 Entries

| Type | Short Descript. |
|---|---|
| C | Field with Alphanumerical Contents (Character Field) |
| N | Field with Numerical Contents (0-9) |
| X | Field with Hexadecimal Contents |
| DDMY | Date in Day, Month, Year Format (with/Without Hyphen) |
| DMDY | Date in Month, Day, Year Format (with/Without Hyphen) |
| DYMD | Date in Year, Month, Day Format (with/Without Hyphen) |
| AMT1 | Amount Field in Form 1234,56 or -1234,56 or 1234,56- |
| AMT2 | Amount Field in Form 1.234,56 or -1.234,56 or 1.234,56- |
| AMT3 | Amount Field in Form 1234.56 or -1234.56 or 1234.56- |
| AMT4 | Amount Field in Form 1,234.56 or -1,234.56 or 1,234.56- |
| PAC0 | Field with Packed Contents Without Decimal Point |
| PAC1 | Field with Packed Contents and 1 Decimal Point |
| PAC2 | Field with Packed Contents and 2 Decimal Points |
| PAC3 | Field with Packed Contents and 3 Decimal Points |
| PAC4 | Field with Packed Contents and 4 Decimal Points |
| PAC5 | Field with Packed Contents and 5 Decimal Points |
| PAC6 | Field with Packed Contents and 6 Decimal Points |
| PAC7 | Field with Packed Contents and 7 Decimal Points |
| PAC8 | Field with Packed Contents and 8 Decimal Points |
| PAC9 | Field with Packed Contents and 9 Decimal Points |

**Figure 9.13** SAP Legacy System Migration Workbench: Possible Types of Source Fields

**Date fields, value fields**

If you define a field as a date field (field types DDMY, DMDY, or DYMD) or value field (field types AMT1, AMT2, AMT3, or AMT4), later when importing the data (Section 9.2.13), you can choose whether you want to convert date values to the internal date format (YYYYMMDD; the 4-digit year number followed by a 2-digit month number followed by a 2-digit day number) and convert value fields to the calculation format (1234.56; without a thousands grouping character and with a period as decimal point).

**Identifying field content**

If a file contains source data for multiple source structures, SAP Legacy System Migration Workbench requires additional information on how to identify a record. For this purpose, a value must be specified in the **Identifying Field Content** field, which specifies which source structure belongs to this record, as shown in Figure 9.12. Identifying field content may only be specified for one field per source structure.



**Selection parameters**

For the fields of structures in the top hierarchy level—in the example, for the fields of the HEAD source structure—a checkmark can be set next to **Selection Parameter for "Import/Convert Data"**. If you select this checkbox, the relevant field is made available as a selection parameter in the **Read data** and **Convert data** steps. This field is usually used for tests to limit the data scope there.

### Defining Source Fields in a Table

Instead of maintaining each source field individually, you can edit all fields for a source structure at once. For this task, position your cursor on a source structure or on a source field that has already been created and click the icon (**Table Maintenance**). A screen similar to the one shown in Figure 9.14 appears.

**Default values**

If you enter the **Field Name**, **Type**, and **Length** fields and confirm by pressing Enter, the field name is taken over.

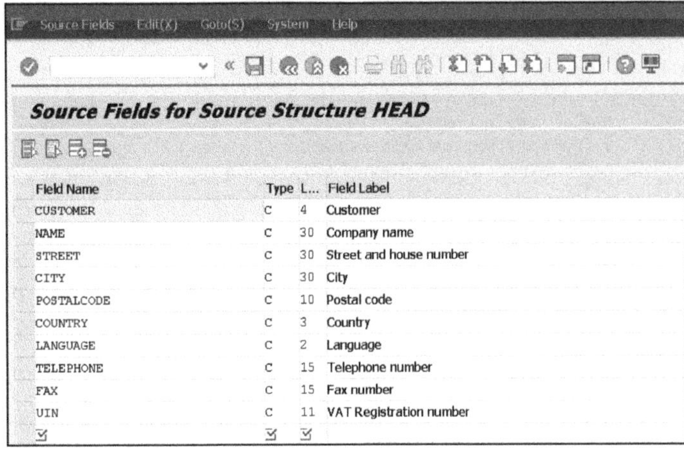


| Field Name | Type | L... | Field Label |
|---|---|---|---|
| CUSTOMER | C | 4 | Customer |
| NAME | C | 30 | Company name |
| STREET | C | 30 | Street and house number |
| CITY | C | 30 | City |
| POSTALCODE | C | 10 | Postal code |
| COUNTRY | C | 3 | Country |
| LANGUAGE | C | 2 | Language |
| TELEPHONE | C | 15 | Telephone number |
| FAX | C | 15 | Fax number |
| UIN | C | 11 | VAT Registration number |

**Figure 9.14** SAP Legacy System Migration Workbench: Defining Source Fields in a Table

### Copying Source Fields from Other Sources

In addition, SAP Legacy System Migration Workbench allows you to copy the descriptions of source fields from other sources to reduce the maintenance work. For this task, position your cursor on a source structure or on a source field that has already been created (shown earlier in Figure 9.11) and click the [icon] icon (**Copy Fields**). In the dialog box shown in Figure 9.15, you'll be asked to select a source.

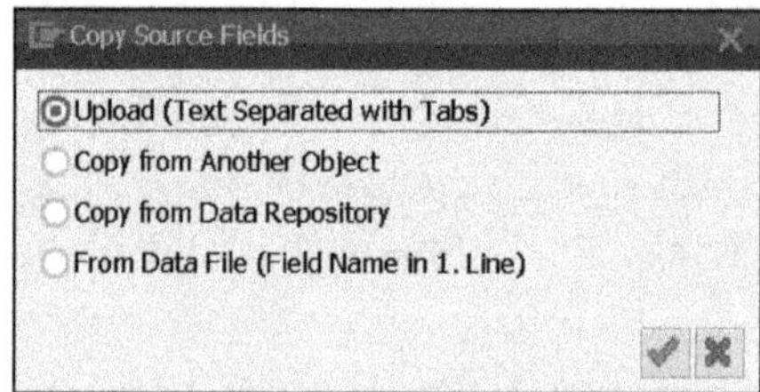


**Figure 9.15** SAP Legacy System Migration Workbench: Copying Source Fields

**Copying structure descriptions**

The following sources are available:

- **Upload (text separated with tabs)**
  In this case, the descriptions of the source fields are to be stored in a text file whose columns are separated by tabs, as shown, for example, in Figure 9.16.

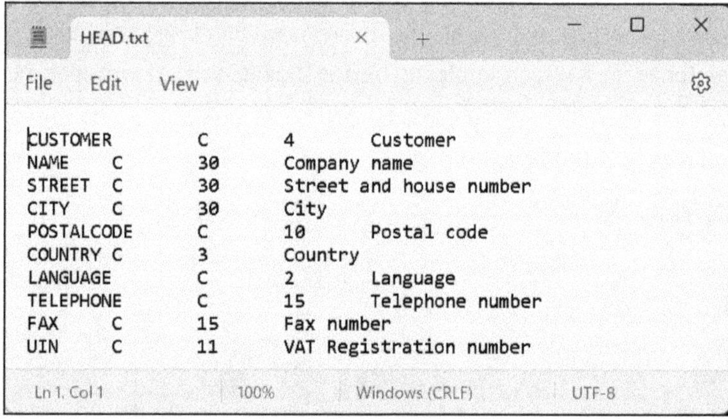


```
CUSTOMER	C	4	Customer
NAME	C	30	Company name
STREET	C	30	Street and house number
CITY	C	30	City
POSTALCODE	C	10	Postal code
COUNTRY	C	3	Country
LANGUAGE	C	2	Language
TELEPHONE	C	15	Telephone number
FAX	C	15	Fax number
UIN	C	11	VAT Registration number
```

**Figure 9.16** SAP Legacy System Migration Workbench: Copying Source Fields from a Text File (Separated Using Tabs)

- **Copy from another object**
  You can copy the source fields from a source structure of another object of SAP Legacy System Migration Workbench. If you choose this variant, in the step that follows, you'll be asked to select the corresponding object (or, more precisely, select a project, subproject, or object).

- **Copy from data repository**
  You can copy the source fields from a structure of the ABAP Dictionary in an SAP system. When using this variant, you're prompted in the step that follows to select the desired structure of the ABAP Dictionary.
- **From data file (Field name in 1. line)**
  You can copy the source fields from a file that contains the associated source data. This file must be in **Text (tab-delimited) format (*.txt)** on the PC and contain the field names in the first line. For all fields, the field type is C, and the field name is assigned as the field label, while the largest length of a field content of the respective field is calculated for the field length.

**Fields with Identical Names**

When defining source fields, you have the option of highlighting fields with the same name by using the corresponding button (shown earlier in Figure 9.11). This option allows you to verify that these fields are actually related in content. If, in the further course of the migration, the source data is available in multiple files, the data records are automatically linked via fields with the same name (Section 9.2.12). You should therefore rename a field that is not related in content to a key field. Otherwise, when you import and convert the data records, fields could be linked incorrectly or not at all.

### 9.2.6 Defining Structure Relations

In Section 9.2.4, we explained that the data from the source system for an object consists of one or more record types, and these record types are referred to as *source structures* in the context of SAP Legacy System Migration Workbench. Consequently, the record types of the target format—the SAP format—are called *target structures*.

**Relations between source and target structures**

In the **Define Structure Relations** step, the relations between source and target structures are defined. The possible target structures are already defined when you select the object type and the import method (shown earlier in Figure 9.8). Our **Customer master** selection specified the format into which the data must be put in order to be processed by the RFBIDE00 batch input program. You can view the result by calling the **Define Structure Relations** step from the navigation screen, as shown in Figure 9.7 (earlier) and Figure 9.17.

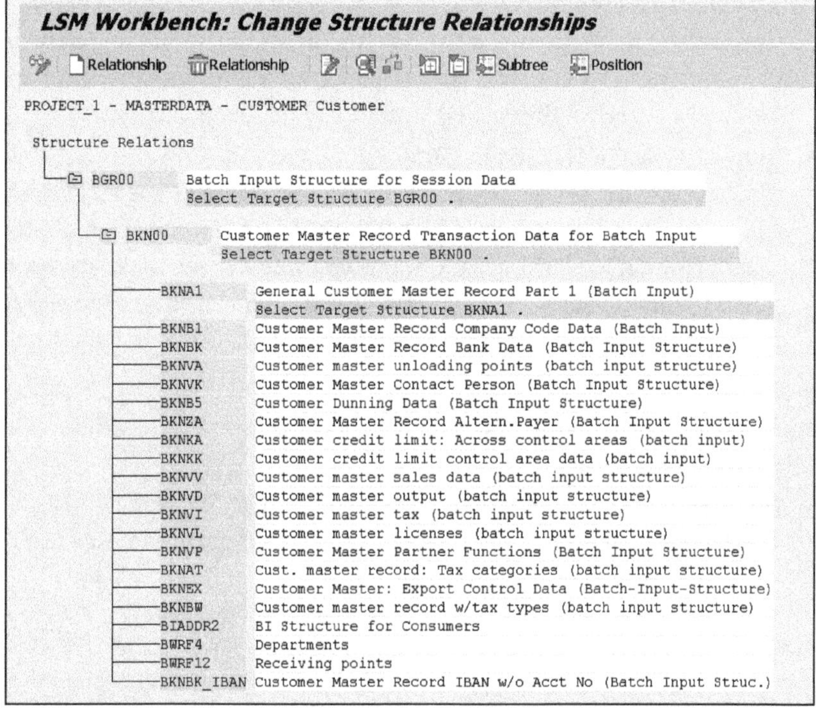


**Figure 9.17** SAP Legacy System Migration Workbench: Defining Structure Relations, Initial State

**Mandatory target structures**

Some target structures that must be selected (*mandatory target structures*). In such a case, you'll receive the message "Select Target Structure <xy>", as shown in Figure 9.18.

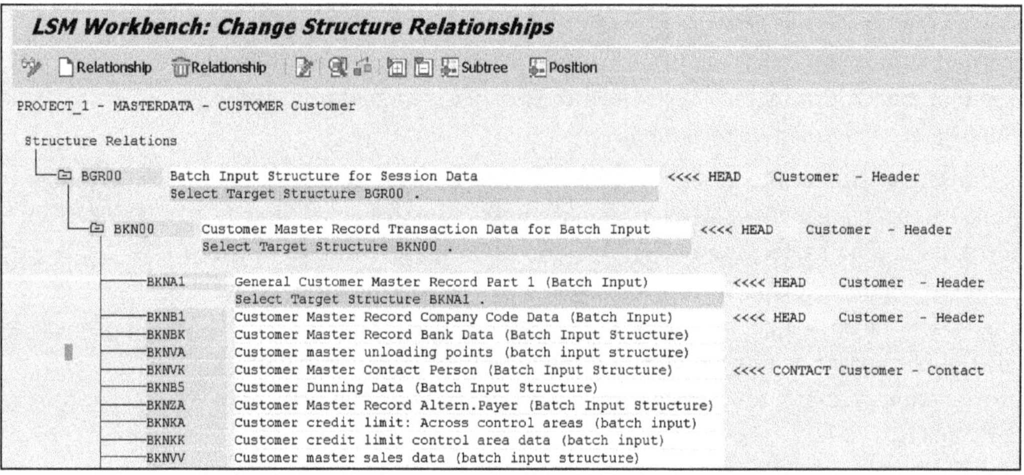


**Figure 9.18** SAP Legacy System Migration Workbench: Defined Structure Relations

To set structural relations, you must position the cursor on a target structure and click the Relationship button (**Create relationship**). A dialog box is displayed offering you the source structures already created for selection. If you want to change the relation, then first remove the already existing relation using the Relationship icon (**Delete relationship**).

Checking structure relations

In addition, you can use the icon (**Check**) to check the structure relations for errors. You'll receive either an error message in the status bar or the message: "There are no errors in the structure relations."

Selecting specific target structures

In our example, we assigned the source structure HEAD to the target structures BGR00, BKN00, BKNA1, and BKNB1, respectively, while the source structure CONTACT is assigned to the target structure BKNVK. The result is shown in Figure 9.18.



Conversely, our definition means that the customers we want to create in the SAP system will *not* contain a lot of information, missing specifically all the information contained in the structures that we have not selected, such as bank data, unloading points, dunning data, alternative payers, etc.

Control records

Finally, we want to point out an important point regarding this work step: Many batch input and direct input programs use a control record called BGR00 or BI000. You should always assign the source structure of the top hierarchy level (header structure) to this record.

### 9.2.7 Defining Field Mapping and Conversion Rules

Two steps in one

In this work step, your task is to assign source fields to the target fields and to determine how the field contents should be converted. This step is also known as *field mapping* and, strictly speaking, two steps are involved:

1. Assigning the source fields to the target fields
2. Determining the conversion rules

However, since the relation between the two substeps is close, they are combined into one step in SAP Legacy System Migration Workbench.

All target fields at a glance

If you call this step from the navigation screen (shown earlier in Figure 9.7), you'll see all the selected target structures and the associated fields displayed in a hierarchical tree structure, as shown in Figure 9.19.

Now that all available target fields are displayed, we can define the assignment of source fields to target fields, initially "on paper." This preparatory step is usually performed by your business departments, sometimes in cooperation with the SAP consultants, if the necessary application knowledge is available there.

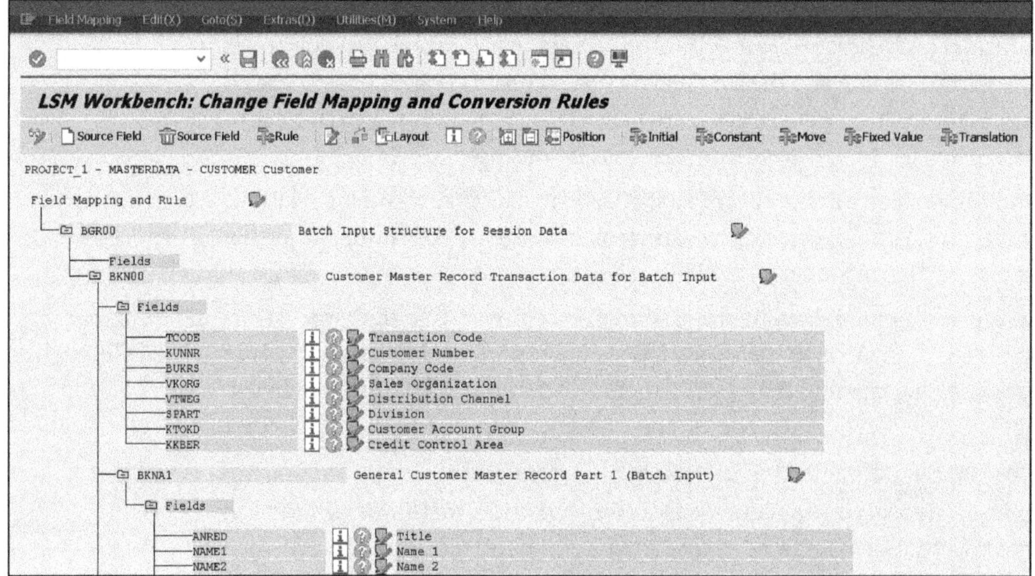


**Figure 9.19** SAP Legacy System Migration Workbench Excerpt: Field Mapping, Initial State

[+]

### Use the Object Overview for Paper Mapping

For this task, the best option is to use the **Object Overview** function, shown earlier in Figure 9.7, which we explain in Section 9.2.18, and export this overview as a table.

[»]

### Number Range Used

You'll probably need to adapt the fixed values and constants used in this example to match the configuration of your system. We used an account group for customers with an external alphanumeric number range.

In our example, we get the result shown in Table 9.2.

| No. | Target Field | Source Field | Conversion Rule |
|---|---|---|---|
| 1 | BKN00-TCODE | - | Constant XD01 |
| 2 | BKN00-KUNNR | HEAD-CUSTOMER | Prefix "D" |
| 3 | BKN00-BUKRS | - | Fixed value FV_BUKRS |

**Table 9.2** Paper Mapping: Field Assignments and Conversion Rules for Customer Master Data

| No. | Target Field | Source Field | Conversion Rule |
|---|---|---|---|
| 4 | BKN00-KTOKD | - | Constant KUNA<br>Choose suitable account group of your system for customers |
| 5 | BKNA1-NAME1 | HEAD-NAME | Transfer (MOVE) |
| 6 | BKNA1-SORTL | HEAD-NAME | Transfer (MOVE); truncated |
| 7 | BKNA1-STRAS | HEAD-STREET | Transfer (MOVE) |
| 8 | BKNA1-ORT01 | HEAD-CITY | Transfer (MOVE) |
| 9 | BKNA1-PSTLZ | HEAD-POASTALCODE | Transfer (MOVE) |
| 10 | BKNA1-LAND1 | HEAD-COUNTRY | Translation |
| 11 | BKNA1-SPRAS | HEAD-LANGUAGE | Translation |
| 12 | BKNA1-TELF1 | HEAD-TELEPHONE | Transfer (MOVE) |
| 13 | BKNA1-TELFX | HEAD-FAX | Transfer (MOVE) |
| 14 | BKNA1-STCEG | HEAD-UIN | Transfer (MOVE) |
| 15 | BKNB1-AKONT | - | Constant 140000<br>(Customers - domestic receivables) |
| 16 | BKNVK-NAME1 | CONTACT-NAME_C | Transfer (MOVE) |
| 17 | BKNVK-TELF1 | CONTACT-TELEPHONE_C | Transfer (MOVE) |
| 18 | BKNVK-ABTNR | CONTACT-DEPARTMENT | Translation |
| 19 | BKNVK-NAMEV | CONTACT-FIRSTNAME | Transfer (MOVE) |
| 20 | BKNVK-ANRED | CONTACT-TITLE | Transfer (MOVE) |
| 21 | BKNVK-PAFKT | CONTACT-POSITION | Translation |

**Table 9.2** Paper Mapping: Field Assignments and Conversion Rules for Customer Master Data (Cont.)

**Fields of the BGR00 control structure**

As an attentive reader, perhaps you noticed that not a single field of the BGR00 control structure is listed in Table 9.2. This omission is due to the fact that all fields of the BGR00 structure are regarded as *technical fields* by SAP Legacy System Migration Workbench and are thus automatically preassigned. For this reason, usually no changes need to be made to these fields.

With this paper mapping, most of the work is already done. Now, this information just needs to be communicated to SAP Legacy System Migration Workbench through the procedure presented in Table 9.3. The sequence number given in the first column refers to the sequence number used earlier in Table 9.2.

| No. | Target Field | Action |
|---|---|---|
| 1 | BKN00-TCODE | ■ As shown in Figure 9.19, position the cursor on the BKN00-TCODE field.<br>■ Then, click the Rule button.<br>■ Select the **Constant** rule in the dialog box shown in Figure 9.20.<br>■ Enter "XD01" in the dialog box that opens and confirm the entry. |
| 2 | BKN00-KUNNR | ■ As shown in Figure 9.19, position the cursor on the BKN00-KUNNR field.<br>■ Click the Source Field button (**Assign Source Field**).<br>■ Select HEAD-CUSTOMER from the list of possible source fields by double-clicking.<br>■ Then, click the **Rule** button. Select the **Prefix** rule in the dialog box shown in Figure 9.20.<br>■ Enter "D" in the dialog box that follows and confirm the entry. |
| 3 | BKN00-BUKRS | ■ As shown in Figure 9.19, position the cursor on the BKN00-BUKRS field.<br>■ Then, click the **Rule** button.<br>■ Select the **Fixed Value (Reusable)*** rule in the dialog box shown in Figure 9.20.<br>■ SAP Legacy System Migration Workbench proposes BUKRS as a name to you. Confirm this proposal.<br>■ In the dialog box that opens next, you can enter the actual value you use for this option (i.e., enter "0001"). |
| 4 | BKN00-KTOKD | Analogous to 1: Constant |
| 5 | BKNA1-NAME1 | ■ As shown in Figure 9.19, position the cursor on the BKNA1-NAME1 field.<br>■ Click the **Assign Source Field** button.<br>■ Select HEAD-NAME from the list of possible source fields by double-clicking. |

**Table 9.3** Implementing a Paper Mapping in SAP Legacy System Migration Workbench

| No. | Target Field | Action |
|---|---|---|
| | | Since SAP Legacy System Migration Workbench automatically selects the **Transfer** (MOVE) rule, no further settings need to be made. |
| 6 | BKNA1-SORTL | Analogous to 5; however, in this case, the source field is longer than the destination field. Thus, the content of the source field gets truncated during the transfer. |
| 7 | BKNA1-STRAS | Analogous to 5: **Transfer** (MOVE) |
| 8 | BKNA1-ORT01 | Analogous to 5: **Transfer** (MOVE) |
| 9 | BKNA1-PSTLZ | Analogous to 5: **Transfer** (MOVE) |
| 10 | BKNA1-LAND1 | ▪ As shown in Figure 9.19, position the cursor on the BKNA1-LAND1 field.<br>▪ Click the **Assign Source Field** button.<br>▪ Select HEAD-COUNTRY from the list of possible source fields by double-clicking.<br>▪ Then, click the **Rule** button.<br>▪ In the dialog box shown in Figure 9.20, select the **Translation (Reusable)** rule.<br>▪ SAP Legacy System Migration Workbench proposes LAND1 as the name. Confirm this proposal. |
| 11 | BKNA1-SPRAS | Analogous to 10: **Translation (Reusable)**<br>Accept the SPRAS proposal |
| 12 | BKNA1-TELF1 | Analogous to 5: **Transfer** (MOVE) |
| 13 | BKNA1-TELFX | Analogous to 5: **Transfer** (MOVE) |
| 14 | BKNA1-STCEG | Analogous to 5: **Transfer** (MOVE) |
| 15 | BKNB1-AKONT | Analogous to 1: **Constant** |
| 16 | BKNVK-NAME1 | Analogous to 5: **Transfer** (MOVE) |
| 17 | BKNVK-TELF1 | Analogous to 5: **Transfer** (MOVE) |
| 18 | BKNVK-ABTNR | Analogous to 10: **Translation (Reusable)**<br>Accept the ABTNR proposal |
| 19 | BKNVK-NAMEV | Analogous to 5: **Transfer** (MOVE) |
| 20 | BKNVK-ANRED | Analogous to 5: **Transfer** (MOVE) |

**Table 9.3** Implementing a Paper Mapping in SAP Legacy System Migration Workbench (Cont.)

| No. | Target Field | Action |
|---|---|---|
| 21 | BKNVK-PAFKT | Analogous to 10: **Translation (Reusable)**<br>Accept the PAFKT proposal |
| * These reusable rules are described in detail in Section 9.2.8. | | |

**Table 9.3** Implementing a Paper Mapping in SAP Legacy System Migration Workbench (Cont.)

When you execute the statements according to Table 9.3, you'll notice that SAP Legacy System Migration Workbench translates your instructions into ABAP code. You also have the flexibility to change or add to the ABAP code as needed. An example would be to replace the leading zero of the area code with the country code in phone numbers, for example "06227 741117" with "+49 6227 741117". In this case, you would replace the generated code BKNA-TELF1 = HEAD_TELEPHONE with:

```
Concatenate '+49' HEAD-TELEPHONE+1 into BKNA1-TELF1 separated
by space
```

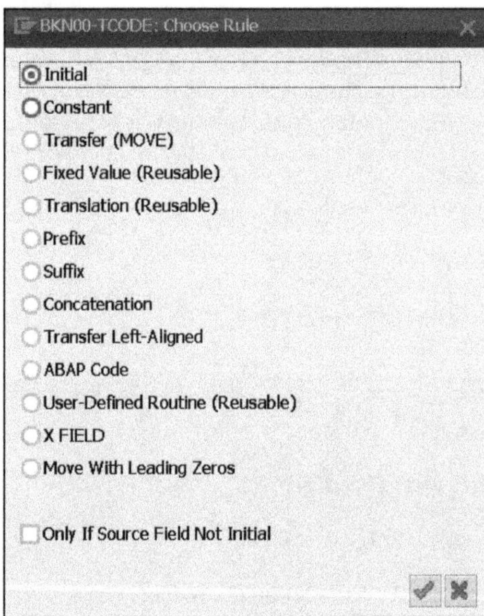


**Figure 9.20** SAP Legacy System Migration Workbench: Predefined Conversion Rules

In our example, however, no adjustments are required.

Summary

We have now worked through all the statements from Table 9.3; the result of our work is shown in Figure 9.21 and Figure 9.22.

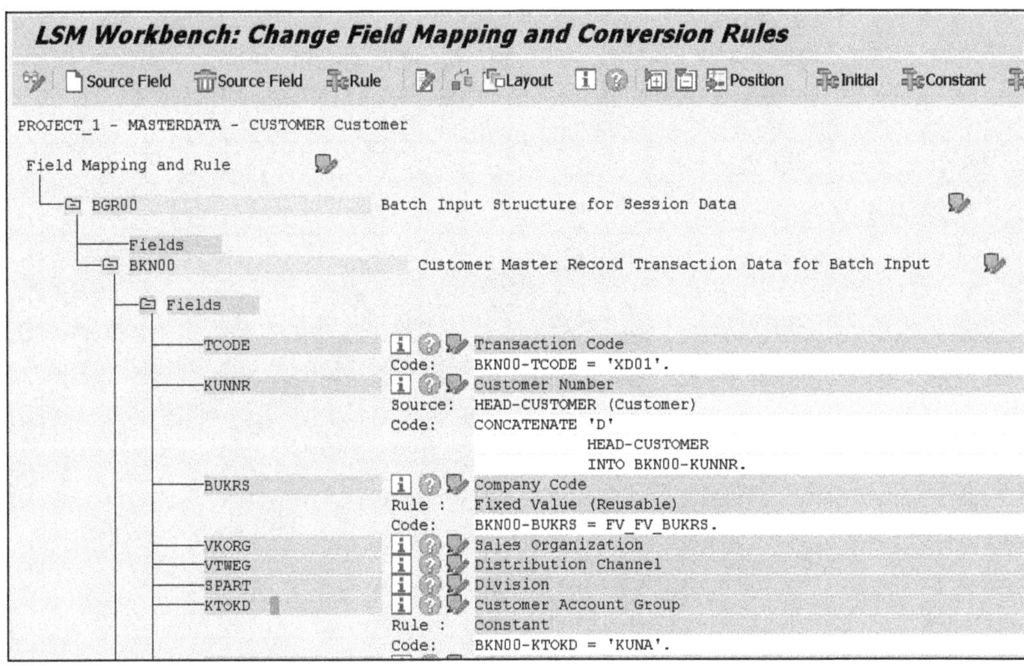


**Figure 9.21** SAP Legacy System Migration Workbench: Field Mapping (Excerpt 1)

Syntax check

You can now directly verify if you have done a good job. For this task, click the icon (**Check syntax**, see Figure 9.22). This step causes SAP Legacy System Migration Workbench to generate the data conversion program and check it for syntactical correctness. If everything was done correctly, the success message appears in the status line: "Conversion program generated successfully." Otherwise, the data conversion program is displayed and clicking the icon (**Check syntax**) displays the error(s). If only predefined rules are used and no custom ABAP code is added, syntax errors are unlikely.

Display variant: do not display initial fields

For the sake of completeness, we want to mention that, as shown in Figure 9.22, all initial fields have been hidden. This adjustment was achieved using the Layout button, which we discuss in detail in Section 9.7.1.

Now, you know the most important functions of the **Define Field Mapping and Conversion Rules** step. A number of additional functions are quite useful, the most important of which will now be described.

Auto field mapping

A very smart feature is the what's called *auto field mapping*. If you select **Extras • Auto-Field Mapping**, SAP Legacy System Migration Workbench will make proposals for mapping source fields to target fields. Using the settings screen shown in Figure 9.23, you can choose whether you want to search for fields with the same or similar name (by hit probability). Furthermore, you do not have to call the function for each field individually but can run through all fields from the cursor position or even all fields of all target structures, with or without confirmation at each individual step. You can choose whether you want to use a reusable rule or a MOVE rule, and whether you want to display proposals only for non-initial fields.

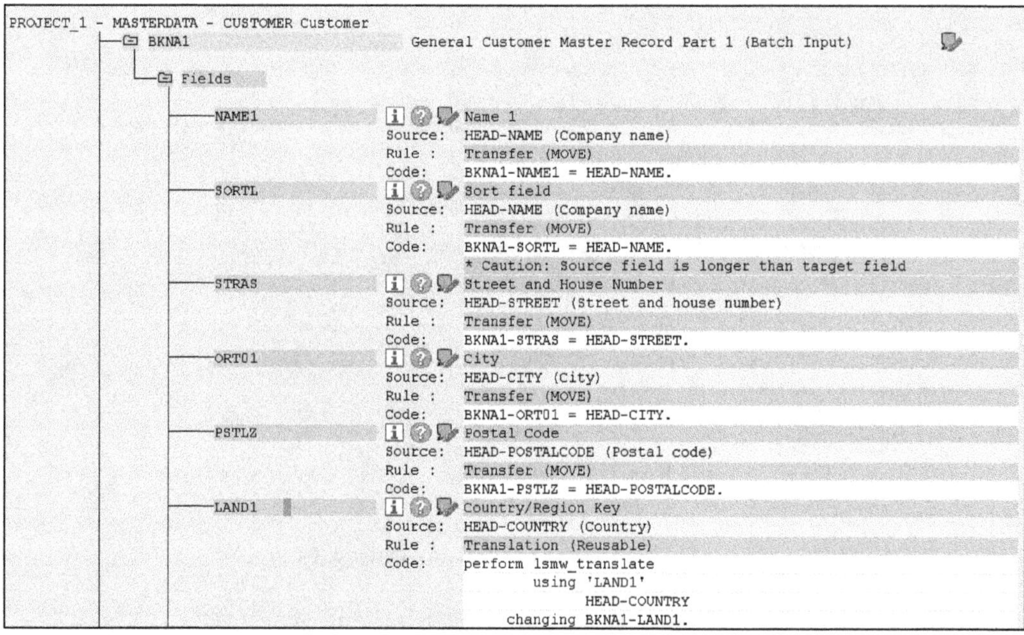


**Figure 9.22** SAP Legacy System Migration Workbench: Field Mapping (Excerpt 2)

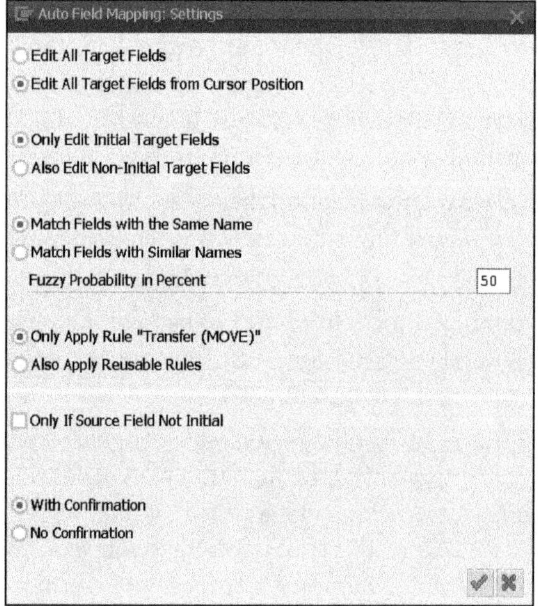


**Figure 9.23** SAP Legacy System Migration Workbench: Auto Field Mapping: Settings

[«]

**Hit Probability**

The hit probability is a measure of the similarity between two words. The more similar two words are, the higher the hit probability. For two identical words, the hit probability is therefore 100%.

If you call this function for target field BKNA1-NAME1, as in our example, you'll be presented with the proposal shown in Figure 9.24.

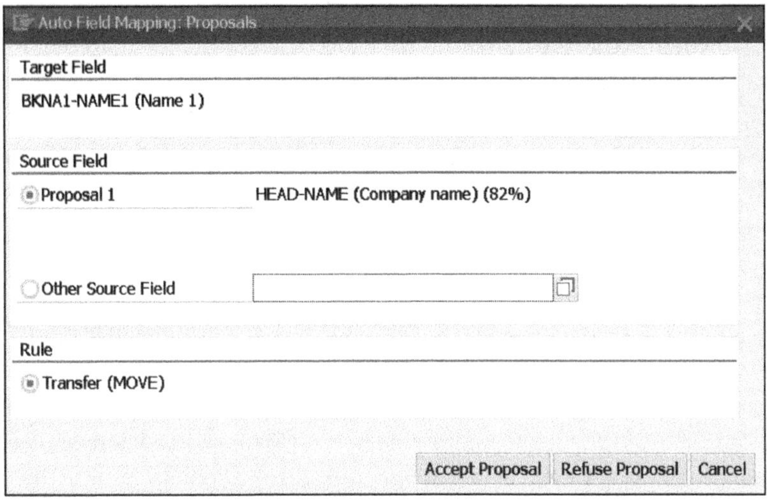

Auto Field Mapping: Proposals

Target Field

BKNA1-NAME1 (Name 1)

Source Field

Proposal 1 HEAD-NAME (Company name) (82%)

Other Source Field

Rule

Transfer (MOVE)

Accept Proposal | Refuse Proposal | Cancel

**Figure 9.24** SAP Legacy System Migration Workbench: Auto Field Mapping, Proposal

If you use a large number of fields and the field names of the source system (more precisely, the source fields) correspond or are similar to the fields of the SAP system (more precisely, the target structures), it makes sense to adapt the field names of the source fields to the field names of the target structure. The auto field mapping function speeds up your work.

Technical fields

Some target fields are preset by the system. These fields are called *technical fields* and are marked with the rule type **Default Settings**. The rules for these fields are not initially displayed in the field mapping but can be displayed using the **Layout** button (Section 9.7.1). Changes to the default assignment can seriously affect the data conversion process under certain circumstances. If you have accidentally changed the default assignment of a field, you can restore it by selecting **Extras • Restore Default**.

The following information is displayed for each target field (shown earlier in Figure 9.21 and Figure 9.22):

- SAP field name
- Any assigned source fields (**Source**)

- Rule type (fixed value, translation, etc.) (**Rule**)
- Coding (**Code**)

**Functions per target and source field**

In addition, the following functions are available for each destination or source field:

- **(Field documentation)**
  A brief description of the target field on which the cursor is located is displayed. The documentation may contain links to further information.
- **(Possible values)**
  In some cases, you get a selection list of all possible values for this target field. Whether the selection list is available depends on the definition of the target field in the Data Dictionary.
- **(Create Documentation)**
  You can maintain the documentation for a field or structure and explain the field assignment and conversion rule in more detail if necessary.
- **Unassigned source fields**
  If you use a large number of fields, you may not remember which source fields have already been assigned and which ones haven't. You can regain this overview by selecting **Extras • Unassigned source fields.**
- **Remove assignment of a source field**
  To remove an assigned source field, you want to position the cursor on a target field in the tree structure and select Source Field (**Delete source field**). If only one source field is assigned, it will be removed. If more than one source field is assigned, you'll get a list of all assigned source fields to choose from and you can select the relevant source field by double-clicking on it.

**Predefined conversion rules**

Finally, for the **Define Field Mapping and Conversion Rules** step, we'll now present the predefined conversion rules in detail, as shown ealier in Figure 9.20:

**Resetting a field**

- **Initial**
  **Initial** deletes the code associated with the target field. Likewise, all source fields associated with the target field are removed. The target field is also assigned an initial value, which can vary depending on the selected import method:
  - **Standard batch input/standard direct input**
    The target field is filled with the value of *nodata character*, which is usually defined in the control record (BGR00, BI000) and preassigned with the "/" character.
  - **Batch input recording**
    The target field is filled with the "/" value as the nodata character.
  - **BAPI, IDoc**
    The ABAP command CLEAR is applied to the target field, which fills character fields with blanks and numeric fields with zeros.

- **Constant**
  A constant value is assigned to the target field.
- **Transfer (MOVE)**
  The transfer of data from the source field to the target field is done using ABAP statement MOVE, which represents, depending on the field type, the type of transfer listed in Table 9.4 in each case.

  In addition, the conversion rules that apply to the ABAP command MOVE are used. For further details, refer to the relevant documentation of the SAP system.

Assignment



| Field Type | Transfer Type |
|---|---|
| C (character) | 1:1 transfer |
| N (numeric) | 1:1 transfer, including any leading zeros |
| Packed field | Unpack into the target field using ABAP statement WRITE ... TO ... |
| Date field | You'll be prompted in a dialog box to decide how to transfer the date field:<br>■ internal format (YYYYMMDD)<br>■ User format (e.g., 01/30/2003) |
| Value field | In case of batch input or direct input, the amount value is formatted according to the user master settings in the output format.<br>In the case of BAPIs or IDocs, the amount value is left in the internal calculation format. |

**Table 9.4** Data Transfer Type Depending on the Field Type

- **Fixed Value (Reusable)**
  The target field is assigned a *fixed value object*, i.e., a variable whose name starts with FV_. In the step **Define Fixed Values, Translations, User-Defined Routines**, this variable is provided with a concrete value. Fixed values are reusable, which means that, unlike constants, they are used in multiple objects in the project. The concrete value must be defined only at one point.
  In our example, we have used a fixed value for the **Company Code** field. Let's suppose you want to migrate data for different company codes one after the other. You can use the associated fixed value in various objects. However, you always have to set the concrete value only at one central point. This is the advantage over a constant. A fixed value is therefore, in a sense, a constant that is valid throughout the project.

- **Translation (Reusable)**
  A code is assigned to the target field, which performs the translation of the field content of the source field using a translation table. The values of this translation table can be entered in the step **Define Fixed Values, Translations, User-Defined Routines** (shown earlier in Figure 9.7). This is described in detail in Section 9.2.8.
- **Prefix**
  You can specify any prefix that will be prepended to the field content of the source field.
- **Suffix**
  You can specify any suffix that will be placed after the field content of the source field.
- **Concatenation**
  You have the possibility to connect two or more source fields and pass them to the destination field.
- **Transfer left aligned**
  The field content is transferred left aligned.

### ABAP Code

If you select this option (or double-click on a target field), you'll be taken to the ABAP Editor. There you can edit generated ABAP code or store your own code. Here, most of the functions of the standard SAP editor are available, such as **Check** (syntax check), **Pretty Printer**, and so on.

### User-Defined Routine (Reusable)

Subroutine written by the user

The system creates the frame of a `FORM` routine (ABAP subroutine) for you with the name prefix `UR_`. This routine is reusable, which means that it can be used in other objects of the project. This routine is different from the ABAP code you can store for a target field.

For all types of reusable rules, SAP Legacy System Migration Workbench suggests one to three possible names, one of which will be recommended by the system. We advise you adopt the proposal. For detailed information on naming, see Section 9.7.4.

When creating user-defined routines, you should note the following:

- Before creating the routine, you need to assign the correct number of source fields (corresponding to the number of input parameters of the routine).
- The source fields must be assigned in the correct order (i.e., in the order of the parameters).

### X Field

X field in IDoc processing

The *X field* is a special function for processing IDocs. In some cases, what's called a *checkbox structure* exists in addition to the data transfer structure (which contains the values to be transferred). This checkbox structure has the same field names as the data transfer structure. However, all fields of this structure are single-digit and are filled with "X" or with blanks. These checkboxes determine whether the corresponding field is transferred from the data transfer structure or not. For an X field, the code shown in Listing 9.1 is generated automatically.

```
IF NOT <field of the data transfer structure> IS INITIAL.
     <field of the checkbox structure> = 'X'.
ELSE.
      <field of the checkbox structure> = ' '.
ENDIF.
```

**Listing 9.1** Code for an X Field



Filling X structures

You can add this code for entire target structures via the menu path **Extras • Complete X structures** (shown earlier in Figure 9.19).

### Only If Source Field Not Initial

In the dialog box with the predefined rules (shown earlier in Figure 9.20) notice the **Only If Source Field Not Initial** checkbox. If this flag is set, the selected rule will be applied only if the corresponding source field has a non-initial value.

## 9.2.8 Defining Fixed Values, Translations, and User-Defined Routines

Reusable rules

We now turn to editing the reusable rules of a project. In our example, we used the following reusable rules: fixed value `FV_BUKRS` for the `Company Code` field and translations for the `Country`, `Language`, `Department` and `Position` fields.

If you call the step **Define Fixed Values, Translations and User-Defined Routines** (shown earlier in Figure 9.7), the screen shown in Figure 9.25 is displayed.

Alternatively, you can also perform maintenance in the **Define Field Mapping and Conversion Rules** step (shown earlier in Figure 9.7) and double-click on a rule to navigate to for maintenance.

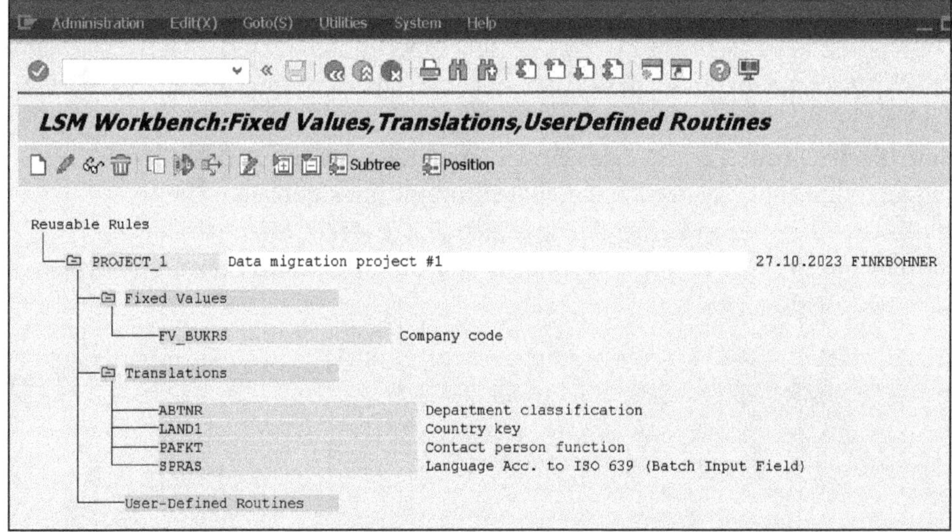


**Figure 9.25** SAP Legacy System Migration Workbench: Reusable Rules

### Fixed Value

Via **Fixed value**, you can specify the length, the type, the case-sensitive flag, and the value in addition to the name.

In our example, we have already assigned the value 0001 for the fixed value FV_BUKRS during the previous step. If the values have not yet been assigned, you can do so now. For this task, position the cursor on the BUKRS entry and click the [edit] icon (**Edit entry**), or alternatively double-click the name of the fixed value. You'll then see the dialog box shown in Figure 9.26.

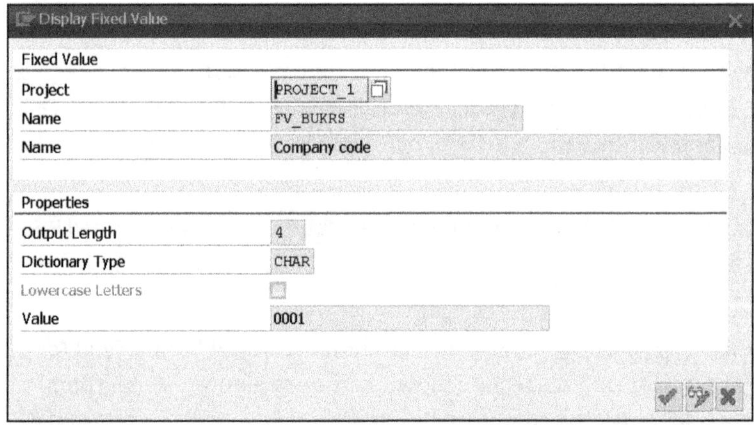


**Figure 9.26** SAP Legacy System Migration Workbench: Fixed Value

**Input help** A very useful and therefore worth mentioning detail is the F4 input help for the value field.

### Translation

In our example, we used the translation technique several times, for example, for the COUNTRY field. Here, the source data contains the entries "GER" (for Germany), "GBR" (for Great Britain), "ESP" (for Spain), and "FRA" (for France). The SAP system, however, expects "DE," "GB," "ES," and "FR." For this reason, the translation shown in Table 9.5 must be performed.

| Old Value | New Value |
|---|---|
| GER | DE |
| GBR | GB |
| ESP | ES |
| FRA | FR |

**Table 9.5** Translation for the COUNTRY Field

This change is exactly what the conversion rule does. The definition of a conversion rule involves several steps, as shown in Figure 9.27 through Figure 9.29:

- **Source Field, Target Field**
  In this context, you can enter information about the source and destination field. If you create a new translation, you must first save the information. Only then can you go to the next tab.

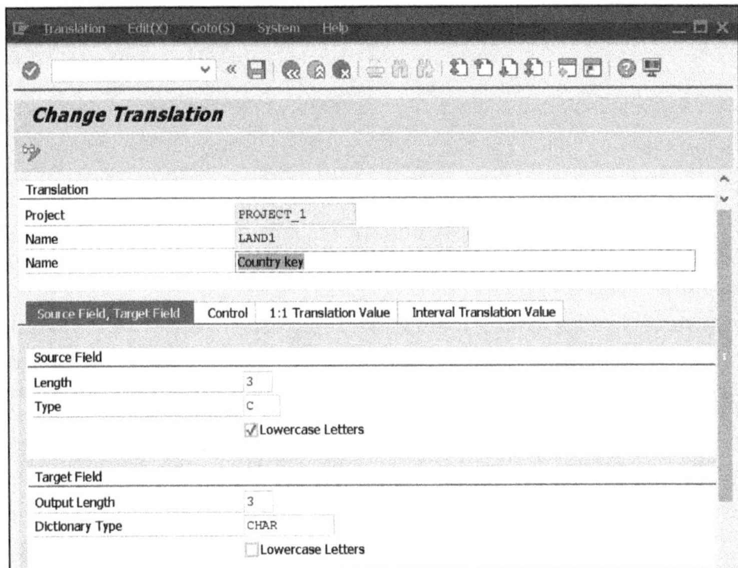


**Figure 9.27** SAP Legacy System Migration Workbench: Translation, Source Field, Target Field

- **Control**
  Specify the type of translation. You can specify in which of the two translation tables a value should be searched first and which alternative should be selected if no matching entry is found. For our example, the suggested settings are sufficient: If the search in the 1:1 translation table is unsuccessful, the target field remains initial. For the automatic value collector to become active, the checkbox **1:1 Translation Table Added to Automatically** must be selected, as shown in Figure 9.28.

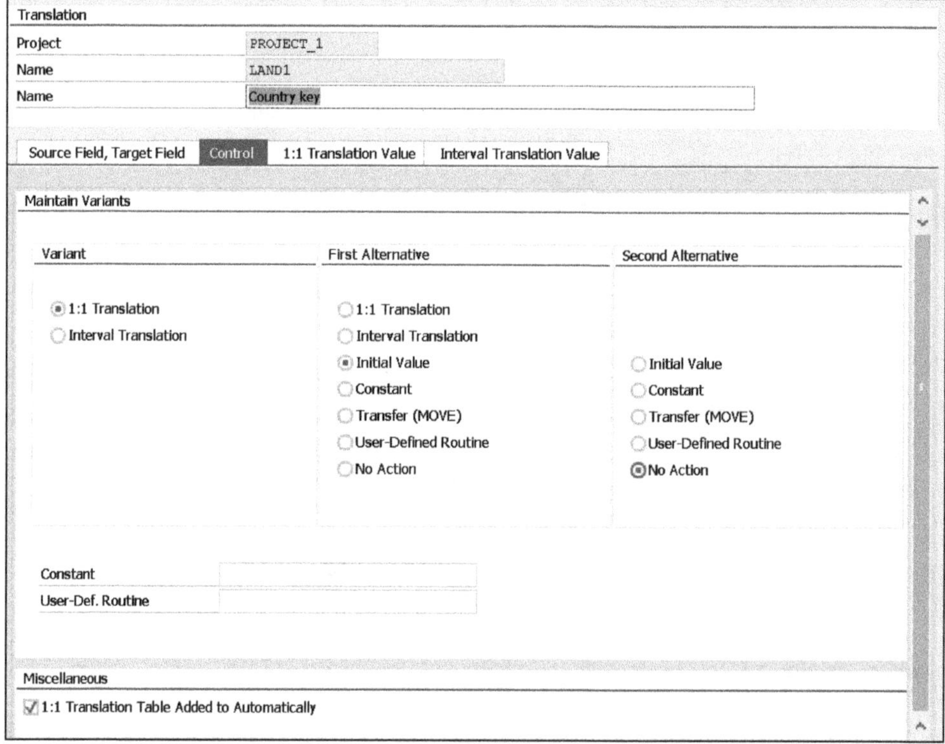


**Figure 9.28** SAP Legacy System Migration Workbench: Translation, Control

- **1:1 Translation Value**
  Specify the value table to be used for the translation. You can also upload the values from a PC file in the **Text (tab-delimited) (*.txt)** format by clicking the icon (**Upload**).

  *Input help and OK indicator*

  Also useful is that the F4 input help is usually available in the **New Value** column. Note that only the values for which the **OK** flag is set are taken into account during the translation.

*Automatic value collector*

- It is also extremely useful that the translation acts as an automatic value collector during data conversion. Thus, all the different values in the source field entries will be collected when you perform the **Convert Data** step (Section 9.2.15). This feature brings advantages that should not be underestimated. From experience, you can be surprised at the entries that have accumulated in the source data over time. To achieve a proper conversion, all these values must be mapped to values that are accepted by the SAP system.

In our example, we thus need a the translation table for the COUNTRY field: In the left-hand column, all entries are collected ("ESP," "FRA," "GBR," and "GER"). Now, you just need to provide the right-hand column with correct values (the [F4] input help makes this easier) and set the **OK** flag. The final result is shown in Figure 9.29.



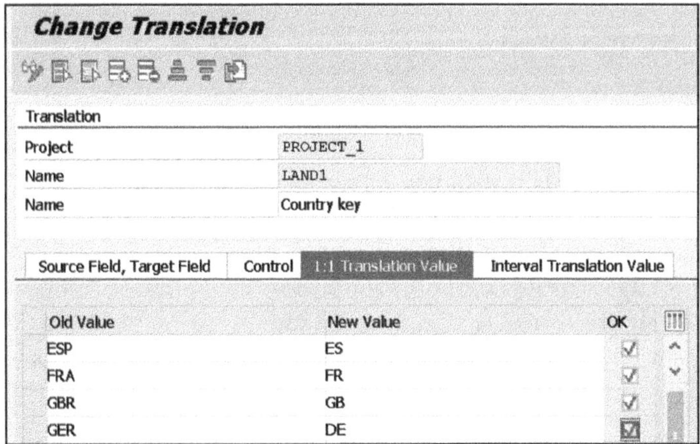


**Figure 9.29** SAP Legacy System Migration Workbench: Translation, 1:1 Translation Values

- **Interval Translation Value**
  In this step, you specify the value table to be used for the translation by intervals. An upload option and [F4] input help are also available here. Furthermore, only the values for which the **OK** indicator is set will be taken into account.

The feature of the translation technique is used relatively rarely. A possible use case is, for example, if multiple accounts from a number interval (old values) are to be combined into one account (new value) within the scope of a chart of accounts cleanup.

### User-Defined Routine

As mentioned earlier, any ABAP code can be added to each target field. If the specific situation requires ABAP code to be used at various points within a project, it is a good idea to use a separate routine.

**Example: editing phone numbers**

Let's suppose there are invalid characters in the phone numbers of your source data that you want to remove. Since phone numbers occur in various places in the SAP system and can also be part of external correspondence, it makes sense to clean up the invalid characters.

On the overview screen of reusable rules (shown earlier in Figure 9.25), position your cursor on the **User-defined Routines** label and click the icon (**Create Entry**). In the dialog box that appears, enter the name and description and confirm these settings with Enter, as shown in Figure 9.30.

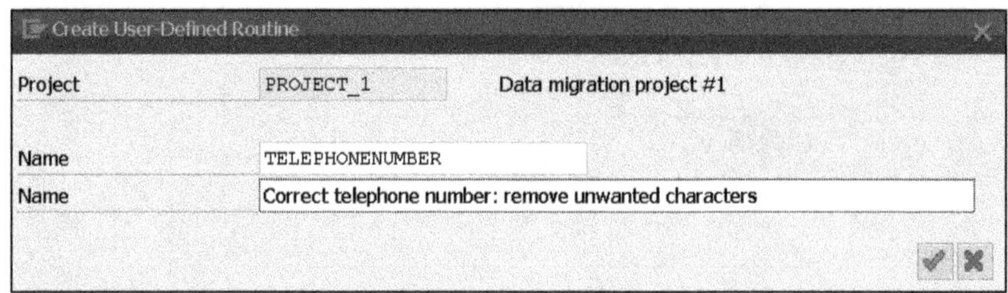


**Figure 9.30** SAP Legacy System Migration Workbench: User-Defined Routine, Name and Designation

**Defining input and output parameters**

Another dialog box follows, asking you for the number of input and output parameters, as shown in Figure 9.31. In our example, we want to pass a phone number to the routine and receive a "clean" phone number. Consequently, we enter a "1" in both fields and confirm with Enter.

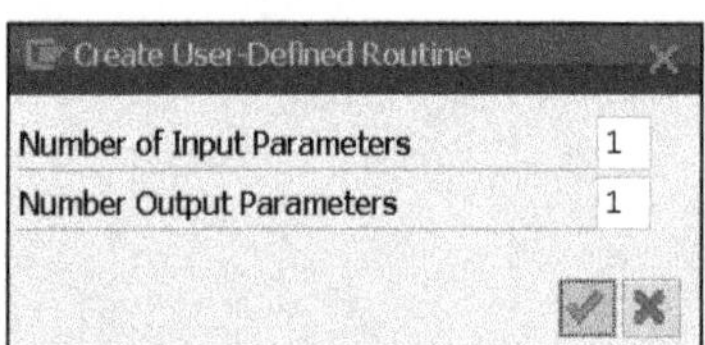


**Figure 9.31** SAP Legacy System Migration Workbench: User-Defined Routine, Number of Parameters

The rule then appears in the tree under **User-Defined Routines**. Select the new rule and click the icon (**Edit Entry**). Then, the ABAP Editor opens, in which SAP Legacy System Migration Workbench already provides a frame that you can fill with your own ABAP code. The main functions (Pretty Printer, syntax check, etc.) that you're familiar with in connection with the ABAP Editor are available, as shown in Figure 9.32.

You can now insert your code; the result is shown in Figure 9.33. The user-defined routine can then be used in the volume of the field mapping in any number of places within the project.

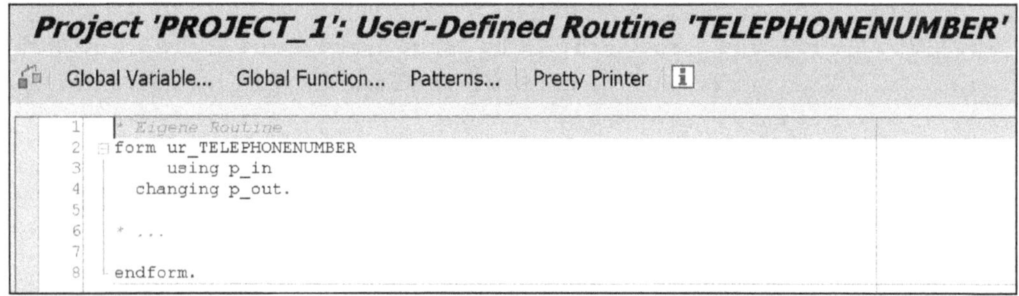


Figure 9.32 SAP Legacy System Migration Workbench: User-Defined Routine, Frame for ABAP Code

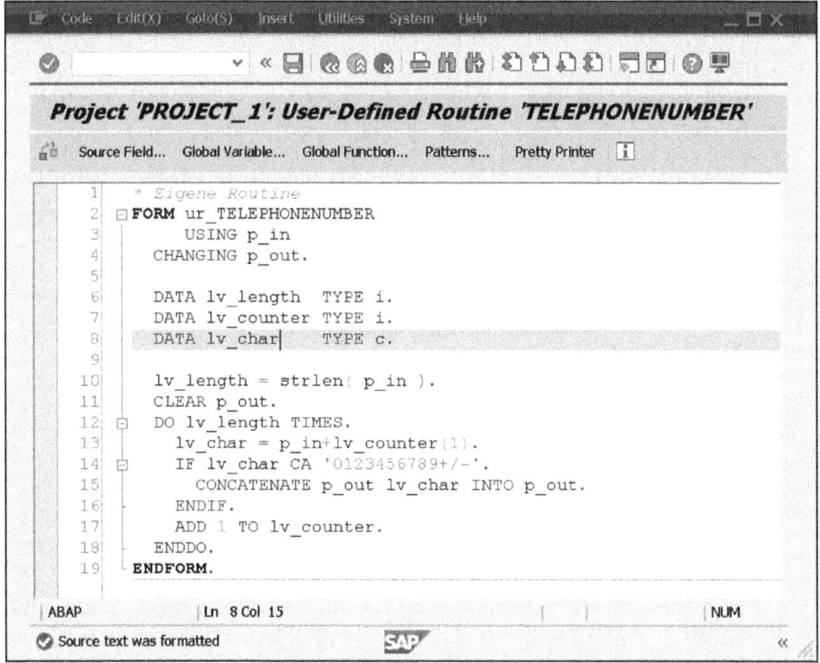


Figure 9.33 SAP Legacy System Migration Workbench: User-Defined Routine, Completed ABAP Code

### 9.2.9 Digression: Files

Before we proceed with the steps of SAP Legacy System Migration Workbench that concern the files, we want to outline the properties by which files can be described and which types of files SAP Legacy System Migration Workbench supports.

Your data can be stored in files in various ways. For a program to process the files correctly, the following information about the files must be available:

Record limit

- **Record end marker or fixed record length**
  Several ways exist to indicate within a file where a record ends and where the next record begins. SAP Legacy System Migration Workbench supports the following three options:
  - **Record End Marker**: Undoubtedly the most common case, this marker is present whenever the file is created with a text editor or exported from a spreadsheet application. We also refer to files with record end markers as *text files*.
  - **Fixed Record Length**: This marker means that each record has the same length in bytes. This case often occurs when data is exported from a mainframe application by means of a program.
  - **Hexadecimal Lth Field (4 Bytes) at Start of Record**: This case occurred in connection with the R/2-R/3 migration.

Structure of the file content

- **Table or sequential file**
  There are basically two possibilities regarding the file content:
  - **Table**: The content of a file can have a table-like character. This is the case when all records in the file have the same structure, that is, when all records in the file belong to the same source structure.
  - **Sequential file**: If the file contains records for different source structures, the question arises how you can determine which records belong together. The only reasonable solution to this problem is to solve it via the sequence. In this case we speak of a *sequential file*.

Field boundary

- **Separator**
  To indicate within a record where a field ends and where the next field begins, the following alternatives are available:
  - **Separator**: One method is to insert an agreed separator between every two fields. Typical separators are the tab character, the comma, and the semicolon. In practice, the use of a comma or semicolon as separators often entails problems because these characters can occur in text fields and in this case lead to unwanted separations.
  - **No Separator**: In this case, SAP Legacy System Migration Workbench places the defined source structure over the record of a file as a template, so to speak, and determines the field delimiters in this way.
- **Field Names at Start of File**
  Possibly the field names are listed at the start of the file. In the case of table-like files, this information can be used for the correct assignment of the file contents to the source fields. In this case, the order of the source fields within the source structure does not necessarily have to match the order of the columns in the file. However, the fields are expected to be separated from each other by a separator character.

- **Character set (code page)**
  If you operate your source system and your SAP system in different operating system environments, the data from the source system may be encoded in a character set that cannot be interpreted by your SAP system. In this case, a character set conversion must be performed. SAP Legacy System Migration Workbench will do this for you automatically. However, you must specify the character set in which the source system data is encoded. A convenient [F4] input help is available for this purpose.

Character set conversion on demand

[+]

**Using Files on the Frontend**

If you use files on the frontend, we recommend that you use files of the **Text (tab-delimited) (*.txt)** format. Files of this type have two advantages:

- They can be generated by all common spreadsheet programs (e.g., Microsoft Excel).
- The tab character does not normally appear in data. A comma or semicolon, on the other hand, may well be included in your data. Thus, if you choose the semicolon as a separator and your data contains a semicolon, this semicolon will be mistakenly interpreted as a separator—with the very likely consequence that the data record in question cannot be interpreted correctly.

If you use a spreadsheet, you should format the cells as text to avoid issues with automatic formatting for numeric values and with the automatic deletion of leading zeros.



### 9.2.10 Specifying Files

We now come to the step where you must describe all the files that will be used in the following steps and make them known to the system.

Files to specify

In detail, the following files are specified:

- The files on the PC and/or SAP application server that contain your source data
- Two internal working files of SAP Legacy System Migration Workbench:
  - The file for the imported data
  - The file for the converted data

In our example, the source data is located in two worksheets of a Microsoft Excel file. Before you can make them known to SAP Legacy System Migration Workbench, you must save each spreadsheet in the **Text (tab-delimited) (*.txt)** format. Assign the file names "Customer_Header.txt" and "Customer_Contact.txt."

[+]

**Delete Blank Lines**

When saving the text files, make sure that you do not export empty lines. You should delete empty lines beforehand.

### Legacy Data on the Frontend (PC)

In this example, we assume that all source data is located on the PC. Proceed by following these steps:

- In the navigation screen (shown earlier in Figure 9.7), call the **Specify Files** step.
- Position the cursor in change mode on the line **Legacy Data - On the PC (Frontend)**, as shown in Figure 9.35.
- Click the icon (**Add Entry**), and you'll see the dialog box shown in Figure 9.34.

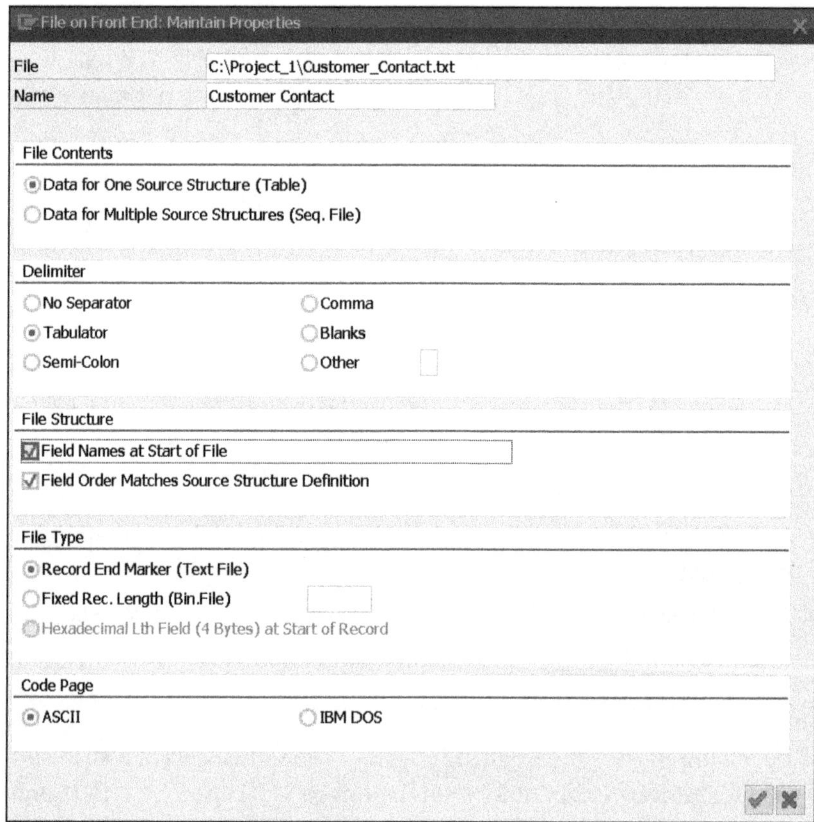


**Figure 9.34** SAP Legacy System Migration Workbench: Specifying a File on the Frontend

- In the dialog box, specify the file path, file name, file description, and other properties corresponding to the first source file. The F4 input help is available for selecting the file path.
- At the end, press Enter and save these settings.

Then, repeat all these steps for the other source files; in the end, you should get the overview shown in Figure 9.35.

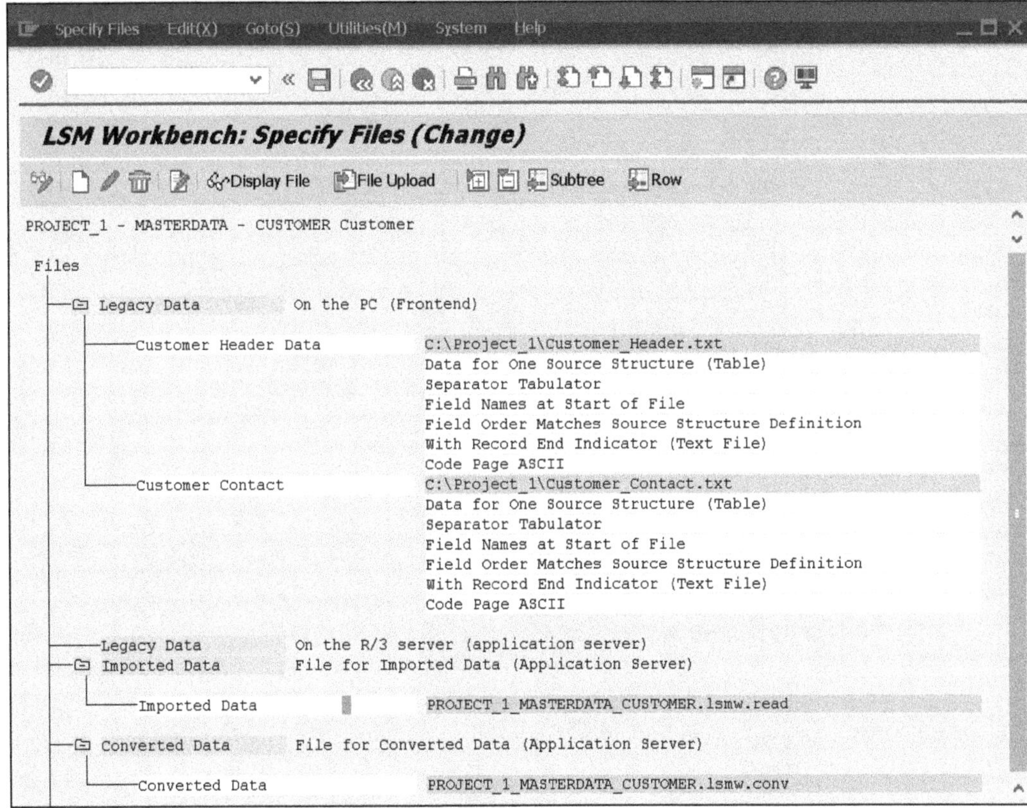


**Figure 9.35** SAP Legacy System Migration Workbench: Specifying Files

### Legacy Data on the SAP Application Server

If you want to use files that are stored on the application server, you can proceed in a similar way to what we described in the previous section:

- In change mode, position the cursor on the line **Legacy Data - On the R/3 server (application server)**, as shown in Figure 9.35.
- When you click the icon (**Add entry**), the dialog box shown in Figure 9.36 will be displayed.
- Again, specify the file path, file name and file designation, and other properties.

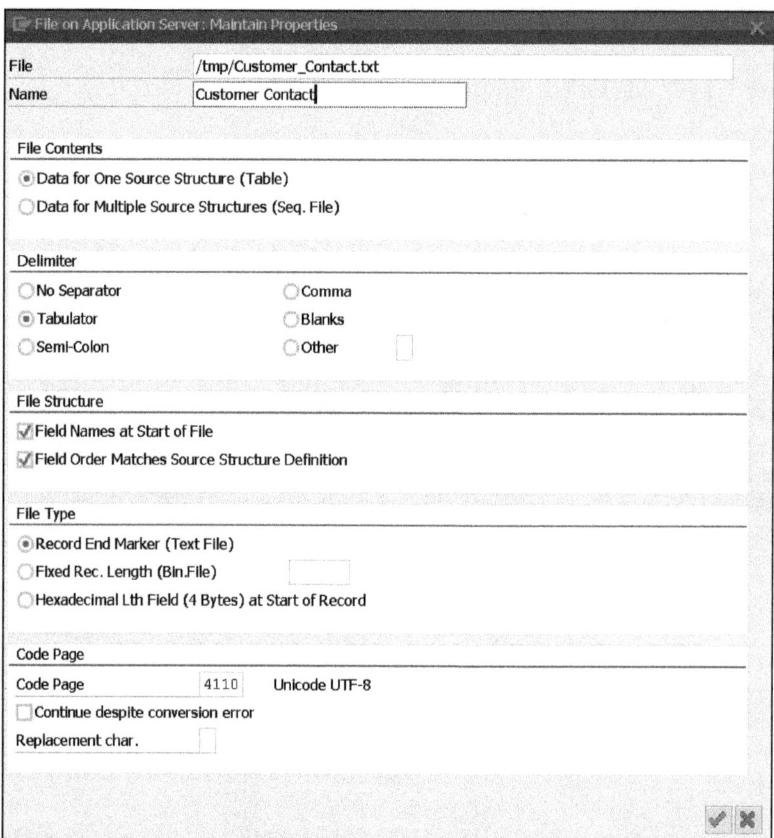


**Figure 9.36** SAP Legacy System Migration Workbench: Specifying Files on the SAP Application Server

**Additional rules regarding the files**

With regard to the files containing the source data, SAP Legacy System Migration Workbench requires compliance with some rules that do not play a role in our example, but which may be important in your specific migration situation. We therefore want to turn to these rules now:

- To the operating system, the SAP system appears under the user ID `<sid>adm`, where `<sid>` represents the three-digit identifier of the SAP system. You must therefore make sure that a read or write permission exists in the selected directory.
- If a file contains data on multiple source structures, the field order in the file must match the order of the source fields defined in SAP Legacy System Migration Workbench.

**Field order or field names**

- If a file contains data for a source structure, either the field order must match the source structure definition, or there must be field names at the beginning of the file, which can be used to assign the columns of the file to the source fields.

- If the file contains record end markers (text file), packed fields are not allowed. In a packed field, not every digit is stored in a separate byte. They are rather stored in a more compact manner.
- If a file contains separators, packed fields are not allowed.
- You can also use files on the frontend and the SAP application server within an object.

Mix of frontend and application server allowed

- A file that contains data for multiple source structures can be assigned to multiple source structures in the subsequent step (assign files). A file that contains data for only one source structure, on the other hand, can be assigned to only one source structure.
- If you use multiple files in an object, the associated source structures must contain fields with the same name. In our example, this is the `CUSTOMER` field. The fields with the same name are used to assign related records. The fields used for this assignment can be highlighted in color when you define the source fields (shown earlier in Figure 9.11) using the **Fields With The Same Name** button.



### File of Imported Data, File of Converted Data

Complete default values

For the two internal work files **File of imported data** and **File of converted data** of SAP Legacy System Migration Workbench, you'll be presented with complete proposals that you can usually accept in their entirety.

The *SAP Home Directory* of the SAP application server or the last directory you used in SAP Legacy System Migration Workbench is selected as the file path.

> **Overview of all SAP Directories**
>
> Transaction AL11 provides an overview of all SAP directories. You can find the physical path of the SAP home directory under the `DIR_HOME` identifier.

The file name is composed of the abbreviations for project, subproject, and object (separated by an underscore in each case) and provided with the file extension *lsmw.read* (for the file of the imported data) or *lsmw.conv* (for the file of the converted data). In our example, this results in the file names shown in Figure 9.37.

Our files are broken down in the following way:

- *PROJECT_1_MASTERDATA_CUSTOMER.lsmw.read* for the file of the imported data
- *PROJECT_1_MASTERDATA_CUSTOMER.lsmw.conv* for the file of the converted data

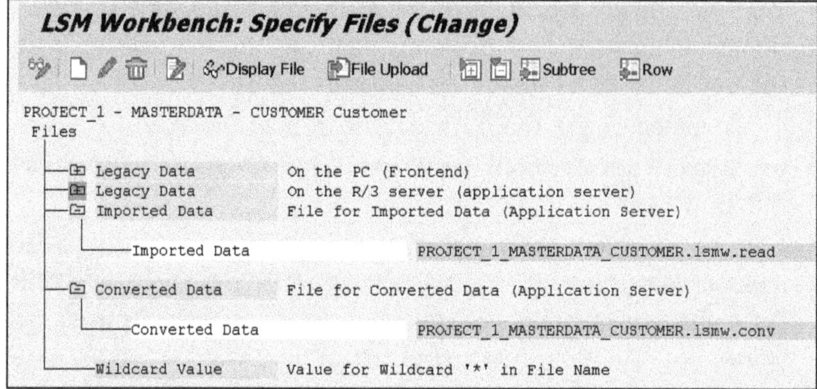


**Figure 9.37** SAP Legacy System Migration Workbench: Specifying Files of the Imported and Converted Data

Of course, you can customize the file names according to your preferences. You can also prefix the file name with a file path. The syntax to follow depends on the operating system of your application server.

[+]

**Data Browser**

An undocumented feature is the [F4] input help for files on the application server. To make it available, you must use Transaction SE16 (Data Browser) to enter an entry in the SAP database table /SAPDMC/LSGCUST with the server name and an "X" to display the input help. However, you need the following authorizations to do so:

- S_TABU_CLI
  - CLIIDMANINT = X
- S_DEVELOP
  - DEVCLASS = /SAPDMC/LSGCUST
  - OBJTYPE = TABL
  - OBJNAME = /SAPDMC/LSGCUST
  - P_GROUP
  - ACTVT = 02

**Logical path, logical file name**

The **Logical Path** and **Logical File Name** fields are not displayed in the **Specify Files** step until the batch input or direct input program to be called in the further course requires this information. [F4] input help is available for the two fields. We strongly recommend that you do not use logical paths and file names that are also used by other applications.

[!]

**Maximum Length of File Names**

The file names on the application server may only be 45 characters long including the path. You'll reach this value pretty quickly. For this reason, it makes sense to save these files in a directory path with only one or very few levels. You should also shorten the file extensions to *.rd* and *.cv*, for example.

### 9.2.11 Using Wildcards in File Names

**Example**

We'll explain the use of wildcards in file names based on an example. Let's assume that the source data is in the following four files:

- File 1: *C:\Project_1\Customer_Header_1.txt*
- File 2: *C:\Project_1\Customer_Contact_1.txt*
- File 3: *C:\Project_1\Customer_Header_2.txt*
- File 4: *C:\Project_1\Customer_Contact_2.txt*

Two files (**1.txt* and **2.txt*) form a *set*; in other words, file 2 contains the item data for the header records in file 1, and file 4, the item data for the header records in file 3. When importing the data, file 1 and file 2 are to be processed first, followed by file 3 and file 4. You can achieve this sequence with the settings for **Wildcard Value** and inserting the wildcard character "*" in the file name in the **Specify Files** step, as shown in Figure 9.38.

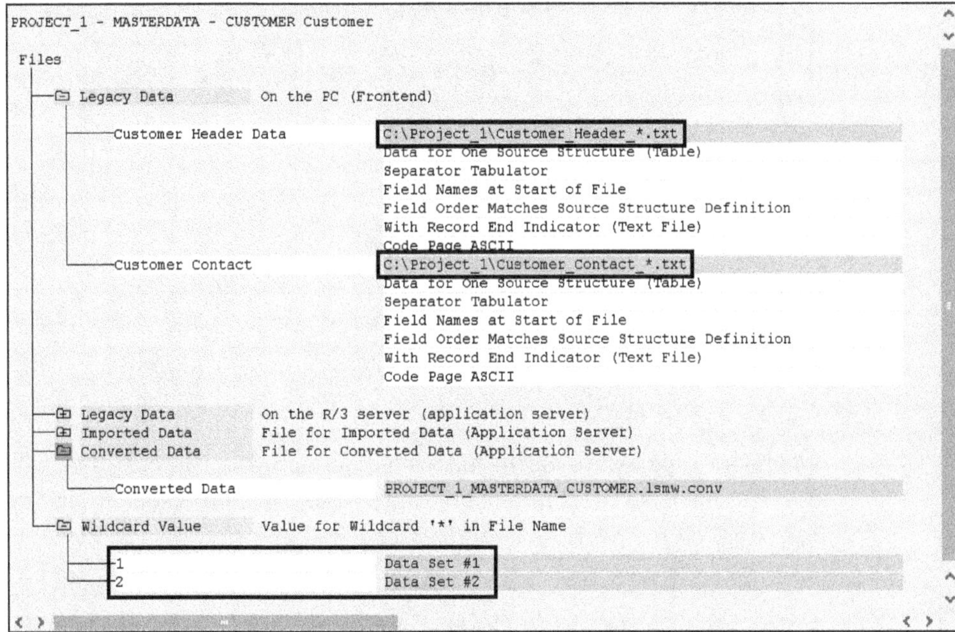


**Figure 9.38** SAP Legacy System Migration Workbench: Using Wildcards in File Names

**Wildcards in all file names**

Note that you can also use wildcards in the file names of the imported and converted data.

### 9.2.12 Assigning Files

In this step, you assign the files you defined in the previous step to the source structures, as shown in Figure 9.39.

In our example, call the **Assign Files** step from the navigation screen (shown earlier in Figure 9.7) and position the cursor on the HEAD source structure. Then, double-click the source structure name or click the Assignment button (**Assign File**). From the list of defined files, select the file that contains the header records. Proceed in the same way with the source structure CONTACT.

**File Names and File Assignment**

Note that, if you subsequently change file names or file properties, the file assignment will be preserved.

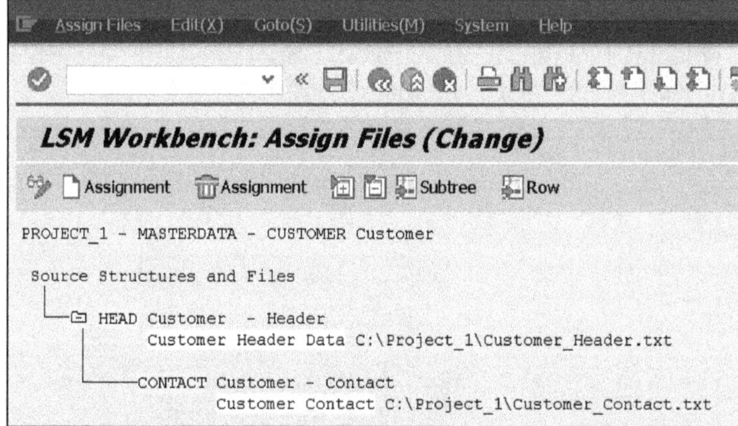


**Figure 9.39** SAP Legacy System Migration Workbench: Assigning Files

### 9.2.13 Importing Data

When you import the data, the files with data from your source system will be converted into a technically uniform format. Based on this format, the next step is the conversion to the SAP format.

**Program regenerated on demand**

In the navigation screen (shown earlier in Figure 9.7), you need to call the **Read Data** step, which will display the screen template shown in Figure 9.40. The system first checks whether the data import program is still up to date. If not, the program will be regenerated automatically.

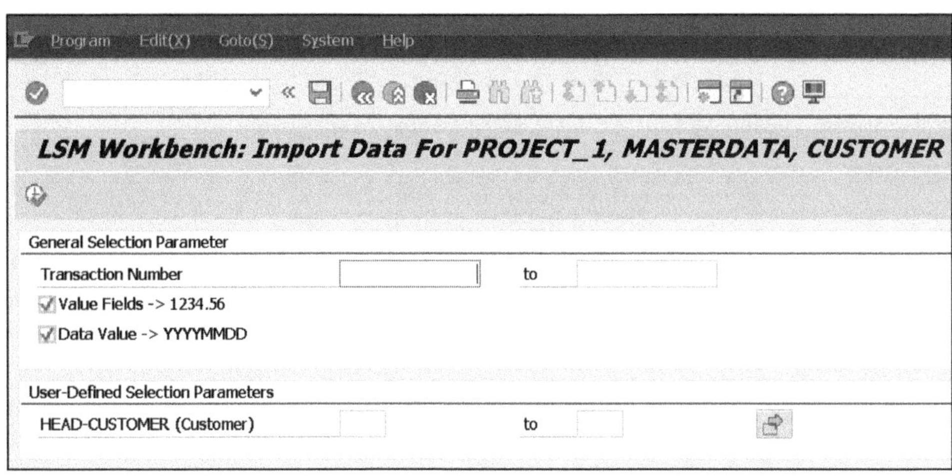


**Figure 9.40** SAP Legacy System Migration Workbench: Importing Data, Selection Screen

Selection parameters

If you want to import all the data that belongs to the object, click **Execute** to start the process. If you want to migrate only a portion of the dataset, you can limit the amount of data you want to migrate in the **General Selection Parameter** area. Make your data selection in the **Transaction Number** field. However, we do not recommend using this selection option during productive migration, as in this case, you run the risk that not all data will be migrated in its entirety.

If you have marked one or more source fields as selection parameters when you defined the source fields (Section 9.2.5), these fields will also be provided as selection parameters. In addition, there are two checkboxes:

- **Value Fields -> 1234.56**
  Value fields are converted to the calculation format (with decimal point).
- **Date Value -> YYYYMMDD**
  Date value fields are converted to the internal SAP format (YYYYMMDD).

Wildcards

If you use the wildcard character (*) in the names of the source data files (see Section 9.2.11) and have defined at least one value for the wildcard, a selection parameter for the wildcard will be generated in the data import program. Typically, this selection parameter behaves in the following sense: If you do not make any input when executing the program, all defined wildcard values will be processed. When you make an input, the matching values will be processed.

In our example, you should accept the default values and click **Execute**.

Log

After the successful completion of the process, you'll see a small summary log, as shown in Figure 9.41, from which you can see whether all records were successfully processed.

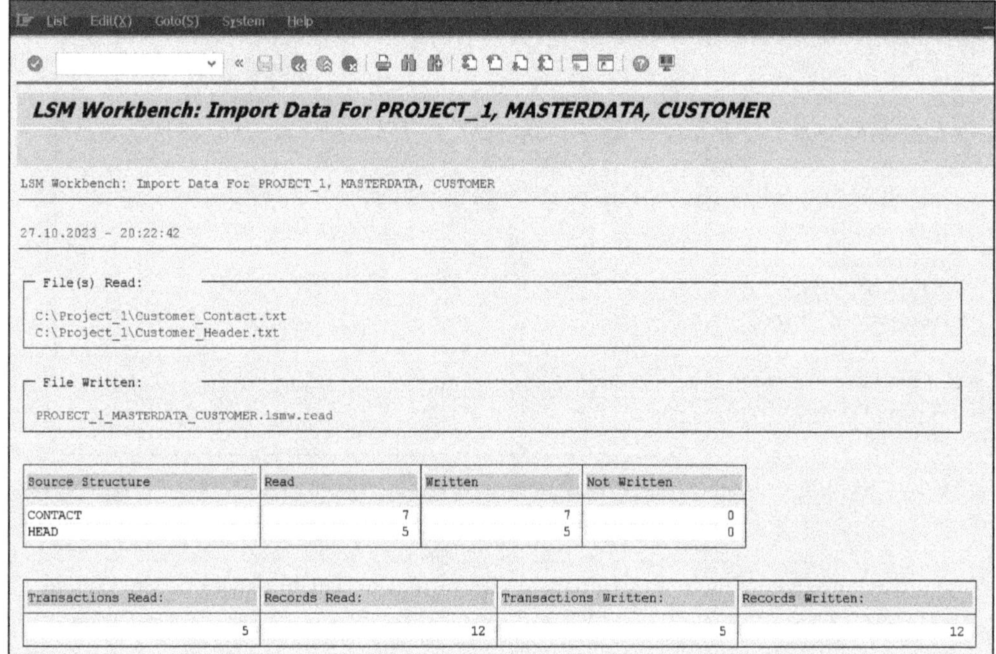


Figure 9.41 SAP Legacy System Migration Workbench: Importing Data, Log

### 9.2.14 Displaying Imported Data

**Color marking** After importing the data, you'll want to look at the result. You can display all or part of the imported data in a table. For the sake of clarity, the individual record types (source structures) are color coded, as shown in Figure 9.42.

Imported Data Edit(X) Goto(S) System Help

**LSM Workbench: Imported Data**

Field Contents

File Contents PROJECT_1_MASTERDATA_CUSTOMER.lsmw.read

| Row | Hierarchy Level | Structure | Cont. |
|---|---|---|---|
| | | File Information | LSMWPROJECT_1 MASTERDATA CUSTOMER QKV 91020231027202240FINKBOHNER |
| 1 | 01 | HEAD | 1001Omega Soft-Hardware Markt Gustav-Jung-Str. 425 Nuernberg 90455 GERD 09256-4548-0 0925 |
| 2 | 02 | CONTACT | 1001 Schmidt Ralf Herr Einkauf |
| 3 | 02 | CONTACT | 1001 Huber Luise Frau Einkauf |
| 4 | 01 | HEAD | 1002Lampen Markt GmbH Auf der Schanz 54 Frankfurt 65936 GERD 069-467653-0 069- |
| 5 | 02 | CONTACT | 1002 Luchs Harald Herr Organisation |
| 6 | 02 | CONTACT | 1002 Wolf Susanne Frau Einkauf |
| 7 | 01 | HEAD | 1003Hever Industrial UK High Wycombe HP12 3TL GBRE 00-170866645 |
| 8 | 02 | CONTACT | 1003 Winchester George Mr Vorstand |
| 9 | 02 | CONTACT | 1003 Smith Judy Ms Finanzen |
| 10 | 01 | HEAD | 1004Editorial Atlántida S.A. Carrer de la Mariana 34, PL. 1Barcelona 8005 ESPS 34-91-556159 |
| 11 | 02 | CONTACT | 1004 Lopez Juan Carlos Sr. Geschäftsführung |
| 12 | 01 | HEAD | 1005Adecom SA ZI Les Echarmeaux Longjumeau 91230 FRAF 0615489712 |

Figure 9.42 SAP Legacy System Migration Workbench: Displaying Imported Data

By clicking on a row, you'll access all the information that is in that line displayed in a clear way one below the other. You'll get the same result if you click on the **Field Contents** button, as shown in Figure 9.43.

**LSM Workbench: Imported Data**

Structure HEAD

| Field Name | Field Text | Field Length | Field Value |
|---|---|---|---|
| CUSTOMER | Customer | 4 | 1001 |
| NAME | Company name | 30 | Omega Soft-Hardware Markt |
| STREET | Street and house number | 30 | Gustav-Jung-Str. 425 |
| CITY | City | 30 | Nuernberg |
| POSTALCODE | Postal code | 10 | 90455 |
| COUNTRY | Country | 3 | GER |
| LANGUAGE | Language | 2 | D |
| TELEPHONE | Telephone number | 15 | 09256-4548-0 |
| FAX | Fax number | 15 | 09256-4548-99 |
| UIN | VAT Registration number | 11 | |

**Figure 9.43** SAP Legacy System Migration Workbench: Displaying Imported Data, Detail

**Detecting errors**

At this point, it may be possible to detect errors that occurred during character set conversion or processing of separators. These errors should of course be cleaned up before further processing.

### 9.2.15 Converting Data

**Conversion according to field assignments and conversion rules**

Once the data from your source system has been converted into a technically uniform format by the import process, this format will now be converted into the SAP format using the field assignments and conversion rules you have defined.

In the navigation screen (shown earlier in Figure 9.7), you need to call the **Convert Data** step to open the selection screen shown in Figure 9.44. In this case, too, the system first checks whether the data conversion program is still up to date. If not, the program will be regenerated automatically.

If you do not want to perform any data selection, you can start the process by clicking the icon (**Execute**). Alternatively, you can define the data selection in the **Transaction Number** field.

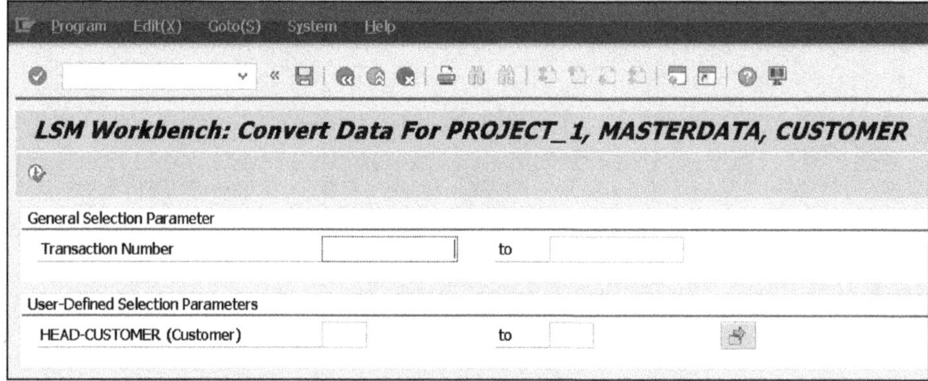


**Figure 9.44** SAP Legacy System Migration Workbench: Converting Data, Selection Screen

**Selection parameters and wildcards**

If you have marked one or more source fields as selection parameters when defining the source fields, these fields will also be offered as selection parameters. If you use a wildcard in the names of the files of the source data and have defined at least one value for it, you'll also get a selection parameter for the wildcard here. If you do not make an entry here, all defined wildcard values will be processed.

**Additional function in case of BAPI, IDoc**

If you have chosen the BAPI or IDoc import method under **Define Object Attributes** (Section 9.2.1), you'll see additional selection parameters in the selection screen of the data conversion program, as shown in Figure 9.45.

When you select **Create File**, a file will be created during the data conversion. If, on the other hand, you select **Create IDocs Directly**, IDocs are collected during the data conversion and then transferred in packages for IDoc generation. You can specify the package size via the **Number of IDocs per Package** parameter. The default setting for this is 50.

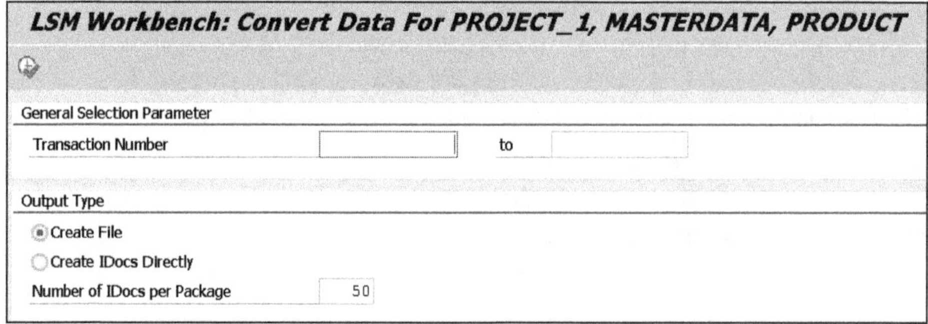


**Figure 9.45** SAP Legacy System Migration Workbench: Data Conversion Program Selection Screen, Additional Selection Parameters for BAPI/IDoc

For our example, simply click the [icon] icon (**Execute**) again to display the log shown in Figure 9.46.

**Log**

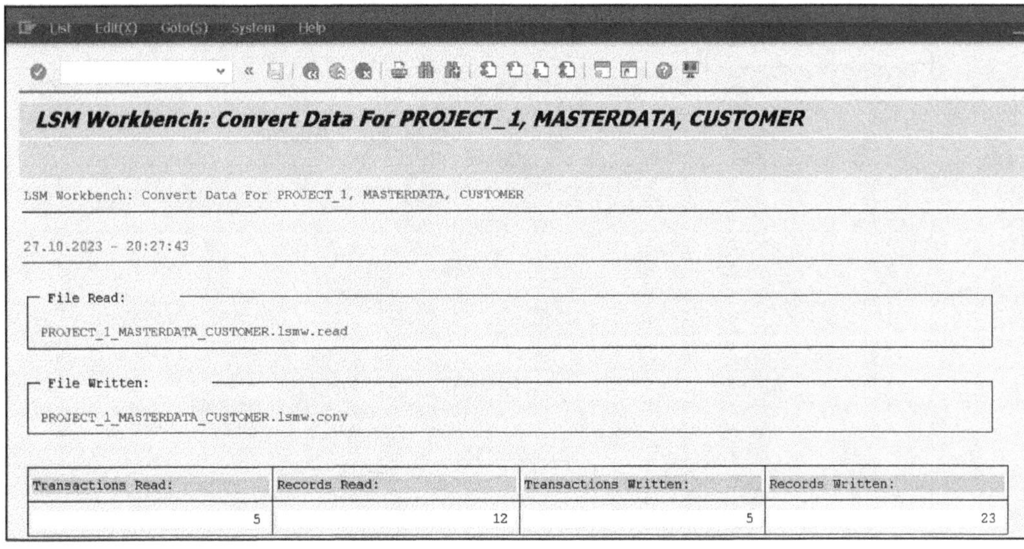


| Transactions Read: | Records Read: | Transactions Written: | Records Written: |
|---|---|---|---|
| 5 | 12 | 5 | 23 |

**Figure 9.46** SAP Legacy System Migration Workbench: Converting Data, Log

### 9.2.16 Displaying Converted Data

**Deviating number of records**

After running the data conversion, you can see the result of the data conversion in the next step—**Display Converted Data**. You have essentially the same options here as in the **Display Imported Data** step (Section 9.2.14). Note that the number of records generally differs from the number of records in the source data. This difference is due to the different structure of source data and SAP format. However, the number of data units (in this case, five customers) should match.

In our example, a BGR00 record is created and one BKN00 record, one BKNA1 record, and one BKNB1 record for each customer. A BKNVK record is generated for each contact record (23 records total), as shown in Figure 9.47.

In this case, too, you can click on a line to display a detailed representation, as shown in Figure 9.48. Some fields have the "/" value (nodata indicator). The batch input program to be called subsequently interprets this value as "No input"—in contrast to a " " input.

**LSM Workbench: Converted Data**

Field Contents

## File Contents PROJECT_1_MASTERDATA_CUSTOMER.lsmw.conv

| Row | Hierarchy Level | Structure | Contents |
|---|---|---|---|
| 1 | 0 | BGR00 | 0CUSTOMER 910FINKBOHNER / |
| 2 | 1 | BKN00 | 1XD01 D1001 0001/ / / KUNA/ |
| 3 | 2 | BKNA1 | 2BKNA1 / Omega Soft-Hardware Markt / / |
| 4 | 2 | BKNB1 | 2BKNB1 0000140000/ / / // / // // / / / // / // |
| 5 | 2 | BKNVK | 2BKNVK // Schmidt 09256-45485 0002Ralf |
| 6 | 2 | BKNVK | 2BKNVK // Huber 09256-45486 0002Luise |
| 7 | 1 | BKN00 | 1XD01 D1002 0001/ / / KUNA/ |
| 8 | 2 | BKNA1 | 2BKNA1 / Lampen Markt GmbH / / |
| 9 | 2 | BKNB1 | 2BKNB1 0000140000/ / / // / // // / / / // / // |
| 10 | 2 | BKNVK | 2BKNVK // Luchs 069-467653-145 0004Harald |
| 11 | 2 | BKNVK | 2BKNVK // Wolf 069-467653-146 0002Susanne |
| 12 | 1 | BKN00 | 1XD01 D1003 0001/ / / KUNA/ |
| 13 | 2 | BKNA1 | 2BKNA1 / Hever Industrial UK / / |
| 14 | 2 | BKNB1 | 2BKNB1 0000140000/ / / // / // // / / / // / // |
| 15 | 2 | BKNVK | 2BKNVK // Winchester 170877755 0001George |
| 16 | 2 | BKNVK | 2BKNVK // Smith 170877756 0009Judy |
| 17 | 1 | BKN00 | 1XD01 D1004 0001/ / / KUNA/ |
| 18 | 2 | BKNA1 | 2BKNA1 / Editorial Atlántida S.A. / / |
| 19 | 2 | BKNB1 | 2BKNB1 0000140000/ / / // / // // / / / // / // |
| 20 | 2 | BKNVK | 2BKNVK // Lopez 34-91-556153 0001Juan Carlos |
| 21 | 1 | BKN00 | 1XD01 D1005 0001/ / / KUNA/ |
| 22 | 2 | BKNA1 | 2BKNA1 / Adecom SA / / |
| 23 | 2 | BKNB1 | 2BKNB1 0000140000/ / / // / // // / / / // / // |

**Figure 9.47** SAP Legacy System Migration Workbench: Displaying Converted Data

**LSM Workbench: Converted Data**

## Structure BKNA1

| Fld Name | Fld Text | Field Length | FldValue |
|---|---|---|---|
| STYPE | Batch Input Interface Record Type | 1 | 2 |
| TBNAM | Table Name | 30 | BKNA1 |
| ANRED | Title | 15 | / |
| NAME1 | Name 1 | 35 | Omega Soft-Hardware Markt |
| NAME2 | Name 2 | 35 | / |
| NAME3 | Name 3 | 35 | / |
| NAME4 | Name 4 | 35 | / |
| SORTL | Sort field | 10 | Omega Soft |
| STRAS | Street and House Number | 35 | Gustav-Jung-Str. 425 |
| PFACH | PO Box | 10 | 90455 |
| ORT01 | City | 35 | Nuernberg |
| PSTLZ | Postal Code | 10 | 90455 |
| ORT02 | District | 35 | / |
| PSTL2 | P.O. Box Postal Code | 10 | 90455 |
| LAND1 | Country/Region Key | 3 | DE |
| REGIO | Region (State, Province, County) | 3 | / |
| SPRAS | Language Acc. to ISO 639 (Batch Input Field) | 2 | D |
| TELX1 | Telex Number | 30 | 09256-4548-99 |
| TELF1 | First telephone number | 16 | 09256-4548-0 |
| TELFX | Fax Number | 31 | 09256-4548-99 |

**Figure 9.48** SAP Legacy System Migration Workbench: Displaying Converted Data, Detail

### 9.2.17 Importing Data

Dependence on import method

The import method selected in the **Define Object Attributes** step (Section 9.2.1) determines which steps you're offered regarding **importing data**, as shown in Table 9.6.

| Import Method | Data Import Step |
|---|---|
| Standard batch input or recording | Create batch input session |
| | Run batch input session |
| Standard direct input | Start direct input session |
| BAPI or IDoc | Start IDoc generation |
| | Start IDoc processing |
| | Create IDoc overview |
| | Start IDoc postprocessing |

**Table 9.6** Import Steps Depending on the Import Method

#### Importing Data via Batch Input or Recordings

In our example (standard batch input), you need to perform the **Create batch input session** step (shown earlier in Figure 9.7). The further course of this step depends strongly on the respective batch input program.

[!]

**RFDIBDEOO/RFBIKROO as of SAP S/4HANA 1809**

For the standard batch input report for transferring customer master data (RFBIDE00) or supplier master data (RFBIKR00), the name **Create Batch Input Session** is somewhat misleading since SAP S/4HANA 1809 because the data is updated directly. After the execution, a dialog box will appear indicating this. The text available for RFBIDE00 via the info button is as follows:

"With release 1809 the customer master data transactions are obsolete for batch input. Running the program will no longer create a batch input session. Instead the customer master data will be created/changed by changing the corresponding business partner data. The business partner data will be updated immediately."

SAP Note 2803681 explains the changes in detail.

The batch input sessions we usually create in this context are typically named the same as SAP Legacy System Migration Workbench object. The global g_groupname variable is assigned the object name while the map name is defined via the default BGR00-GROUP = g_groupname. assignment.

Note that there can be any number of batch input sessions with the same name. As mentioned, in the present example, no sessions are created, but the data is updated directly. The name of the file with the converted data is already suggested. The RFBIDE00 program brings some messages and then returns to the starting point. To save yourself a click storm of dialog boxes, you should choose the **Log** option instead of the **Dialog Box** option. After that, click the icon (**Execute**).

**Final step: run batch input session**

In the case of your own recordings or other batch input programs, you must then execute the **Run batch input session** step (shown earlier in Figure 9.7). Here, a jump is made to the standard SAP Transaction SM35. However, only the batch input sessions for the selected object are displayed. Note the following: If you have also used the name of the object in other projects or subprojects, batch input sessions from these objects may also be displayed. You have been familiarized with running batch input sessions in Chapter 7. For this reason, we'll not go into more detail at this point.

### Importing Data via Direct Input

If you have selected the direct input import method (see Table 9.6), in the **Start direct input session** step, then—depending on the object type—you either directly call the standard direct input program pertaining to the object or you're given the option to choose between direct input program and direct input transaction.

### Importing Data via BAPIs or IDocs

If you have selected BAPI or IDoc as the import method, the import essentially takes place in three steps:

1. **Start IDoc generation**
   First, the file of the converted data is imported. The information packages contained therein are stored in IDoc format in the SAP database—but not in the database of the associated application. The system assigns a number to each IDoc. Then, the file of the converted data gets deleted.
2. **Start IDoc processing**
   The IDocs created in the first step are transferred to the associated application program. This application program checks the data and, if necessary, posts it to the application's database.
3. **Create IDoc overview**
   You get a status overview where you can drill down to the individual IDoc.

Note the following: The **Start IDoc generation** step will be omitted if you selected the **Create IDocs Directly** option during the data conversion (shown earlier in Figure 9.45).

**Shortcut: generating IDocs directly**

It depends on the settings of the ALE-EDI Customizing whether the second step (**Start IDoc Processing**) is triggered automatically or not. The essential setting for this trigger is stored in the partner profile (for a partner and a message type) (Section 9.2.3). There it is defined whether the processing of the IDocs should be done immediately or with the help of a background program. For more details, see Section 9.5.

[+]

**Inbound IDoc Processing**

When processing inbound IDocs, you should consider the following:

- For the partner profiles created automatically by SAP Legacy System Migration Workbench, the setting is **Trigger by Background Program**. You can change this setting manually at any time.
- When processing inbound IDocs, *work items* are created in the standard system. These elements of the SAP workflow are generally mot needed during data migration. SAP Note 2031151 describes how you can suppress the creation of work items.

Our example, which took us through the main functions of SAP Legacy System Migration Workbench, is now complete.

### 9.2.18 Object Overview

Before we finally finish this extended example, we'll show you how you can get an overview of all the definitions you have made in connection with the CUSTOMER object.

For this task, you need to return to the overview screen of the work steps (shown earlier in Figure 9.7) and click the Object Overview button (**Object Overview**). The result is shown in Figure 9.49, Figure 9.50, and Figure 9.51.

In addition to the overview in list format, an overview in table format is also available (**Overview in Table Format** button). This overview is suitable as a template for "mapping on paper" (Section 9.2.7).

**Reusable rules**

In addition, you can list all the reusable rules actually used in the object (**Overview of Reusable Rules** button). For the purpose of converting the data, you may find it useful that the content of the conversion table is output.

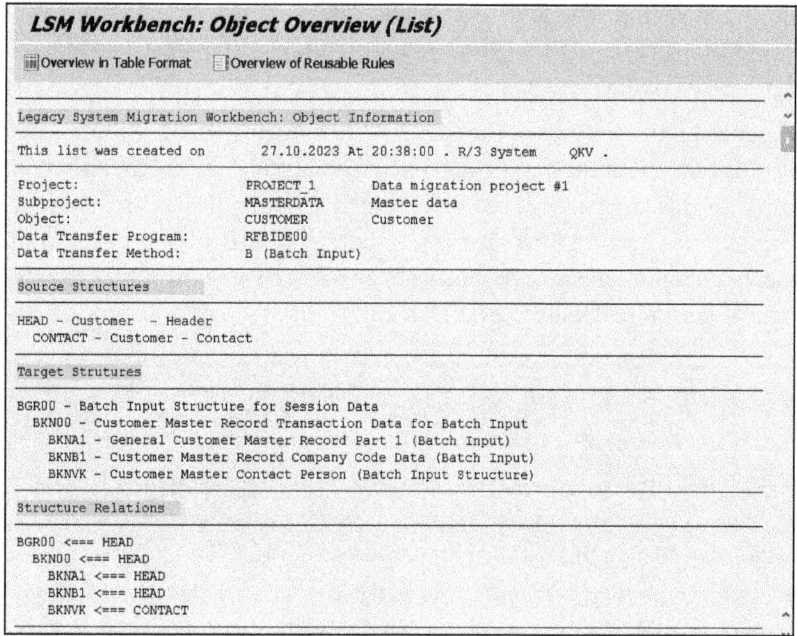

LSM Workbench: Object Overview (List)

Overview in Table Format | Overview of Reusable Rules

```
Legacy System Migration Workbench: Object Information

This list was created on      27.10.2023 At 20:38:00 . R/3 System    QKV .

Project:                      PROJECT_1        Data migration project #1
Subproject:                   MASTERDATA       Master data
Object:                       CUSTOMER         Customer
Data Transfer Program:        RFBIDE00
Data Transfer Method:         B (Batch Input)

Source Structures

HEAD - Customer  - Header
  CONTACT - Customer - Contact

Target Strutures

BGR00 - Batch Input Structure for Session Data
  BKN00 - Customer Master Record Transaction Data for Batch Input
    BKNA1 - General Customer Master Record Part 1 (Batch Input)
    BKNB1 - Customer Master Record Company Code Data (Batch Input)
    BKNVK - Customer Master Contact Person (Batch Input Structure)

Structure Relations

BGR00 <=== HEAD
  BKN00 <=== HEAD
    BKNA1 <=== HEAD
    BKNB1 <=== HEAD
    BKNVK <=== CONTACT
```

**Figure 9.49** Object Overview in List Format: General Data, Source Structures and Target Structures

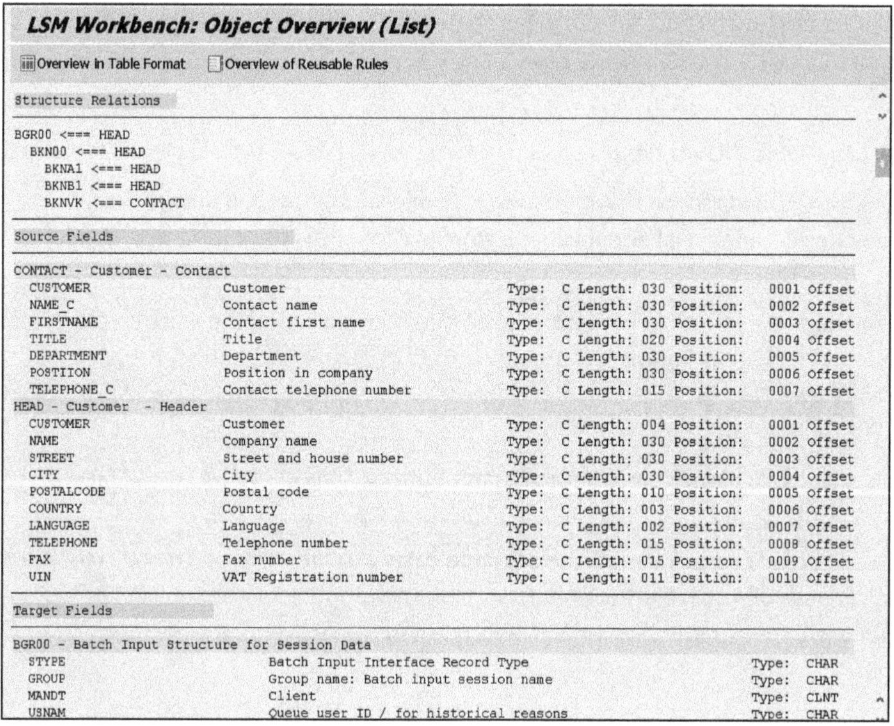

LSM Workbench: Object Overview (List)

Overview in Table Format | Overview of Reusable Rules

```
Structure Relations

BGR00 <=== HEAD
  BKN00 <=== HEAD
    BKNA1 <=== HEAD
    BKNB1 <=== HEAD
    BKNVK <=== CONTACT

Source Fields

CONTACT - Customer - Contact
  CUSTOMER      Customer                      Type:  C Length: 030 Position:   0001 Offset
  NAME_C        Contact name                  Type:  C Length: 030 Position:   0002 Offset
  FIRSTNAME     Contact first name            Type:  C Length: 030 Position:   0003 Offset
  TITLE         Title                         Type:  C Length: 020 Position:   0004 Offset
  DEPARTMENT    Department                    Type:  C Length: 030 Position:   0005 Offset
  POSTION       Position in company           Type:  C Length: 030 Position:   0006 Offset
  TELEPHONE_C   Contact telephone number      Type:  C Length: 015 Position:   0007 Offset
HEAD - Customer  - Header
  CUSTOMER      Customer                      Type:  C Length: 004 Position:   0001 Offset
  NAME          Company name                  Type:  C Length: 030 Position:   0002 Offset
  STREET        Street and house number       Type:  C Length: 030 Position:   0003 Offset
  CITY          City                          Type:  C Length: 030 Position:   0004 Offset
  POSTALCODE    Postal code                   Type:  C Length: 010 Position:   0005 Offset
  COUNTRY       Country                       Type:  C Length: 003 Position:   0006 Offset
  LANGUAGE      Language                      Type:  C Length: 002 Position:   0007 Offset
  TELEPHONE     Telephone number              Type:  C Length: 015 Position:   0008 Offset
  FAX           Fax number                    Type:  C Length: 015 Position:   0009 Offset
  UIN           VAT Registration number       Type:  C Length: 011 Position:   0010 Offset

Target Fields

BGR00 - Batch Input Structure for Session Data
  STYPE         Batch Input Interface Record Type              Type:  CHAR
  GROUP         Group name: Batch input session name           Type:  CHAR
  MANDT         Client                                         Type:  CLNT
  USNAM         Queue user ID / for historical reasons         Type:  CHAR
```

**Figure 9.50** Object Overview in List Format: Structure Relations and Source Fields

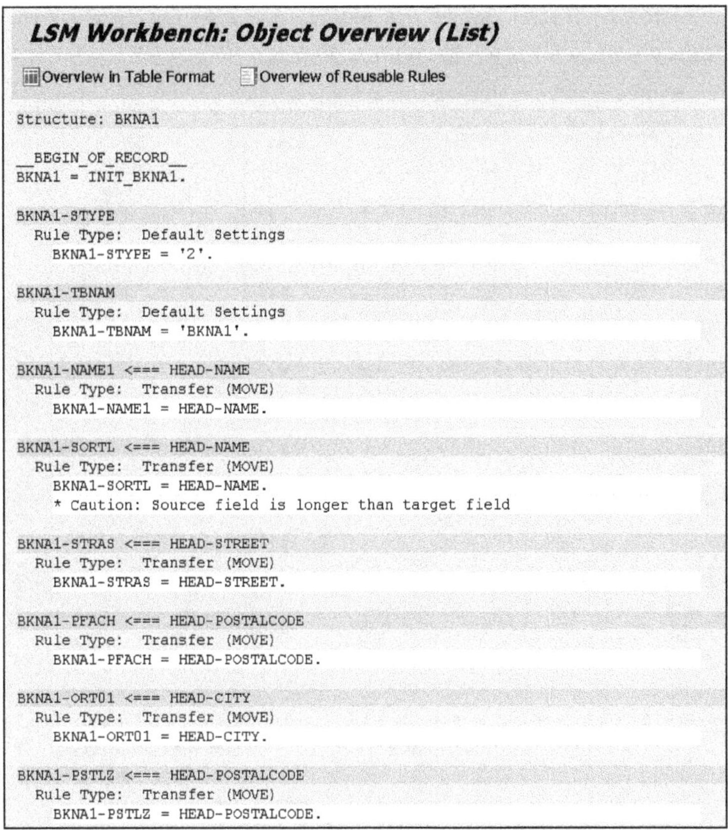


**Figure 9.51** Object Overview in List Format: Conversion Rules (Excerpt)

## 9.3 Recordings

In Section 9.2, we guided you through the central functions of SAP Legacy System Migration Workbench using the example "Migrating Customer Master Data Using Standard Batch Input." In this section, we'll use another example to show you how SAP Legacy System Migration Workbench is integrated into the recording function of the SAP system. You already got to know the transaction recorder in Chapter 7. Of course, we do not present this second example in as much detail as the first. Rather, we focus on where the approach differs from the approach in the first example.

**Example: creating a general ledger account**

The example we have chosen is called "Creating a G/L Account." Now, you wonder why we don't perform this task using the standard batch input program RFBISA00 for the transfer of general ledger accounts. Of course, this would also be possible. However, since our source data has very few fields, hopefully by the end of this section, we've convinced you that the recording function will get you there faster in this case.

Let's assume our task is to create general ledger accounts in the SAP system based on the source data shown in Figure 9.52.

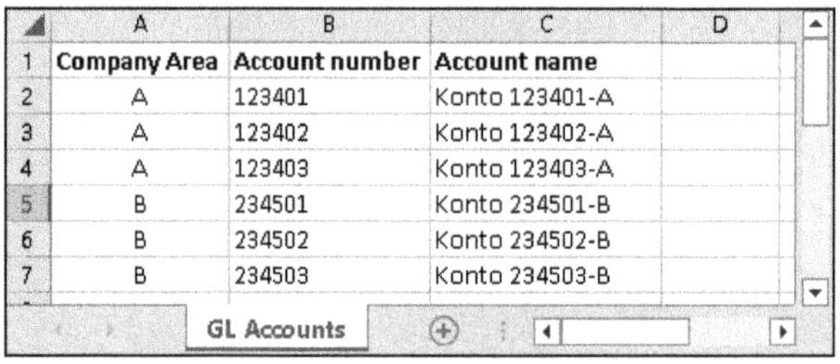

| | A | B | C | D |
|---|---|---|---|---|
| 1 | Company Area | Account number | Account name | |
| 2 | A | 123401 | Konto 123401-A | |
| 3 | A | 123402 | Konto 123402-A | |
| 4 | A | 123403 | Konto 123403-A | |
| 5 | B | 234501 | Konto 234501-B | |
| 6 | B | 234502 | Konto 234502-B | |
| 7 | B | 234503 | Konto 234503-B | |

GL Accounts

**Figure 9.52** Example: Migrating General Ledger Account Master Data (Excerpt)

We further assume that we have already familiarized ourselves with the associated Transaction FS00 and therefore know which data needs to be entered where, as shown in Table 9.7.

| Area/Tab in Transaction FS00 | Field Name | Input |
|---|---|---|
| **Header** | **G/L account** | Account number field from source data |
| | **Company code** | To be derived by converting from the Company Area field of the source data:<br>A→0001<br>B→1000 |
| **Type/Designation** | **Short text** | Account name field from the source data |
| | **G/L account type** | "X" = stock account |
| | **Account group** | SAKO |
| | **Group account number** | "10101011" |
| **Control data** | **Balances in local currency only** | "X" |
| **Entry/Bank/Interest** | **Field status group** | "G001" |

**Table 9.7** Creating a General Ledger Account: Mapping on Paper

With this step, everything is ready. We can now get to work with SAP Legacy System Migration Workbench. The group account number has been set constant here to keep the example as simple as possible.

### 9.3.1 Creating and Postprocessing a Recording

No simulation

In the start screen (shown earlier in Figure 9.4), select **Goto • Recordings**. The overview screen appears, which is still empty and in which all recordings for the current project will be listed, as shown in Figure 9.53.

Note that a recording always belongs to a project. In addition, it is worth remembering that, when a recording is created, the corresponding transaction is actually executed and not just simulated. We therefore create the recording using an account that is not listed in our source data.

[+]

**Creating Recordings Using Transaction SM35**

Unfortunately, the use of recordings created directly in Transaction SM35 is not possible here because SAP Legacy System Migration Workbench requires additional information that a recording created in Transaction SM35 does not contain.

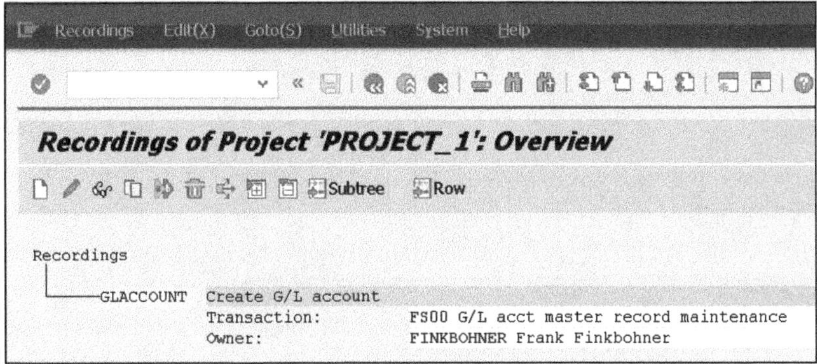


**Figure 9.53** SAP Legacy System Migration Workbench: Recordings

In the recordings overview screen, click the [icon] icon (**Create Recording**). In the dialog box that appears, as shown in Figure 9.54, enter the abbreviation GLACCOUNT and the name "Create G/L account" for the recording.

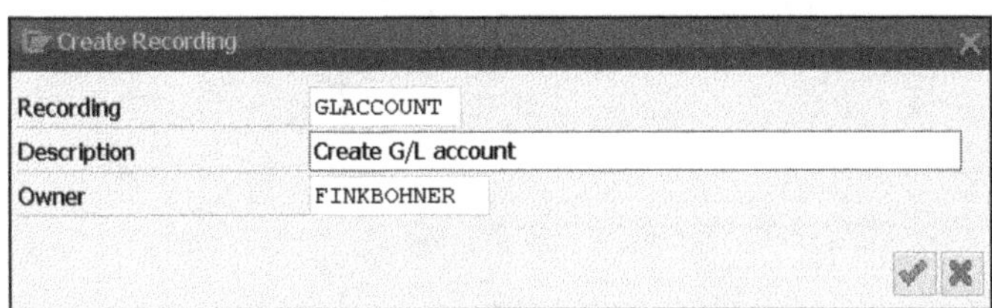


**Figure 9.54** SAP Legacy System Migration Workbench: Creating a Recording

Then, click the [icon] icon (**Next**). You'll now be prompted to enter Transaction FS00, as shown in Figure 9.55.

**Figure 9.55** SAP Legacy System Migration Workbench: Entering the Transaction Code for the Recording

Enter the transaction code and confirm your entry. This step will take you to Transaction FS00, which you fill in with general ledger account 123492. Then, click the icon (**Create**) and edit the general ledger account, as shown in Figure 9.56, Figure 9.57, Figure 9.58, and Figure 9.59.

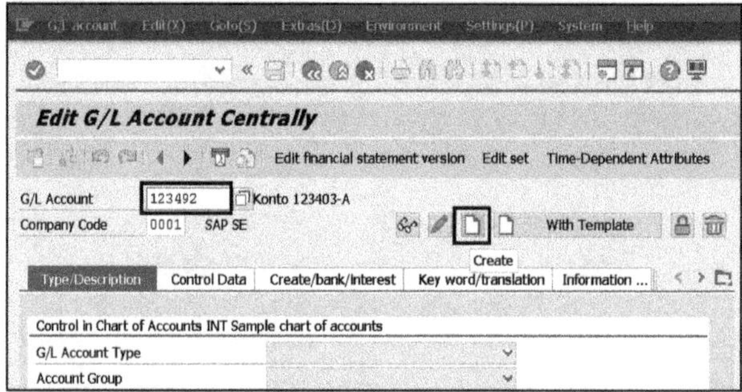


**Figure 9.56** SAP Legacy System Migration Workbench: Recording, Transaction FS00 (Image 1)

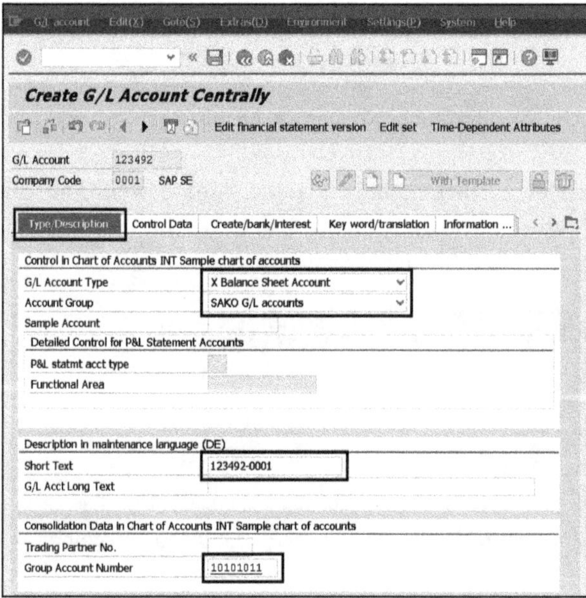


**Figure 9.57** SAP Legacy System Migration Workbench: Recording, Transaction FS00 (Image 2)

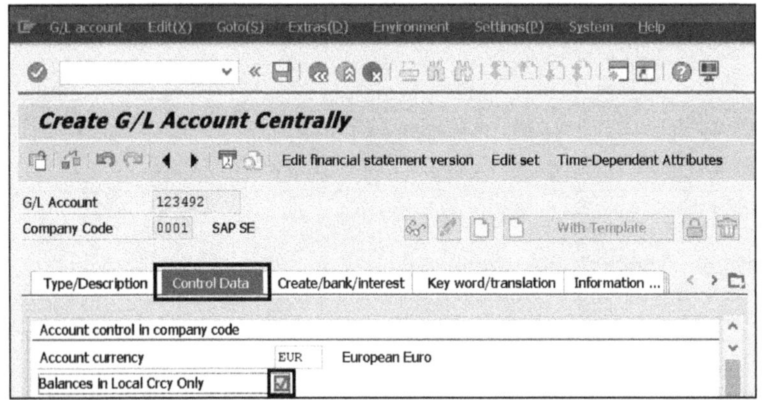


**Figure 9.58** SAP Legacy System Migration Workbench: Recording, Transaction FS00 (Image 3)

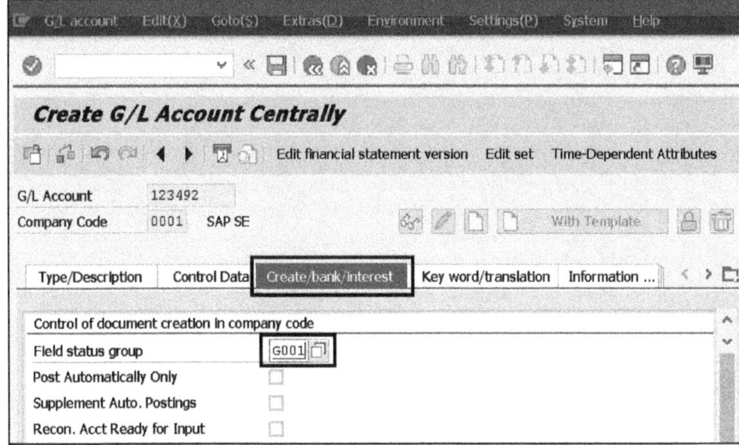


**Figure 9.59** SAP Legacy System Migration Workbench: Recording, Transaction FS00 (Image 4)

Saving the newly recorded account will end the recording, and SAP Legacy System Migration Workbench takes control again.

**Tree-like representation of the recording**

This leads us to a screen in which the recording is shown with its technical components, as shown in Figure 9.60. The representation corresponds to the structure as you already got to know it in Chapter 7. The tree-like structure is organized by transaction (FS00), screens (SAPLGL_ACCOUNT_MASTER_MAINTAIN 2001, etc.) and fields (GLACCOUNT_SCREEN_KEY-SAKNR, etc.).

In the center column, the colored fields contain the values you used when creating the recording or the values suggested to us by the SAP system. The right-hand column is empty, indicating that you need to rework here. If you trace this in the system, you'll clearly see three columns based on the color scheme. In the black and white rendering, the right-hand column is not highlighted clearly enough.

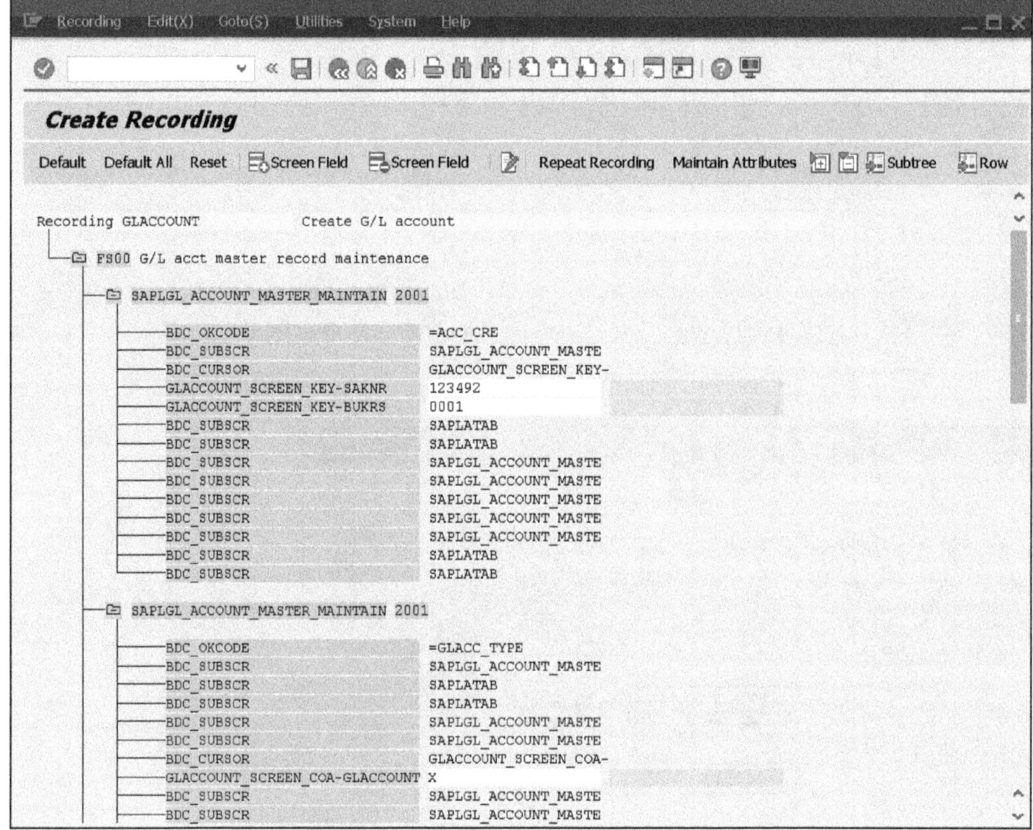


**Figure 9.60** SAP Legacy System Migration Workbench Excerpt: Result of the Recording of Transaction FS00

**Defining field names of the target structure**

From Section 9.2 you already know that SAP Legacy System Migration Workbench requires a target structure to which the source data will be mapped in the further course. A structure requires field names, and you must now assign these field names. For this task, position the cursor on the first input field in the list—the GLACCOUNT_SCREEN_KEY-SAKNR screen field with the value 123492—and double-click it. A dialog box appears asking you to assign a name and label for the field, as shown in Figure 9.61. This information will appear later in the target structure. Enter "GLACCOUNT" in the first **Name** field and "G/L account number" in the second **Name** field, delete the default value and click **Next**.

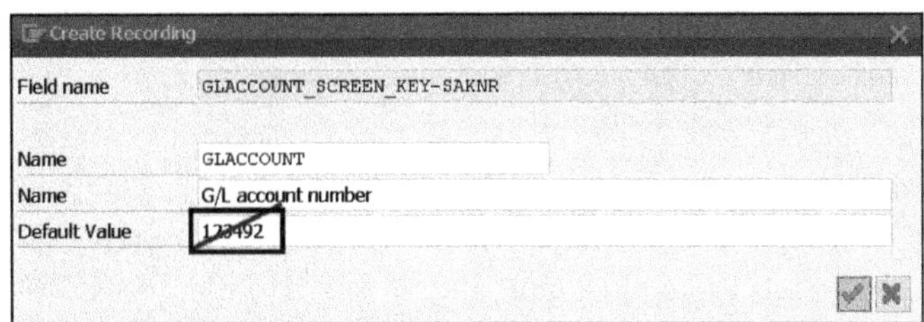


**Figure 9.61** SAP Legacy System Migration Workbench: Recording, Defining Field Names

**Naming the fields to be supplied**

Proceed in the same way with all input fields that you want to supply with field values from the source data. For all other fields, leave the name and label blank. If name and label are empty for a field, SAP Legacy System Migration Workbench automatically adopts the stored default value.



Therefore, you only define names and labels for three fields: GLACCOUNT, COMPANYCODE, and SHORTTEXT Figure 9.62 shows a section of the designation made. The SHORTTEXT field is assigned to the GLACCOUNT_SCREEN_COA-TXT20_ML field in the SAPLGL_ACCOUNT_MASTER_MAINTAIN 2001 screen.

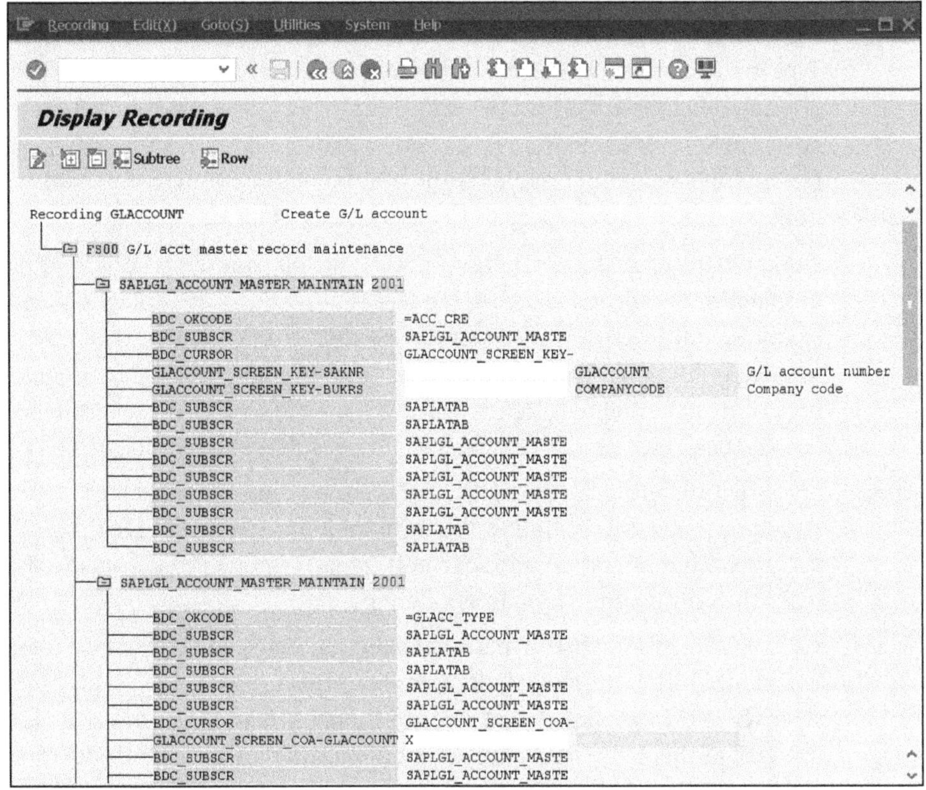


**Figure 9.62** SAP Legacy System Migration Workbench Excerpt: Result of Recording Transaction FS00 (Field Names and Labels)

For technical reasons, at least one field must be assigned a name and a label. Save the result and return to the overview screen of all recordings of the project (shown earlier in Figure 9.53).

"Default" shortcut

Before we continue, we want to show you a "shortcut" for the last step. In the display in Figure 9.60, where you have just post-processed the recording, you'll find the **Default value** or **Default all** buttons. If you use this, SAP Legacy System Migration Workbench automatically assigns names and labels for the field on which the cursor is positioned, or for all fields, which are derived from the underlying screen fields. The advantage is speed. The disadvantage is that the field names are not always meaningful. In addition, it can happen that in this way, two fields are assigned the same name, with the consequence that only one field with this name is available in the field mapping.

### 9.3.2 Using Recordings

Defining source structure and source fields

Now you can already use the recording in SAP Legacy System Migration Workbench. For this task, create an object named "GLACCOUNTS" with the description "G/L account master data" in project PROJECT_1 and subproject MASTERDATA. In the **Define Object Attributes** step, you want to select the **Object Type item and then the Batch Input Recording** import method. Then use the F4 input help to select the GLACCOUNT recording, as shown in Figure 9.63. In addition, SAP Legacy System Migration Workbench provides the option to put multiple recordings one after the other to apply two or more transactions sequentially to a data record. However, this feature is rarely used.

Similar to the procedure in Section 9.2, we now continue with the definition of source structure and source fields. This is not particularly difficult in our example, since we use only one source structure and three source fields. (Generally speaking: When using the recording technique, only one source structure can be defined). Define an ACCOUNT source structure with the fields COMPANYCODE, GLACCOUNT and DESCRIPTION. The result is shown in Figure 9.64.

Structure relations

The structure relations are trivial, since they are only a source and a target structure. If you call the **Define Structure Relations** step, you'll see that SAP Legacy System Migration Workbench is "smart" enough to automatically make the only possible assignment right here.

Field mapping and conversion rules

We now come to the maintenance of field mapping and conversion rules. This step, which is normally quite complex, is also very simple in this case, since we only have three target fields to supply. The result is shown in Figure 9.65. At this stage, we want to point out again that all other fields from the recording are automatically filled with the default values stored in the recording.

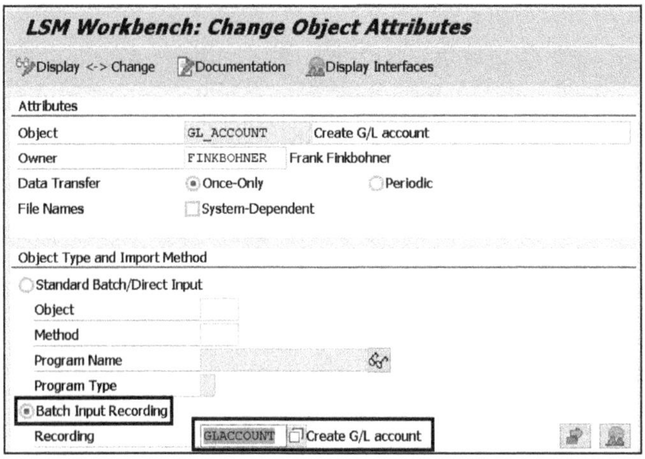

LSM Workbench: Change Object Attributes

Display <-> Change | Documentation | Display Interfaces

Attributes

Object: GL_ACCOUNT Create G/L account

Owner: FINKBOHNER Frank Finkbohner

Data Transfer: Once-Only / Periodic

File Names: System-Dependent

Object Type and Import Method

Standard Batch/Direct Input

Object

Method

Program Name

Program Type

Batch Input Recording

Recording: GLACCOUNT Create G/L account

**Figure 9.63** SAP Legacy System Migration Workbench: Recording, Changing Object Attributes

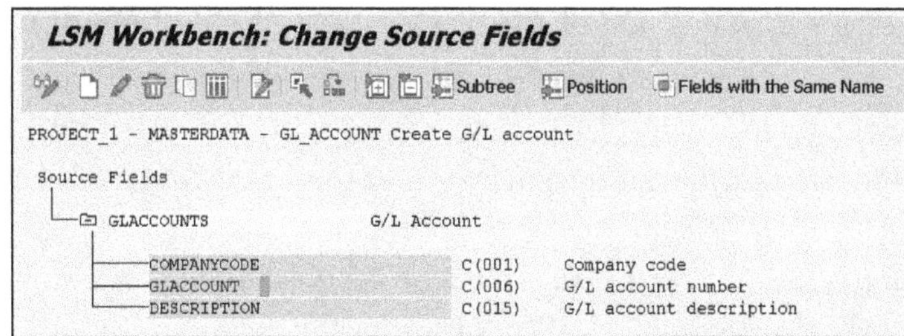


**Figure 9.64** SAP Legacy System Migration Workbench: Recording, Source Fields

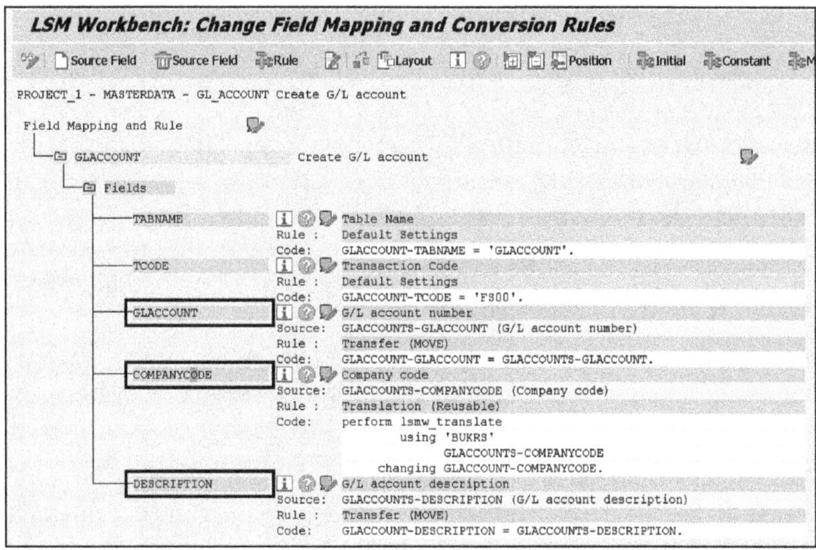


**Figure 9.65** SAP Legacy System Migration Workbench: Recording, Field Mapping

**Additional steps**

The last steps can be done quickly:

- Specify the copy of the Microsoft Excel file in the **Text (tab-delimited) format (*.txt)** as the file of the source data and assign it.
- Import the data and display the imported data.
- Convert the data and display the converted data.
- Maintain the translation for the company codes.
- Convert the data again.
- Create a batch input session and run it.

**Applicability of the recording method**

This example was supposed to demonstrate to you that the recording method in SAP Legacy System Migration Workbench quickly leads to the goal, provided that it is applicable.

However, under what conditions is the recording method applicable?

- It must be possible to represent the source data in a single table, and in such a way that one row corresponds to one transaction. Remember our first example from Section 9.2? There we had one variable per customer, i.e., any number of contacts. In this case, the recording method is not applicable.
- For each data record, the transaction must pass through the same screen sequence.
- In SAP S/4HANA, you must make sure that the transaction is still supported by the system.

## 9.4 Transporting Projects

SAP Legacy System Migration Workbench allows you to transport the data for a project both using the SAP transport system and via download and upload between two SAP systems. The programs generated by SAP Legacy System Migration Workbench are not transported, but are regenerated in the target system.

### 9.4.1 Creating a Change Request

At the push of a button, you can create an SAP change request that contains all the information about an SAP Legacy System Migration Workbench project.

This SAP change request can be exported and imported using the usual means of the SAP Change and Transport system. You can find this function on the initial screen (shown earlier in Figure 9.4) under **Extras • Create Change Request**.

**SAP Change and Transport system**

If you transport data for a project in the way described here, you can always track who transported what to where and when in the SAP Change and Transport system.

Keep the following in mind:

- When importing such a change request, the entire project is first deleted in the target system and then recreated.
- When you export the transport request, the current version is exported and not the version at the time the transport request was created.
- Since the entire project will be exported, you should make sure that the project has a "clean" state at the time of export.



### 9.4.2 Exporting a Project

In the initial screen (shown earlier in Figure 9.4), select **Extras • Export Project**. This first displays the structure tree of the selected project, as shown in Figure 9.66. Using [Select / Deselect] you can select or deselect parts of the project for the export. If you left-click on a section, such as Fixed values, all fixed values will be deselected or selected. Parts selected for export are marked with [icon]. After that, click the **Export** [F8] button. This will create a text file. Of course, the corresponding documentation for the selected elements will also be exported.

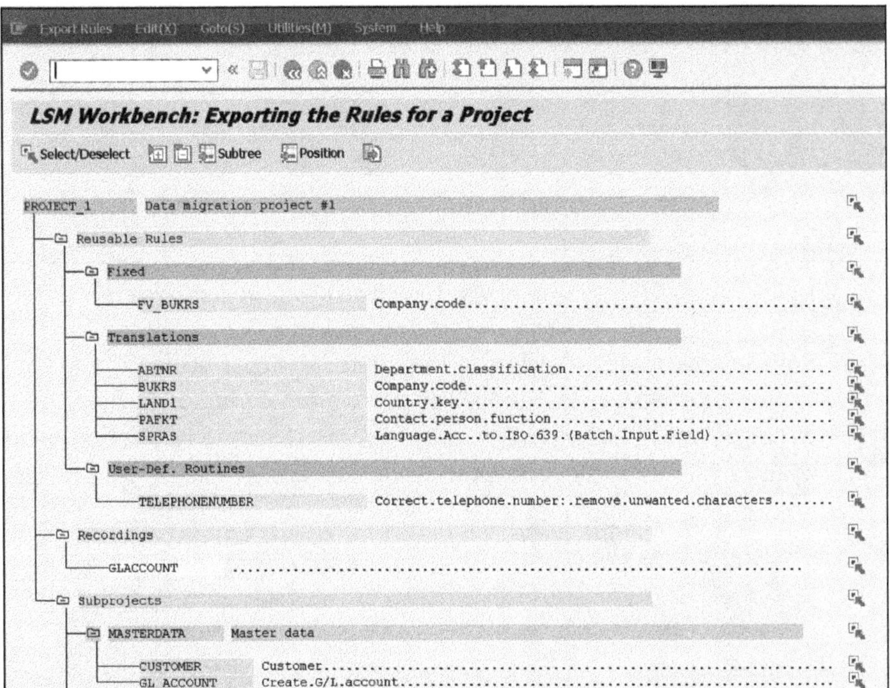


**Figure 9.66** SAP Legacy System Migration Workbench: Exporting a Project

### 9.4.3 Importing a Project

The exported project can be imported into another SAP system. For this task, select **Extras • Import Project** in the initial screen of the other SAP system. You'll be prompted to specify the name of the text file. The file then gets imported and the content gets analyzed. Once the analysis is complete, you'll receive a list of the identified subprojects and objects, as shown in Figure 9.67.

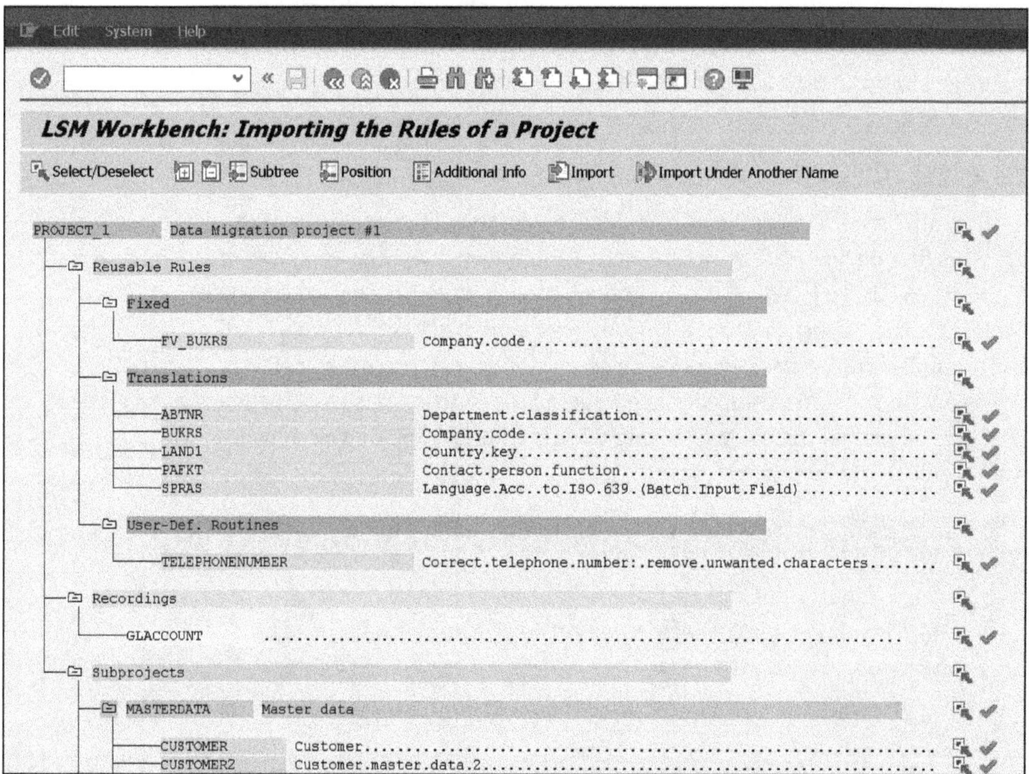


**Figure 9.67** SAP Legacy System Migration Workbench: Importing a Project

**Overwriting existing elements**

You can now tick the objects to be imported. Existing data of the project are marked by check marks; they will be overwritten during the import. Overwriting a project that already exists in the target system can be prevented by using the **Import Under Another Name** function.

**Clients During the Import**

Note that when importing the settings for inbound IDoc processing (Section 9.5), the source client is not automatically replaced by the target client. However, you can manually adjust the client in the text file generated during export before importing. You can find the relevant entry in the text file in Section "[Project Information]" in line "/SAPDMC/LSMCUST".

## 9.5 Preparatory Measures for Using Inbound IDoc Processing

*IDocs* (Intermediate Document) were developed for the exchange of messages between different systems (e.g., between two SAP systems). A detailed treatment of this topic would undoubtedly go beyond the scope of this book. However, you'll find some useful information on ALE and IDoc in Chapter 8.



We have already indicated several times that SAP Legacy System Migration Workbench supports this import method. However, in order to use it, some presets and preparations are required. You can find these settings summarized in the initial SAP Legacy System Migration Workbench screen, as shown in Figure 9.4) under **Settings • Inbound IDoc Processing**, as shown in Figure 9.68. These settings must be made per client and project.

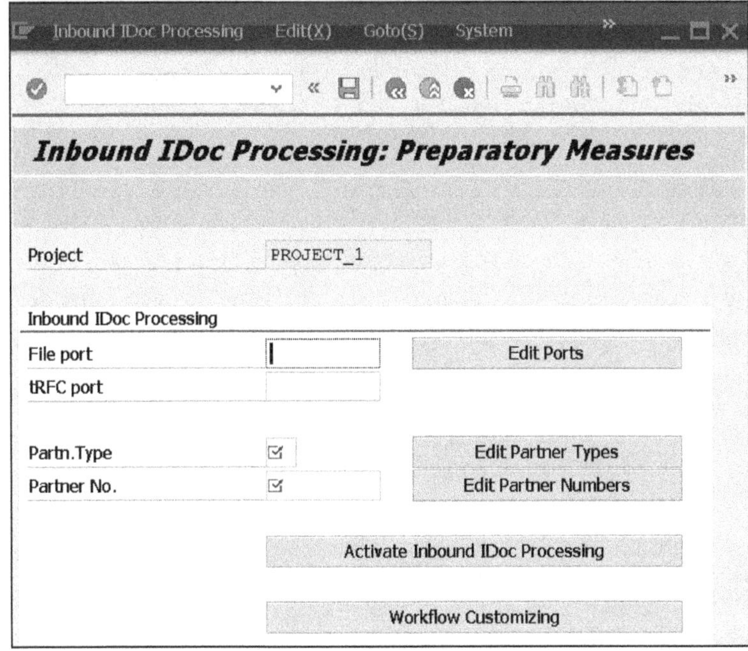


**Figure 9.68** SAP Legacy System Migration Workbench: Settings for Inbound IDoc Processing

In detail, you must perform the following steps:

1. **File port**
   First, a file port must exist for the transfer of the file. If necessary, create a port of type **File** via **Edit Ports** by positioning the cursor on **File** and selecting **Create**. For this task, you must be in change mode. We recommend the settings from Table 9.8.

| Field | Value |
|---|---|
| **Port** | LSMW |
| **Description** | Legacy System Migration Workbench |
| **Version** | IDoc record types SAP Release 4.x |
| **Outbound file** | Enter a file path that your system administrator can tell you and a file name (e.g., *filelsmw*). |

**Table 9.8** Inbound IDoc Processing: File Port

[»]

**Custom Directory for Data Migration**

It has proven to be very helpful if you're assigned separate directories for data migration. Attention should also be paid to the maximum storage size of the directory. Files for data migration often require several hundred megabytes. So the directory should have a few gigabytes of free space.

2. **tRFC Port**
   In addition, you can specify a tRFC port. This port is required if you do not want to create a file during data conversion, but want to transfer the data in packages directly to inbound IDoc processing. The recommended settings are listed in Table 9.9.

| Field | Value |
|---|---|
| **Port** | <Ask your administrator about the naming convention in your company>. |
| **Version** | IDoc record types SAP Release 4.x |
| **RFC Destination** | <Name of the SAP system> |
| **Description** | Legacy System Migration Workbench |

**Table 9.9** Inbound IDoc Processing: tRFC Port

3. **Partner Type**
   Define a partner type or select a partner type. We recommend setting **Partner Type** to **US**.
4. **Partner Number**
   Define a partner number or select a partner number. We recommend the settings shown in Table 9.10.

| Field | Value/Comment |
|---|---|
| **Partner Number** | LSMW |
| **Partner Type** | US |
| **Classification/ Partner Status** | A = Active, partner profile is ready for use |
| **Type** | US |
| **Processor** | <Your user ID> |
| **Language** | DE or EN |

**Table 9.10** Inbound IDoc Processing: Partner Number

5. **Inbound IDoc Processing**
   Activate inbound IDoc processing via **Activate Inbound IDoc Processing**. Confirm the prompt with **Yes**. This must be done once per system.
6. **Workflow Customizing**
   Verify the workflow customizing. This must also be done once per system.

The following entries of the workflow runtime system must have a green checkmark, as shown in Figure 9.69:

- **Configure RFC Destination**
- **Edit System Administrator for Workflow**
- **Classify Decision Task as General**
- **Activate Send to Objects and HR Objects**

You can start automatic customizing for this purpose. You should then set the **Schedule Background Jobs SAP Business Workflow** item to **Not Scheduled**. Otherwise, the SAP system would repeatedly attempt to post incorrect IDocs that arise during data migration.

[«]

**Automatic Workflow Customizing**

You can also start the automatic workflow customizing afterwards via Transaction SWU3. SAP Note 2366252 describes this transaction in detail.

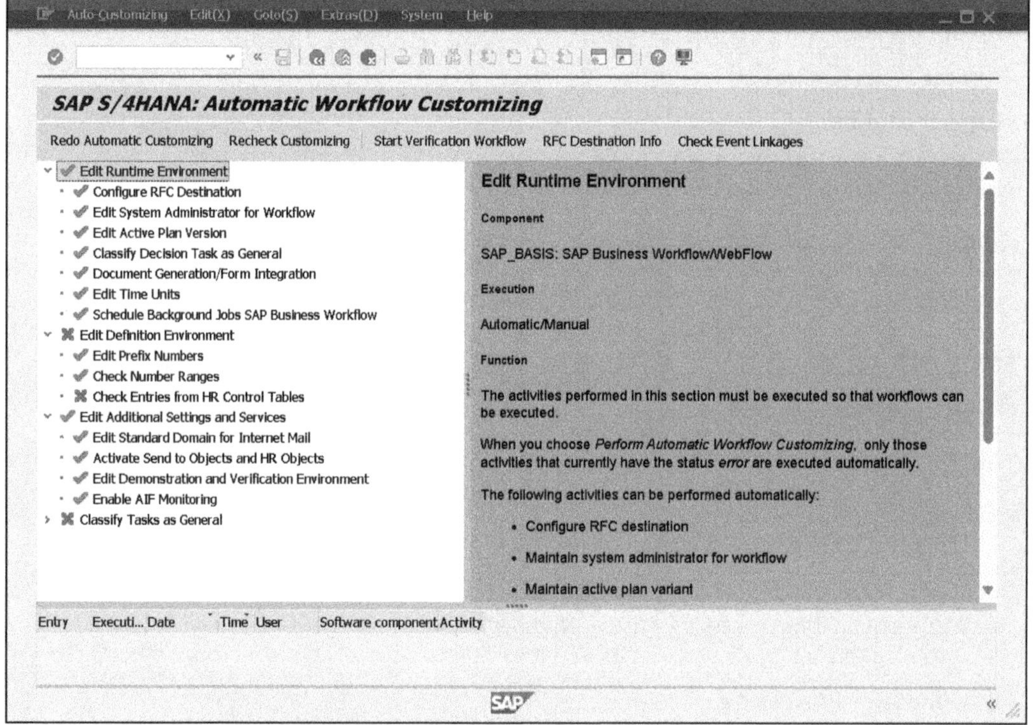


**Figure 9.69** Workflow Customizing

## 9.6 SAP Legacy System Migration Workbench Conversion Program

In the following sections, we describe the structure of SAP Legacy System Migration Workbench conversion program. If you know what the generated conversion program will look like in the end and where individual pieces of code in your field mapping will be generated to, you can use that code in a much more targeted and refined way. This knowledge is helpful if you use your own code or routines in your field mapping to be able to map more complex mappings.

### 9.6.1 Displaying Additional Menus

**Personal settings**

As described in Section 9.2.2, you can compose your own user menu. To make it easier for you to follow and check the tips below, you should include the additional **Generate Conversion Program** and **Display Conversion Program** steps in your user menu. For this task, open the **User Menu** window and activate the mentioned steps, as shown in Figure 9.70.

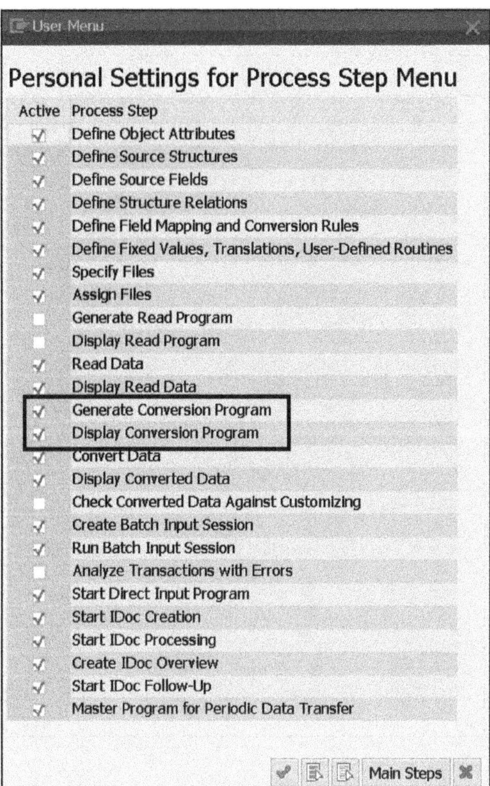


**Figure 9.70** Personal Settings: Displaying the Conversion Program

### 9.6.2 Customizing the Display Variant

Customize the display variant in the field mapping as described in Section 9.7.1, and activate all listed options as in Figure 9.72.

### 9.6.3 Structure of the Conversion Program

Based on the definitions and the field mapping, SAP Legacy System Migration Workbench generates a temporary local ABAP report in the `/1CADMC/` namespace, such as `/1CADMC/SAP_LSMW_CONV_00000003`, when the **Convert Data** step is called. The number at the end of the report gets incremented for each new conversion program that's generated locally.

[!]

> **Never Change Import or Conversion Programs**
>
> Since import and conversion programs are local programs, they cannot be transported to other systems. They are regenerated in each system, assigned a new number and thus a new name. Changes to the code of these programs are deleted again each time they are regenerated.

Calling the program

Call the additional menu item **Display Conversion Program**, as shown in Figure 9.71, to display the conversion program.

```
Process Step
Define Object Attributes
Define Source Structures
Define Source Fields
Define Structure Relations
Define Field Mapping and Conversion Rules
Define Fixed Values, Translations, User-Defined Routines
Specify Files
Assign Files
Read Data
Display Read Data
Generate Conversion Program
Display Conversion Program
Convert Data
Display Converted Data
Create Batch Input Session
Run Batch Input Session
```

**Figure 9.71** "Display Conversion Program" Work Step

Structure of the report

The report is structured as shown in Table 9.11.

| Program section | What happens here? |
|---|---|
| `Data declaration` | Data definitions and integration of standard SAP Legacy System Migration Workbench. |
| `__GLOBAL_DATA__` | Custom data declarations and code at label `__GLOBAL_DATA__`. The best place to put your own selection screen parameters, data declaration, and initialization. |
| `Selection parameters` | Definition of the standard selection screen for the implementation report. |
| `Initialization` | Definition of some global variables. |
| `Start-of-Selection` | Here the generated program components are called. |

**Table 9.11** Program Sections of SAP Legacy System Migration Workbench Conversion Program

| Program section | What happens here? |
|---|---|
| PERFORM determine_wildcardvalues | Import the defined wildcard values depending on the so_wldcd selection parameter. All subsequent program sections are then called within a LOOP..ENDLOOP loop using these wildcard values (only available if wildcard values have been defined under **Specify Files**). |
| PERFORM initializations | Get file paths and initialize other global variables. Import further object-dependent values and open imported files. |
| PERFORM reset_counters_and_flags | Standard routine for resetting counters and flags. |
| PERFORM reset_tables | Standard routine for resetting the internal tables. |
| __BEGIN_OF_ PROCESSING__ | Custom code at processing time BEGIN_OF_PROCCESSING (only available if the code was defined within that event). |
| PERFORM execute_data_conversion<br>PERFORM convert_transaction<br>PERFORM convert_0001<br>PERFORM convert_<nnnn> | Generated routine that performs the conversion.<br>Within this routine, **convert_transaction** gets called for field mapping. For each target structure another converted routine—**convert_<nnnn>**—is called here, which contains the field mapping for this target structure. |
| PERFORM final_message | Standard routine for displaying the counters and the standard protocol. |
| PERFORM close_files | Standard routine for closing the used files. |
| __END_OF_PROCESSING__ | Custom code at processing time END_OF_PROCESSING (only available if the code was defined within that event). |
| PERFORM write_protocol | Standard routine for determining the runtime. |

**Table 9.11** Program Sections of SAP Legacy System Migration Workbench Conversion Program (Cont.)

| Program section | What happens here? |
|---|---|
| FORM Routines | Integration of standard SAP Legacy System Migration Workbench subroutines (FORM routines). |
| Generated form routines | Generated Form routines and custom routines per SAP Legacy System Migration Workbench object. |

**Table 9.11** Program Sections of SAP Legacy System Migration Workbench Conversion Program (Cont.)

## 9.7 Functions for Advanced Users

In this section, we present some expert features that will be of interest to you as you become more deeply involved with SAP Legacy System Migration Workbench:

- Display variant and processing times
- Global variables
- Global functions
- Naming for reusable rules

### 9.7.1 Display Variant and Processing Times

In the **Define Field Mapping and Conversion Rules** step (shown earlier in Figure 9.7), click on the **Hide and Display Fields/Code** Layout button; the dialog box **Determine Layout** appears, as shown in Figure 9.72.

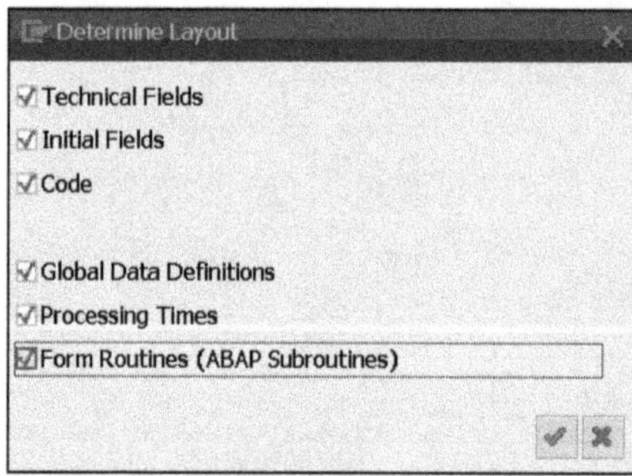


**Figure 9.72** SAP Legacy System Migration Workbench: Determining the Display Variant

This feature is especially useful for experienced users who want to modify their field mapping. You can determine which information is displayed:

- **Technical Fields**
  Technical fields are target fields for which SAP Legacy System Migration Workbench proposes a conversion rule (e.g., constant). As a rule, no changes need to be made here.
- **Initial Fields**
  Initial fields are the fields that have no source field and no conversion rule assigned to them. This selection is used in particular to compress the display. However, it is not recommended until you have already identified the target fields that are relevant to you.
- **Code**
  If you check **Code**, the ABAP code implicitly formulated by SAP Legacy System Migration Workbench or explicitly formulated by you gets displayed, if available.
- **Global Data Definitions**
  If **Global Data Definitions** is checked, the `__GLOBAL_DATA__` label will be displayed. There you can define variables, structures, tables, etc. that you want to use in your own code in the field mapping. Thus, all data definitions are bundled in a central location, which improves clarity.
- **Processing Times**
  **Processing Times** (events) enables you to insert your own code at specific processing times. The following processing times are available, as shown in Table 9.12.



| Processing Time | Default Assignment |
|---|---|
| `__BEGIN_OF_PROCESSING__` | Before the processing of the data starts |
| `__BEGIN_OF_TRANSACTION__` | Before the processing of the data for a transaction starts |
| `__BEGIN_OF_RECORD__` | Before applying the conversion rules to a source structure |
| `__END_OF_RECORD__` | After applying the conversion rules to a source structure<br>Default assignment: `Transfer_record`. |
| `__END_OF_TRANSACTION__` | After the end of the processing of transaction<br>Default assignment:<br>`Transfer_transaction`. |
| `__END_OF_PROCESSING__` | After the end of the processing of all transactions |

**Table 9.12** Processing Times

- **Form Routines**
  Finally, you can display the `__FORM_ROUTINES__` label for **Form Routines (ABAP Subroutines)**. Here you can define ABAP subroutines that you want to use in your own code in the field mapping or include includes with `FORM` routines.

[!]

**Display Variants Are Neither Exported Nor Transported**

Display variant settings are neither exported nor transported. So, you should not be alarmed if your entire code seems to have "disappeared" in the field mapping of a freshly imported project. Don't panic, this is only the display variant that is still set to the initial values. Simply re-enable all the settings you want, and the seemingly lost code becomes visible again.

### 9.7.2 Global Variables

SAP Legacy System Migration Workbench uses a number of global variables internally. You can use these variables in your own ABAP code within SAP Legacy System Migration Workbench. The variables that can be used in your ABAP code are shown in Table 9.13.

| Global Variable | Description |
|---|---|
| `g_project` | Current project |
| `g_subproj` | Current subproject |
| `g_object` | Current object |
| `g_record` | Current target structure |
| `g_cnt_records_read` | Number of records imported so far |
| `g_cnt_records_skipped` | Number of records skipped so far |
| `g_cnt_records_transferred` | Number of records transferred to the output file so far |
| `g_cnt_transactions_read` | Number of transactions imported so far |
| `g_cnt_transactions_skipped` | Number of transactions skipped so far |
| `g_cnt_transactions_transferred` | Number of transactions transferred to the output file so far |
| `g_cnt_transactions_group` | Number of transactions contained in the current batch input session |

**Table 9.13** Global Variables of SAP Legacy System Migration Workbench

| Global Variable | Description |
|---|---|
| g_userid | User ID |
| g_groupname | Name of the current batch input session |
| g_groupnr | Sequence number of the current batch input session |

**Table 9.13** Global Variables of SAP Legacy System Migration Workbench (Cont.)

### 9.7.3 Global Functions



SAP Legacy System Migration Workbench provides a set of functions that can be used anywhere in the ABAP code. These functions enable you to influence the flow of the data conversion program—in some cases considerably. For this reason, you should use these functions very carefully. The functions that are available are shown in Table 9.14.

| Global Function | Description |
|---|---|
| transfer_record. | The current record (for the current target structure) is transferred to the output buffer (see also Section 9.8.4). |
| transfer_this_record '...'. | A record of another target structure is transferred to the output buffer. The name of the target structure must be specified as an argument in quotation marks. |
| at_first_transfer_record. | The current record is transferred to the output buffer if it is the first transaction. |
| on_change_transfer_record. | The current record is transferred to the output buffer if it has changed since the last record (see also Section 9.9.2). |
| transfer_transaction. | The current transaction is written to the output file. This transfers all records of the output buffer to the output file. |
| skip_record. | The current record is not transferred to the output buffer (Section 9.8.2). |
| skip_transaction. | The current transaction is not written to the output file (Section 9.8.3). |

**Table 9.14** Global Functions of SAP Legacy System Migration Workbench

### 9.7.4 Naming for Reusable Rules

You have become familiar with reusable rules as rules that can be used project-wide, that is, in all objects of a project. Reusable rules are fixed values, translations, and user-defined routines.

**Up to three proposed names**

When you assign a reusable rule to a target field, the system proposes up to three different names. To understand these naming suggestions, we need to look at the definition of data objects in the SAP system.

**Data objects in the SAP system**

The definition of data objects in the SAP system takes place in three stages:

1. **Domain**
   On the lowest level, technical properties are defined, such as the field type, field length, value table, or fixed value.
2. **Data element**
   On the second level, based on a domain and its properties, "semantic" properties are defined, such as language-dependent texts, documentation, etc.
3. **Field**
   At the top level, based on a data element, properties of the field are defined in the context of a structure or a table, such as foreign key relations, input helps, etc.

In particular, thus, a domain generally has multiple data elements that reference the domain. A data element usually has multiple fields that reference this element.

**Accepting the recommendation**

As a rule, it is recommended you accept the name proposed by the system. An exception to this recommendation is when the domain is of a very general nature, such as CHAR1 or XFIELD. If you were to use the name of the domain in such a case, the reusable rule might not be usable with another field with the same domain since that field might have a completely different meaning. This naming procedure keeps the number of conversion rules small and ensures consistency in data conversion. Table 9.15 shows an example.

| No. | Field | Data Element | Domain | Description |
|---|---|---|---|---|
| 1 | BUKRS | BUKRS | BUKRS | Company code |
| 2 | CO_ CODE | CO_CODE | BUKRS | Company code |

**Table 9.15** Naming

Both fields have the description *company code*. The field names are different, but the domain is the same. Accordingly, both fields should be supplied with the same fixed value or the same conversion or the same user-defined routine.

## 9.8 Application Examples for Advanced Users

In this section, we describe some techniques that you probably won't need very often. However, if you have a corresponding use case, the solutions we present in this section might save you a considerable amount of work.

### 9.8.1 Determining the Transaction Code at Runtime

Let's suppose you want to transfer records, some of which are already created in the system. For the existing records, you want to call the respective change transaction instead of the entry transaction. Using the example of the customer master, we'll explain the solution.



**Creating or changing**

First, you must determine the situation. In our example customer master, check whether an entry with the customer number in question exists in table KNA1. In detail, you would carry out the following steps:

1. Add the code shown in Listing 9.2 into __GLOBAL_DATA__ (Section 9.7.2).

   ```
   TABLES: KNA1.
   ```

   **Listing 9.2** Data Definition in Global Data for Customer Number

2. Insert the ABAP code shown in Listing 9.3 into field BKN00-TCODE.

   ```
   SELECT count(*) FROM KNA1
     WHERE kunnr = <old_customer_number>.
   IF sy-dbcnt = 0.
     bkn00-tcode = 'XD01'.
   ELSE.
     bkn00-tcode = 'XD02'.
   ENDIF.
   ```

   **Listing 9.3** ABAP Code for Finding the Customer Number

This code calls Transaction XD01 for records to be inserted and calls Transaction XD02 for records to be changed. Replace the <old_customer_number> placeholder with the source field of your defined source structure that contains the customer number to be checked, for example, HEAD-CUSTOMER.

### 9.8.2 Skipping a Record

You may not want to transfer certain data records from your source system to your SAP system. Of course, you can also perform this filtering process already when exporting the data from the source system. Now, we want to show you how you can also do this conveniently in SAP Legacy System Migration Workbench. When we want to skip a data record, we don't want it converted and transferred to the output file. The solution is simple: Formulate the corresponding condition in ABAP and include it in the following ABAP statements:

```
IF <condition>.
  skip_record.
ENDIF.
```

Within the code, these ABAP statements can be inserted into any field of the relevant target structure. Note that this code skips only one record (for example, a contact for a customer) and not the entire data object (e.g., customers).

### 9.8.3 Skipping All Data Records of a Transaction

If you skip all records of a data object (a transaction) depending on a certain condition, you must insert the following ABAP statements anywhere in the field mapping:

```
IF <condition>.
  skip_transaction.
ENDIF.
```

To avoid unnecessary further code, it often makes sense to insert an EXIT command after skip_transaction.

### 9.8.4 Duplicating a Record

**Example** Let's suppose you want (or need) to create two or more target records from a single source record. For example, in the source system, your customer master consists of a record that contains, among other things, the FIRSTNAME, NAME, and TELEPHONE fields for two contacts, as shown in Figure 9.73. In the SAP system, on the other hand, a BKNVK record must be filled for each contact person.

You can solve this problem by including the code from Listing 9.4 in the _END_OF_RECORD processing event for the BKNVK target structure.

```
transfer_record.
BKNVK = INIT_BKNVK.
BKNVK-STYPE = '2'.
BKNVK-TBNAM = 'BKNVK'.
BKNVK-NAME1 = HEAD-NAME2.
BKNVK-TELF1 = HEAD-TELEPHONE2.
BKNVK-NAMEV = HEAD-FIRSTNAME2.
transfer_record.
```

**Listing 9.4** Duplicating a Record

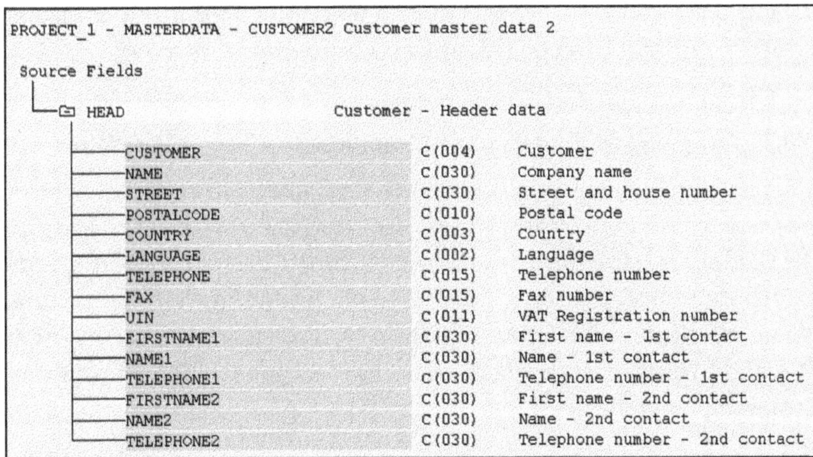

PROJECT_1 - MASTERDATA - CUSTOMER2 Customer master data 2

Source Fields

HEAD Customer - Header data

| Field | Type | Description |
|---|---|---|
| CUSTOMER | C(004) | Customer |
| NAME | C(030) | Company name |
| STREET | C(030) | Street and house number |
| POSTALCODE | C(010) | Postal code |
| COUNTRY | C(003) | Country |
| LANGUAGE | C(002) | Language |
| TELEPHONE | C(015) | Telephone number |
| FAX | C(015) | Fax number |
| UIN | C(011) | VAT Registration number |
| FIRSTNAME1 | C(030) | First name - 1st contact |
| NAME1 | C(030) | Name - 1st contact |
| TELEPHONE1 | C(030) | Telephone number - 1st contact |
| FIRSTNAME2 | C(030) | First name - 2nd contact |
| NAME2 | C(030) | Name - 2nd contact |
| TELEPHONE2 | C(030) | Telephone number - 2nd contact |

**Figure 9.73** SAP Legacy System Migration Workbench: All Fields in a Source Structure



The corresponding set of rules then is shown in Figure 9.74.

```
PROJECT_1 - MASTERDATA - CUSTOMER2 Customer master data 2
  BKNVK                  Customer Master Contact Person (Batch Input Structure)
    Fields
      __BEGIN_OF_RECORD__  Before Using Conversion Rules
                           Rule :  Default Settings
                           Code:   BKNVK = INIT_BKNVK.
      STYPE                Batch Input Interface Record Type
                           Rule :  Default Settings
                           Code:   BKNVK-STYPE = '2'.
      TBNAM                Table Name
                           Rule :  Default Settings
                           Code:   BKNVK-TBNAM = 'BKNVK'.
      NAME1                Name 1
                           Source: HEAD-NAME1 (Name - 1st contact)
                           Rule :  Transfer (MOVE)
                           Code:   BKNVK-NAME1 = HEAD-NAME1.
      TELF1                First telephone number
                           Source: HEAD-TELEPHONE1 (Telephone number - 1st contact)
                           Rule :  Transfer (MOVE)
                           Code:   BKNVK-TELF1 = HEAD-TELEPHONE1.
                                   * Caution: Source field is longer than target field
      NAMEV                First Name
                           Source: HEAD-FIRSTNAME1 (First name - 1st contact)
                           Rule :  Transfer (MOVE)
                           Code:   BKNVK-NAMEV = HEAD-FIRSTNAME1.
      __END_OF_RECORD__    After Using Conversion Rules
                           Rule :  Default Settings Modified
                           Code:   transfer_record.
                                   bknvk = init_bknvk.
                                   bknvk-stype = '2'.
                                   bknvk-tbnam = 'BKNVK'.
                                   bknvk-name1 = head-name2.
                                   bknvk-telf1 = head-telephone2.
                                   bknvk-namev = head-firstname1.
                                   transfer_record.
```

**Figure 9.74** SAP Legacy System Migration Workbench: Duplicating a Record

**Processing time: __END_OF_RECORD__**

The crucial place is the code at the __END_OF_RECORD__ processing time. The first statement (transfer_record) transfers the BKNVK record with the data of the first contact person. Then, the BKNVK record is initialized, filled with the data of the second contact person, and written by means of the second `transfer_record` statement. This process creates two `BKNVK` records.

### 9.8.5 Assigning Multiple Source Structures to a Target Structure

**Assigning source structures**

You may want to assign multiple source structures to a target structure. In this case, you should create your source structures as usual. Then, assign the child source structure to the target structure. This step makes the fields from both source structures available to the fields of the target structure. Figure 9.75 shows this relationship. The ABAP code that SAP Legacy System Migration Workbench generates from the structure relations is shown in the lower part of the figure.

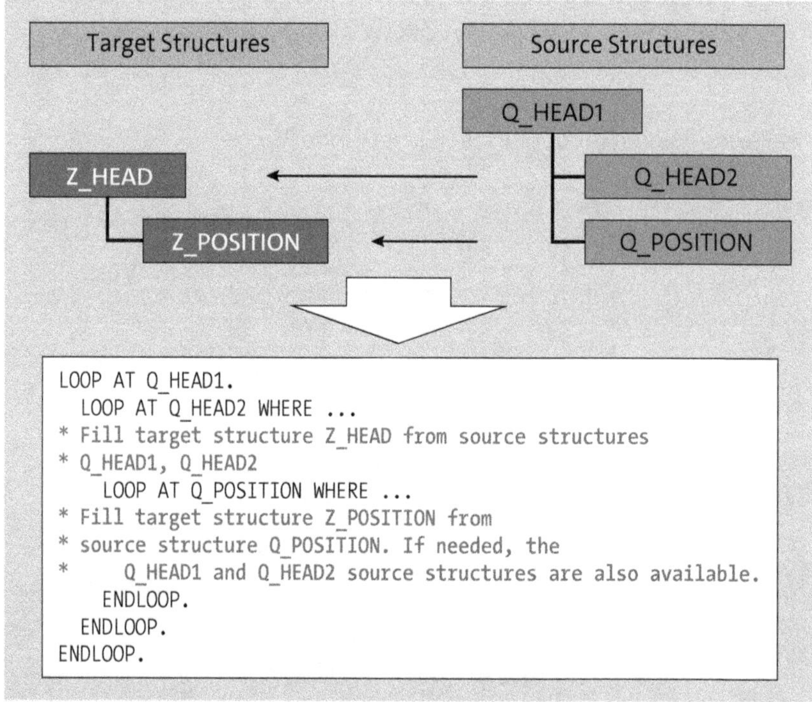


**Figure 9.75** SAP Legacy System Migration Workbench: Structure Relation, Multiple Source Structures to One Target Structure

### 9.8.6 Custom Selection Parameters in the Conversion Program

**Migrating Product/Material Master Data**

The following phased rollout scenario applies to Section 9.8.6, see Section 9.8.7, and Section 9.8.8.

Let's say you need to gradually roll out your materials management processes in plants 1000, 1010, and 1100 of your company. The material masters are stored in a source system. Some of the plants use different material masters, so for the first go-live for plant 1000, only the material masters used by this plant need to be extracted. For this task, you have defined wildcards in the file definitions that reflect the plant (i.e., 1000, 1010, and 1100).

For the transfer of these material masters, we can use IDoc type MATMAS07.

**Defining selection parameters**

You can influence the behavior of your conversion program via your own selection parameters in the global data definition _GLOBAL_DATA. The simplest solution is to simply insert a selection parameter. However, the corresponding text will then be missing from the selection screen. In the following code example, you create an additional *selection parameter* for the plant. This parameter is then used to load all imported values for this plant. To use your own selection parameters (PARAMETERS) or selection options (SELECT-OPTIONS), including text elements in the conversion program, you must follow these steps:

**Including code in the global data definition**

1. Insert the sample code shown in Listing 9.5 into the global data definition of the field mapping, _GLOBAL_DATA, as shown in Figure 9.77. The text elements are defined via the INITIALIZATION event.

```
SELECTION-SCREEN BEGIN OF BLOCK a WITH FRAME TITLE t_b1.
SELECTION-SCREEN BEGIN OF LINE.
SELECTION-SCREEN COMMENT (30) t_s1 FOR FIELD pa_werk.
SELECTION-SCREEN POSITION 32.
PARAMETERS: pa_werk TYPE werks_d.
SELECTION-SCREEN COMMENT (30) t_c1.
SELECTION-SCREEN END OF LINE.
SELECTION-SCREEN END OF BLOCK a.
INITIALIZATION.
 t_b1 = 'Own selection parameter'. "block-title"
 t_s1 = 'Target plant'. "1st Selection parameter"
 t_c1 = '<- Select your plant'.
```

**Listing 9.5** Defining Selection Parameters for the Target Plant

2. Assign the parameter to the E1MARCM-WERKS field, as shown in Figure 9.81 and Listing 9.6.

```
E1MARCM-WERKS = pa_werk.
```

**Listing 9.6** Assigning Selection Parameters

The code from the global data definition _GLOBAL_DATA) is inserted between the **data declaration** and the **selection parameter** in the conversion program. Although the implementation program now contains two (!) **INITIALIZATION** events, the ABAP compiler eventually merges them into one event.

**Input help** Your conversion program now has an additional selection parameter for the target plant including an input help, as shown in Figure 9.76.

[+]

**Using Includes**

You can also swap out everything to an include and then include that. Using an include has advantages if common parameters are involved that you want to use in other objects as well. Changes to the include will then take effect in all conversion programs.

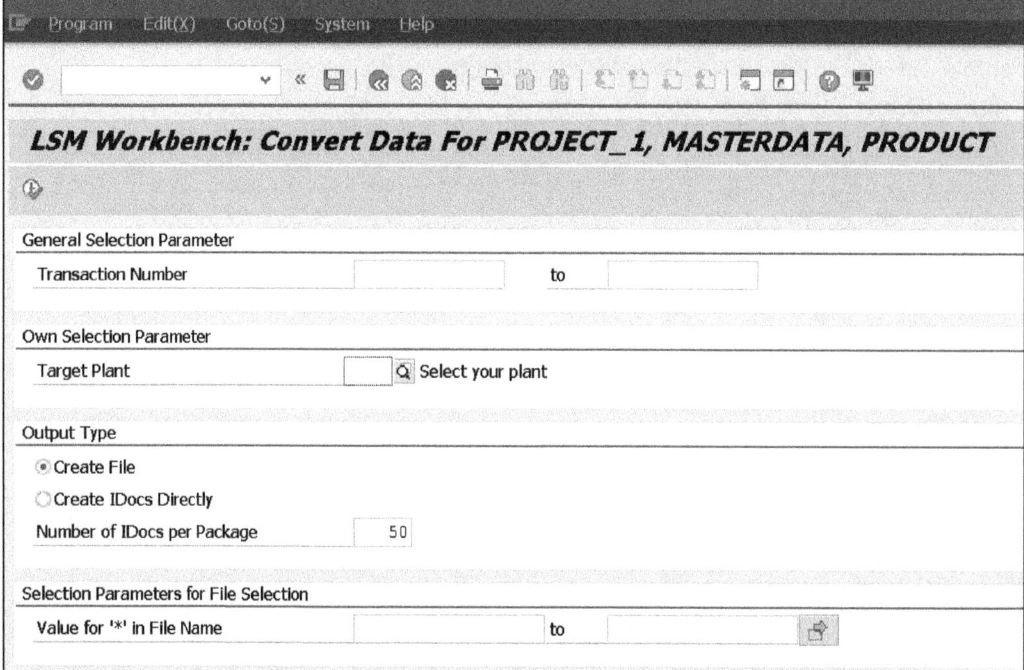


**Figure 9.76** Own Parameter for the Target Plant

[»]

**Phased Rollouts**

In phased rollouts, you use data migration objects more often. In most cases, the assignment of the source fields to the target fields does not change. However, individual conversion rules may change depending on the organizational unit. Additional code will then be included in the field mapping, and the different mapping rules are executed via CASE..ENDCASE constructs, for example.

### 9.8.7 Initializing IDoc Structures with Nodata Flags

In many IDocs, as with records and batch input, the so-called *nodata character* is used. This character is used to tell the program that a field does not contain a value ("no data"); that is, it should not be overwritten with a blank value. As a rule, the nodata character is the slash (/).

**g_nodata variable**

By default, SAP Legacy System Migration Workbench generates a structure variable init_<target_structure>, which can be used to populate the fields of a target structure with nodata characters using the FORM routine initialize_with_nodata. The value of the global **g_nodata** variable is used for this purpose. This variable is not pre-filled by SAP Legacy System Migration Workbench during data transfers via IDoc and must therefore be set before you call the routine(s).



In our sample scenario, all fields of IDoc type MATAMAS07 must be prefilled with the nodata character "/". We use the target structures E1MARAM, E1MAKTM, and (in Section 9.8.8) also E1MARCM. Empty fields would cause existing field values to be overwritten and thus deleted.

**Filling the structure variable**

The code for filling the structure variables init_e1maram and init_e1maktm is inserted in the BEGIN_OF_PROCESSING block, as shown in Listing 9.7. The name of the target structure E1MARAM or E1MAKTM and then the name of the structure variable init_**e1maram** or init_**e1maktm** are passed first as transfer parameters for the routine.

```
__BEGIN_OF_PROCESSING__
g_nodata = '/'.
PERFORM initialize_with_nodata USING
          'E1MARAM' init_e1maram.
PERFORM initialize_with_nodata USING
          'E1MAKTM' init_e1maktm.
```

**Listing 9.7** Filling the Structure Variables with Nodata Characters

**Initializing fields**

The fields of the structure are then initialized in the respective _BEGIN_OF_RECORD_ block of the structure, as shown in Listing 9.8.

```
* Target structure E1MARAM
__BEGIN_OF_RECORD__
g_edidd_segnam = 'E2MARAM006'.
g_edidd_segnum = '0001'.
g_edidd_psgnum = '000000'.
```

```
g_edidd_hlevel = '02'.
clear E1MARAM.
E1MARAM = INIT_E1MARAM.

* Target structure E1MAKTM
__BEGIN_OF_RECORD__
g_edidd_segnam = 'E2MAKTM001'.
g_edidd_segnum = '0004'.
g_edidd_psgnum = '000001'.
g_edidd_hlevel = '03'.
clear E1MAKTM.
E1MAKTM = INIT_E1MAKTM.
```

**Listing 9.8** Initializing a Target Structure with Nodata

You can see what the entire thing looks like in the field mapping shown in Figure 9.77. The entries in BEGIN_OF_PROCESSING are inserted between the program sections reset_tables and execute_data_conversion.

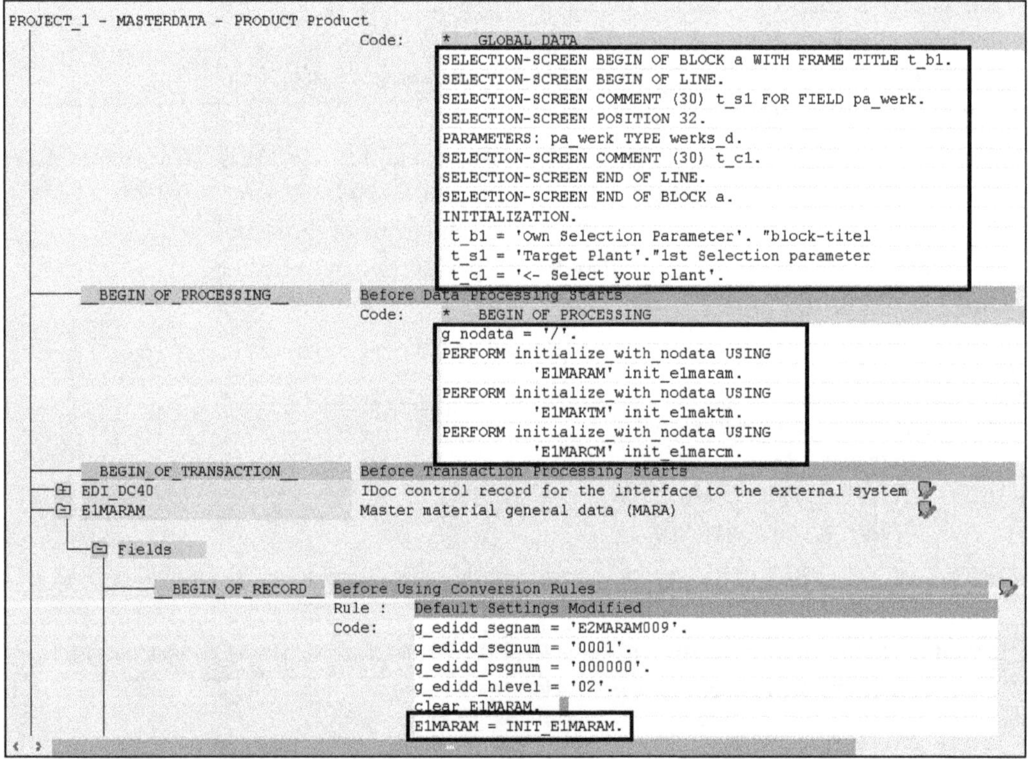


**Figure 9.77** Initializing a Target Structure with Nodata

### 9.8.8 Assigning Multiple Source Structures to a Target Structure with Substructures

The procedure described in Section 9.8.5 only works if you want to access the data of a subordinate structure. If you need to access different substructures (i.e., different data sources) simultaneously in the field mapping of a target structure, the procedure will not work. In Chapter 11, we describe how you can solve such problems using, for example, external tools such as Microsoft Excel without additional code. However, with the help of your own code within the field mapping, you can also solve these problems within SAP Legacy System Migration Workbench. We explore how this feature works in this section. For this purpose, let's first extend our data transfer scenario from the previous sections.



[Ex]

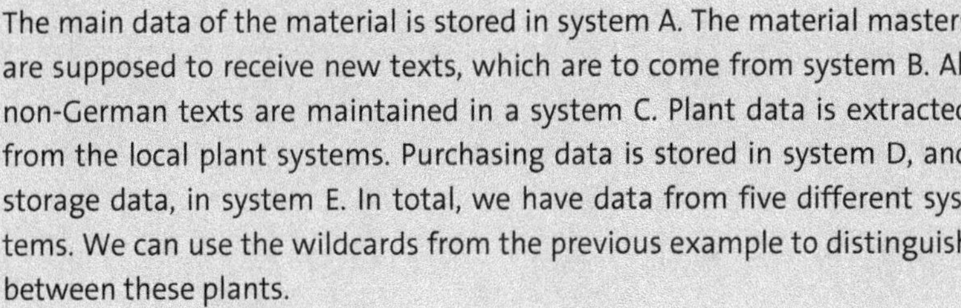

**Data from Multiple Data sources for Different Fields of a Target Structure**

The main data of the material is stored in system A. The material masters are supposed to receive new texts, which are to come from system B. All non-German texts are maintained in a system C. Plant data is extracted from the local plant systems. Purchasing data is stored in system D, and storage data, in system E. In total, we have data from five different systems. We can use the wildcards from the previous example to distinguish between these plants.

**Fields of the source structure**

The source structures contain the following fields:

- **BASIC_DATA_MATERIAL**
  MATERIALNUMBER
- **MATERIALTEXT1**
  MATERIALNUMBER, TEXT (in English)
- **MATERIALTEXT2**
  MATERIALNUMBER, LANGUAGE, TEXT2 (in non-English language)
- **PLANT_DATA1**
  MATERIALNUMBER, PURCHASINGGROUP
- **PLANT_DATA2**
  MATERIALNUMBER, MAX_STORAGE_TIME, EH_MAX_LZ

The source structures should be logically assigned to the relevant target structures, as shown in Figure 9.78.

**Assigning a source structure**

Since we now have multiple structures on one level, we cannot use the procedure described in Section 9.8.5. For this reason, we can assign only one source structure to the target structure at a time, as shown in Figure 9.79.

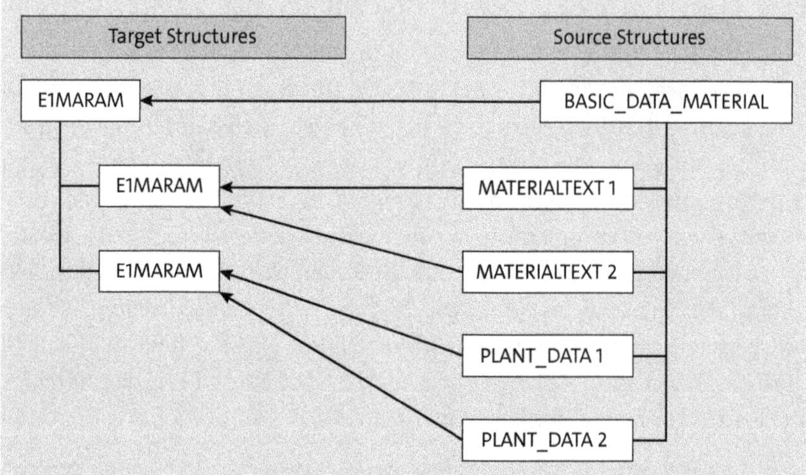


**Figure 9.78** One Main Structure, Multiple Substructures

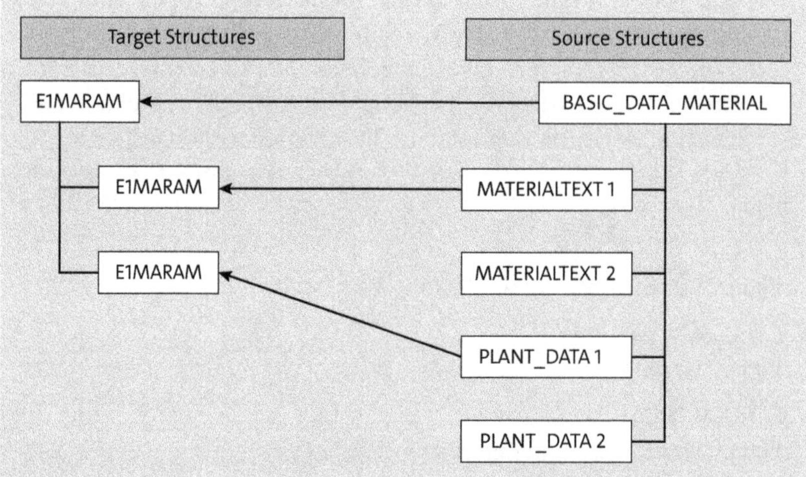


**Figure 9.79** Assigning a Source Structure to the Target Structures

**Importing data**

If the data is imported via the **Read Data** step, the screen shown in Figure 9.80 appears.

SAP Legacy System Migration Workbench generates, for each source structure, an internal structure named *LSMW_<name of source structure>* and an internal table with the name of the source structure. Although we have only assigned the **PLANT_DATA1** source structure to the **E1MARCM** target structure in our example, the data of the other source structure (**PLANT_DATA2**) can also be accessed in the internal **PLANT_DATA2** table at runtime.

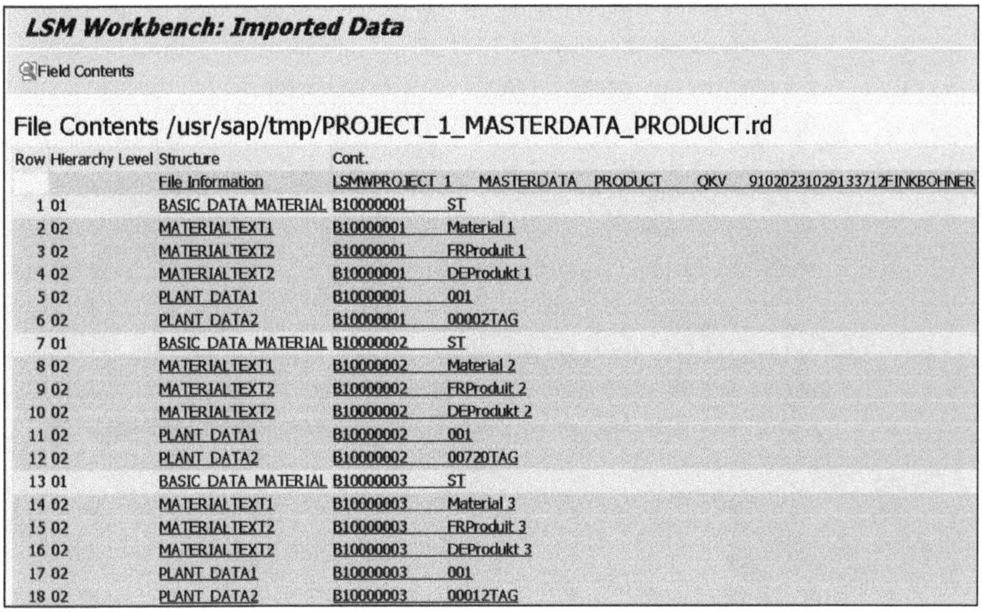

**LSM Workbench: Imported Data**

Field Contents

File Contents /usr/sap/tmp/PROJECT_1_MASTERDATA_PRODUCT.rd

| Row | Hierarchy Level | Structure | Cont. |
|---|---|---|---|
| | | File Information | LSMWPROJECT_1 MASTERDATA PRODUCT QKV 91020231029133712FINKBOHNER |
| 1 | 01 | BASIC_DATA_MATERIAL | B10000001 ST |
| 2 | 02 | MATERIALTEXT1 | B10000001 Material 1 |
| 3 | 02 | MATERIALTEXT2 | B10000001 FRProduit 1 |
| 4 | 02 | MATERIALTEXT2 | B10000001 DEProdukt 1 |
| 5 | 02 | PLANT_DATA1 | B10000001 001 |
| 6 | 02 | PLANT_DATA2 | B10000001 00002TAG |
| 7 | 01 | BASIC_DATA_MATERIAL | B10000002 ST |
| 8 | 02 | MATERIALTEXT1 | B10000002 Material 2 |
| 9 | 02 | MATERIALTEXT2 | B10000002 FRProduit 2 |
| 10 | 02 | MATERIALTEXT2 | B10000002 DEProdukt 2 |
| 11 | 02 | PLANT_DATA1 | B10000002 001 |
| 12 | 02 | PLANT_DATA2 | B10000002 00720TAG |
| 13 | 01 | BASIC_DATA_MATERIAL | B10000003 ST |
| 14 | 02 | MATERIALTEXT1 | B10000003 Material 3 |
| 15 | 02 | MATERIALTEXT2 | B10000003 FRProduit 3 |
| 16 | 02 | MATERIALTEXT2 | B10000003 DEProdukt 3 |
| 17 | 02 | PLANT_DATA1 | B10000003 001 |
| 18 | 02 | PLANT_DATA2 | B10000003 00012TAG |

**Figure 9.80** Data Imported From Five Source Systems (Excerpt)

**Initializing the target structure**

In addition to the code listed in Section 9.8.7, add the following code from Listing 9.9 to _BEGIN_OF_PROCESSING_. This code initializes the **E1MARCM** target structure with nodata characters.

```
__BEGIN_OF_PROCESSING__
PERFORM initialize_with_nodata USING
            'E1MARCM' init_e1marcm.
```

**Listing 9.9** Filling a Structure Variable for E1MARCM

In the _BEGIN_OF_RECORD block of the **E1MARCM** target structure, you must assign the initial **INIT_E1MARCM** structure filled in **_BEGIN_OF_PROCESSING**, as shown in Listing 9.10.

```
__BEGIN_OF_RECORD
g_edidd_segnam = 'E2MARCM005'.
g_edidd_segnum = '0004'.
g_edidd_psgnum = '000001'.
g_edidd_hlevel = '03'.
clear E1MARCM.
E1MARCM = INIT_E1MARCM.
```

**Listing 9.10** Assigning the Initial Structure

**Filling storage values**

In our example, we define a local variable with the structure of **PLANT_DATA2** using **LIKE LSMW_PLANT_DATA2**. Then, using the READ command, we import the data from the internal **PLANT_DATA2** table into the local **ls_werks2** variable. The **materialnummer** key field of the internal **PLANT_DATA2** table is compared with the current value of the **materialnummer** field of the same name of the top-level structure **BASIC_DATA_MATERIAL**. If a value was found, the corresponding storage values are written to the associated target fields. If no value was found, the values are filled again with the nodata character as a precaution. Insert the code in the field mapping of field E1MARCM-MAXLZ, as shown in Listing 9.11 and Figure 9.81.

```
*--- E1MARCM-MAXLZ
DATA ls_werk2 LIKE lsmw_werkdaten2.
READ TABLE werksdaten2 INTO ls_werk2
  WITH KEY materialnummer = basic_data_material-materialnummer.
IF sy-subrc = 0.
  e1marcm-maxlz = ls_werk2-max_lagerungszeit.
  e1marcm-lzeih = ls_werk2-eh_max_lz.
ELSE.
  e1marcm-maxlz = g_nodata.
  e1marcm-lzeih = g_nodata.
ENDIF.
```

**Listing 9.11** Code for Filling the Storage Values

[!]

**Maintenance Status MATMAS07**

To make sure that the fields for the purchasing group and the storage period can also be written, you must fill in the maintenance status field (E1MARCM-PSTAT) with the value "EL" for the purchasing and storage view. Theoretically, an automatic detection of the maintenance status exists, but it does not always work correctly. If a field is not filled in, only a warning message appears. For this reason, it is always important to also look at the warnings in the log when migrating data using MATMAS07. Otherwise, you may overlook fields that have not been loaded because they are not ready for input due to a lack of maintenance status.

**Duplicating the target structure**

In the next example, we now fill the same fields of a target structure with different data sources. The target structure is therefore duplicated.

**Data from Multiple Data Sources for the Same Fields of a Target Structure**

For the material texts we have a similar problem. What complicates matters even more is that now we have two source structures of the same level for the same fields. The MATERIALTEXT1 source structure contains the

English texts, and the MATERIALTEXT2 source structure contains the material texts of other languages for the same material.

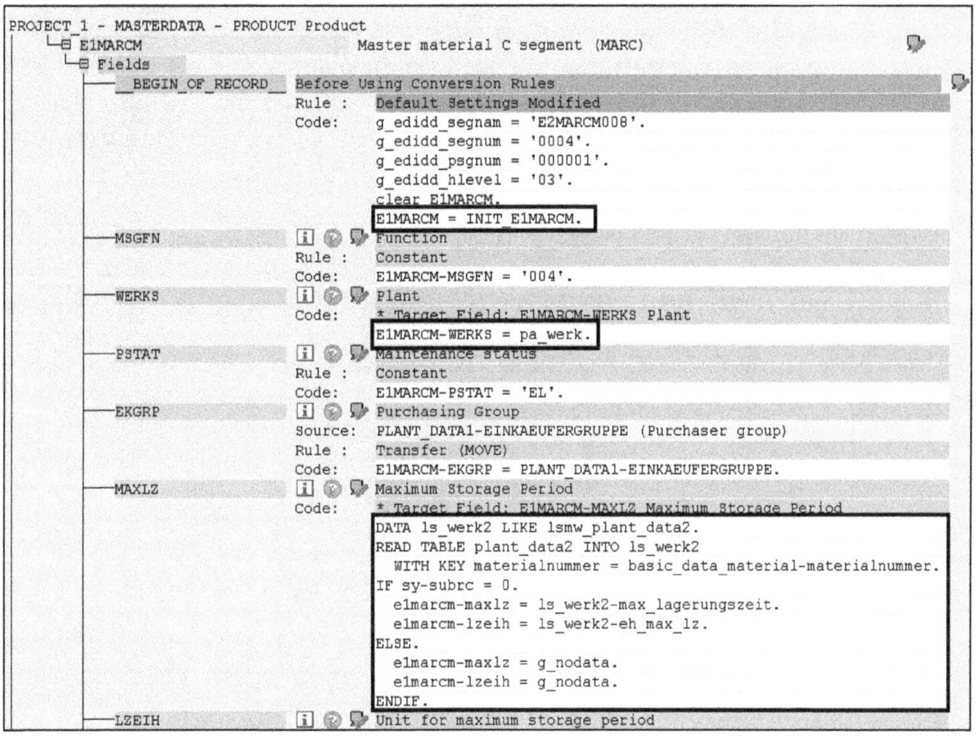

```
PROJECT_1 - MASTERDATA - PRODUCT Product
  E1MARCM                         Master material C segment (MARC)
    Fields
      __BEGIN_OF_RECORD__ Before Using Conversion Rules
                          Rule :  Default Settings Modified
                          Code:   g_edidd_segnam = 'E2MARCM008'.
                                  g_edidd_segnum = '0004'.
                                  g_edidd_psgnum = '000001'.
                                  g_edidd_hlevel = '03'.
                                  clear E1MARCM.
                                  E1MARCM = INIT_E1MARCM.
      MSGFN                       Function
                          Rule :  Constant
                          Code:   E1MARCM-MSGFN = '004'.
      WERKS                       Plant
                          Code:   * Target Field: E1MARCM-WERKS Plant
                                  E1MARCM-WERKS = pa_werk.
      PSTAT                       Maintenance Status
                          Rule :  Constant
                          Code:   E1MARCM-PSTAT = 'EL'.
      EKGRP                       Purchasing Group
                          Source: PLANT_DATA1-EINKAEUFERGRUPPE (Purchaser group)
                          Rule :  Transfer (MOVE)
                          Code:   E1MARCM-EKGRP = PLANT_DATA1-EINKAEUFERGRUPPE.
      MAXLZ                       Maximum Storage Period
                          Code:   * Target Field: E1MARCM-MAXLZ Maximum Storage Period
                                  DATA ls_werk2 LIKE lsmw_plant_data2.
                                  READ TABLE plant_data2 INTO ls_werk2
                                    WITH KEY materialnummer = basic_data_material-materialnummer.
                                  IF sy-subrc = 0.
                                    e1marcm-maxlz = ls_werk2-max_lagerungszeit.
                                    e1marcm-lzeih = ls_werk2-eh_max_lz.
                                  ELSE.
                                    e1marcm-maxlz = g_nodata.
                                    e1marcm-lzeih = g_nodata.
                                  ENDIF.
      LZEIH                       Unit for maximum storage period
```

**Figure 9.81** Importing Source Structures of the Same Level (READ)

The source structure and source field definition are shown in Figure 9.78. We only assign the source file of the English texts to the target structure, as shown in Figure 9.79.

**Assigning fields**

In the actual field mapping, you then assign the corresponding fields of the **MATERIALTEXT1** structure to the individual target fields. Since these texts are English texts, we can preassign the constant **'EN'** to field **E1MAKTM-SPRAS_ISO**, as shown in Listing 9.12.

```
* --- E1MAKTMAKTX
E1MAKTM-MAKTX = MATERIALTEXT1-TEXT.
* --- E1MAKTM-SPRAS_ISO
E1MAKTM-SPRAS_ISO = 'EN'.
```

**Listing 9.12** Assigning the "Materialtext1" Structure

We then insert the code for the second source file at the _END_OF_RECORD point below the global function macro **transfer_record**.

**Importing the second structure**

Similar to the plant data, we define a local variable for the **MATERIALTEXT2** structure and insert a loop over all values of the internal table **material text2**. Within this loop, we fill in the corresponding fields with the values of the source structure (**MATERIALTEXT2**) and write away the corresponding record of the target structure using the **transfer_record** macro. Since at runtime only the respective values of the material number of the current source data record are in the internal **MATERIALTEXT2** table, a **WHERE** condition in the loop can be omitted. To be on the safe side, you can add a **WHERE** condition, as shown in Figure 9.82 and Listing 9.13.

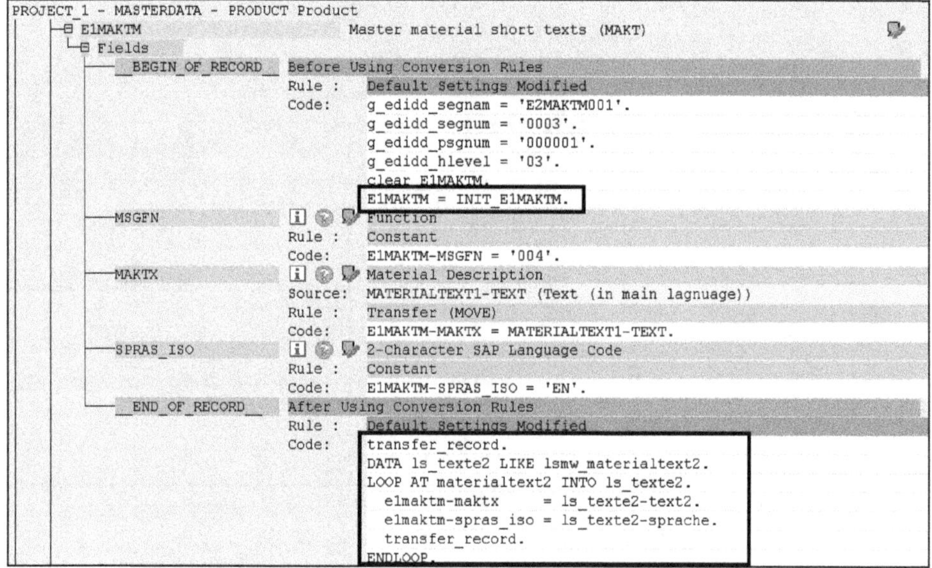


**Figure 9.82** Importing Source Structures of the Same Level (Loop)

```
*--- __END_OF_RECORD__
transfer_record.
DATA ls_texte2 LIKE lsmw_materialtext2.
LOOP AT materialtext2 INTO ls_texte2.
  e1maktm-maktx     = ls_texte2-text2.
  e1maktm-spras_iso = ls_texte2-sprache.
  transfer_record.
ENDLOOP.
```

**Listing 9.13** Sample Code for Importing the Second Structure

After converting the data using the **Convert Data** step, the data will appear as shown in Figure 9.83. As requested, the corresponding text segments for English and the other languages were created for each material, as well as a work segment whose data was supplied from two files.

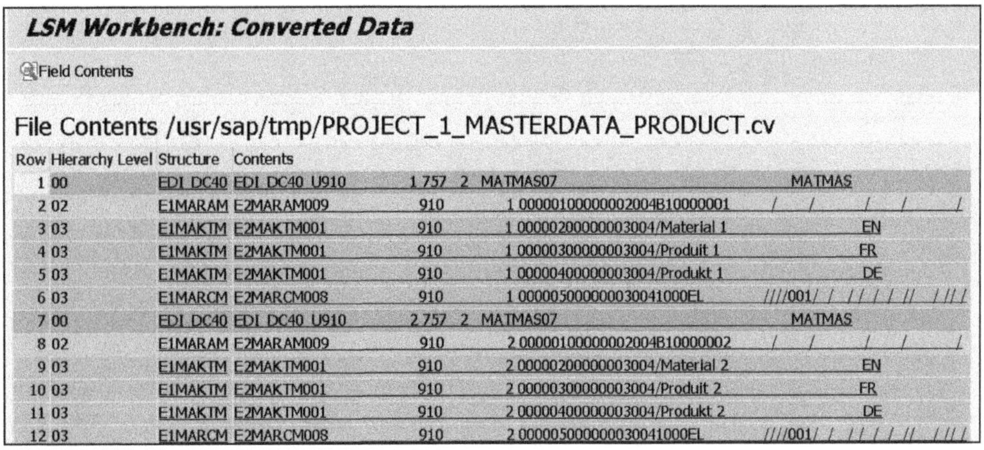

**LSM Workbench: Converted Data**

Field Contents

File Contents /usr/sap/tmp/PROJECT_1_MASTERDATA_PRODUCT.cv

| Row | Hierarchy Level | Structure | Contents |
|---|---|---|---|
| 1 | 00 | EDI_DC40 | EDI_DC40_U910 1 757 2 MATMAS07 MATMAS |
| 2 | 02 | E1MARAM | E2MARAM009 910 1 00000100000002004B10000001 / / / / / |
| 3 | 03 | E1MAKTM | E2MAKTM001 910 1 00000200000003004/Material 1 EN |
| 4 | 03 | E1MAKTM | E2MAKTM001 910 1 00000300000003004/Produit 1 FR |
| 5 | 03 | E1MAKTM | E2MAKTM001 910 1 00000400000003004/Produkt 1 DE |
| 6 | 03 | E1MARCM | E2MARCM008 910 1 000005000000030041000EL ////001/ / / / / / // //// |
| 7 | 00 | EDI_DC40 | EDI_DC40_U910 2 757 2 MATMAS07 MATMAS |
| 8 | 02 | E1MARAM | E2MARAM009 910 2 00000100000002004B10000002 / / / / / |
| 9 | 03 | E1MAKTM | E2MAKTM001 910 2 00000200000003004/Material 2 EN |
| 10 | 03 | E1MAKTM | E2MAKTM001 910 2 00000300000003004/Produit 2 FR |
| 11 | 03 | E1MAKTM | E2MAKTM001 910 2 00000400000003004/Produkt 2 DE |
| 12 | 03 | E1MARCM | E2MARCM008 910 2 000005000000030041000EL ////001/ / / / / / // //// |

**Figure 9.83** Converted Data from Five Source Systems (Excerpt)

## 9.9 Long Texts

**Multiline texts**

With regard to long texts, we're now turning to a subject that is often neglected, but that always causes trouble when handled improperly. When speaking of *long texts*, we are referring to multiline texts. Long texts for an object usually need to be migrated separately because a central repository for long texts is used in the SAP system.

**Object type 0001**

In SAP Legacy System Migration Workbench, the object type with the identifier 0001 is available for the transfer of long texts. However, for this object to be available to you, you may need to run program /SAPDMC/SAP_LSMW_SXDA_TEXTS once. But before we describe the migration of long texts, we first want to explore how long texts are stored in an SAP system.

### 9.9.1 Long Texts in the SAP System

In the SAP system, long texts are stored in SAPscript format in a text pool which consists of the STXH tables for the header and STXL tables for the rows. The key of a long text is made up of four components, as shown in Table 9.16.

| Key Field | Meaning | Example | Length | Check Table |
|---|---|---|---|---|
| OBJECT | Application object | MVKE = Material sales texts | 10 | TTXOB, TTXOT |
| ID | Text ID | 0001 | 4 | TTXID, TTXIT |

**Table 9.16** Components of a Long Text Key

| Key Field | Meaning | Example | Length | Check Table |
|---|---|---|---|---|
| NAME | Actual text key | Material number (18 digits) + Sales organization (four digits) + Distribution channel (two digits) | 70 | (None) |
| SPRAS | Language | DE | 1-2 | T002 |

**Table 9.16** Components of a Long Text Key (Cont.)

**Text key: no uniform rule**

No standard rule governs the structure of the actual text key Name. It is important that numeric keys are padded with leading numbers and that missing optional key components are replaced with the right number of spaces. To determine the values for OBJECT and ID as well as the structure of Name for a specific text type, you must follow these steps:

1. Display a text of the relevant text type (e.g., material sales text) and go to the editor.
2. In the editor, you can use the menu path **Goto • Header** to display the desired information in the **Text name** field.

### 9.9.2 Target Structures and Field Assignment

**Two target structures**

Let's now look at how you can use object type 0001 for migrating long texts. This object has the following two target structures:

- **/SAPDMC/LTXTH: Long text header**
  - STYPE: Record type (technical field, value = 1)
  - OBJECT: Application object
  - NAME: Text name
  - ID: Text ID
  - SPRAS: Language
- **/SAPDMC/LTXTL: Long text line**
  - STYPE: Record type (technical field, value = 2)
  - TEXTFORMAT: SAPscript format key (2 digits) (e.g., "*" for standard paragraph)
  - TEXTLINE: Text line

The TEXTFORMAT field is used to store formatting information for the text. If you want to transfer the text in a one-to-one relationship, you should always enter the SAPscript format key for the standard paragraph "*" in this field. Figure 9.84, Figure 9.85, Figure 9.86, and Figure 9.87 show what an object in SAP Legacy System Migration Workbench could look like based on the example of material sales texts.

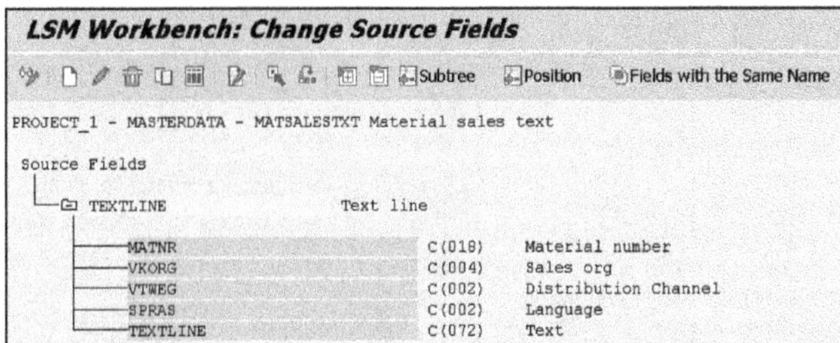


**Figure 9.84** SAP Legacy System Migration Workbench: Long Text Source Fields



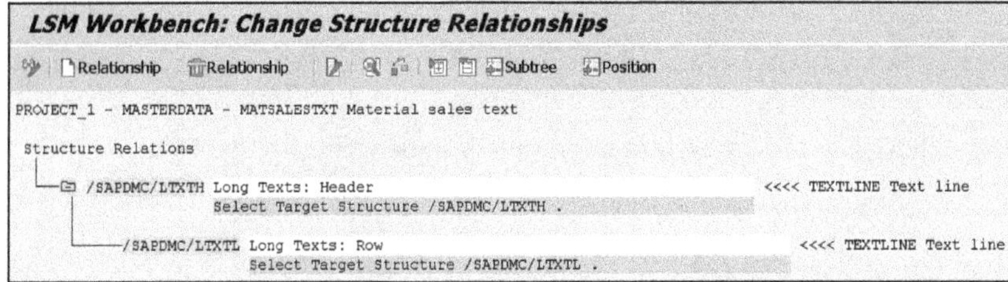


**Figure 9.85** SAP Legacy System Migration Workbench: Long Text Structure Relations

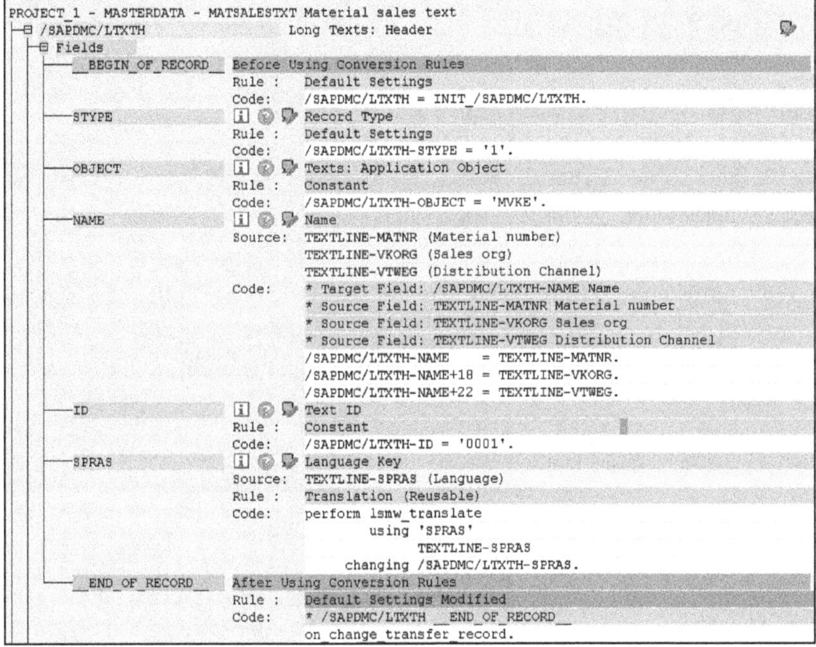


**Figure 9.86** SAP Legacy System Migration Workbench: Long Text Field Mapping (Part 1)

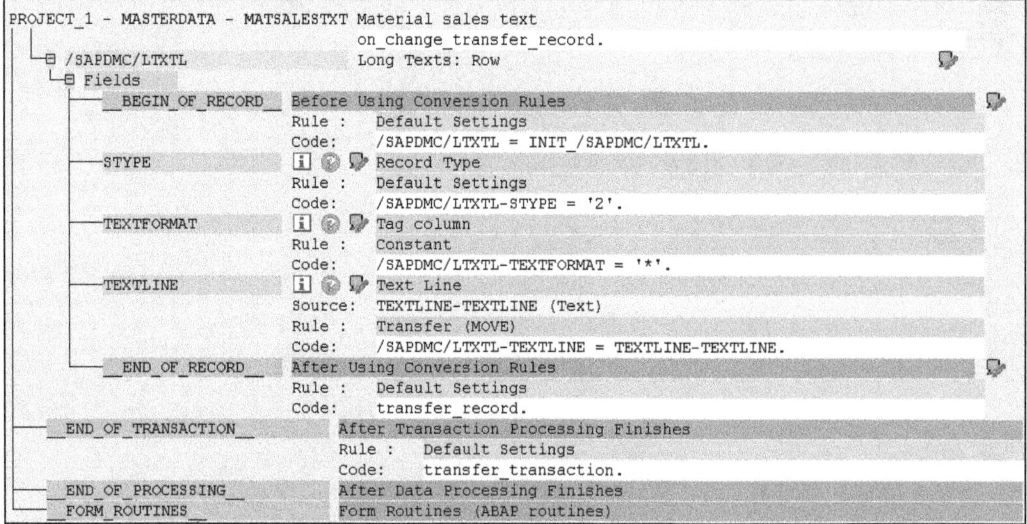


**Figure 9.87** SAP Legacy System Migration Workbench: Long Text Field Mapping (Part 2)

**Long text step by step**

In detail, you would perform the following steps:

1. **Define object**
   Create an object with the abbreviation MATVERTTXT and the name "Material sales texts" in PROJECT_1 and in the MATERIALS subproject.
2. **Define object attributes**
   On the **Object type and Import method** screen section, select **Standard batch/direct input**, object 0001, and method 0001.
3. **Define source structures**
   Define a source structure with the abbreviation TEXTLINE and the name "Text line."
4. **Define source fields**
   Five fields are required: MATNR (material number, length 18), VKORG (sales organization, length 4), VTWEG (distribution channel, length 2), SPRAS (language, length 2), and TEXTLINE (text content, length 72). All fields are of type C.
5. **Define structure relations**
   The single source structure TEXTLINE is assigned to the two target structures: /SAPDMC/LTXTH and /SAPDMC/LTXTL.
6. **Define field mapping and conversion rules**
   The actions to be performed are shown in Table 9.17.

| Target Field | Action |
|---|---|
| /SAPDMC/LTXTH-OBJECT | Assign constant MVKE |
| /SAPDMC/LTXTH-NAME | Assign source field TEXTLINE-MATNR, VKORG, VTWEG<br>Double-click on /SAPDMC/LTXTH-NAME |
| | Enter ABAP code:<br>/SAPDMC/LTXTH-NAME = TEXTLINE-MATNR.<br>/SAPDMC/LTXTH-NAME+18 = TEXTLINE-VKORG.<br>/SAPDMC/LTXTH-NAME+22 = TEXTLINE-VTWEG. |
| /SAPDMC/LTXTH-ID | Assign constant 0001 |
| /SAPDMC/LTXTH-SPRAS | Assign source field TEXTLINE-SPRAS |
| /SAPDMC/LTXTL-TEXTFORMAT | Assign constant * |
| /SAPDMC/LTXTL-TEXTLINE | Assign source field TEXTLINE-TEXTLINE |

**Table 9.17** Overview of Actions to Be Performed on Target Fields

If we leave it at that, a highly undesirable effect will occur: A text header is created for each text line. Thus, each text line is a separate text. To prevent this problem, we must reach into our "SAP Legacy System Migration Workbench bag of tricks" again (Section 9.7.3) and try the following steps:

- Click the **Hide and Display Fields/Code** button and check **Processing Times**.
- Various labels (including __END_OF_RECORD__) now appear at the end of each target structure. Double-click this label at the end of the /SAPDMC/LTXTH target structure. The ABAP editor opens, in which you need to add `on_change_transfer_record`.

This statement ensures that the text header will be transferred only if it has changed compared to the preceding record, which is exactly what we want to achieve. For more detailed information, refer to Section 9.7.

### 9.9.3 Importing Long Texts

**Import via direct input**

Texts are imported into the SAP system using the direct input technique. The corresponding direct input program /SAPDMC/SAP_LSMW_IMPORT_TEXTS can be called conveniently from SAP Legacy System Migration Workbench via **Start Direct Input Program**.

**Postprocessing**

However, imported long texts are then sometimes not readable from the corresponding application, although they are correctly stored in the database. This problem arises due to the fact that some applications have fields in the master data that contain the language of the long text, or in which it is noted whether a long text exists. These fields are not filled in by the direct input program because this program applies to all applications, and no information is available at runtime about which application a text belongs to.

Two possible solutions to this problem include the following:

- The relevant field is supplied by a report to be written by the user, which is called after the long texts have been imported.
- When defining the conversion rules of the associated object (Section 9.2.7), you must fill the relevant field, if any, in the appropriate manner.

[»]

**Additional Material**

You can download all the examples listed in this chapter as SAP Legacy System Migration Workbench project exports (text file) via this book's webpage (*https://www.sap-press.com/5817/*) and import them into your system for your own use (Section 9.4.3) for modification and adaptation. The use of these sample programs is without guarantee and at your own risk. On this webpage, you'll also find other useful sample programs that did not make it into the print edition due to space limitations.

## 9.10 Summary

In this chapter, we showed you how SAP Legacy System Migration Workbench represents a tool that offers extensive support in dealing with your data migration requirements. For more information, *https://community.sap.com* will take you to SAP Community, where you can search for the term "LSMW."

Chapter 10

# Preparing Data Migration Using Cloud Integration

*Cloud Integration is SAP's middleware solution for integrating cloud and on-premise applications. In this chapter, we provide an overview of how you can use Cloud Integration in the context of a data migration. For this purpose, we'll introduce you to the secure connection of an Secure File Transfer Protocol (SFTP) server. You'll also find helpful courses of action for implementing interfaces in Cloud Integration.*



This chapter familiarizes you with the use of *Cloud Integration* in the context of data migrations. In this chapter, we'll show you how to best design your integration processes in Cloud Integration. In this way, we'll put you in a position where you can develop your own *integration processes* as part of your data migrations and implement them according to your requirements. For this task, we'll provide step-by-step instructions on how to establish a secure connection to an *on-premise SFTP server* using the *cloud connector*. We'll also explore some examples of how to build the basic framework of your configuration in Cloud Integration and how to extend this basic framework based on common data preparation requirements.

The steps we describe in this chapter are aligned with SAP Business Technology Platform (SAP BTP), Cloud Foundry environment. If you're use to SAP BTP, Neo environment, you can partially implement some of these steps in an analogous manner in your environment.

Cloud Integration is part of *SAP Integration Suite*, SAP's integration platform as a service (iPaaS) product. With its broad product portfolio, SAP Integration Suite enables the smooth integrations of on-premise and cloud-based applications and processes. SAP-managed tools and preconfigured content are available for this purpose. Cloud Integration was formerly known as *SAP Cloud Platform Integration*, *SAP HANA Cloud Integration*, and *SAP HANA Cloud Platform, integration service*. Cloud Integration is thus the successor to the long-established *SAP Process Orchestration* middleware. SAP offers out-of-the-box integrations, harmonized access to popular third-party cloud applications, and artificial intelligence (AI)-powered cloud and hybrid integration development as part of Cloud Integration. But above all, Cloud Integration is designed to empower you to define, document, and manage your integration strategy through a systematic approach.

## 10.1 Cloud Integration in the Context of Data Migration

In the context of data migration, Cloud Integration is used to prepare existing but incomplete *XML documents* according to your requirements or to enrich them with additional data from other XML documents. Furthermore, Cloud Integration can transform XML documents into the desired target format or to perform data splitting.

**Using standard migration methods**

In addition, Cloud Integration does not perform value mappings as part of data migrations. The goal should be to prepare the XML documents in such a way that you can use them with the standard migration methods described in this book, without requiring any manual data preparation. We call this approach the *file-to-file approach*.

## 10.2 Choosing the File-to-File Approach

**Cloud Integration as a help tool**

In this section, you'll learn about the file-to-file approach without value mapping since you already administer other systems for your data migration and prepare them for migration, and the migration tools are configured as well. For this reason, the logic should only exist in one system, which allows you a better overview of the system landscape, especially in the case of complicated and lengthy migrations. Cloud Integration is a helpful tool to prepare your existing XML documents for a data migration. Thus, the logic of the migration remains in one system and is not extended to systems that are otherwise not involved.

## 10.3 Preparation

So you can process your context-related data using Cloud Integration through the file-to-file approach, some preparation is required. First, the SFTP server must be made accessible with a secure connection in order to both retrieve data from it and store the prepared data on it again. For this task, see Section 10.3.2 for an explanation of how the cloud connector comes into play. The cloud connector is necessary if you perform data transfers using an SFTP server located locally on your network (which is also referred to as "on-premise").

Then, depending on your requirements, you can perform *message splitting* and a *data merge*. On this basis, you can weigh whether any of these scenarios should play a role during the migration process.

### 10.3.1 Configuring a Secure Connection to the SFTP Server

The cloud connector enables you to access internal systems via controllable access. By accessing the internal SFTP server, you enable a data exchange between Cloud Integration and the internal network.

> **SFTP Server in the Cloud**
>
> If you host an SFTP server in the cloud, the cloud connector is not required.

[«]

**What is the cloud connector?**

A Java application provided by SAP, the cloud connector is often installed in a *demilitarized zone (DMZ)*. This application serves as a connection point between *SAP BTP* and your on-premise systems in a network. The cloud connector is easy to set up with a simple configuration of systems integrated with SAP BTP. Because it is frequently installed in a DMZ, the cloud connector acts as an on-premise agent in a secure part of the network and, in a sense, serves as a reverse proxy between the local network and SAP BTP.



You can restrict the extent to which cloud applications are allowed to access the disclosed systems and resources. To establish connection availability, the cloud connector is capable of automatically restoring interrupted connections. You can even use multiple cloud connectors to implement a high availability setup. Figure 10.1 shows an example architecture with several connectivity options and cloud connector communication flows.

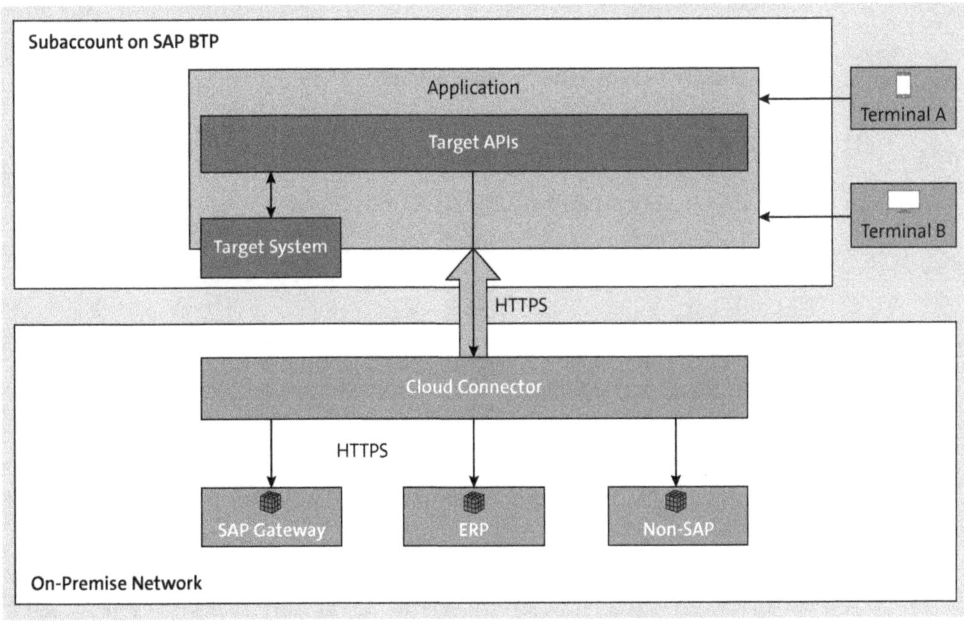


**Figure 10.1** Cloud Connector: Architecture Diagram (Source: SAP)

No opening of ports or using reverse proxies

The advantages of using the cloud connector come into play when considering opening ports in a firewall or using reverse proxies to access internal on-premise systems. When using the cloud connector, a company's internal firewall does not need to be configured for external access to SAP BTP. In addition, no changes are required for permitted outgoing connections.

Broad protocol portfolio

In addition, the cloud connector supports a wide portfolio of protocols, which it offers for communication purposes. This portfolio includes not only the HTTP protocol, but also the remote function call (RFC) protocol, for example. The RFC protocol allows native access to ABAP systems by calling function modules. Furthermore, on-premise databases and business intelligence tools can be connected to SAP HANA databases in the cloud.

Enabling principal propagation

To propagate identities of the calling user, the cloud connector provides the *principal propagation* feature. Via principal propagation, cloud users can be redirected to the on-premise system in a secure manner. If the cloud connector is commissioned in the DMZ of your network, note that firewall activations must be carried out for the IP addresses of the cloud connector. An overview is available at SAP Help Portal, where you'll also find additional notes on installation and system requirements: *http://s-prs.co/v581720*.

### Basic Configuration of the Cloud Connector

Once the cloud connector has been installed, you can access it in your browser via the following URL: *http://<hostname>:8443*. In this link, the hostname is the name of the machine on which you installed the cloud connector. If you want to open the cloud connector using the browser of a virtual machine, you must use *localhost* as the host name.

[»]

**Default Port of the Cloud Connector**

Port 8443 is the default port of the cloud connector. Deviations are possible if you configure a different port during the installation process. This alternative port should then be used in the host URL according to your customization.

Now, on the logon screen, enter "Administrator" and "manage" (both case sensitive) for `<User Name>/<Password>`. When you log on for the first time, you'll be prompted to change the password. Follow this prompt and change this password because the default password is the same for each installed cloud connector. Thus, a unique personal admin password is not generated. Finally, in the same dialog box as the password change, you must specify which instance the cloud connector is. Select **Master** if you've installed a single cloud connector. If you decide to build a failover or high availability architecture, you should also select **Master** for the first cloud connector.

For additional cloud connectors to be associated with a master instance, select **Shadow**. In both cases, you need to confirm your input by clicking **Save** in the upper right. For more information on failover and high availability architectures, see *http://s-prs.co/v581721.*

For now, you've completed the basic configuration. If you encounter inexplicable error messages during this process or get stuck, you can usually find help on SAP Help Portal, in the cloud connector section, and also in some SAP blog posts.

### Authorizations on SAP Business Technology Platform

**Subaccount and authorizations**

The cloud connector is now accessible, and you can move on to coupling SAP BTP to it in next. For all subsequent steps, you must have access to the *SAP BTP cockpit*, and you also need the authorization to view the *subaccount* in the cockpit. A subaccount of SAP BTP allows you to organize the *global account* according to the project specifications specified in your company in terms of users, authorizations, and licenses. A subaccount is always bound to a region. The physical server of your subaccount is located in this region. To ensure that you and your S-user also have the necessary authorizations to connect the subaccount, an overview of the *roles* or *role collections* that are required on the corresponding environment of your SAP BTP is provided in Table 10.1.

**Feature set**

If the *feature set* is unknown to you, navigate to the SAP BTP cockpit. In the top-right corner of the screen, there is a dropdown menu with your first name. Click on the **About** information box to open a window in the middle of your browser. The **Cloud Management Tools** item lists which feature set you have. Proceed with the authorization assignment according to Table 10.1.

| Environment | Required Role/Role Collection |
|---|---|
| SAP BTP, Cloud Foundry environment, Feature Set A | Either the user is a *member* of the global account to which the subaccount belongs, or the user has been assigned the **Security Administrator** role. |
| SAP BTP, Cloud Foundry environment, Feature Set B | Add one of the following default role collections to the user (all of which include the **cloud connector Administrator** role):<br>▪ **Subaccount Administrator**<br>▪ **Cloud Connector Administrator**<br>▪ **Connectivity and Destination Administrator** |

**Table 10.1** Authorizations Depending on the Environment and Feature Set (Source: SAP)

| Environment | Required Role/Role Collection |
|---|---|
| | Alternatively, a custom role collection that includes the **Cloud Connector Administrator** role can be assigned to the user. |
| SAP BTP, Neo environment | Add at least one of the following roles to the user:<br>▪ **Cloud Connector Admin**<br>▪ **Administrator**<br>Alternatively, the user can be assigned a custom role that includes the manageSCCTunnels authorization. |

**Table 10.1** Authorizations Depending on the Environment and Feature Set (Source: SAP) (Cont.)

### Connecting Subaccounts

**Making the subaccount known in the cloud connector**

With the authorizations from the previous section, you're now equipped to connect the SAP BTP subaccount that you want to use for SAP Integration Suite. For this task, click the hamburger icon in the top-left corner of the image. Then, click the **Connector** button. To add a subaccount, click the **+Add Subaccount** button in the upper right, which will open an input form on your screen. Insert the relevant information in the corresponding buttons:

- **Region**: Select the region of your subaccount in this dropdown menu.
- **Subaccount**: Referring to the subaccount ID, you can find this information in the SAP BTP cockpit under the navigation menu item **Overview**. There, you'll find the relevant metadata for your subaccount under the **General** tab. Copy the subaccount ID and paste it into the **Subaccount** field in the cloud connector.
- **Display Name**: This display name can be freely chosen by you and can be seen in the cloud connector after creation.
- **Subaccount User**: In this area, you should use the email address of the S-user to whom you added the authorizations in the previous step. This information is not stored and is only used for a one-off creation.
- **Password**: Enter the password for the corresponding S-user.
- **Location ID**: The location ID is used in the cloud connector architecture to uniquely identify the cloud connector. If you have connected your subaccount to additional cloud connectors, it is essential to specify a location ID. This ID can be freely chosen but should best reflect the location of the cloud connector. Later, the location ID will be used to specify in the Cloud Integration interfaces at which cloud connector the connection to the SFTP server can be found and thus also established.
- **Description**: Any description can be freely entered by you.

Save your entries by clicking the **Save** button on the form. Upon confirmation, the subaccount will appear in the **Subaccount Dashboard** on your screen.

### Connecting an SFTP Server

Making the SFTP server known in cloud connector

At this point, you've overcome most of the hurdles and can move on to the actual connection of the SFTP server. For this task, click again on hamburger icon in the top-left corner of the screen. Click the **Cloud To On-Premise** button. In the dropdown menu above the button, you'll find the display name of the subaccount you entered when you created it. If this name does not match the subaccount you're looking for, click the selection icon. At the top of the screen, you'll see a selection field labeled **Subaccount**. Select the correct subaccount and then click the **Cloud To On-Premise** button in the navigation menu. In the new window, click the plus button on the right side of the screen. An input form will then open. Go through the steps as follows:



1. First you need to select the **backend type**. Since this is an SFTP server, select **Non-SAP System**. Click **Next** to confirm the selection.
2. Select **TCP** as the protocol. Click **Next** to confirm this selection.
3. Next, specify the **Internal host** and **Port** of the on-premise SFTP server. Click **Next** to confirm your entries.
4. In the form that follows, you can specify any **Virtual host** and **Port**. These values will be used later in your interfaces. This information is later distributed to the group of people who are not supposed to know the actual connection information of the SFTP server; the cloud connector dynamically converts the virtual host and port to the internal connection information. Click **Next** to confirm your entry.
5. Enter an optional description for the connection. Click **Next** to confirm your entry.
6. At this point, you'll receive a summary of your information. Select the **Check Internal Host** checkbox. Clicking **Finish** now performs a connection test to the connected system at the same time. You'll see the result as a green or red information field on the **Check Result** tab of your connections table.

Testing the connection to the SFTP server

If setting up a connection fails, check the internal connection data to the on-premise system. By clicking the pencil icon, you can edit this connection information, and when you click on the calendar/magnifying glass icon, you can test the connection test. Once the connection has been successfully established, you've created a secure connection between SAP BTP and your on-premise system. Now, you can exit the cloud connector interface.

Viewing released systems in the SAP BTP cockpit

Connected cloud connectors and released systems can also be viewed in the SAP BTP cockpit of your subaccount. For this task, click the hamburger icon in the SAP BTP cockpit of your subaccount and then select **Cloud connectors** from the **Connectivity** dropdown list. You'll see all connected instances of **Cloud Connectors** and the provided backend systems.

### 10.3.2 Message Splitting or Data Merge

Understanding the benefits of message splitting

In the context of Cloud Integration, *message splitting* refers to splitting the message into multiple individual messages. This splitting process can also occur in groups, which means that blocks can be formed.

An example scenario would an *application programming interface (API)* can only ever accept 500 items per call. However, the original message contains 10,000 items. Message splitting can solve this problem by grouping messages into 500-item blocks and then managing these blocks as individual messages and thus as individual calls. In the context of data migration, sometimes certain messages do not contain more than 10,000 items. However, especially in the case of data migrations, there may well be more than 10,000 items in a message. In such a case, you must perform message splitting. A simple example of message splitting is shown in Table 10.2.

**Original Message**

```
<customerList>
  <customers>
    <customerNumber>0001</customerNumber>
    <customerName>John Doe</customerName>
  </customers>
  <customers>
    <customerNumber>0002</customerNumber>
    <customerName>Jenny Smith</customerName>
  </customers>
</customerList >
```

**Result with Message Splitting**

```
<!-- Message 1 -->
<customerList>
  <customers>
    <customerNumber>0001</customerNumber>
    <customerName>John Doe</customerName>
  </customers>
</customerList>
```

**Table 10.2** Representation of a Simple Message Splitting

```
<!-- Message 2 -->
<customerList>
  <customers>
    <customerNumber>0002</customerNumber>
    <customerName>Jenny Smith</customerName>
  </customers>
</customerList>
```

**Table 10.2** Representation of a Simple Message Splitting (Cont.)

In Section 10.4.4, we'll show you in detail how to implement message splitting.

**Understanding the benefits of a data merge**

By *data merge*, we refer to a message whose content gets *merged* with another message. This data is mostly supplementary data that is not present in the original file. In the context of data migration, in some cases, you may have a message that does not contain enough information because this necessary information is reserved for another file. Then, you need to merge the information from both messages to get the full benefit of combining them. For this purpose, you can use the data merge functionality. To learn how to implement a data merge in Cloud Integration, see Section 10.4.3.



## 10.4 Implementing the Interface

At this point, you've made the SFTP server accessible to the cloud. This section focuses on a sample implementation of the interface to prepare your data for data migration. For this task, you'll learn about the basic framework of this type of interface and how to implement message splitting and data merges. Furthermore, we'll describe the basic principles of mapping, with the help of which you can transform the old datasets into another target format and prepare them for data migration.

We hope to provide consultants and developers actively involved in implementing the file-to-file approach with key information on preparing for data migrations using Cloud Integration. However, key users who are not directly involved in an implementation are also welcome to read this section. In Section 10.5, we present some options so you can make adjustments and modify configurations in the already implemented objects presented in this chapter.

### 10.4.1 Building the Basic Structure

In this section, you'll learn how to use flow steps to build a solid foundation for your interfaces in a data migration context. For this task, we turn our attention again to the file-to-file approach.

**Prerequisites for developing interfaces**

Before you can start developing your interfaces, a few prerequisites must be met. You must ensure that your user has the necessary authorizations to create *packages*, *integration flows* (*iFlows*), and *credentials*. Again, you can maintain authorizations in the SAP BTP cockpit. To jumpstart with all the required permissions, the administration role for Cloud Integration can be added. If you strive for more detailed role management, you can learn more about available roles through their descriptions at *http://s-prs.co/v581722*.

#### Creating Packages in Cloud Integration

**Creating a package**

Next, you must create a package in Cloud Integration. For this task, first navigate to the Cloud Integration interface, which you can access via the SAP BTP cockpit. Next, click the hamburger icon and then select **Instances and Subscriptions**. In the center of your screen, click **Integration Suite** to be redirected to the capacity overview page. Once this page, click the **Cloud Integration** tile. Then, click the hamburger icon, click the pencil icon, and select **Integrations**.

Now, you can manage your packages. Click the **Create** button in the top-right corner to create a new package. Enter a name in the input field in accordance with the naming convention defined in your company.

Finally, you must fill in the *description* of the package. You've just completed the minimum requirements for package creation; save your package by clicking **Save**. Then, click the **Artifacts** tab for your package. From the **Add** dropdown menu, select **Integration Flow**. Enter a name in the **Name** field according to your company's naming convention. Confirm your entries by clicking **OK**. With these steps, you have now fulfilled all the requirements for package creation. In the next step, we'll show you how to retrieve data from the SFTP server and also store it back in the SFTP server.

#### Essential Flow Steps and Functionalities

**Sender and receiver channel**

In the following section, you'll learn the essential flow steps in Cloud Integration for building the basic structure of an iFlow for a data migration. In this context, we'll mainly deal with the *sender channel* and *receiver channel* components.

**Externalized parameters**

We'll also explore how to make the configuration more dynamic by using *externalized parameters*. Finally, we'll introduce you to the mapping flow step. Let's start with a concrete diagram of a basic iFlow structure, shown in Figure 10.2.

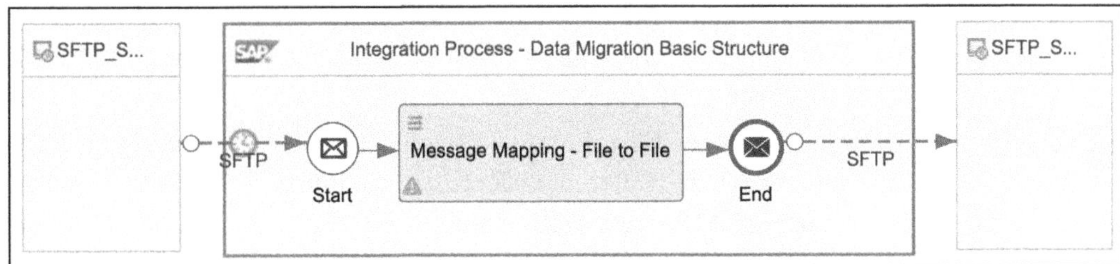


**Figure 10.2** Basic Structure of the iFlow

Standard flow steps in the newly created iFlow

To call the iFlow we created earlier, click the button with the name of your iFlow. You can now see the standard flow steps, which are predefined when you created a new iFlow. These steps include the integration process in the middle, in which the flow steps for processing are integrated in the structure of a message flow from left to right. To the left of the integration process rectangle, you can see the object of the sender and on the opposite side is the object of the receiver. Within the integration process, you have a start message event and an end message event. (These specific behaviors and the various start and end events go beyond the scope of this book.) The configuration of the flow steps is described in more detail in the following sections. We proceed from left to right and thus focus first on the sender channel, then turn to message mapping, and finally move to the receiver channel.

### Creating Trust with the SFTP Server

Known host file

For a *Secure Shell (SSH)*-based communication, the Cloud Integration tenant requires the *host key* of the SFTP server. This information is required because Cloud Integration can only establish connections to *trusted* SFTP servers. To establish this trust for secure SSH communication, the *known host file* is used. The host key of the SFTP server must be added to the known host file. The host key itself can be downloaded from the SFTP server, or you can get it from your SFTP server administrator.

Host key via Cloud Integration

Alternatively, you can also copy the host key via Cloud Integration. For this task, click the hamburger icon ☰ in the top left and select the bar icon (**Monitor**). Then, in the dropdown menu, select **Integrations.** The Cloud Integration monitoring overview page opens in the middle of the screen. The third row contains the tiles with the **Manage Security** heading. Click the **Connection Test** tile and then click the **SSH** tab. In the input fields, specify the necessary connection information by following these steps:

1. First, in the **Host** input field, enter the virtual host name that you configured in the cloud connector. If you have a cloud SFTP server, enter the host name of your SFTP server.

2. In the **Port** input field, enter the virtual port you have configured in the cloud connector. If you have a cloud SFTP server, enter the port of your SFTP server.
3. From the **Proxy Type** dropdown menu, select **On-Premise** if you have connected your SFTP server via the cloud connector or select **Internet** if you use an SFTP server in the cloud.
4. You can keep the authentication method with the **None** setting.
5. Confirm your entries by clicking the **Send** button.

Cloud Integration now tries to ping the server without authenticating. Then, in the right half of the screen, you'll find the response to your request. If the request was successful, you'll find some information about your SFTP server. Below that is a button labeled **Copy Host Key**. Click this button to copy the host key.

**Updating the host file**

Then, navigate again via the hamburger icon ☰ in the top-left corner and then click on the eye icon 👁 (**Monitor**). In the dropdown menu, select **Integrations**. In the middle of the screen, the Cloud Integration monitoring function overview page opens. The third row contains tiles under the **Manage Security** heading. Click on the **Security Material** tile and check if you already have a security material with named `known.hosts`. If so, download this file by clicking the **Download** button to the right of the table entry.

**Recreating the known host file**

If no table entry with this name exists, create a new text file called *known.hosts* on your computer and then delete the *.txt* file name extension from its name. Open this file in any text editor and paste the host key you just copied as a new line into this file. Each host key must be on a separate line in the file. For a new line, press [Enter]. Pressing [Ctrl] + [V], you can paste the host key into the new line, provided that your copy of the host key is still in your clipboard. Save the file by pressing [Ctrl] + [S]. You'll then need to upload the file to the Cloud Integration tenant again. For this task, click on the **Upload** dropdown menu in the still open window in the upper-right corner. Click **Known Host (SSH)** and, in the dialog box that opens, select the file you just saved. After the upload, the file deploys automatically, and you have successfully made your SFTP server known to Cloud Integration.

### Configuring the Sender Channel in the Basic Structure

**Configuring the sender channel**

Now, you must return to your iFlow. To configure the sender channel, navigate to and click on the sender component, as shown in Figure 10.3.

Three icons will appear to the right of the object. Place your mouse cursor on arrow icon, drag and drop the arrow to the start message event, and release the left mouse button. This action will open a dialog box listing all the sender channel types. Scroll down and select **SFTP** as the channel type. The SFTP label can now be recognized by the **dashed arrow**, which means you have successfully created the sender channel. To configure it now, double-click on it. At the bottom of the screen, a detailed overview of the channel opens, which represents the basis for connecting your SFTP server. The detailed overview contains four tabs: **General**, **Source**, **Processing**, and **Scheduler**.

**Figure 10.3** Selecting the Sender Channel

**Storing specific information**

The **General** tab contains the adapter-specific information and is for your information only. The **Source** tab stores the directory and file name you want to read from the directory. Furthermore, you must enter the connection information to the SFTP server here. Enter your specific information in the **Directory** and **File Name** fields. You can use wildcards for file names. Such wildcards are marked with an asterisk (*). Then, go to the connection parameters, where you must enter, in the **Address** field, the virtual host and virtual port of the SFTP server you configured in Section 10.3.1 in the cloud connector. Enter the relevant information according to the schema *<hostname>:<port>*. If you have a cloud SFTP server, which differs from the connectivity we present via the cloud connector, you must specify the host name and port of your SFTP server. For connectivity via the cloud connector, select **Proxy Type On-Premise** from the dropdown menu. If you have set up the cloud connector with a location ID, then you want to enter it in the **Location ID** field that appears. If you have a cloud SFTP server, keep **Internet** as the **proxy type** configuration. In the **Authentication** dropdown menu, we recommend keeping the **User Name/Password** selection for simplicity.

> **Other Authentication Methods**
>
> To explore the other authentication methods, we recommend reading the SAP blog post by Mandy Krimmel (2017) available at *http://s-prs.co/v581723*.

Then, you must specify the name of the security material where the credentials to the SFTP server are stored.

Creating credentials

If you have not created any credentials, click the hamburger icon in the top left and select the bar icon (**Monitor**). There, select **Integrations** in the dropdown menu and click **Manage Security** in the third line. Click the **Security Material** tile and **Create** in the upper right-hand corner, then select **User Credentials** as the authentication type. In the input form that appears, specify a name for the user credential.

Storing credentials in the channel

You also need to specify the **User** with permission on the SFTP server and also the **Password**. You'll have to enter the password again. Confirm your entries by clicking the **Deploy** button. Your newly created security material now appears in the table. Copy or memorize the name and navigate back to your iFlow. Click on the sender channel and then, on the **Source** tab and in the **Credential Name** input field, enter the name of the security material you just created. You can set the other four connection parameters as you wish. We recommend keeping the default parameters. Now, click on the **Processing** tab to configure the processing parameters. You can keep the default parameters, but we recommend, in the **Post Processing** dropdown list, selecting **Move File** and then specifying an archiving directory on your SFTP server. This choice ensures that you still have the file for archiving or further use even after the file has been processed. If a message processing error occurs while you process the file, you won't lose all your work.

Post processing

*Post-processing* does not take effect until processing is successfully completed. Thus, if the process fails, the file will remain on your SFTP server. Now, navigate to the **Scheduler** tab where you can configure the polling feature. *Polling* stands for a cyclic request of data, in this case, files with the file names you configured from the configured directory from the SFTP server. Thus, after deploying the integration process, Cloud Integration automatically requests data from the SFTP server using the configured SFTP sender channel.

Providing iFlow

You can deploy the iFlow by clicking the **Deploy** button in the upper-right corner of the screen. Only after the first deployment of the iFlow will the integration process become active and automatically request the data from the SFTP server in cycles. The basis for polling cycles is the **Scheduler** tab interval, as defined by you. You should therefore set values that you agree with; the setting is up to you and you alone. This step completes the configuration of the sender channel, and you should definitely save your adjustments by clicking the **Save** button in the upper right-hand corner.

### Configuring the Receiver Channel in the Basic Structure

The next step is to configure the receiver channel. As shown in Figure 10.4, you need to navigate to the end message event and hold down the arrow symbol using the left mouse button.

**Configuring the receiver channel**

Then, drag and drop your mouse pointer to the receiver component and release the left mouse button. In the dialog box that opens, scroll to the bottom and select **SFTP** from the list of available adapter types. As you already know from creating the sender channel, you must double-click on the dotted line of the receiver channel.



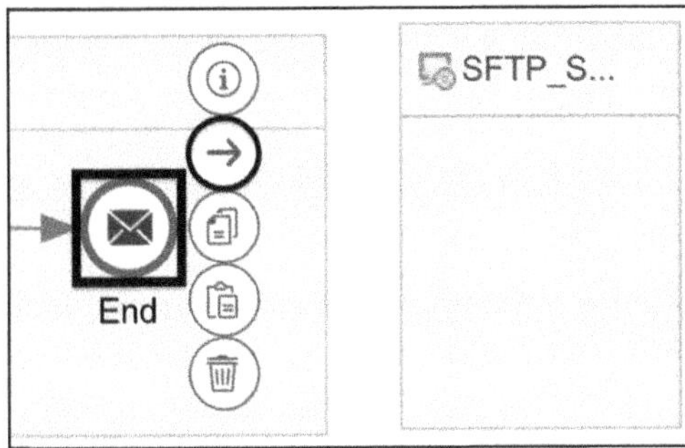


**Figure 10.4** Selecting the End Message

The detailed view of the channel then opens at the bottom of the screen with three tabs: The **General** tab contains adapter-specific information and is for your information only. The **Target** tab is used to configure the file access parameters and connection parameters, similar to the sender channel. In the **Directory** field, you must specify the directory where you want to store the files after the transformation. Proceed according to the schema *parent directory/child directory*. For the **File Name** input field, specify the file name you want to use when saving. If you leave this parameter field empty, the content of the `CamelFileName` header will be used as the file name.

[«]

**CamelFileName Header in Case of SFTP Sender Channel**

Since you use an SFTP sender channel, this header will be set automatically. Thus, the file name in the repository is identical to the file name the file had when it was fetched by Cloud Integration. This approach has the advantage of keeping track of which files you store. The prerequisite for this feature is that you have named your files in a comprehensible way.

**Timestamp**

When you click **Append Timestamp**, a timestamp will automatically be appended to the file name. If you have left the **File Name** input field empty, the file name will contain only a timestamp. If you want to have the file name set dynamically via the `CamelFileName` header, you must deselect the **Append Timestamp** checkbox. By appending a timestamp, files can be assigned to a corresponding date after they have been processed. Now, go to the connection parameters and, for **Address**, enter the virtual host and virtual port of the SFTP server you configured in Section 10.3.1 in the cloud connector. Enter the relevant information according to the schema *<host-name>:<port>*.

Note the type of SFTP server

If you have a cloud SFTP server in your cloud connector, other than the connectivity shown here, you must specify the host name and port of your SFTP server. For connectivity via the cloud connector, select **Proxy Type On-Premise** from the dropdown menu. If you have set up the cloud connector with a location ID, then you want to enter this ID in the **Location ID** field that appears. If you have a cloud SFTP server, keep the **Internet** proxy type configuration. You can set the other four connection parameters as you wish. We recommend that you keep the default parameters.

Storing credentials in the channel

In the dropdown menu for selecting authentication, we recommend that you keep the default **User Name/Password** selection. For information on other authentication methods, refer to the SAP blog post by Mandy Krimmel mentioned earlier. Then, you must specify the name of the security material where the credentials to the SFTP server are stored. If you use the same SFTP server as for the sender channel, you can also enter the name of the security material.

Creating credentials

If you have not created credentials or are using another SFTP server for storing data, navigate via the hamburger icon to the bar button (**Monitor**), select **Integrations** from the dropdown menu, and check the third line named **Manage Security**. Click the **Security Material** tile, then click **Create** in the upper right, and then select **User Credentials** as the authentication type. Enter a name for the user credential in the input form that appears. Also specify the user with permission for the SFTP server and the password. You'll have to enter the password again. Confirm your entries by clicking the **Deploy** button. Your newly created security material now appears in the table. Copy or memorize the name and navigate back to your iFlow.

Click on the receiver channel and then select the **Target** tab. Enter the name of your just created security material in the **Credential Name** input field. Click the **Processing** tab to configure the processing parameters. We recommend that you keep the default parameters.

**More Information about these Parameters**

For more information about these parameters, refer to SAP Help Portal, specifically the following link: *http://s-prs.co/v581724*.

Now, you've finished configuring the receiver channel, and you must save these adjustments by clicking the **Save** button in the top-right corner.

### Getting to Know and Using Externalized Parameters

**Externalized parameters**

You've just learned how to configure channels. In the next step, we'll show you how to make these parameters configurable. In the Cloud Integration environment, these types of parameters are referred to as *externalized parameters*. These parameters can be configured after deployment, without putting their iFlows into edit mode. The main advantage of this approach is that you can, for example, make the host and port of the SFTP server configurable. If you change the SFTP server (keeping records of this change of course), you can adjust the connection information. Furthermore, the *externalize* feature has the advantage that these parameters can be reused in the *scope* of the iFlow after they have been created once. To create an externalized parameter, navigate to an input field, such as the **Address** field of the sender channel, for example.



**Defining externalized parameters**

Define an *externalized parameter* in the format `{{parameter_1}}` and press `Enter` to confirm. With the entry confirmed, you can set the default value by double-clicking on this parameter. If this parameter is not configured, the default value is used in any case. Should you wish to adjust the parameter, save your iFlow by clicking the **Save** button in the top-right corner of the screen, if you have not already done so.

**Subsequent configuration of parameters**

Next, click the **Cancel** button next to it, and the **Configure** button will appear, which you should click now. A popup window opens where you can find all the externalized parameters that have been created in the iFlow. If you make any adjustments, save these changes by clicking the **Save** button in the bottom-right corner of the dialog box. If you want to reuse the parameter, for example, use the same SFTP server in the sender and receiver channels, navigate to the SFTP receiver channel, and click the **Target** tab. Navigate to the **Address** input field and enter "{{". The created externalized parameters will then appear below the field as suggestions. Click the parameter you created for the address of the SFTP server and confirm your choice by pressing `Enter`. After making an adjustment in iFlow, don't forget to save your changes by clicking the **Save** button in the top-right corner of the screen. With these parameters, you can build iFlows that are as user specific as possible.

[+]

> **Transporting iFlows with Externalized Parameters**
>
> When making changes or using transports to another environment (*stage*), you can adjust the changing parameters without having to put the iFlow into edit mode.

**Message mapping step**

Let's now turn our attention to the last flow step in the basic structure, the *message mapping step*. This flow step allows you to design structural mappings between two different structures in a graphical mapping editor, which will be briefly introduced in the following sections. Various standard mapping functions are available for this purpose. In Cloud Integration, message mapping supports the following file types:

- *XML Schema Definition (XSD) files*
- *OData V2/V4 metadata files*, with the *.edmx* or *.xml* file extension
- *Web Service Description Language (WSDL)*
- *Swagger/Open API Specifications* as JavaScript Object Notion (JSON) files

Since you're dealing with XML files in the data migration context, we'll focus on the use of XSD files. First, let's create the message mapping step. In Section 10.4.2, we'll describe how you can use the flow step. We'll also define the concept behind XSD and where you can get this kind of file.

**Necessity of Message Mapping**

Let's now look at a quick example to help you understand the need for message mapping. Imagine your employee data is stored in system A. Each employee in system A has a birth date in the format yyyy-MM-dd (y = year, M = month, d = day) as well as a last name and a first name. Now, you want this data to be available in system B as well. However, system B requires the date of birth in the format dd:MM:yyyy and the last name and first name consolidated into one field. Message mapping can help you meet these requirements.

To create the message mapping step, you must navigate to the arrow between the start and end message events in your iFlow, as shown in Figure 10.5.

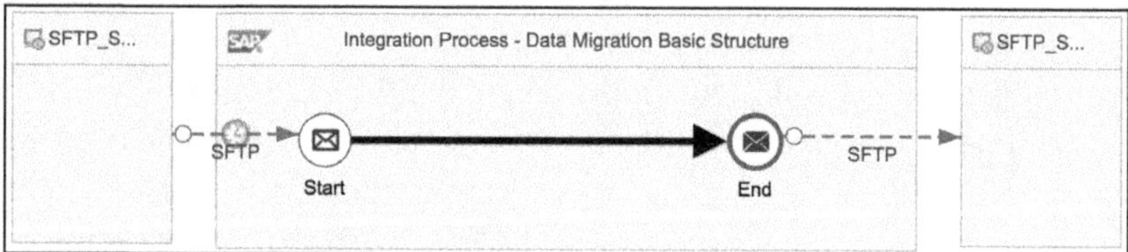


**Figure 10.5** Creating Message Mapping

Double-click on the arrow and click the plus sign icon that is displayed. As shown in Figure 10.6, select the **Message Mapping** checkbox in the dialog box or search for "Message Mapping" in the search box.

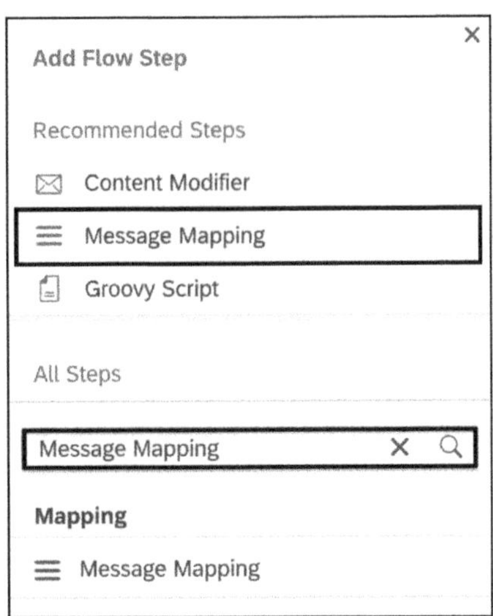


**Figure 10.6** Adding a Message Mapping Step



**Using message mapping**

The flow step then appears in your integration process between the start and end message events. To create a new message mapping click the **Create** button and specify a name for the message mapping. Confirm your input by clicking the **Create** button, which opens the editor of the message mapping. In Section 10.4.2, you'll learn how to proceed further. For now, click the **OK** button in the top-right corner of the screen. Click the **Later** button in the error dialog box that appears and save the iFlow by clicking **Save** to save your changes for the time being. To use your iFlow, you still need to deploy it, which you can do by clicking the **Deploy** button located in the top right. If, contrary to expectations, this button is not visible, you're still in edit mode. You can close edit mode by clicking the **Cancel** button in the top right. You should then also see the **Deploy** button. In the following section, we'll familiarize you with the message mapping graphical editor so that you can create your own mappings for your data migrations on your own.

### 10.4.2 Implementing Message Mapping

In this section, we'll introduce you to the basic principles of message mapping. In the previous section, you integrated the message mapping step into your basic structure. The files you come into contact with during data migration are XML documents. To define and validate XML documents, you must use the XML Schema Definition (XSD). These XSD files represent the basis of message mapping for us, as already briefly explained in the previous section.

**What Is XSD?**

XSD is a recommendation of the *World Wide Web Consortium (W3C)* to define structures of XML documents. Both the type of structure of XML documents and the elements they contain can be validated. An XML schema is usually stored as an XSD file with the *.xsd* file extension.

**Creating an XDS file**

If you have an XML document, you can create an XSD file based on it. You can use different software for this purpose, such as Visual Studio, for example. Also, many providers on the internet can help you create XSD files from your XML documents.

**Implementing field mapping**

You need an XSD file both for the source file, which is the file that is fetched from your SFTP server by Cloud Integration, and for the target file, which is the file that is to be placed back on the SFTP server after processing. Once you have these files, navigate to your message mapping step and double-click it. Click the **Processing** tab and then click the resource you created in the previous section, which opens the message mapping editor.

**Source structure in message mapping**

The message mapping editor is divided into three sections: The upper part is divided into the source structure on the left and the target structure on the right. The lower part includes the section where you'll perform the mapping. First, navigate to the text highlighted in blue (**Add source message**) in the source structure area. In the window that opens, click **Upload from File System** in the lower left-hand corner. This step opens the file explorer. Locate your XSD file for your source file and double-click on the file to open it automatically in Cloud Integration.

**Target structure in message mapping**

To add the XSD file for the target structure, click the **Add target message** link and follow the same procedure to select the file. Now that you have the two structures, you can start mapping the fields. Drag and drop individual fields from the target structure to the lower area of the screen. The source fields from your source structure that now need to be mapped to the target field can also be dragged and dropped into the editor. In the **Mapping Expression** area, some selection fields appear on the left. These fields reflect the standard mapping functions. Pick the functions you need and drag and drop them into the area as well. To implement a field mapping, shown in Figure 10.7, map the first and last name of the source structure, concatenated with a space as a separator to the name as the target field. Additionally, you need to drag and drop the *customer* parent node from the source structure to the target structure and also to the *customer* parent node. This step enables the names for each customer to be generated in the target structure by transferring the context. The procedure for message mapping is always the same.

To simulate your performed field mapping, click the **Simulate** button in the top right. In the window that opens, click the blue **Browse** button and select a sample message from the file explorer that matches the source structure.

**Simulating message mapping**

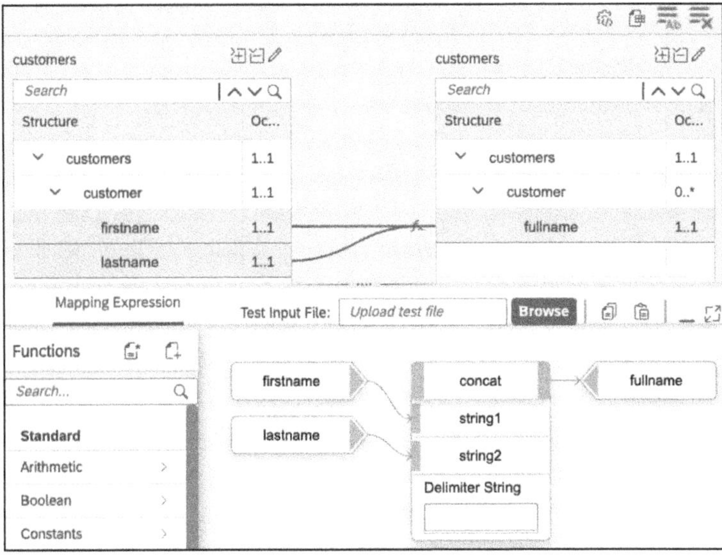


**Figure 10.7** Example of a Field Mapping

To start the simulation, click the **Test** button in the bottom-right corner of the window. After a short waiting period, if no errors have been detected, the target structure is located on the right side of the window. To close the simulation window, click the **Close** button.

**Provisioning message mapping**

When you're done with implementing the field mapping, click the **OK** button in top right. Then, click the **Save** button to save your changes and click the **Deploy** button to deploy your iFlow. After deployment, the SFTP sender channel actively polls the directory configured in the channel. After a short time, you'll see the processed files in the storage directory of your SFTP server.

**Dynamic message mapping**

During your implementation project, you may find that you need to re-create a message mapping over and over again in order to reuse it in your integration processes. In this section, we'll explain how you should deal with this type of situation. With the October 2022 release of Cloud Integration, you can now reuse message mappings, the steps necessary for which we'll present next.

The basic structure of our iFlow will look like the structure shown in Figure 10.8, after our adjustments.

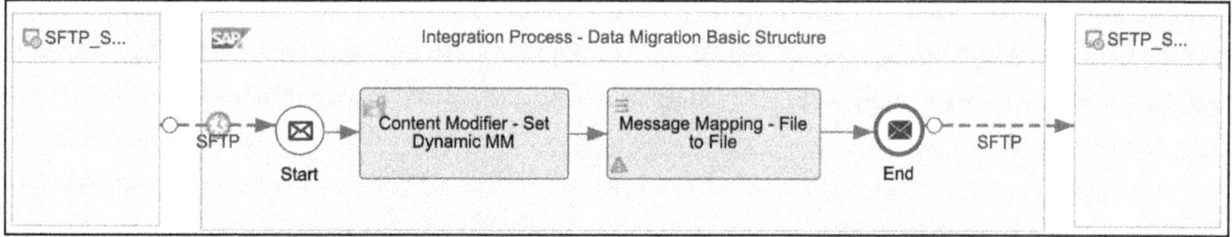


**Figure 10.8** Basic Structure of Dynamic Message Mapping

To be able to reuse a message mapping, it must be created as a separate artifact. For this task, navigate to the package overview. Put the package into edit mode by clicking the **Edit** button in top right. To create the message mapping artifact, from the **Add** dropdown menu, select **Message Mapping**. A form will open where you can specify the name of your message mapping. Confirm your entries by clicking the **OK** button. After that step, the artifact is included in the overview.

In the next step, navigate to the artifact. The message mapping editor you're already familiar with opens. Additionally, on the left of this editor, you'll see an overview of **Resources** that are included. You can perform message mapping in the same way as described in Section 10.4.2.

To deploy and actively use the artifact, save it as a version using the **Save-as-version** button in top right. Then, click the **Deploy** button. At this point, you have created the message mapping as its own artifact. To reuse it in the different iFlows in your package, however, you still need to perform a number of steps.

First, you need the identifier of the message mapping artifact. For this purpose, click on the icon in the **Actions** column of the artifacts list. Then, click **View metadata**. The property you need is the name of the ID. Save the ID name because you'll need this information in subsequent steps.

Then, go to the iFlow in which you want to include the message mapping (Section 10.4.1). Use the structure shown in Figure 10.8 as a guide for the following steps. Put your iFlow into edit mode by clicking the **Edit** button in the upper-right corner. Now, insert a content modifier flow step between the **Start Message** event and the **Message Mapping** step and double-click on the content modifier step to adjust the configuration. Navigate to the **Exchange Property** tab and add a new property by clicking **Add**. The **Action** and **Source Type** columns remain. Specify any **Name** for the property an enter *ref:<message mapping artifact ID>* in the **Source Value** field. You'll need the name of the property in the next step: Double-click on the **Message Mapping** step. On the **Processing** tab, change the **Reference Type** to **Dynamic**. Then enter `${property.<property name>}` in the **Expression** field. At this point, you have created a link to the property from the content modifier. The message mapping referenced in the property is thus executed in the integration process at runtime. Save the iFlow by clicking **Save** or **Save-as-version**. Then, deploy the iFlow by clicking the **Deploy** button.

At this point, you've stored the message mapping dynamically in your iFlow. Of course, you can connect this step with the external parameters described in Section 10.4.1.

### 10.4.3 Implementing a Data Merge

Building on the basic structure, in this section, we want to turn to extending the data merge mentioned earlier. A data merge is required if, for example, the employee names are in one file but the employees' places of residence are in another file, and you now need them in a single XML document. The data merge realized by means of the **Poll Enrich** step helps you to achieve this goal. As shown in Figure 10.9, the flow step makes use of an SFTP sender, connected to a sender component.

Configuring the Poll Enrich step

To implement the data merge, click the button shown in Figure 10.10 in the menu bar and click **Sender** in the selection.

Then, click the arrow between the start message event and the message mapping step of your basic structure and click the plus button. In the window that opens, search for "Poll Enrich" in the search bar and confirm your entry by clicking on the **Poll Enrich** option that appears.

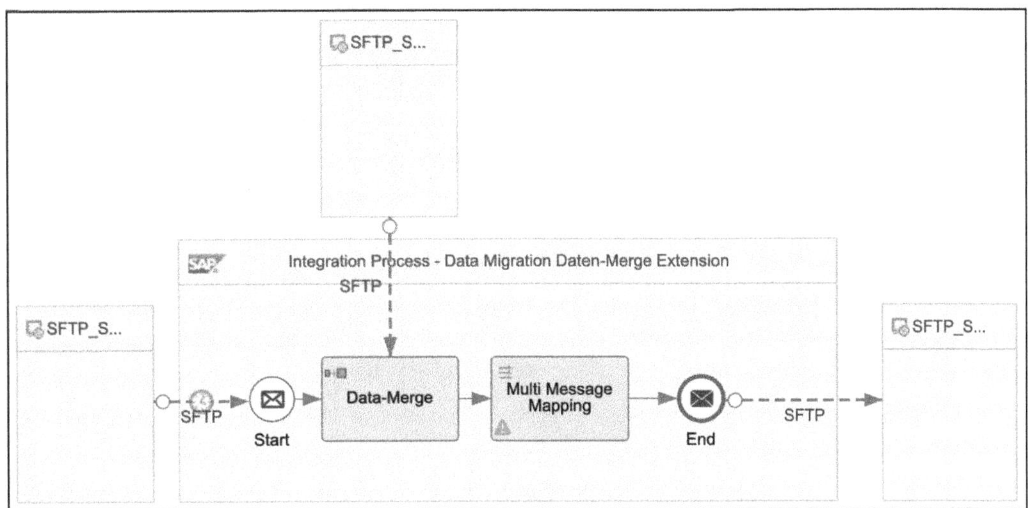


**Figure 10.9** iFlow with Data Merge

**Figure 10.10** Selection Banner

Double-click the **Poll Enrich** step, navigate to the **Processing** tab, and select **Aggregation Algorithm Combine XML** in the field. This step ensures that the two XML documents (from the SFTP sender channel in the basic structure and from the Poll Enrich step created afterwards) are merged into a single message. To create the second SFTP sender channel for the data merge functionality, click on the sender block you just created, drag and drop the arrow to the Poll Enrich step, and then release. In the window that opens, click on the **SFTP** item. Configure the channel in the same way as you did for the sender channel in the basic structure. Remember to change the directory and file name if necessary, should they differ from the actual directory and file name. If you have configured externalized parameters, you can reuse them now. Save your changes by clicking **Save** in the upper right-hand corner of the screen.

**Multi-mapping function**

Then, navigate to your message mapping flow step. Double-click on it, click the **Processing** tab, and select your already created mapping under **Resource**. The XML documents were assembled by the Poll Enrich step. You now need to cover these XML documents in the mapping editor as well. For this purpose, the XSD files of the two source files serve both as the main file and as a file added by the Poll Enrich step for a data merge. In the upper right, click the **Add source message** link; then click **Upload from File System** in your file explorer and locate the XSD file.

**Enriching the source structure**

If you've already added an XSD file and want to extend it only with a second XSD file, click the pencil icon in the source message area. Then, click the plus button in the **Source Message** area and insert the additional XSD file using the same method as for the first XSD file. This time, select the XSD file used for the data merge. After making a selection, you'll see two messages at the source message. Confirm your changes by clicking **OK** in the top-right corner of the screen. Now, create a field mapping as described in Section 10.4.2. Confirm your mapping by clicking **OK** in the top right.

Table 10.3 shows a sample run of the data merge function.

| Main Document |
|---|

```xml
<?xml version="1.0" encoding="UTF-8"?>
<customers>
  <customer>
    <firstname>John</firstname>
    <familyname>Doe</familyname>
  </customer>
  <customer>
    <firstname>Jenny</firstname>
```

**Table 10.3** Example of a Data Merge

| |
|---|
| `<familyname>Smith</familyname>`<br>`</customer>`<br>`</customers>` |
| **Poll Enrich Document** |
| `<?xml version="1.0" encoding="UTF-8"?>`<br>`<addresses>`<br>`<address>`<br>`<city>Boston</city>`<br>`<zip>02108</zip>`<br>`<street>Blue Hills Parkway</street>`<br>`<housenumber>13</housenumber>`<br>`</address>`<br>`<address>`<br>`<city>Chicago</city>`<br>`<zip>60131</zip>`<br>`<street>LaSalle Street</street>`<br>`<housenumber>7</housenumber>`<br>`</address>`<br>`</addresses>` |
| **XML Document Combined by Poll Enrich Step** |
| `<?xml version='1.0' encoding='UTF-8'?>`<br>`<multimap:Messages xmlns:multimap="http://sap.com/xi/XI/SplitAnd-Merge">`<br>`<multimap:Message1>`<br>`<customers>`<br>`<customer>`<br>`<firstname>John</firstname>`<br>`<familyname>Doe</familyname>`<br>`</customer>`<br>`<customer>`<br>`<firstname>Jenny</firstname>`<br>`<familyname>Smith</familyname>`<br>`</customer>`<br>`</customers>`<br>`</multimap:Message1>`<br>`<multimap:Message2>`<br>`<addresses>`<br>`<address>`<br>`<city>Boston</city>`<br>`<zip>02108</zip>`<br>`<street>LaSalle Street</street>` |

**Table 10.3** Example of a Data Merge (Cont.)

| |
|---|
| <pre>        <housenumber>13</housenumber><br>      </address><br>      <address><br>        <city>Chicago</city><br>        <zip>60131</zip><br>        <street>LaSalle Street</street><br>        <housenumber>7</housenumber><br>      </address><br>    </addresses><br>  </multimap:Message2><br></multimap:Messages></pre> |
| **XML Document after Multi-Message Mapping** |
| <pre><?xml version="1.0" encoding="UTF-8"?><br><ns0:Messages xmlns:ns0="http://sap.com/xi/XI/SplitAndMerge"><br>  <ns0:Message1><br>    <customers><br>      <customer><br>        <name>John Doe</name><br>        <zip>02108</zip> <city>Boston</city><br>        <street>Blue Hills Parkway</street><br>        <housenumber>13</housenumber><br>      </customer><br>      <customer><br>        <name>Jenny Smith</name><br>        <zip>60131</zip><br>        <city>Chicago</city><br>        <street>LaSalle Street</street><br>        <housenumber>7</housenumber><br>      </customer><br>    </customers><br>  </ns0:Message1><br></ns0:Messages></pre> |

**Table 10.3** Example of a Data Merge (Cont.)

You can save your adjustments by clicking on the **Save** button in the upper right-hand corner of your iFlow. To use the iFlow, click on the **Deploy** button in the upper right corner of the screen.

[»]

**File Name**

If you have left the **File Name** field empty in the receiver channel, the file name of the document that has been included in the message flow by the Poll Enrich step will be taken. If you prefer a different file name, you can change it in the receiver channel. Keep the following in mind: If you process more messages and you have not taken the already processed documents from the storage directory, there is a risk that they will be overwritten. First, remove the already processed files from the storage directory, or change the name of the file that will be added to the message flow by the Poll Enrich step.

### 10.4.4 Implementing Message Splitting

In this section, we want to focus on extending message splitting. Reminder: During your data migration, you may be faced with a requirement that only a certain maximum number of elements be contained in an XML document so that you can process it. In this context, message splitting provides a solution if there are more than this maximum number of elements in the source file. There are a number of splitter flow steps available in Cloud Integration; we'll focus on the **General Splitter** step. In Figure 10.11 you can see the extension of the basic structure with message splitting.

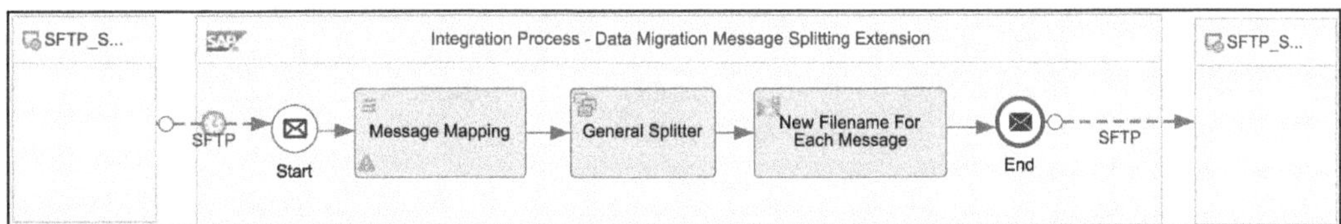


**Figure 10.11** iFlow with Message Splitting

**Configuring the general splitter**

To implement message splitting, go to your iFlow with the basic structure. Click the arrow between the message mapping step and the end message event, then click the plus button. In the window that opens, type "General Splitter" in the search bar and then click on the name in the selection. A content modifier is also required.

To add this content modifier, proceed in the same way. Click the arrow between the general splitter step you just created and the end message event. Click the plus button, search for Content Modifier in the window that opens, and confirm by clicking the name in the list. Finally, configure the general splitter by double-clicking on the step. The flow step contains the **General** and **Processing** tabs. The **General** tab is for your information. Click on the **Processing** tab. Since you use XML files, select **Expression Type XPath** from the dropdown menu. Then enter your XPath expression, on the basis of which you want to perform the splitting. You can specify the path as a relative path or also as an absolute path. The relative path starts with two slashes (//) followed by the element name, which may be anywhere in the XML file. The absolute path starts with a single slash (/) followed by the *root* *node* of the XML file. This root node can be further refined by following the format: `/root node/parent node/child node/[...]`.

In the **Grouping** input field, you can insert the number of elements that can be grouped into one XML document. For example, if you have a requirement that there be a maximum of 10,000 entries in an XML document, enter the number 10,000 here. To clarify the splitting process, you can take another look at Table 10.2. You can keep the other parameters with the default values. Having done this, you have finished the configuration of the splitter. Save your changes by clicking the **Save** button in the upper right-hand corner of the screen.

**Configuring the content modifier**

Next, you need to configure the content modifier. The content modifier is needed because the filename in the splitting process causes problems during file storage on the SFTP server. Recall again the setting of your receiver channel with regard to the file name. You have most likely left this blank or assigned your own name to it. Now, however, after the splitting process, there are multiple files that all have the same file name according to the current configuration and will thus be overwritten. To prevent this, you must implement the content modifier after the splitter step. After the splitting process, each message has its own *properties* during further processing and also its own *header* data. We want to take advantage of this by manipulating the `CamelFileName` header, which is automatically referenced when there is an empty file name in the receiver channel and contains the original file name on the SFTP server. For this task, navigate to the content modifier and click on it. Go to the **Message Header** tab. On the right there is an **Add** button; click on it. Now enter the following data in the new line that appears:

- **Action**: Select **Create** because we want to overwrite the existing `CamelFileName` header.
- **Source Type**: Select **Expression** because we'll access another header that contains the split index.
- **Source Value**: Enter the following here: "${header.CamelSplitIndex}_${header.CamelFileName}". In this way, we'll manage to overwrite the header. Let's assume that the file name is *Test*, and the file extension is *.xml* at the time it is requested. The file still has so many entries that the splitter splits it into two files. Then, after storing them on the SFTP server, the file names are *0_Test.xml* and *1_Test.xml*. *0* or *1* stands for the index of the splitting operation and *Test.xml* was already our file name; we used this again. This will give us different file names, and the files will not be overwritten when saved. The prerequisite for this is that in the receiver channel you have left the configuration of the file name empty.
- **Data Type**: Since this is a *string*, type the following: "java.lang.String".

This means that you have successfully implemented message splitting. Using the content modifier you have prevented split messages from overwriting each other on the SFTP server. To save your changes click the **Save** button in the upper right corner of the screen. If you want to use your iFlow, it must be deployed via the **Deploy** button.

When processing a new file, the split index starts again at 0. If files have the same name, these files are potentially overwritten when they are stored after processing.

[+]

**Combining Data Merge and Message Splitting**

We want to point out here that you can also combine message splitting and data merge. You can integrate the message splitting steps into the data merge steps. For this task, create the message splitting steps after the data merge steps in a common iFlow.



## 10.5 Options for Action for Key Users

This section is specifically intended for key users who do not actively develop iFlows. We'll describe actions that you can carry out yourself as a key user in order to support the development and consultants working in your company or to exert influence yourself, e.g., on the data storage. For this purpose, we'll describe the distribution of roles, the use of monitoring and the configuration of external parameters.

### 10.5.1 Distribution of Roles

**Assigning a role collection**

In order for staff in consulting and development to gain access and authorizations for Cloud Integration, they must be assigned the necessary roles in the SAP BTP cockpit. These roles are mostly found in predefined *role collections*.

You can assign the role collections to a user. For this task, go to the hamburger icon ☰ in the SAP BTP cockpit, open the **Security** dropdown menu and click **Users**. In the middle of the screen a table opens which contains all created users of the subaccount. You can search for the relevant user via the search bar. Alternatively, just browse the table. To add a user to a role collection, click on the user in the table. On the right-hand side of the screen, the user information opens. Scroll down to **Role Collections** and click the dots icon. Then click **Assign Role Collection**. In the dialog that opens, select the required role collection. For a detailed listing of the role collection and the included roles for Cloud Integration, check the following link: *http://s-prs.co/v581725*.

Once you have selected the role collection, click the blue **Assign Role Collection** button in the lower right-hand corner of the dialog. In this way, you have assigned the desired roles to a user.

**Creating a user**

If you need to create a user first, the following steps will help you: To create a new user, go to the hamburger icon in the SAP BTP cockpit and open the **Security** dropdown menu. Click **Users**. In the top right-hand corner of your screen, click the blue **Create** button.

In the dialog that opens, specify the user name. The user name is the email address stored for the S-user, if you use the default *Identity Provider* of SAP. You also store this email address in the **Email** field. In the **Identity Provider** dropdown menu, you can keep the setting.

If you use a different identity provider in your company, then you need to enter the user's information required for this provider. Click on **Create** in the bottom right-hand corner of the dialog to save your entries.

You have now successfully authorized a new user on the subaccount.

### 10.5.2 Using Monitoring

As a key user, one of your tasks could be to monitor the message history. We'll now describe how you can perform simple activities in the *Monitoring* section of Cloud Integration. First, you need to go to SAP Integration Suite, which you can access via the SAP BTP cockpit. For this task, click on the hamburger icon and then on **Instances and Subscriptions**. There you'll find SAP Integration Suite in the center of your screen. Click on the name. This will redirect you to the central SAP Integration Suite page. Select the hamburger icon, and then click the bar icon. After that, click **Integrations** in the dropdown menu that opens. Cloud Integration Monitoring opens in the middle of your screen.

**Log-level setting and integration content**

The tiles we would like to describe in detail are the first ones in each of the first and second rows.

Click the tile labeled **All** in the second row. All iFlows deployed on Cloud Integration are located in this section. On the left you can see a list of iFlows with a *deployment status*. The deployment status provides information about what state the iFlow is in. On the right you can see the specific information of the iFlow.

Click on the **Log Configuration** tab and you'll be redirected to the iFlow's *log level*. If the log level is set to **Info**, only a small log file of the message history will be created.

However, if you want to be able to trace every single step in message processing for testing or troubleshooting purposes, you must set the log level in the dropdown menu to **Trace**. For all messages that occur within the coming ten minutes this log level is decisive. If messages get processed, you can look into the trace in message monitoring and follow each individual step of message processing. The log level automatically resets to the previously configured setting after ten minutes. The trace of a message can be called for one hour after it has been recorded.

Message monitoring

You can access the entire *message monitoring* by clicking on the first tile of the first row. The message monitoring screen is divided into three sections.

- The upper section contains the tools that allow you to filter for a specific message or artifact.
- Below that in the left-hand section, you can see all the messages that have been processed by Cloud Integration.
- On the right, there is the specific information of a message. There you'll also find the corresponding error message in case of an incorrect message.



If you have set the trace of an iFlow, as explained in the previous section, you'll find the **Trace** button on the specific message. For this task, scroll to the end of the message and click on the blue **Trace** button. A new window will open in your browser with the view of the trace. You can find information on the navigation in this window and the individual meanings in the following link: *http://s-prs.co/v581726*.

You have now mastered the first tasks in the monitoring section of Cloud Integration. If you ever get stuck, it's best to contact your development or consulting team.

### 10.5.3 Configuring Parameters

In this section, you'll learn how to make an adjustment to the parameters of an iFlow by performing a task during data migration.

First you need to locate the package that contains the iFlow and find out which iFlow it specifically is. You can then click on the pencil icon in the SAP Integration Suite interface and then click **Integrations**. Locate the package and click on it. Then navigate to the **Artefacts** tab. There you'll see the artifacts of a package, to which an iFlow belongs as well. In the list, you need to find the iFlow you want to edit.

Configuring external parameters

In the **Actions** column, click on the icon and then on the **Configure** button. If a dialog box opens stating that no attributes are available for a configuration, then no external parameters have been created in the iFlow. We'll explain later how you can proceed in this case. If a dialog window opens instead, in which you can configure parameters, then external parameters are available in the iFlow. The type is set to **All Parameters** by default. Below that you can see the parameters that can be configured in the iFlow. That you should find the parameters you want to change. If you do not find a parameter, it is either not an external parameter or it has a different type. To select a different type, in the **Type** dropdown menu, find the flow step where the parameter is located and edit it according to your requirements. When you have finished editing, click the **Save** button in the bottom right-hand corner to save your changes. By clicking the **Deploy** button, you also deploy the changes you have made at runtime.

**Configuring parameters in the iFlow**

If, as described, you find the message that there are no configurable attributes, you have two options. If you're confident, you can adjust the parameters in the iFlow yourself or ask a developer or consultant to do so. Navigate to the iFlow you want to edit. For this task, click on it in the artifact overview of the package. Click the **Edit** button in the top right-hand corner of the screen to put the iFlow into edit mode. Find the flow step in which you need to make the adjustment. If you need to make changes to the connection data to the SFTP server, you can use the configurations in Section 10.4.1 as a guide. Then save your changes by clicking the **Save-as-Version** button. After that, click the **Deploy** button to deploy the changes. The **Save-as-Version** button allows you to define a new version of the iFlow. If, contrary to your expectations, the adjustments result in an error, you can reset the version accordingly. For this task, navigate to the artifacts overview. Find the iFlow from the list and click on the number highlighted in blue in the **Version** column. This opens the **version history** of the artifact. If you want to restore a previous version, find it and hover your mouse over the line of the version you want to restore. A clock icon appears next to the line, click on it. In the dialog box that opens, click the **OK** button in the bottom right-hand corner. You have successfully restored a previous version of the artifact. If you want to restore the newer version, follow the same procedure. In most cases, the previously active version is located as the second entry in the version history.

## 10.6 Summary

You can quickly implement iFlows to process your XML documents for data migrations using Cloud Integration, although some infrastructure requirements must be met first so that you can communicate with an on-premise SFTP server via the cloud, for example. Furthermore, some settings in Cloud Integration are required to establish the trust between Cloud Integration and an SFTP server.

Then, you learned about the basic structure of processing simple XML documents. Finally, you learned about extensions involving message splitting and data merges to handle common XML document requirements. In these ways, document processing capabilities can convert your XML documents into the format required for your data migration.

We hope we've given you some insights and explored some avenues for action so that, when you decide to implement field mapping using Cloud Integration, this chapter will support you in the best possible way.

Chapter 11

# Methods to Avoid Programming

*In this chapter, we'll introduce you to simple but effective methods so you can do without programming in your data migrations and thus save time and money.*

It goes without saying that you want to avoid time-consuming programming as part of the data migration process: In most cases, these generated programs would be used exactly once. In many cases, programming can be bypassed by combining the capabilities of the data migration methods presented in this book with the capabilities of common office applications. This chapter introduces you to some common use cases.



## 11.1 Critical Point: Data Conversion

If programming is performed as part of the data migration process, it is usually done in the following two phases:

- **Exporting the source data from the source system**
  If the source system does not provide a connection or a function for exporting the data, you'll have to develop (or need to have developed) appropriate programs for this purpose. From a cost perspective, this expense is usually not critical since experts for your source system are usually in-house, and therefore, no cost-intensive external support is required.
- **Converting source data to SAP data format**
  When you convert the source data to the SAP data format, the situation is different: For this task, you need both knowledge of the structure of the SAP data format and possibly also knowledge of the ABAP programming language. However, at the time of SAP implementation, both skills are not always sufficiently available.

In this chapter, we focus on the second set of problems: converting the source data to the SAP data format.

## 11.2 Methods in the Context of Data Conversion

With the methods described below for avoiding programming during data conversion, we want to show you how to transfer your source data to the SAP system largely without programming.

How can you achieve this goal? Many of the tools presented in this book include preconceived rules such as assignments, remappings, fixed values, and so on. These elements usually cover the majority of all conversions that occur. In this chapter, we'll show you how to use methods in Microsoft Excel to prepare your source data in such a way that the entire conversion can be performed without programming.

**Customizing structures and field contents**

Basically, two problem areas exist in this context, that is, situations that cannot (or cannot elegantly) be handled using built-in conversion functions. These areas include the following:

- Modifying structures
- Modifying field contents

We'll address these points in detail in the following sections.

### 11.2.1 Modifying Structures

Structures must be modified if the distribution of data to tables in the source system differs from their distribution in the SAP system.

**Unproblematic flat structures in the source system**

If the structure of the source system is "flatter" than that of the SAP system (i.e., if certain data in the source system is located in a table that is distributed across several tables in the SAP system), then you don't have a problem. Most tools allow you to assign the same structure of the source system (the source structure) to multiple structures of the SAP system (the target structure). An example of this case was described when were loaded customer master data in Chapter 9. The associated relationships among the structures are shown again in Figure 11.1.

If, for example, the data of the customer header in the source system is distributed across two structures, as shown in our example in Figure 11.2, assigning them properly is more problematic.

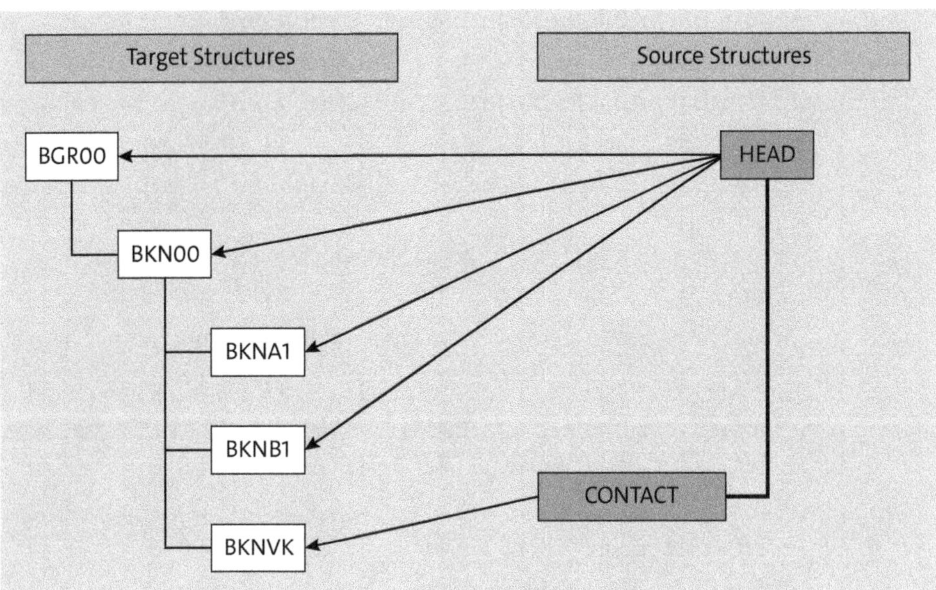


**Figure 11.1** Source Structures Flatter than Target Structures: "Customers" Example

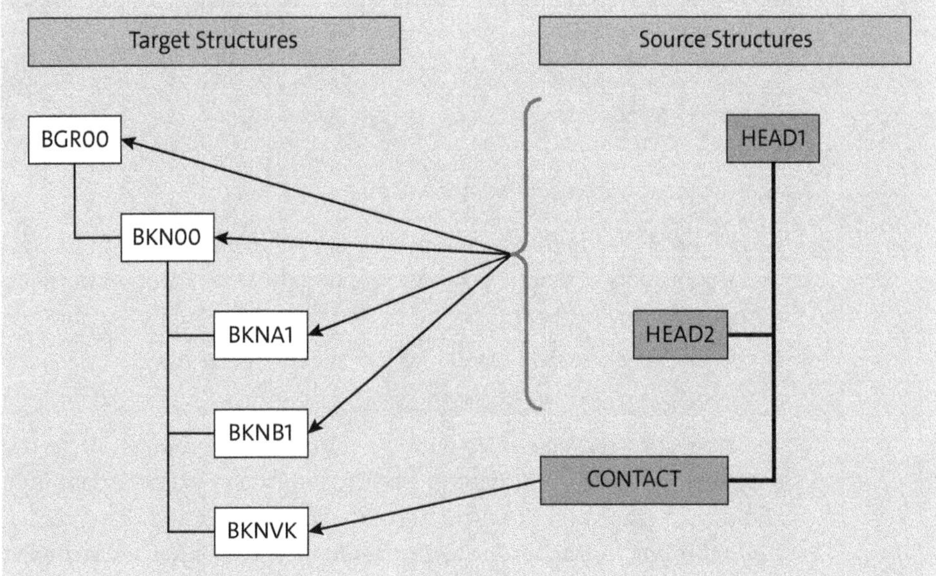


**Figure 11.2** Source Structures Deeper than Target Structures: "Customers" Example

How can you fix this problem? Microsoft Excel offers a simple solution.

**Joins with Microsoft Excel**

The basic idea is to combine the two structures HEADER1 and HEADER2 into one structure. In database language, this action is referred to as a *join*.

So, let's assume that the fields of the HEADER structure from our example are distributed to the HEADER1 and HEADER2 structures in the following way:

- **HEADER1**
  CUSTOMER, NAME, STREET, CITY, POSTALCODE, and COUNTRY
- **HEADER2**
  CUSTOMER, LANGUAGE, TELEPHONE, FAX, and UIN (tax ID number)

Using Microsoft Excel, you can easily create a new structure that contains all the fields of both structures. In this case, we can use the somewhat more complex VLOOKUP function from Microsoft Excel as a join.

[»]

**VLOOKUP Function**

The VLOOKUP function follows the syntax VLOOKUP(Lookup_value;Table_array;Col_index_num;[Range_ lookup])
Make sure that the comparison table (matrix) is sorted by the key and that the key column has the same formatting as the cells of the search criterion in the cell to be filled.

(A detailed description of the VLOOKUP function can also be found in the Microsoft Excel help feature.)

Now, in Microsoft Excel, follow these steps:

1. Import the HEADER1 and HEADER2 tables as separate worksheets.
2. Add the headers of worksheet HEADER2, excluding the customer number, to the HEADER1 worksheet.
3. In cell G2 of the last field of HEADER1, add the following VLOOKUP rule:
   - =VLOOKUP(A2;Header2!A:E;**2**;FALSE)
4. The rule links the customer number from HEADER1 and cell A2 to the search range (Header2!A:E) in the HEADER2 worksheet via columns A to E and returns the value from the second column (2) of the search range (i.e., column B in this case). The last value—FALSE—causes the searched customer number in HEADER2 to match HEADER1 in value and formatting.

5. Copy the rule to all other cells from H2 to J2 and count up the column number of the search range whose value is to be returned:
   - H2: `=VLOOKUP(A2;Header2!A:E;`**`3`**`;FALSE)`
   - I2: `=VLOOKUP(A2;Header2!A:E;`**`4`**`;FALSE)`
   - J2: `=VLOOKUP(A2;Header2!A:E;`**`5`**`;FALSE)`
6. As shown in Figure 11.3, all values from `HEADER2` are now also visible in `HEADER1`.
7. Save the file under the name `HEADER1_HEADER2`.
8. Export the result in the appropriate format for your data migration tool.
9. Process this file with your data migration tool.

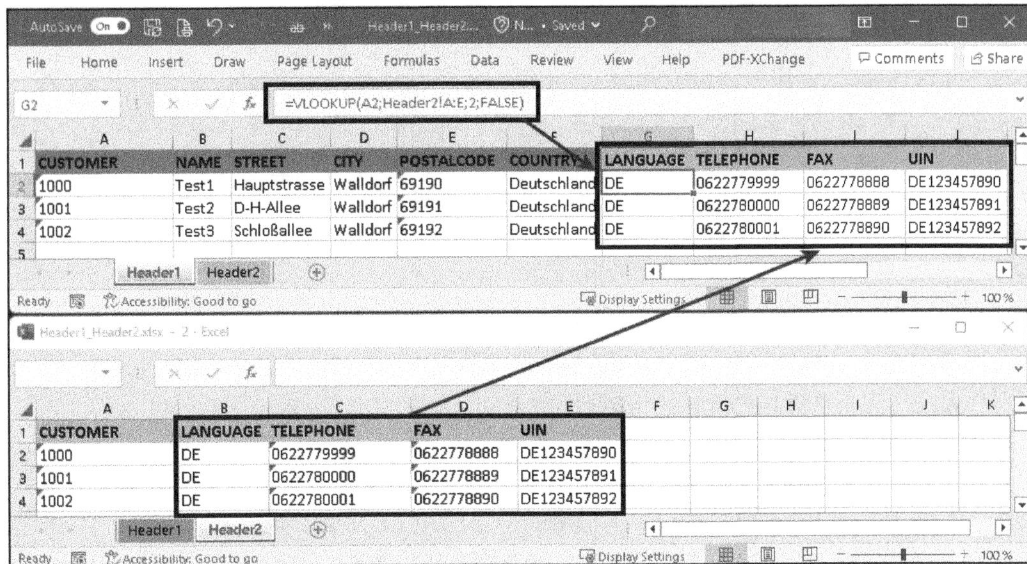


**Figure 11.3** Microsoft Excel: Join of HEADER1 and HEADER2

In this way, the structure relations are simplified, as shown in Figure 11.4. The complex case (shown earlier in Figure 11.2) can now be transformed into a situation whose structure relations correspond to those shown earlier in Figure 11.1 and can consequently be handled by the data migration tool.

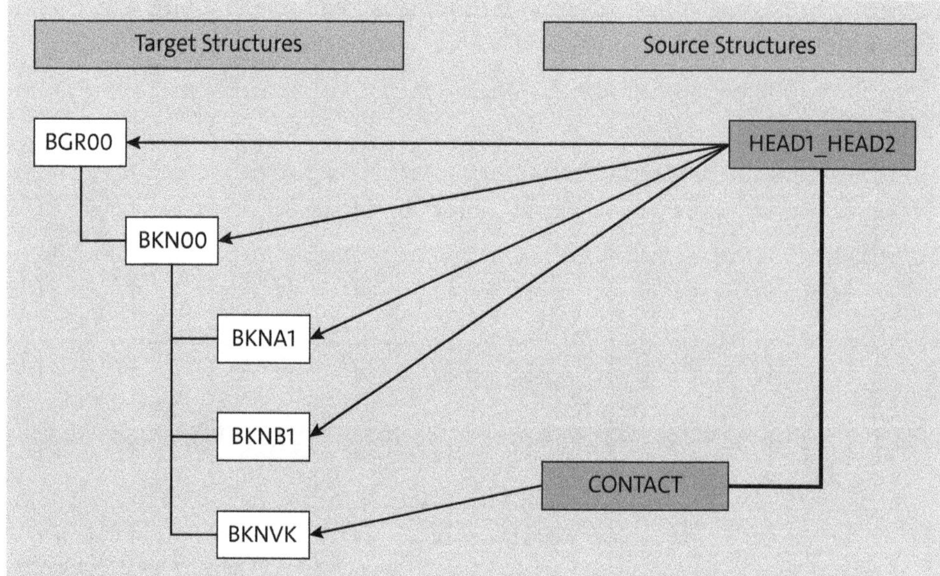


**Figure 11.4** Simplified Structure Relations after Join between HEADER1 and HEADER2

### 11.2.2 Modifying Field Contents

In this section, we'll consider some cases in which necessary adjustments to field contents can be made without programming in Microsoft Excel.

#### Date Formats

Date values occur in just about all data objects. Depending on which import method has been chosen, the SAP system expects the date values in a certain format:

- **Batch input**
  With the batch input method, a date value is usually expected in the format defined in the SAP user ID settings (e.g., `DD.MM.YYYY`). Reminder: If you follow the menu path **System • User Defaults • User Data** from the initial SAP screen and click the **Fixed Values** tab on the following screen template, you'll reach the input screen shown in Figure 11.5. On this screen, you can specify, among other things, how date values should be displayed.
- **Business Application Programming Interface (BAPI) and IDoc**
  With the BAPI and IDoc import method, a date value is normally requested in the internal format, that is, `YYYYMMDD`.

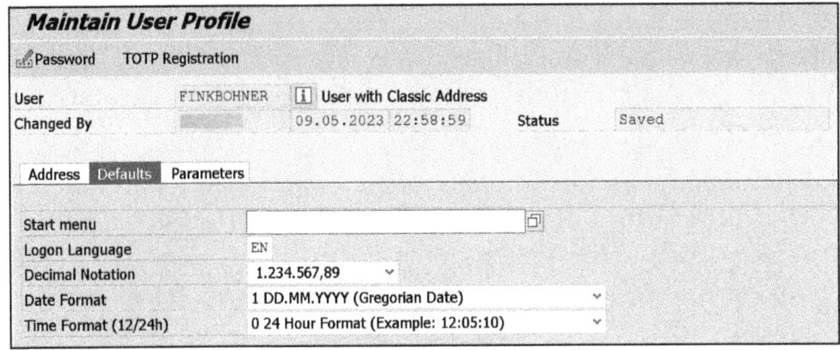


**Figure 11.5** SAP System: Defining User Defaults

But how do you transform a date value, for example, from DD.MM.YYYY to YYYYMMDD? For this task, follow these steps:



1. In Microsoft Excel, select **Format Cells** from the context menu by right-clicking in the relevant table column. Then, click the **Number** tab.
2. Select the **Custom** menu item on the left and enter the value "YYYYM-MDD" in the **Type** field, as shown in Figure 11.6.
3. If you now save the file in **Text (tab-delimited) (*.txt)** format, the date values will be output in YYYYMMDD format.

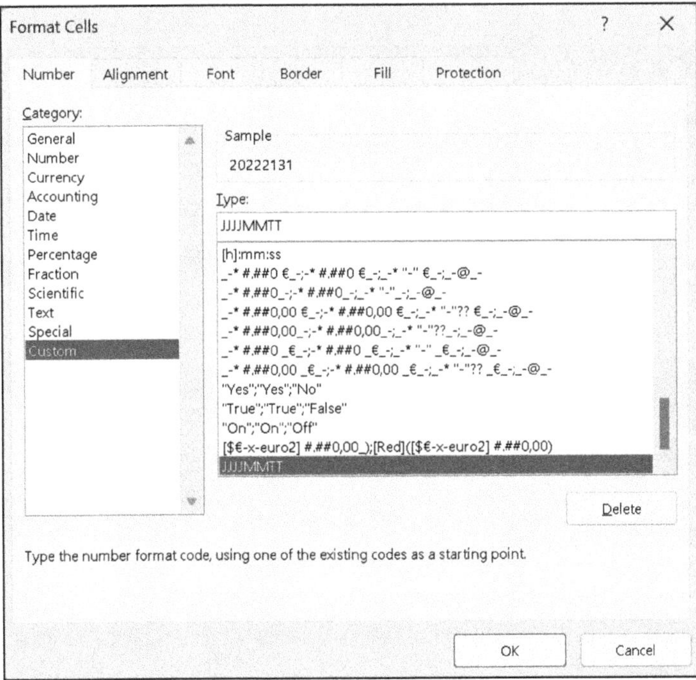


**Figure 11.6** Selecting Date Format YYYYMMDD

### Number Formats

Like date values, numerical values occur in almost all data objects. Depending on the import method, the SAP system also expects a certain format for numerical values in the following ways:

- **Batch input**
  With the batch input technique, a numerical value is usually expected—similar to date values—in the format as it exists in the SAP user ID settings. The example `123,456.78`, shown earlier in Figure 11.5, has a thousands separator and uses a dot as the decimal point.
- **BAPI and IDoc**
  With the BAPI and IDoc import method, a numerical value is normally expected in the internal format `123456.78` (i.e., without a thousands separator but still with a dot as the decimal point).

Now, how do you transform a numerical value of, say, 123,456.78 into 123456.78? This conversion is quite simple. Follow these steps:

**Converting the number format**

1. In Microsoft Excel, select **Format Cells** from the context menu by right-clicking in the relevant table column. The dialog box shown in Figure 11.7 appears.

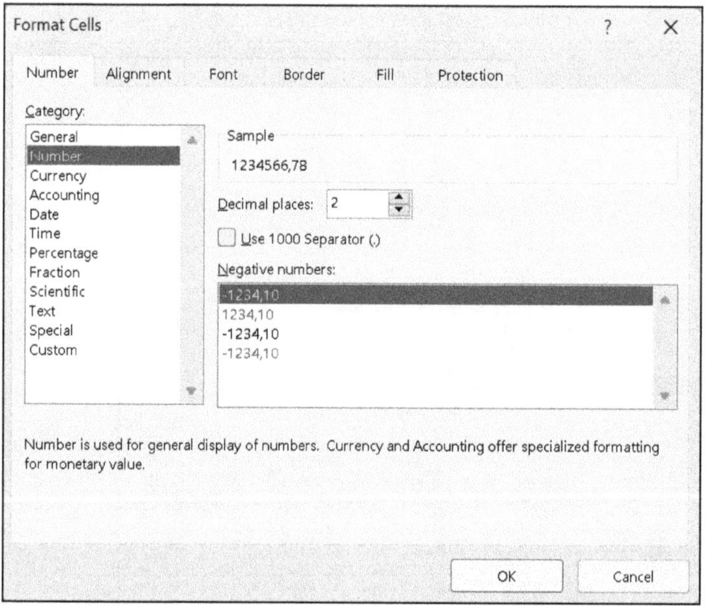


**Figure 11.7** Selecting the Number Format without Thousands Separator

2. Click the **Number** tab.
3. Select the **Number** menu item on the left and deselect the **Use 1000 separator (,)** checkbox.

4. If you now save the file in **Text (tab-delimited) (*.txt)** format, the numerical values will be output in the desired format (i.e., 123456.78).

You can change the decimal point from a dot to a comma by calling—while Microsoft Excel is open—the Region settings from the Windows Control Panel and making the appropriate changes under **Additional settings**, as shown in Figure 11.8. The data will then be immediately updated in Microsoft Excel.



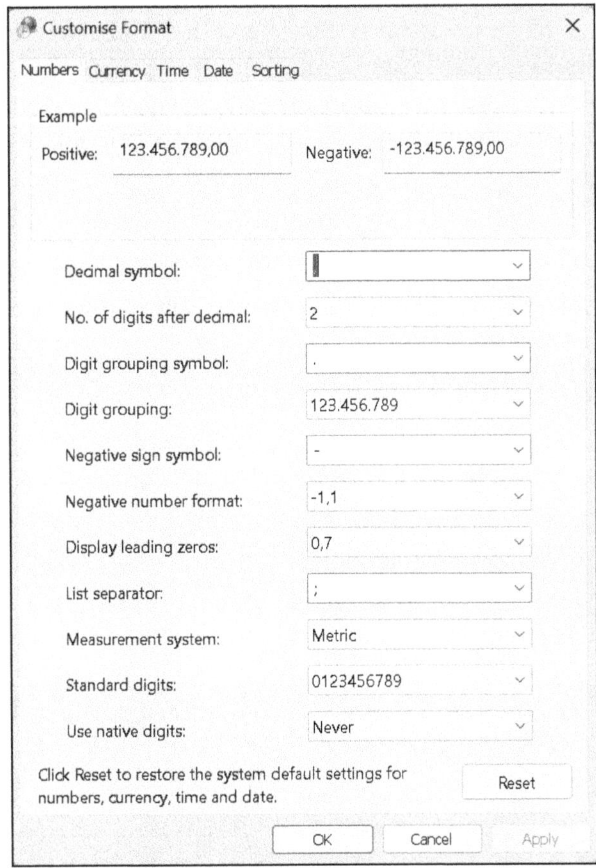


**Figure 11.8** Regional Settings: Selecting the Decimal Symbol

### 11.2.3 Access Data of the SAP System

During the data conversion, you might need to access the data of the SAP system.

**Historically grown customer numbers**

Let's assume you want to clean up your historically grown customer numbers as part of the SAP implementation, and for this reason, you decide to have your SAP system assign new customer numbers during the data migration. Accordingly, you select the what's called the *internal number assignment* option. To avoid losing the reference between old and new customer numbers, you should store the old customer number in a specially designated table field in the SAP system.

However, if you also want to transfer customer open items, which usually only know the old customer number, you must replace the old customer number with the new one in these open items, as the former is for information purposes only for the SAP system. In other words, the old number does not represent an account that can be posted to.

**Reference between old and new customer numbers**

Note, however, that the reference between the old and the new customer numbers is only known to the SAP system, more precisely, known only to SAP table KNB1. To create this reference during the data conversion of your documents, you can either program the activity or proceed with the following steps:

- You must load the section of the SAP table that creates the reference between the old and new customer numbers onto your workstation.
- Then, replace the old customer number with the new one in the document data using Microsoft Excel.

**Extracting data using the Data Browser**

Let's consider these steps in a bit more detail:

1. Migrate your customer master data to the SAP system using the internal number assignment. Store the old customer number in field KNB1-ALTKN.
2. In the SAP system, start Transaction SE16 (Data Browser), enter "KNB1" in the **Table Name** field, and press [Enter].
3. On the following selection screen, as shown in Figure 11.9, select all relevant data records via fields KUNNR and BUKRS.

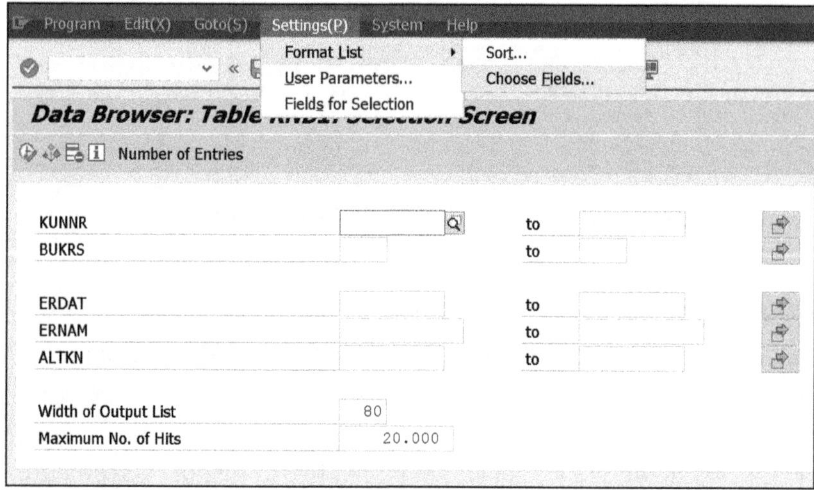


**Figure 11.9** Transaction SE16 (Data Browser)

4. Open the menu path **Settings • Format List • Choose Fields** and select the fields MANDT, KUNNR, and ALTKN in the dialog box that follows. Confirm your selection by clicking **Next**.
5. Then, select **Settings • User Parameters** and, in the dialog box that opens, select the **ALV Grid Display** option. Enter a sufficiently large number in the **Maximum Number of Hits** field, as shown in Figure 11.10, and close the window.

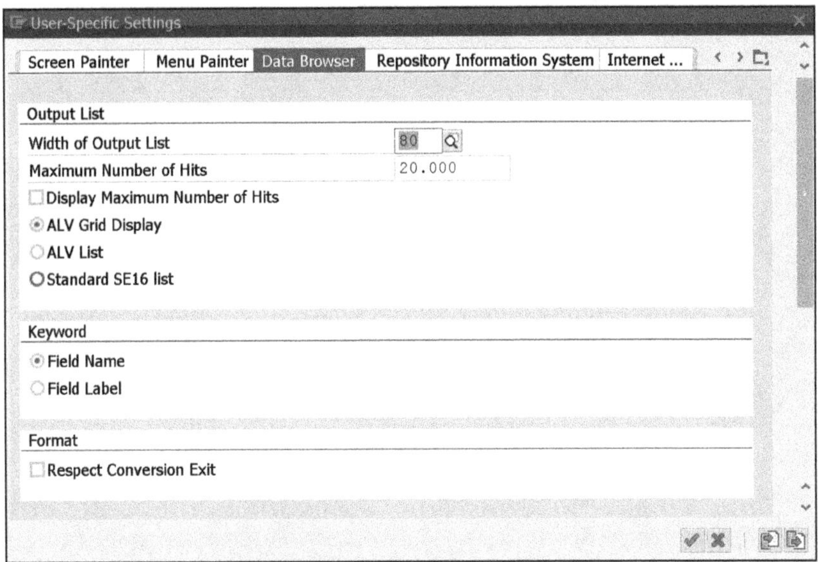


**Figure 11.10** Transaction SE16: ALV Grid Display

6. If you now select **Execute** [F8], you'll get the relevant data in a table.
7. In the displayed ALV list, follow the menu path **Table Entry • List • Export • Spreadsheet**. In the dialog box that opens, as shown in Figure 11.11, select the value **xlsx-TT Spreadsheet(*.xlsx)** in the **Format** field and select the **LOCAL_FILE Local** option in the **Destination** dropdown list.

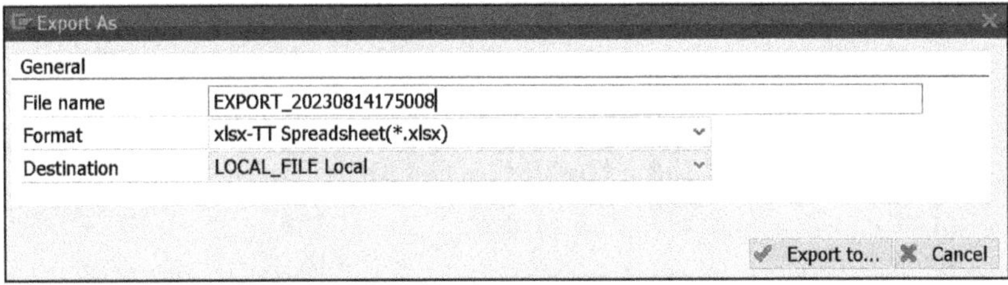


**Figure 11.11** Transaction SE16: Saving the File

8. After entering a file name with the *.xlsx* extension, click **Export to...**.

9. The data is then transferred to your workstation computer. Microsoft Excel opens automatically. You may need to make some corrections, reformat the cells as text, and resave the file in a different format. This file should look something like the spreadsheet shown in Figure 11.12.

| | A | B | C | D | E | F | G | H |
|---|---|---|---|---|---|---|---|---|
| 1 | MANDT | KUNNR | ALTKN | | | | | |
| 2 | 910 | 0000013260 | 556677 | | | | | |
| 3 | 910 | 0000013262 | 556678 | | | | | |
| 4 | 910 | 0000013264 | 556679 | | | | | |
| 5 | 910 | 0000013270 | 556680 | | | | | |
| 6 | 910 | 0000013276 | 556681 | | | | | |
| 7 | 910 | 0000013277 | 556682 | | | | | |
| 8 | 910 | 0000014479 | 556683 | | | | | |

CustomerNo_Old_New

**Figure 11.12** Table with New and Old Customer Numbers

**Copying Data Directly**

As an alternative to exporting to a Microsoft Excel file, you can select all the relevant columns in the ALV grid and copy-paste them directly to an open Notepad file. For this task, follow these steps:

1. Scroll to the end of the list in the ALV grid by pressing [Page Down]. Doing so serves to fill the buffer of the ALV. For very large lists, the records are not loaded into the ALV buffer until they are displayed.
2. Mark the relevant columns in the heading line by clicking the left mouse button while pressing down the [Ctrl] key. Alternatively, you can press [Ctrl] + [A].
3. Move the mouse over a column heading and open the context menu via the right mouse button.
4. Select the **Copy Text** menu item or press [Ctrl] + [C].
5. Open a Microsoft Notepad or other text editor and paste the copied text or press [Ctrl] + [V].
6. The text editor now contains the values from the table, separated by tabs.
7. If necessary, add a header with the field names.

**Importing Extracted Table Contents Using IDocs**

Since IDocs are used to exchange data between SAP systems, the IDoc structures often correspond to the tables in which the data is stored in the system. Thus, if you extract the tables into files, they can often be mapped one-to-one to the IDoc structures afterwards, for instance, in SAP Legacy System Migration Workbench or via a rapid data transfer using SAP Data Services.

## 11.3 Summary

You should always ensure that the use of these methods is thoroughly documented, which is the only way to avoid forgetting a step during the productive transfer.

In addition, you must remember that the methods we described cannot be easily incorporated into automated processes, such as periodic data migrations.

Nevertheless, we hope that this chapter has made it clear that there are several situations in which you might be tempted at first to resort to programming. However, by using Microsoft Excel skillfully, you can sometimes avoid costly programming.

# Chapter 12
# Related Areas

*In this chapter, we cover topics that are not directly related to the core topic of this book but are relevant for a broad and general understanding of data migrations.*

12

The focus of this book is avowedly on the data migration procedures provided by SAP, which are most often used as part of a new implementation. However, for a comprehensive presentation on the topic of migration, we thought it appropriate to also mention other methods, such as database migrations and system conversions, in a brief outline. These special methods, which are part of the brownfield migration strategy described in Chapter 1, have been successfully used by countless customers over the past decades and are virtually the gold standard if you want to adapt your system architecture or quickly migrate your existing systems to the SAP HANA and SAP S/4HANA world. In Section 12.1, we provide a brief introduction to the topic of database migration, while in Section 12.2 we'll introduce you to system conversion. For both topics, we'll provide links to further information.

## 12.1 Database Migration

So far, we have only talked about migrating data, that is, moving data records from one system to another. However, sometimes, an SAP system must be migrated from one database to another. In the following sections, we'll briefly describe some basic principles behind such a database migration and then explain them using two sample scenarios.

### 12.1.1 Basic Principles

When an SAP system is to be migrated from one database to another, then we are referring to a *database migration* or *DB migration*. Using the SAP tools for operating system (OS)/DB migration, which can copy classic SAP systems with and without a change in operating system, a special data migration is possible in certain use cases. For this reason, we'll briefly outline the procedure using an ABAP-based SAP system in this section.

**Using a backup** Databases can be copied and duplicated by using a previously created backup to restart a new system. In addition to all traditional SAP databases, this capability also includes SAP HANA as a database. Some databases support restoring such a database backup even on a new operating system. However, in our case, we'll focus only on the migration of the database itself.

**Tools** The requirements for a database migration are similar to the requirements for a data migration; you want to perform this migration quickly and easily. For this purpose, SAP also provides additional and technically simple tools.

**SAP HANA** SAP HANA's in-memory technology in particular allows you to benefit from the potential of big data and, for example, improve the performance of reporting and analyses in your SAP BW system.

### 12.1.2 Sample Database Migration

**Migration to SAP HANA** In the following sections, we describe the database migration of an existing SAP Business Suite system from any database to SAP HANA as an example. For such a change of database, three different options and three different tools can come into play, as shown in Figure 12.1:

- **Option 1: New installation**
  This option is the most expensive, as a completely new system gets built and then the data must be loaded into the new system by means of a data migration.

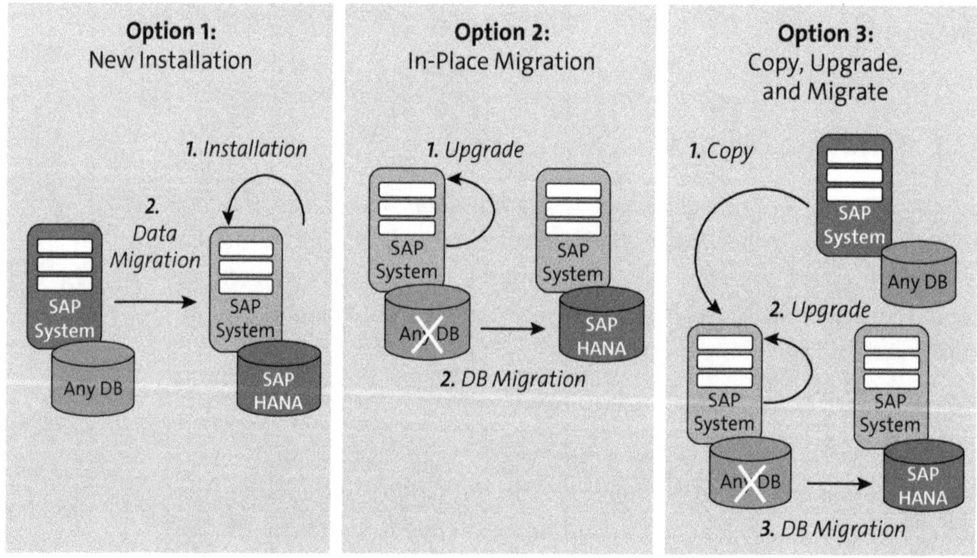


**Figure 12.1** Options for a Database Migration

- **Option 2: In-place migration**
  This option enables you to completely migrate a system to the new SAP HANA database by way of a database migration. This option avoids changes to your system landscape because the system ID, IP address, host name, etc. of your system remain the same. After the upgrade, only the underlying database has been replaced.
- **Option 3: Copy, upgrade, and migrate**
  With this option, an existing system is first copied to another system, and then option 2 is performed.

You have two possible scenarios at this point. Let's look at each scenario next.

### Scenario 1: Migration Using Software Update Manager and Software Provisioning Manager

The first scenario can be described as the classic database migration. As shown in Figure 12.2, the *Software Update Manager (SUM)* is first used to update the SAP system to the latest *Enhancement Package (EHP)*.

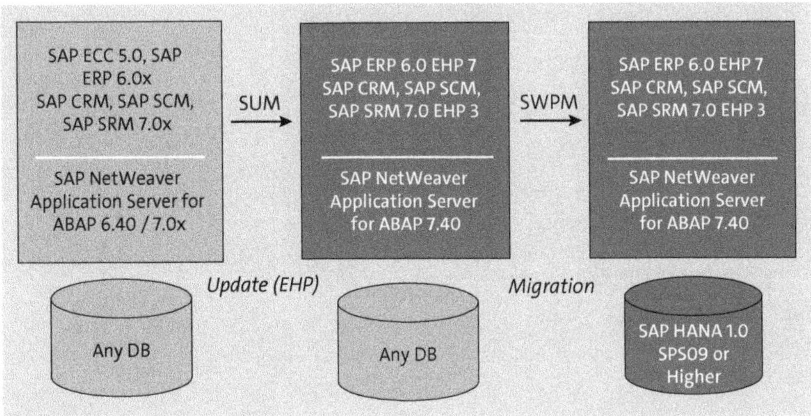


**Figure 12.2** Scenario 1: Update and Migration Using SUM and Software Provisioning Manager (SWPM)

The underlying database remains unchanged. The second step consists of copying (migrating) your database to SAP HANA using the *SWPM*.

[«]

**Software Update Manager (SUM)**

SUM is a tool that can be used for a variety of maintenance scenarios, such as updates, upgrades, service pack imports, system conversions, etc. For more information on SUM, go to *http://support.sap.com/sltoolset*. Navigate to the **System Maintenance** section or go directly to the following link: *http://s-prs.co/v581715*.

[»]

**Software Provisioning Manager (SWPM)**

Software Provisioning Manager (SWPM) is the successor to the SAP tool, SAPinst. You can use SWPM to install, copy, transform, split, rename, and uninstall products based on SAP NetWeaver Application Server ABAP (SAP NetWeaver AS ABP) or SAP NetWeaver Application Server JAVA (SAP NetWeaver AS Java) or based on SAP S/4HANA Foundation.

To find out which SWPM version is the right one for your systems and for more information about SWPM, go to *https://support.sap.com/sltoolset* in the **System Provisioning** section, or follow this link: *http://s-prs.co/v581716*.

### Scenario 2: Migration with Database Migration Option (DMO) for Software Update Manager (SUM)

Rapid migration

In this scenario, SUM is used with the *database migration option (DMO)*. This method enables a rapid migration with reduced effort and shorter downtimes of the SAP system during the database migration: Both the upgrade and the database migration are performed using only one tool, as shown in Figure 12.3.

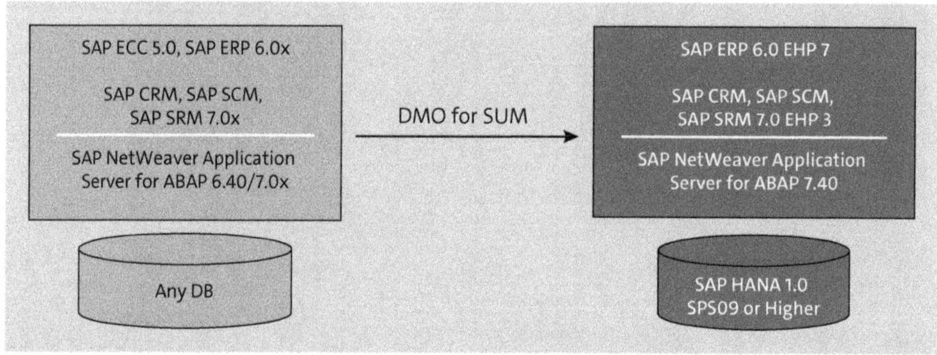


**Figure 12.3** Scenario 2: Migration with DMO and SUM

Sequence

An outline of the individual phases of a migration with DMO is shown in Figure 12.4.

The first step is to prepare the SAP HANA database. The second step is to update the SAP system to the current EHP status. In the third step, the downtime starts, and the system's database connection is redirected. After that, the system's data gets migrated to the new SAP HANA database and, if necessary, the data types are converted. After that, the upgrade will finish, and you can restart your SAP HANA-based system. Your old database will still be available throughout the process and afterwards. It remains essentially unchanged. This method thus enables a simple fallback mechanism.

[«]

**Data Migration Option (DMO) for SUM**

DMO is an option of SUM, which combines the upgrade of the system with the database migration. In contrast to the classic database migration, this approach can also be referred to as a one-step migration. For more information on DMO, see, for example, *https://support.sap.com/sltoolset*. Navigate to the **System Maintenance** section or access this feature directly at *http://s-prs.co/v581717*. Scroll down to the **Database Migration** item.

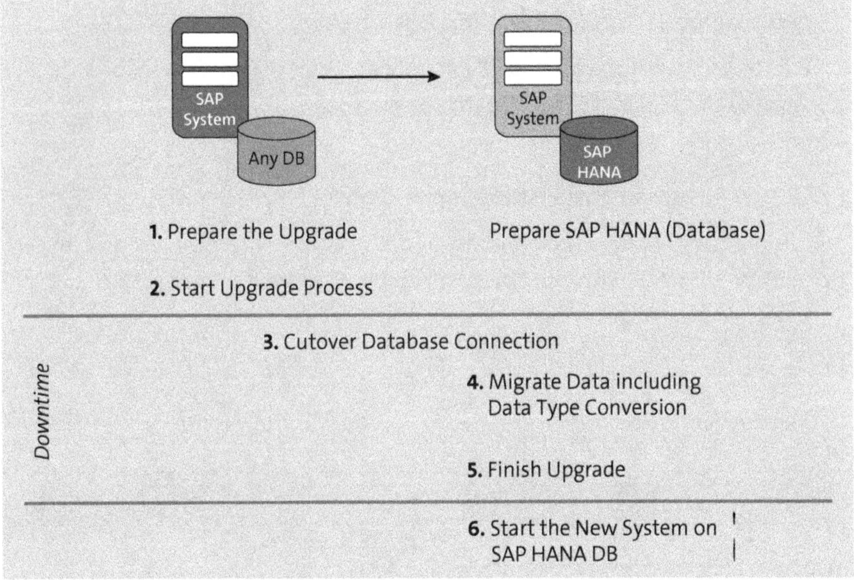


**Figure 12.4** Phases of Database Migration with DMO

## 12.2 System Conversion

A system conversion enables you to not only change your database to an SAP HANA database, but also to convert your entire SAP ERP 6.0 system to an SAP S/4HANA system. This scenario is common today to migrate a legacy SAP ERP system to a new SAP S/4HANA system.

### 12.2.1 Technical Conditions

For the technical conditions for the system conversion, see the SAP Note for your SAP S/4HANA version. For SAP S/4HANA 2022, you want SAP Note 3145277 in the **Conversion Information** section. For other release versions, the best option is to search for "SAP S/4HANA 2022: Release Information Note" and replace "2022" with the relevant release. For a system conversion to SAP S/4HANA 2022, the minimum Support Package (SP) level of the underlying SAP NetWeaver release must be one of the following:

- SAP NetWeaver 7.0 SP17
- SAP EHP1 for SAP NetWeaver 7.0 SP05
- SAP EHP2 for SAP NetWeaver 7.0 SP06
- SAP EHP3 for SAP NetWeaver 7.0 SP01
- SAP NetWeaver 7.4 SP02

If your system is on a lower product version, you must first upgrade to one of the supported product versions listed here.

For the conversion, you must be running at least SUM 2.0 SP 15. Be sure you always use SUM at the latest patch level.

### 12.2.2 Phases of the System Conversion

In general, a system conversion is divided into the prepare and realize phases. Figure 12.5 shows the basic steps of a system conversion and the tools used in the process.

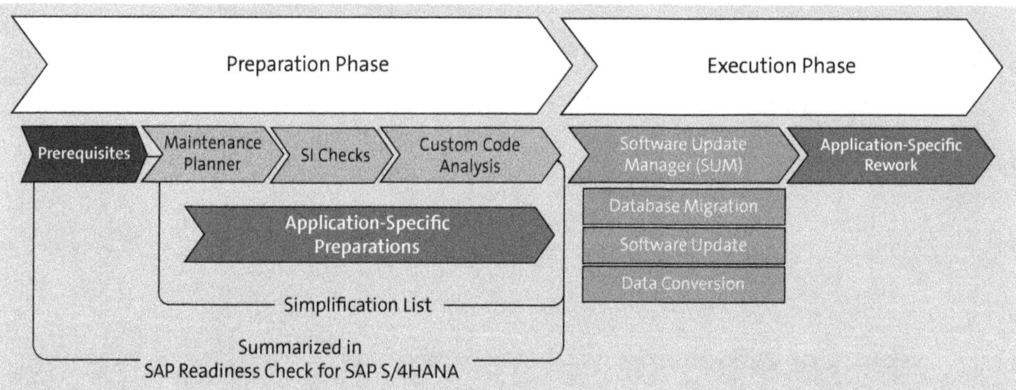


**Figure 12.5** Steps and Tools of an SAP S/4HANA System Conversion

**Prepare phase**

In the *prepare phase*, you must check if the system you want to convert meets system requirements. Systems that do not meet the requirements must be upgraded to the respective version, if possible. This rule also applies to non-Unicode systems, which must first be converted to Unicode. The *SAP Readiness Check* supports you in this respect. It clearly presents the results of the maintenance planner, simplification item check, simplification list, and custom code analysis tools. In addition, some application-specific and cross-application preparations are necessary. You may need to make numerous adjustments in the source system before you can start the system conversion process.

**Realize phase**

In the *realize phase*, your source system gets technically converted using the Software Update Manager (SUM). This step involves bringing the database onto SAP HANA, converting the software to SAP S/4HANA, and converting or transferring application data into the formats and tables relevant for SAP S/4HANA. All this is done in a one-step process.

After using SUM, you must customize your modifications and extensions (custom code) via Transaction SPDD, Transaction SPAU, and Transaction SPAU_ENH. Furthermore, any issues that arise due to SAP S/4HANA simplifications must be resolved. You can use the ABAP Test Cockpit (ATC) to find issues. Perform a check run using the `S4HANA_READINESS` check variant locally in your converted system. After that check, you still need to perform additional application-specific post-processing work, however.

[+]

**Data Transition Validation Tool (DTV)**

The Data Transition Validation Tool is available with SAP S/4HANA 2021 and later. You can use this tool to check your data after the conversion process. For this task, you must call the validation reports provided by SAP. You can save a lot of time, however, since you no longer need to check each individual document. For more information on DTV, check out the central DTV blog "Data transition validation the tool to validate business data after conversion" by Boris Rubarth, which you can find at SAP Community via the following link: *http://s-prs.co/v581718*.

**More Information about System Conversions**

For information on a system conversion to SAP S/4HANA, the conversion guide in SAP Help Portal for SAP S/4HANA is available at *https://help.sap.com/s4hana*. The guide is available for SAP S/4HANA 2022 via the following link: *http://s-prs.co/v581719*.

In general, you should also observe the following SAP Notes:

- SAP Note 3226469 (Preparation)
- SAP Note 3206747 (SUM)
- SAP Note 3226548 (Follow-Up Note)
- SAP Note 3230844 (Restriction Note)

A very good and detailed description can also be found in the book *Migration to SAP S/4HANA* by Densborn, Finkbohner, Höft, Mathäß, Klöß, and Rubarth (SAP PRESS, 2024).

## 12.3 Summary

This chapter has provided a brief outline on the topics of database migrations and system conversions. You learned what a database migration and a system conversion are and what different scenarios and options are available.

With this book, you have gained deep insight and helpful information and tips that should help bring success to your data migration projects involving SAP S/4HANA, SAP SuccessFactors, SAP Ariba, and SAP Business ByDesign. For your next migration project, we wish you easy planning, consideration of all possible issues, and a successful outcome! Hopefully, this book will be a valuable companion for you in that process.

# Appendix A
# Glossary of Terms

### ABAP

Advanced Business Application Programming. Fourth-generation programming language developed by SAP for developing application programs.

### ABAP Dictionary

The ABAP Dictionary describes the logical structure of application development objects and their mapping in the structures of the underlying relational database.

All components of the runtime environment, such as application programs or database interfaces, get their information about these objects directly from the ABAP Dictionary.

The ABAP Dictionary is fully integrated into the ABAP Workbench.

### Basis release

The release of the Basis implementation underlying the SAP system.

### Batch input

Interface that allows you to transfer large amounts of data into an SAP system. The initial transfer of legacy data and the periodic import of external data can be performed via batch input.

### Batch input session

Summary of a sequence of transactions that have been provided with user data by a program. These transactions are stored as a batch. The batch input session can be played back at a later time. Database changes do not take place until the session gets played back. This procedure allows large amounts of data to be imported into the SAP system in a short period of time.

### Brownfield

Migration strategy for migrating into an existing system landscape.

### Business Application Programming Interface (BAPI)

Standardized programming interface that allows external access to the business processes and data of the SAP system. BAPIs are defined in the Business Object Repository (BOR) as methods of SAP business objects and provide an object-oriented view of the business components of the SAP system. BAPIs are implemented and stored as remote function call (RFC)-enabled function modules in the Function Builder of the ABAP Workbench.

### Business object

Business data unit such as customer master, material master, financial document.

### Client

A self-contained unit within an SAP system in terms of commercial law, organization, and data technology with separate master records and an independent set of tables.

### Cloud Integration

Cloud Integration is a cloud-based integration platform that helps organizations seamlessly integrate applications in the cloud and on-premise.

### Constant

A value that is set when the fields of a program are initialized and cannot be changed at runtime. In ABAP, constants are declared using the `CONSTANTS` statement.

### Conversion rule

Rule used to translate source data into target data.

### Converting the data

Data conversion is the structural or value manipulation of data according to certain rules to be defined.

### CSV file

*Comma-separated values* or *character-separated values* files A CSV file is a text file used to exchange data. The data is simply structured, and the values are separated from each other by a character such as a comma, a semicolon, etc.

### Customizing

Settings that the system administrator must make during system implementation. In Customizing, you can adapt enterprise-neutral functionality delivered to the specific business requirements of your company. Customizing is mandatory before the system goes live and is carried out in the SAP system using the Implementation Guide (IMG).

### Data merge

The merging of two different messages into a single message.

### Data migration

Transfer of business data (master data and transaction data) from any system into an SAP system.

### Data migration object

Description of a data object and other properties that are important for data migration: the structure of the data object in the legacy system and in the SAP system and the mapping that merges the two structures.

### Data migration option (DMO) for SUM

Data migration option is an option of Software Update Manager (SUM) that combines a system upgrade with a database migration. In contrast to the classic database migration, this option can also be referred to as a *one-step migration*.

### Data object

→ Business object.

### Data Services

SAP Data Services combines the products for data quality management and data integration. SAP Data Services serves as an extract, transform, and load (ETL) tool that processes and cleanses both structured and unstructured data from SAP and non-SAP data sources.

### Data Transfer Workbench

Tool for managing and organizing data transfer projects. Can be used in combination with SAP Legacy System Migration Workbench.

### Direct input

Data migration procedure in which the data to be transferred is checked directly, that is, bypassing the dialog box transaction, and written to the database of the SAP system.

### DMO

→ Data migration option for SUM.

### Document

Evidence of a business transaction. A distinction is made between original documents, such as incoming invoices, bank statements or copies of outgoing invoices, and computerized documents. The latter can be, for example, accounting vouchers, sample vouchers, or recurring entry vouchers. The accounting document reflects the original document in the system. All other IT documents mentioned serve as entry aids for the original documents.

### Domain

Object that describes the technical properties, such as the data type and length, of a field. The definition of a domain can also specify a range of values that contains the valid data values for the fields that reference that domain. Different technically and professionally similar fields can be combined by a domain. Fields that reference the same domain are changed at the same time when the domain is changed. This approach ensures the consistency of these fields.

### Dynpro

Dynamic program. A dynpro (screen) consists of a screen template and the underlying flow logic. The main components of a screen are attributes (e.g., screen number, next screen); layout (e.g., arrangement of texts, fields, and other elements); field attributes (e.g., definition of the properties of the individual fields); and the flow logic (calls of the relevant ABAP modules).

### EIM

Enterprise information management. Software solution for data integration and quality, master data management, event processing, content management, and data governance.

**Elementary data type**

ABAP data type. ABAP has eight predefined elementary data types: C (alphanumeric characters), D (date), F (floating point number), I (integer), N (numeric text), P (packed number), T (time), and X (hexadecimal).

The data types D, F, I, and T are predefined in every respect, while the data types C, N, P, and X can be additionally specified.

**ETL**

Process for data extraction, transformation, and loading.

**Excel**

→ Microsoft Excel.

**Field mapping**

Assignment of source fields to target fields specifying the conversion rules.

**File-to-file approach**

Using Cloud Integration to prepare data for migration using standard migration methods.

**File Transfer Protocol (FTP)**

Network protocol for transferring files through networks.

**Frontend**

The workstation computer—usually a PC.

**Function module**

Function for general use. Function modules are external subroutines written in ABAP that are developed in the function builder. They are stored there in a central function library and can thus be called by all ABAP programs. In this way, redundancies in code can be avoided, and programming can be made more effective. In contrast to FORM routines, function modules have clearly defined interfaces.

**Greenfield**

Migration strategy for migrating to a new system landscape.

### Historical data

Data that is no longer productively required and can therefore be archived or deleted in compliance with statutory retention periods.

### IDoc (Intermediate Document)

SAP format in which business messages are transmitted. An IDoc (or IDoc message) refers to the concrete message, while an IDoc type refers to its structure.

### Integration Flow (iFlow)

A graphical representation of the flow and processing of messages between two or more participants using an integration runtime platform that ensures successful communication.

### Internal table

Data structure that exists only during the runtime of a program. An internal table is one of two structured data types in ABAP. It consists of any number of identically structured table rows with or without a header row. The header row corresponds to a field bar and serves as a work area for the internal table. The data type of the rows can be elementary or structured.

### Legacy data

Business data to be transferred from the legacy system to the SAP system.

### Legacy system

Business management IT system that is to be completely or partially replaced in the course of introducing an SAP system. In the course of this replacement, data from this system is to be transferred to the SAP system.

### Loop

Sequence of program statements that are executed repeatedly. This action can be occur either over a certain number of repetitions or until a certain condition becomes true or false.

### Mandatory field

Input field in which data must be entered. A mandatory field is usually indicated by a question mark. Screen templates can only be processed successfully if all mandatory fields are filled in with data. In contrast to mandatory fields, an entry is not required for optional fields.

### Mapping

Assignment of fields between a legacy IT system and a succeeding system. This step is necessary to transfer the field contents of the legacy system to the succeeding system as part of the data migration.

### Master data

Data that remains unchanged over a longer period of time. Master data contains information that is needed again and again in the same way. The master data of a cost center can be mentioned as an example, which contains the name, the responsible person, the associated hierarchy area, and further information.

### Microsoft Excel

Microsoft Excel is a spreadsheet software for data organization, analysis, and visualization in the Microsoft Office suite.

### Migrate your data

→ SAP S/4HANA migration cockpit.

### Migration

Used synonymously with data migration.

### Migration cockpit

→ SAP S/4HANA migration cockpit.

### Migration object

→ Data migration object.

### Migration Object Modeler

→ SAP S/4HANA migration object modeler.

### Migration Workbench

A tool developed by SAP, specialized in transferring data between SAP systems. Part of SAP Landscape Transformation.

### Model your migration object

→ SAP S/4HANA migration object modeler.

**Modeling migration objects**

→ SAP S/4HANA migration object modeler.

**New implementation**

→ Greenfield.

**News splitting**

Split messages into individual groups using a splitter, thereby creating smaller messages.

**Non-SAP system**

→ Legacy system.

**Object**

→ Data migration object.

**ODBC**

Open Database Connectivity. Standardized database interface for connecting relational databases via the SQL database language.

**On demand**

Software delivery model that avoids hardware investments and reduces operating costs. The application is operated at the provider and provided to the customer as a service.

**Optional field**

Input field in which data can be entered. In contrast to optional fields, an entry is required in mandatory fields.

**Parameters**

Data required for the execution of a method, such as a function block or a subroutine (import parameters) or returned by the method after it has been called (export parameters). With the definition of the parameters the interface of the method call gets determined.

**Primary window**

Main window of an application. Primary windows are windows where the main action of the application takes place. Within a mode, there is always exactly one primary window on the screen at a time. Primary windows can send secondary windows or be replaced by other primary windows.

### Rapid data migration

A special SAP Rapid Deployment solutions package for migrating legacy data to SAP target systems. The various packages available are based on SAP Data Services as software and provide the content for mapping and data validation as well as for improving data quality.

### RFC

Remote function call. SAP's own protocol for calling function modules in remote systems.

### SAP application server

Part of the SAP client server architecture. Designates the computers on which the application logic of the SAP system runs.

### SAP Ariba

SAP Ariba is a cloud-based purchasing and supply chain platform for optimizing procurement processes, supplier management, and payment processing. It can be integrated with other SAP products and enables a centralized collaboration with suppliers.

### SAP Best Practices

SAP Best Practices are prepackaged, industry-specific solutions that include proven business processes and implementation methodologies to simplify and accelerate the implementation of SAP software for enterprises.

### SAP Business Technology Platform (SAP BTP)

SAP BTP a cloud-based platform from SAP that helps companies develop, integrate, and manage applications. It provides a variety of services including databases, integration, analytics, and mobile development. SAP BTP can be integrated with other SAP products and also provides APIs to integrate other applications into the platform.

### SAP ERP Human Capital Management (SAP ERP HCM)

SAP ERP HCM is on-premise software that helps companies streamline and automate their HR processes, including HR administration, payroll, time management, and talent management.

### SAP Legacy System Migration Workbench

Tool based on SAP technology for the one-off and periodic transfer of data from legacy systems to SAP systems.

**SAP Rapid Deployment solutions**

Rapid Deployment Solution. Packaged SAP software with predefined content. A service tailored to this package is also offered by SAP.

**SAP S/4HANA migration cockpit**

Data migration tool in SAP S/4HANA and SAP S/4HANA Cloud until SAP S/4HANA 2020 accessed as a Web Dynpro transaction (Transaction LTMC). Since SAP S/4HANA 2020 and in SAP S/4HANA Cloud, you could use SAP Fiori app Migrate Your Data Migration Cockpit.

**SAP S/4HANA migration object modeler**

Modeling tool in SAP S/4HANA for modifying delivered standard migration objects, creating your own migration objects, and creating your own selections for the direct transfer approach for the → SAP S/4HANA migration cockpit. Limited functionality is available in the SAP Fiori app Model Migration Objects - Migration Object Modeler in SAP S/4HANA Cloud.

**SAP SuccessFactors**

SAP SuccessFactors is a suite of cloud-based human capital management (HCM) applications that helps organizations manage people processes, talent management, and improve the employee experience.

**SAP SuccessFactors Employee Central**

SAP SuccessFactors Employee Central is a module of SAP SuccessFactors and focuses on managing central employee data and HR processes.

**Secure File Transfer Protocol (SFTP) server**

An SFTP server is a server that enables a secure file transfer between a client and a server using SFTP, which is an encrypted alternative to the FTP protocol and ensures the secure transfer of files over the internet. The SFTP server provides a central storage platform for exchanging data between multiple users or systems.

**Selective data transition**

A selective transition of data from a source system to a productive target system.

**Sequential file**

File with sequential organization. The data records within a sequential file are stored one after another without a key term.

### Software Update Manager (SUM)

SUM is a tool for upgrading SAP systems and importing Enhancement Packages (EHP) or Support Packages (SPs).

### Software Provisioning Manager (SWPM)

SWPM is the successor to the SAP tool SAPInst. You can use SWPM to install, copy, transform, split, rename, and uninstall products based on SAP NetWeaver Application Server (SAP NetWeaver AS).

### Standard SAP interface

An interface to the SAP system of type batch input, direct input, BAPI, or IDoc.

### Status bar

Element of the graphical user interface (UI). The status bar contains an output field to which the SAP system outputs messages triggered by the primary window. Other fields provide information about the status of the system. The status bar is located at the bottom of the primary window and extends across its entire width.

### Subroutine

Module of a program that can be called by multiple programs. Subroutines are used to write frequently used program parts only once. Data can be passed explicitly from and to subroutines.

### System conversion

Converting an SAP ERP system to an SAP S/4HANA system → Brownfield.

### Table type file

A file in which all data records have the same structure.

### Test configuration

Combination of test script and test data.

### Test script

Auditable unit of the SAP system. A test script often maps a specific business process or process chain.

### Text file

A file containing text information whose records are separated by a record end marker.

### Transaction

An application. To open to the initial screen of an application, you can navigate through the menu hierarchy or enter a 4-digit transaction code in the command field. Using the transaction code saves navigating through the various menus and takes a user directly to the initial screen.

### Transaction code

Sequence of alphanumeric characters that names a transaction in the SAP system. To call a transaction, enter the transaction code in the command field and press Enter.

### Transaction data

Transaction-related data that is short lived and assigned to specific master data. Individual posting documents are referred to as transaction data. For example, transaction data relating to the development of sales can be assigned to the master data of a supplier. The total turnover of a supplier is composed of the data of the individual business transactions (the transaction data).

### Translation

Value-based manipulation of a field content using a table that specifies which old value is to be replaced by which new value.

### Transport request

Document for copying corrections between different types of systems. A transport request captures released corrections. After the release of a transport request, the transport is carried out. For example, you can transport corrections from a development system to a consolidation system.

### Value mapping

Conversion table between values from the legacy system and SAP values.

### Variable

Placeholder used to store and address data under a specific name in a specific format. Variables differ by their name, type, length, and structure.

### XML file

Standing for *Extensible Markup Language*, XML enables the structured representation of data and is a common method of exchanging data between different systems and applications. This small database exists in the format of a text file. XML files are much larger than CSV files due to their structure (→ CSV file). For example, while you can store the data of a complex SAP S/4HANA business object, such as a product, in a single XML file, you would need to distribute the same data into dozens of CSV files. Conversely, these dozens of CSV files can consume less disk space than the XML file.

### XML Schema Definition (XSD)

XSD is a language used to define XML document structures and data types. It enables the data validation and interoperability support between different systems using XML.

# Appendix B
# The Authors

**Frank Finkbohner** is SAP Product Owner for the development of predefined SAP S/4HANA Cloud data migration content for the SAP S/4HANA migration cockpit. He has worked for SAP since 1999, 13 years of which were spent in consulting, where he supported customers in the development and enhancement of ABAP applications, among other things. Frank focused on the area of data migration already at an early stage. Within the scope of numerous customer projects, he used the entire range of tools and techniques for SAP data migration, including SAP Legacy System Migration Workbench, batch input, Business Application Programming Interface (BAPI), IDocs, and rapid data migration. As a manager of data migration projects, Frank has developed his own method of structuring migration projects into work packages (SEAMAP) and has also successfully applied it in his customer projects. After moving to SAP Development in 2012, he was Project Manager and Delivery Manager for SAP Rapid Deployment solution Rapid Data Load for SAP Hybris Marketing until 2016. Since 2015, he has been a product owner responsible for the development of SAP S/4HANA Cloud data migration content. Frank is also the author of several SAP PRESS books on data migration. Frank Finkbohner earned his diploma in computer science at the University of Applied Sciences in Fulda, Germany and is also a graduate industrial engineer from studies at Esslingen University of Applied Sciences, Germany. You can contact Frank Finkbohner via LinkedIn: *https://www. linkedin.com/in/frankfinkbohner*.

**Martina Höft** works in product management for SAP S/4HANA in the area of data migration and data transformation with a focus on the SAP S/4HANA migration cockpit. Martina has been with SAP since 2011 and has held various positions in consulting and product management. Previously, she worked for many years as a senior consultant for various consulting firms in numerous national and international projects. Her focus was on SAP Customer Relationship Management (SAP CRM) and SAP Sales and Distribution. Martina Höft is a trained office administrator and studied economics.

**Dr. Michael Roth** oversees data migration and data extraction for the SAP Business ByDesign midmarket cloud solution at SAP, most recently as a Product Manager and before that as a consultant in the Platinum Engineering Team. He studied mathematics and computer science at the Technical University of Munich and joined SAP in 1994.

**Jonas Kinold** works as Head of Competence Center SAP Integration Services and certified SAP integration consultant for RealCore Group. During his training, he quickly found his way to his focus in the area of SAP integration, in which he is still active today. In 2022, he completed a bachelor's degree in business information systems part time. In his area of responsibility, Jonas carries out projects related to SAP Process Integration/SAP Process Orchestration and the SAP Integration Suite.

**Wolfgang Kuchelmeister** was born in Mengen, Germany, in 1968. He studied mechanical engineering at the Georg Simon Ohm University of Applied Sciences in Nuremberg from 1990 to 1996. In 2000, he and other founding members established Pentos AG in Munich, one of the leading SAP HXM consulting firms. Since 2011, he has been working in the SAP SuccessFactors Employee Central area. Since 2017, Wolfgang has been responsible for data integration topics and corresponding data migrations in SAP SuccessFactors.

**Lukas Widera** has been working at SAP Deutschland SE & Co.KG since 2018 as a Technical Solution Architect in the Global Presales Team for SAP procurement solutions. Prior to that, he was a project manager for rapid deployment solutions in the area of integration at SAP SE since 2010.

# Index

## D

## E



## F

## G



## H



## I



## J



## K



## L

## T